Textbook of Microbial Ecology

Textbook of Microbial Ecology

Edited by Nigel Hogan

SYRAWOOD
PUBLISHING HOUSE
New York

Published by Syrawood Publishing House,
750 Third Avenue, 9th Floor,
New York, NY 10017, USA
www.syrawoodpublishinghouse.com

Textbook of Microbial Ecology
Edited by Nigel Hogan

International Standard Book Number: 978-1-68286-544-6 (Hardback)

Cataloging-in-Publication Data

Textbook of microbial ecology / edited by Nigel Hogan.
p. cm.
Includes bibliographical references and index.
ISBN 978-1-68286-544-6
1. Microbial ecology. 2. Microbiology. I. Hogan, Nigel.
QR100 .T49 2018
579.17--dc23

Printed in China.

TABLE OF CONTENTS

PREFACE

This book aims to highlight the current researches and provides a platform to further the scope of innovations in this area. This book is a product of the combined efforts of many researchers and scientists, after going through thorough studies and analysis from different parts of the world. The objective of this book is to provide the readers with the latest information of the field.

Microbial ecology studies the interaction of microorganisms with each other, their hosts as well as with the environment. This book on microbial ecology seeks to understand microbial diversity and biogeochemical processes. Microbes are often used for industrial processes such as fermentation, etc. Thus the study of microbes is significant for all areas of science and technology. A number of latest researches have been included to keep the readers up-to-date with the global concepts in this area of study. The readers would gain knowledge that would broaden their perspective about microbial ecology.

I would like to express my sincere thanks to the authors for their dedicated efforts in the completion of this book. I acknowledge the efforts of the publisher for providing constant support. Lastly, I would like to thank my family for their support in all academic endeavors.

Editor

1

Dynamics of the Microbiota in Response to Host Infection

Clara Belzer[1,2¤a], Georg K. Gerber[1,2,3], Guus Roeselers[4¤b], Mary Delaney[1,3], Andrea DuBois[1,3], Qing Liu[1,3], Vera Belavusava[1,3], Vladimir Yeliseyev[1,3], Andres Houseman[5¤c], Andrew Onderdonk[1,2,3], Colleen Cavanaugh[4], Lynn Bry[1,2,3]*

1 Department of Pathology, Brigham & Women's Hospital, Boston, Massachusetts, United States of America, 2 Department of Pathology, Harvard Medical School, Boston, Massachusetts, United States of America, 3 Center for Clinical and Translational Metagenomics, Brigham & Women's Hospital, Harvard Digestive Diseases Center, Harvard Medical School, Boston, Massachusetts, United States of America, 4 Department of Organismic and Evolutionary Biology, Harvard University, Cambridge, Massachusetts, United States of America, 5 Department of Biostatistics, Harvard School of Public Health, Boston, Massachusetts, United States of America

Abstract

Longitudinal studies of the microbiota are important for discovering changes in microbial communities that affect the host. The complexity of these ecosystems requires rigorous integrated experimental and computational methods to identify temporal signatures that promote physiologic or pathophysiologic responses *in vivo*. Employing a murine model of infectious colitis with the pathogen *Citrobacter rodentium*, we generated a 2-month time-series of 16S rDNA gene profiles, and quantitatively cultured commensals, from multiple intestinal sites in infected and uninfected mice. We developed a computational framework to discover time-varying signatures for individual taxa, and to automatically group signatures to identify microbial sub-communities within the larger gut ecosystem that demonstrate common behaviors. Application of this model to the 16S rDNA dataset revealed dynamic alterations in the microbiota at multiple levels of resolution, from effects on systems-level metrics to changes across anatomic sites for individual taxa and species. These analyses revealed unique, time-dependent microbial signatures associated with host responses at different stages of colitis. Signatures included a *Mucispirillum* OTU associated with early disruption of the colonic surface mucus layer, prior to the onset of symptomatic colitis, and members of the Clostridiales and Lactobacillales that increased with successful resolution of inflammation, after clearance of the pathogen. Quantitative culture data validated findings for predominant species, further refining and strengthening model predictions. These findings provide new insights into the complex behaviors found within host ecosystems, and define several time-dependent microbial signatures that may be leveraged in studies of other infectious or inflammatory conditions.

Editor: Yolanda Sanz, Instutite of Agrochemistry and Food Technology, Spain

Funding: Work supported by a pilot feasibility grant from the Harvard Digestive Diseases Center (NIH Award #P30-DK034854; CMC, LB), R01-HD061916 (LB, CB), Rubicon grant from the Netherlands Organization for Scientific Research, NWO (GR), Stanley L. Robbins Memorial Research award (GKG), and Harvard Catalyst, The Harvard Clinical and Translational Science Center (AH; NIH Award #UL1 RR 025758 and financial contributions from Harvard University and its affiliated academic health care centers). The funders had no role in study design, data collection and analysis, decision to publish, or preparation of the manuscript.

* Email: lbry@partners.org

These authors contributed equally to this work.

¤a Current address: Laboratory of Microbiology, Wageningen University, Wageningen, The Netherlands
¤b Current address: Microbiology and Systems Biology, TNO, Zeist, The Netherlands
¤c Current address: Department of Biostatistics, College of Public Health and Human Sciences, Oregon State University, Corvallis, Oregon, United States of America

Introduction

Large-scale characterization of the host's microbiota has been enabled by recent innovations in sequencing technologies [1] and computational methods [2–4]. These developments have provided initial insights into the microbiota's association with normal physiology and disease [5–11]. Longitudinal studies are particularly valuable for unraveling causal interactions among the host and microbial inhabitants. However, studies of these ecosystems over time require new analytic approaches to fully explore their extraordinarily complex dynamics and identify signatures relevant to host outcomes [12,13].

We used a mouse model of inflammatory colitis, caused by the attaching and effacing pathogen *Citrobacter rodentium* [14,15], to investigate dynamic changes in microbial communities relative to a defined perturbation in the host. Prior studies have identified alterations in the gut flora at the height of acute infection [16–19], supporting our hypothesis that commensal populations change dynamically before and after onset of host symptoms, and may thus play important roles at different stages of disease. However, the kinetics of these changes are neither known, nor characterized. This experimental model thus provides a valuable system in which to discover the complex behaviors of the microbiota, across gut locations, and at different stages of host disease.

In immunocompetent mice *C. rodentium* infection follows four distinct stages [20]: 1) early colonization (≈1–6 days post-challenge), during which the pathogen establishes a small free-living reservoir in the cecum and ileum and initiates adherent infection in the distal colon [20,21], 2) symptomatic infection (≈7–17 days), characterized by epithelial hyperplasia, influx of host immune cells, colitis, and development of early adaptive antibody responses [14,22], 3) resolution (≈17–25 days), during which pathogen-specific IgG responses evolve and the pathogen is cleared from the host [23], and 4) convalescence (≈26–62 days), during which tissue damage is repaired.

We used high-throughput 16S rDNA gene sequencing to broadly characterize the microbiota over the course of infection. We also employed quantitative culture of the pathogen and predominant commensals to provide a complementary, non-nucleic acid based dataset. This approach allowed evaluation of the ecosystems under study at progressively finer levels of resolution, starting with systems-level properties, such as diversity and time to recovery, progressing to sets of taxonomic units, and lastly incorporating quantitative culture data to identify changes at the level of individual species.

Longitudinal analyses of complex microbial ecosystems present several computational challenges. First, the numbers of time-points and replicates collected from the host population(s) are frequently small due to sampling logistics and experimental costs. Second, biologic and analytic factors cause high amounts of noise. Third, limited sequencing depth, constrained by cost, and combined with relative rarity of certain organisms, can lead to low sequence counts for some taxa. Although a number of existing computational tools readily compare ecosystems' taxonomic compositions and diversities [2–4], or abundances of taxa between conditions [24], they have not been designed to analyze time-dependent changes in taxa relative to perturbations in the ecosystem. Tools for analyzing microarray time-series data [25–28] are also suboptimal for this latter application as their underlying algorithms do not model the characteristics of high-throughput sequencing data or microbiome data.

To address these challenges we extended a computational model that we recently developed, Microbial Counts Trajectories Infinite Mixture Model Engine (MC-TIMME) [29] to enable analysis of longitudinal changes in the microbiota during a host infection. MC-TIMME represents a new approach to analyzing microbiome time-series data, employing nonparametric Bayesian methods and continuous-time models of dynamics coupled with an error model tailored for high-throughput sequencing data. In prior work, we introduced the algorithm and applied it to a publicly available dataset measuring the microbiota of human subjects exposed to sequential antibiotic exposures [12]. In that work, we showed that MC-TIMME accurately inferred time-varying signatures for individual taxa while simultaneously compressing similar signatures into groups. Our method further identified a number of new features in the dataset that had not been found using standard analysis techniques. These new findings included characterization of relaxation time distribution, or the kinetics of ecosystems' return to baseline or new levels after introduced perturbations, and discovery of consensus signature groups (CSGs), which represent sets of reference OTUs within or among subjects that share common behaviors over the time-series. In the present work, we extend MC-TIMME with a new, flexible model of dynamics to capture behavior of the microbiota during an ongoing host infection and introduce methods for incorporating complementary data sources into analyses, including quantitative culture data.

Application of the extended version of MC-TIMME to our datasets of 16S rDNA gene signatures and quantitatively cultured isolates from mice infected with *C. rodentium* enabled study of host microbial ecosystems during an infection at progressively finer levels of resolution. First, we analyzed time-dependent changes in systems-level properties of the intestinal ecosystems and found substantial differences across anatomic sites. Second, we used Consensus Signature Group (CSG) analyses to characterize the range of time-varying signatures observed in the microbiota subsequent to host infection with the pathogen. Third, we generated time-maps to temporally order CSG dynamics and to visualize coordinate and cascading changes contributed by individual taxa across intestinal sites. Lastly, we incorporated quantitative culture data for the pathogen and predominant commensals to validate and refine model predictions of the dynamics observed, providing the final component of our ecosystem-to-species level discovery of temporal dynamics in the microbiota during a host infection.

Results

High-throughput longitudinal profiling of the microbiota during host infection

To explore the dynamic effects of *C. rodentium* infection on intestinal ecosystems, samples from ileum, cecum and distal colon were collected from infected mice and uninfected controls, at days 3, 7, 10, 14, 21, 28 and 62 post-challenge with the pathogen. Each time point consisted of biological replicates for the infected or uninfected groups. Samples were subjected to massively parallel 16S rDNA gene sequencing and quantitative culture using media selective for the pathogen and predominant commensals (Fig. 1A, Tables 1, 2, Table S1, Datasets S1, S2).

We calculated Good's coverage estimator [30] (Table 1) for sequenced samples to determine if sequencing coverage was equivalent between infected and uninfected mice, particularly in distal colon where large pathogen burdens (maximum of ≈99% of sequencing reads in the colon, and ≈4% of sequencing reads in cecum and ileum) could impact ability to detect shifts in the underlying microbiota. Average coverage was lowest in the more microbiologically diverse cecal samples at ≈85%, and highest, at ≈95–98% for colon and ileum. However, infected and uninfected mice demonstrated comparable coverage at each location, indicating that introduction of the pathogen did not prevent detection of underlying commensal populations in infected mice.

Robust inference of time-varying signatures of taxa

We applied an extended version of MC-TIMME to our dataset to infer time-varying signatures of taxa. Replicated time-series of counts, either of Operational Taxonomic Units (OTUs) derived from sequencing data, or of individual species measured by quantitative culture, served as input to MC-TIMME (Fig. 1B). MC-TIMME uses nonparametric Bayesian methods to simultaneously estimate the number of signatures, the shapes of signatures, and assignments of taxa to signatures (Fig. 1C). Taxa may follow different signatures in distinct anatomic sites and in the infected and uninfected states, which our algorithm automatically detects [29]. From the 16S rDNA data, 210 OTUs had sufficient counts for analysis. Of these, MC-TIMME identified 45 OTUs with detectable changes in response to infection in at least one intestinal site. For the culture data, out of 7 predominant commensal species determined to have sufficient counts for analysis across biological replicates, all demonstrated detectable changes in response to infection in at least one site. The inferred

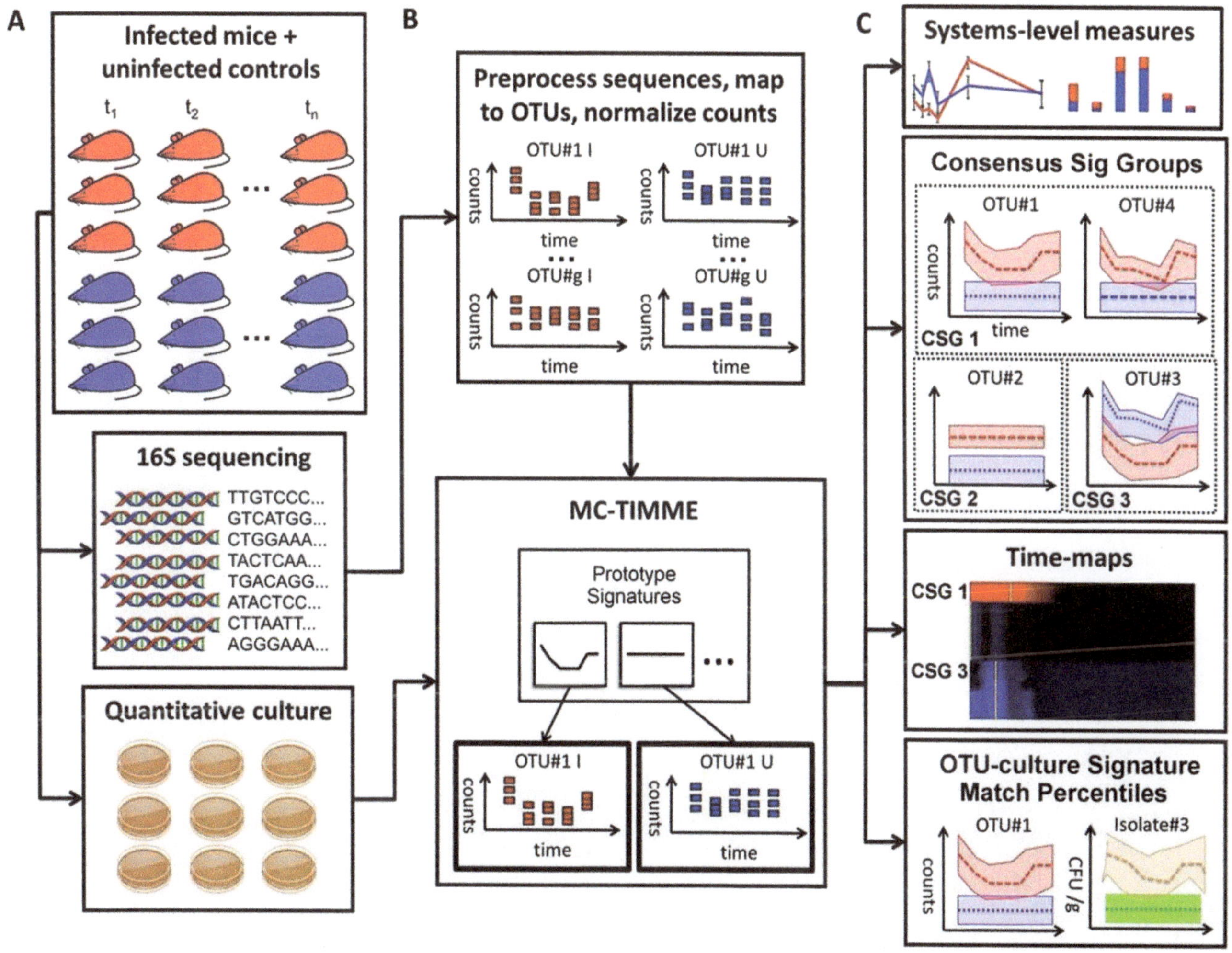

Figure 1. Experimental and computational framework. (A) Experimental model, with cohorts of infected mice (red) or uninfected age-matched controls (blue), from which ileum, cecum and distal colon samples were collected. Samples were subjected to high-throughput 16S rDNA sequencing as well as quantitative culture for the pathogen and predominant commensals. **(B)** Computational pipeline begins with preprocessing and clustering of 16S rDNA sequences into Operational Taxonomic Units (OTUs). Normalized OTU or culture counts serve as input to MC-TIMME, which simultaneously infers the number, shapes, and assignments of taxa to prototype signatures. **(C)** MC-TIMME outputs summarize dynamic changes in microbial communities across intestinal sites at multiple levels of detail. Systems measures detect large-scale changes in microbial community structure and dynamics. Consensus Signature Groups (CSGs) compress OTUs into sets with members exhibiting similar behaviors over time. Time-maps compactly visualize microbiota dynamics in tissues, organizing CSGs by their times of maximal change to reveal cascades of coordinate alterations. Signature Match Percentiles (SMPs) identify taxa for which sequence and culture-derived signatures have strong correspondences.

trajectories formed the basis for subsequent analyses at ecosystems-to-species levels of detail (Fig. 1C).

Ecosystems-level measures: dynamics of microbiota diversity and ecosystem recovery times

We used inferred signatures to estimate an ecosystem-level measure of microbiologic diversity, Shannon entropy [31], and to detect changes in this measure over the infection. In both groups of mice, the Shannon entropy measure indicated that cecum had the highest levels of diversity over the time series (Fig. 2C), followed by colon (Fig. 2E), and ileum (Fig. 2A). Diversity decreased with infection across all gut locations, with nadirs occurring over days 7–14, the period of symptomatic infection. Of note, the most profound decreases occurred in distal colon at the primary site of infection. However, by 2 months post-challenge, overall diversity in the infected mice returned to that of uninfected mice at all gut locations.

We next evaluated the time it took for each ecosystem to return to baseline after introduction of the pathogen. To quantify this duration, we developed a new measure, the Microbiota Recovery Time (MRT). The MRT is defined as the latest time-point post-challenge with *C. rodentium* for which microbial communities from infected mice and controls exhibit no detectable differences ($>$ 95% similarity). For the MRT, the calculated similarity between microbial communities takes into account the relative taxa frequencies derived from their inferred signatures (Protocol S1). The estimated MRT was 62 days for ileum, 14 days for cecum, and 62 days for colon. This analysis highlighted very different patterns of change among gut locations over the course of the infection as well as the time required for recovery to a stable state. Ileum (Fig. 2B) showed a biphasic pattern with prominent

Table 1. Summary of sequenced samples and Good's coverage estimates.

	T	N (ctrl)	N (inf)	Reads (ctrl) K	Reads (inf) K	Coverage (ctrl)	Coverage (inf)
Ileum	3	3	3	2.9–3.6	0.8–3.6	98–100%	98–99%
	7	3	3	3.1–4.0	3.7–4.3	95–99%	98–99%
	10	3	3	2.2–3.3	1.9–3.9	98–99%	97–99%
	14	3	2	3.4–4.5	4.7–5.6	97–99%	99–100%
	28	3	3	2.7–2.9	0.8–3.0	96–98%	92–96%
	62	3	3	4.2–5.8	3.8–4.2	96–98%	97–99%
Cecum	3	3	4	1.0–1.3	1.1–2.4	84–98%	88–93%
	7	3	3	0.7–0.8	0.7–1.2	70–76%	83–88%
	10	3	3	2.0–2.6	2.3–2.6	88–92%	92–95%
	14	3	3	1.1–1.3	1.0–1.3	77–79%	74–78%
	28	3	3	2.0–4.4	1.7–2.3	87–90%	85–89%
	62	3	3	1.8–2.1	1.6–1.8	81–85%	79–82%
Colon	3	3	3	3.5–4.6	3.4–3.7	93–100%	97–100%
	7	2	2	2.8–3.9	2.6–2.9	95–99%	99–100%
	10	3	3	0.8–2.5	2.4–3.6	90–96%	98–99%
	14	1	2	2.8	2.8–2.9	96%	99–100%
	28	4	3	1.9–3.2	2.3–2.5	94–98%	87–91%
	62	3	2	1.0–6.9	1.5–6.1	94–98%	96–99%

Numbers of reads are listed in thousands, and are after all preprocessing quality control and filtering steps. The average number of reads across all samples was ≈2,700. Good's coverage is a nonparametric estimate of the proportion of classes (i.e., OTUs) that are observed in a sample out of the total number of classes inferred to be present in the population. Labels in the top table row: T = time-point, ctrl = control, uninfected mice; inf = infected mice.

Table 2. Species identified by quantitative culture.

Species	Selective Media	Growth Conditions	Identification	Threshold of detection	16S rDNA gene ID
Citrobacter rodentium	MAC	Aeroboic[1], 24 hr	API-20E panel and full 16S sequence.	100 CFU/g	*Citrobacter rodentium*
Enterobacter hormachei	MAC	Aeroboic[1], 24 hr	API-20E panel and full 16S sequence.	100 CFU/g	*Enterobacter hormachei*
Lactobacillus johnsonii	CNA	Anaerobic[2], 48–72 hr.	Biochemical typing and 16S sequence	500 CFU/g	*Lactobacillus johnsonii*
Lactobacillus murinus	BKV and CNA	Anaerobic[2], 48–72 hr.	Biochemical typing and 16S sequence	100 CFU/g	*Lactobacillus murinus*
Lactobacillus reuteri	CNA	Anaerobic[2], 48–72 hr.	Biochemical typing and 16S sequence	500 CFU/g	*Lactobacillus reuteri*
Proteus vulgaris	MAC+TET	Aeroboic[1], 24 hr	API-20E panel and full 16S sequence.	100 CFU/g	*Proteus vulgaris*

Agar media used: MAC (MacConkey agar) for selection of enteric and bile-resistant non-fermenter species; MAC+TET (MacConkey agar with 10 μg/mL of Tetracycline) to select for *Proteus vulgaris*. CAN (Colistin Naladixic Acid agar with 5% sheep's blood on a Columbia agar base) for suppression of enteric species; BKV (Brucella-Kanamycin-Vancomycin agar) for selection of aminoglycoside+vancomycin-resistant Lactobacilli;

[1]Aerobic incubation conditions were in 5% CO_2 humidified atmosphere at 37°C; plates read at 24 and 48 hours of incubation.

[2]Anaerobic conditions were in a Coy anaerobic chamber at 37°C. Plates were read in the chamber at 72 hours of incubation.

increases of taxa abundances during early infection followed by decreases during acute infection. The microbiologically diverse cecum showed little detectable change throughout infection (Fig. 2D), while colon showed uniform decreases in early infection that resolved after the recovery phase (Fig. 2F).

Consensus Signature Groups: defining patterns of commensal responses to infection

We next used Consensus Signature Groups (CSGs) to categorize the types of dynamic changes that occurred in the gut microbiota after introduction of the pathogen (Figs 3,4). A Consensus Signature Group represents a set of taxa that share similar dynamics within a tissue, providing a means to identify common behaviors among taxa regardless of their phylogenetic relationships. All 45 OTUs identified by MC-TIMME as having detectable changes in response to infection in at least one intestinal site were assigned to CSGs.

Figure 3 illustrates representative signatures discovered by our method for individual OTUs. These OTUs belong to CSGs that exemplify predominant patterns of change in the ecosystems studied. As expected, the pathogen *Citrobacter rodentium* in colon (OTU#6; Fig. 3A), shows a rapid increase with maximal change during acute infection, and returns to baseline by the resolution phase. In contrast, *Mucispirillum* in colon (OTU#1; Fig. 3B), demonstrates a rapid decrease during early infection. This decreased exhibits a prolonged delay to recovery, beyond the period of pathogen clearance between days 21–28. Of note, *Parabacteroides* in colon (OTU#8; Fig. 3C) also decreased during early infection, but returns to baseline by the recovery phase when the pathogen has been cleared. *Lactobacillus* (OTU#3; Fig. 3E) in ileum shows an increase that occurs during acute infection, and then quickly returns to baseline by the recovery phase, whereas a second *Lactobacillus* (OTU#13; Fig. 3F) shows an immediate and prolonged decrease in the ileum. In contrast, *Clostridium* (OTU#24) in ileum (Fig. 3G) and cecum (Fig. 3H), shows a very delayed increase that persists into the convalescent phase.

Of note, many taxa demonstrated no detectable differences between infected and uninfected mice, highlighting the fact that profound changes in host microenvironments may preferentially affect select taxa. For instance, *Parabacteroides* (OTU#8), which showed changes in the colon (Fig. 3C), had no detectable changes at its predominant site of residence in the cecum (Fig. 3D). Other examples of taxa that showed no differences between infected and uninfected mice include a member of family Lachnospiraceae in colon (OTU#10; Fig. 3I) and a member of family Clostridiaceae in ileum (OTU#4; Fig. 3J), which further classified as segmented filamentous bacteria (SFB) by the RDP Sequence Match tool [32].

The analysis of Consensus Signature Groups suggested that a substantial number of taxa in each tissue responded to infection in a coordinate manner over the 2-month period. To quantify this effect, we developed a measure, Consensus Signature Group Coordination (CSGC), which is one minus the ratio of the number of CSGs identified in a tissue to the number of taxa present in that location across the time-series. CSGC thus measures the degree of coordination of the changes among taxa within the microbial tissue ecosystem. Higher CSGC values indicate more coordinated behavior, while lower values indicate more varied responses. The CSGC values for our dataset were 83% in ileum, 92% in cecum, and 90% in colon. These values indicate highly coordinated changes in the microbiota within all intestinal ecosystems during infection, but with ileum showing notably more varied changes than cecum or colon.

Time-maps: identifying tissue-specific, cascading changes among commensal taxa

We created time-maps to visualize interrelations among the dynamic responses observed for taxa within intestinal ecosystems (Figs 4A, 4B, 4C). Time-maps revealed cascading and coordinate changes across all gut locations studied, identifying distinct patterns within each tissue.

In ileum (Fig. 4A), groups of OTUs coordinately increased or decreased in response to infection, with the majority showing two waves of peak responses. The first peak occurred early during infection, over days 3–7 post-challenge (CSG IL1 and IL6), while the second wave occurred during acute symptomatic infection over days 10–14 post-challenge (CSGs IL2, IL3, IL7–IL10). Overall, dominant effects in infected mice were seen in the orders Clostridiales and Lactobacillales within phylum Firmicutes,

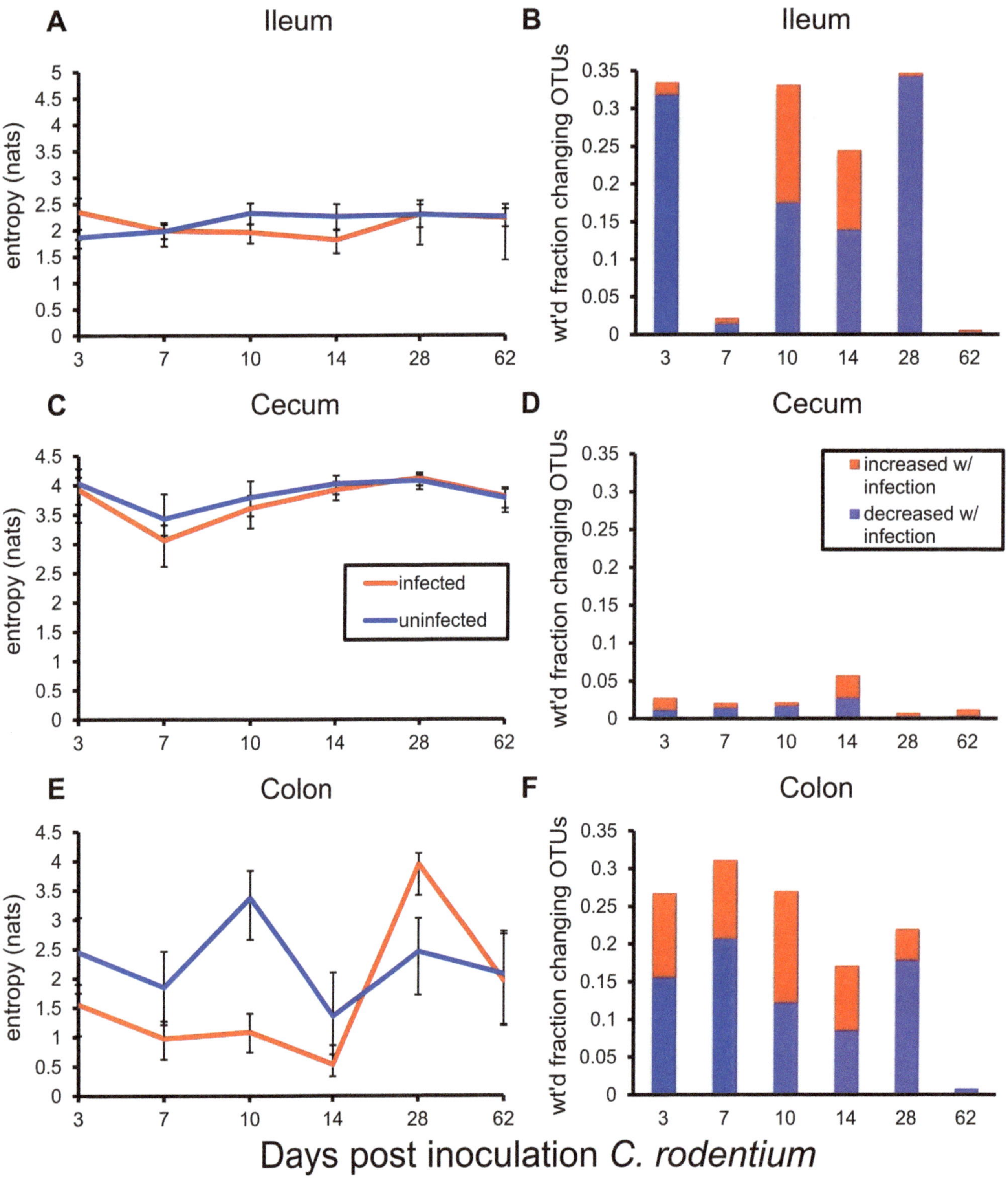

Figure 2. Systems-level measures of microbiota diversity dynamics and tissue recovery times in response to infection. (**A, C, E**) Shannon entropy calculated from inferred Operational Taxonomic Unit (OTU) signatures measures dynamic changes in the diversity of the microbial ecosystems in each tissue. Red line = entropy of infected tissue ecosystem; blue line = entropy for controls. Vertical bars denote 95% credible intervals. (**B, D, F**) The Microbiota Recovery Time (MRT) in each tissue measures the latest time-point post-challenge with the pathogen for which microbial communities from infected mice and controls exhibit >95% similarity overall. Red bars = weighted measure of detected changes in taxa increasing with infection; blue bars = corresponding measure for decreasing taxa. The weighted measure takes into account relative taxa abundances in both infected and uninfected cohorts. The recovery time was 62 days for ileum, 14 days for cecum, and 62 days for colon.

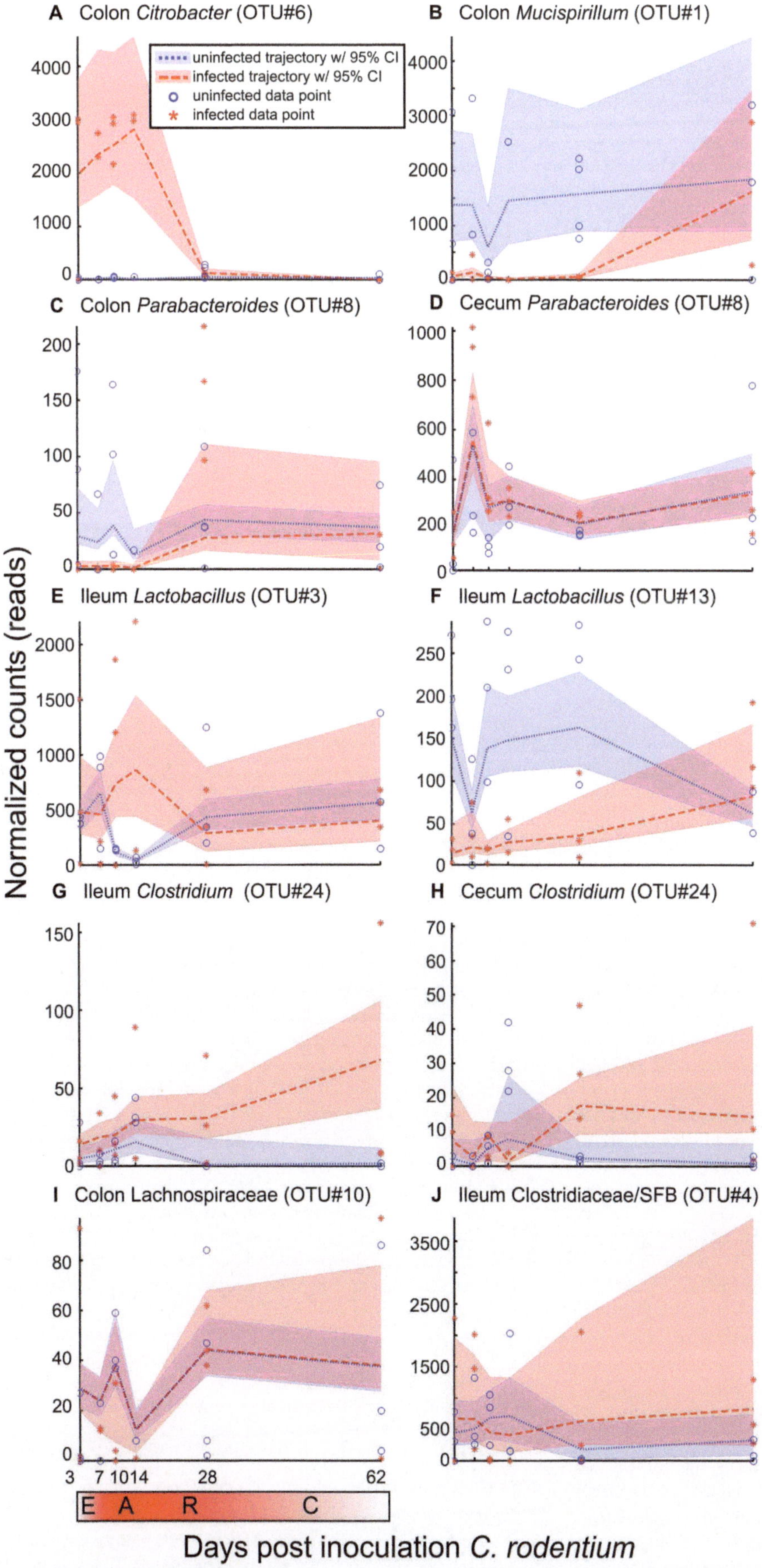
A Colon Citrobacter (OTU#6)
B Colon Mucispirillum (OTU#1)
uninfected trajectory w/ 95% CI
infected trajectory w/ 95% CI
uninfected data point
infected data point
C Colon Parabacteroides (OTU#8)
D Cecum Parabacteroides (OTU#8)
E Ileum Lactobacillus (OTU#3)
F Ileum Lactobacillus (OTU#13)
G Ileum Clostridium (OTU#24)
H Cecum Clostridium (OTU#24)
I Colon Lachnospiraceae (OTU#10)
J Ileum Clostridiaceae/SFB (OTU#4)
Normalized counts (reads)
3 7 10 14 28 62
E A R C
Days post inoculation C. rodentium

Figure 3. Consensus Signature Groups systematically characterize patterns of time-dependent microbiota changes in response to infection. Consensus Signature Groups (CSGs) represent sets of taxa that share similar dynamics within a tissue, providing a means to identify common behaviors among taxa regardless of their phylogenetic relationships. Representative signatures of individual Operational Taxonomic Units (OTUs) from CSGs are shown. Horizontal axis indicates days post-inoculation with the pathogen; vertical axis shows normalized sequencing counts for the OTU. Dashed or dotted lines indicate median signature shapes for OTUs. Shaded regions indicate 95% credible intervals for signatures; regions of overlap indicate time-periods during which changes were not detected. Phases of infection are E = early, A = acute, R = recovery, C = convalescence. **(A)** The pathogen, *Citrobacter rodentium* (OTU#6) in colon. **(B)** *Mucispirillum* (OTU#1) in colon, rapidly decreases and does not return to baseline until the convalescent phase. **(C)** *Parabacteroides* (OTU#8) in colon, decreases during early infection, but returns to baseline by the recovery phase. **(D)** *Parabacteroides* (OTU#8) in cecum had no detectable change between cohorts. **(E–F)** Two *Lactobacilli* in ileum, showing different dynamics: OTU#3 increases during acute infection, while OTU#13 decreases. **(G–H)** *Clostridium* (OTU#24) in ileum and cecum, has a delayed increase that persists into the convalescent phase. **(I–J)** Representative OTUs in colon and ileum showing no detectable changes between cohorts.

though many OTU classifying at the same taxonomic level exhibited quite different dynamics. In infected mice, OTU associated with the genus *Lactobacillus* both increased (OUT#3, #10) and decreased (OTU#13, #14), while members of the *Clostridiales* increased (OTU#33, *Coprobacillus*; OTU# 46, #60, Clostridiales; and OTU#69, *Dorea*) and also decreased (OTU#11, Clostridiales).

In cecum (Fig. 4B), although the *Citrobacter* signature comprised $<\approx 5\%$ of total reads per sample, even at the height of infection, numerous OTUs demonstrated altered trajectories in infected mice. Members of family Lachnospiraceae both increased (OTU#29) and decreased (OTUs #117 and #017) in infected mice as compared to uninfected controls. Within the *Clostridiales* OTUs #045 and #071 (CE2) increased in infected mice only over the first 7 days of infection. In contrast, OTU#67 increased over early and acute stages of infection (CSGs CE4), while OTUs corresponding to genus *Clostridium* and that typed taxonomically to other members of the Clostridiales (OTU#24, 131 and 294), increased after the pathogen's clearance (CSGs CE5, CE6, CE8). Among the Lachnospiraceae, OTU #29 (CE3) increased over acute infection, while OTU#117 (CE7) declined during early stages of colonization, as did OTU #017 (CE9). In contrast, these taxa showed nominal variation over the course of infection in uninfected mice.

In colon (Fig. 4C), other than the pathogen, OTUs with detectable changes showed profound decreases in infected mice, with many CSGs exhibiting a time of maximal decrease by day 3 (CSGs CO2–CO5), prior to the onset of symptomatic infection. Affected taxa included *Anaerostipes* (OTU#56) and members of the Lachnospiraceae (OTUs#82, 103, 122 and 153). A second wave of affected CSGs showed maximal decrease by day 10 during symptomatic infection (CSGs CO6–CO8). Both waves of affected CSGs largely recovered by the time of pathogen clearance at day 28. Genus *Mucispirillum* in colon (OTU#1; CSG CO9) was an exception, exhibiting a sustained decrease into the convalescent phase, beyond *C. rodentium*'s clearance by day 28. Of the tissues studied, the colon demonstrated the most phylogenetically diverse set of OTUs that changed in response to infection, including OTUs classifying at the genus level as *Mucispirillum* (phylum Deferribacteres), *Robinsoniella*, *Lactobacillus*, *Turicibacter* (phylum Firmicutes) and *Parabacteroides* (phylum Bacteroidetes).

Validating predictions to the species level with complementary data sources

We used quantitative culture of predominant species to validate dynamics of corresponding taxa identified with high-throughput sequencing. Sequence and culture-based datasets provide complementary information. With sequence-based counts the prevalence of a given OTU must be interpreted relative to the total counts for all OTUs in the ecosystem. In contrast, quantitative culture normalizes counts for a species relative to the input mass of tissue. By measuring viable organisms, culture-based analyses are not confounded by the presence of nucleic acid signatures from large numbers of dying organisms, which could occur with pathogen clearance. Thus, although culture based methods cannot be used to broadly characterize a complex ecosystem, selective use provides an alternate and sensitive method for measuring changes in defined species. The presence of common signals in both data sources for a given taxon, and corresponding species, provides stronger evidence that the signatures reflect the underlying dynamics *in vivo*.

In addition to sequence and culture-based dynamics detected for *C. rodentium* (OTU#6; Fig 5), MC-TIMME detected altered ileal or cecal dynamics in response to infection for five OTUs, (#2, 3, 9, 13 and 14), which the RDP Classifier [33] classified to the genus *Lactobacillus*. These OTUs proved resolvable to the species level with the RDP Sequence Match tool (34). OTU#2 classified as *L. johnsonii*, OTUs#3 and 9 as *L. murinus*, and OTUs#13 and 14 as *L. reuteri*. Interestingly, quantitative culture of predominant organisms identified these same species, and we applied MC-TIMME to generate signatures based on the culture counts (Fig 6).

To evaluate the similarity of signatures inferred from sequence and culture-based datasets, we calculated the rank of the match between the culture-based signature and the corresponding OTU signature, relative to all other OTU signatures in the system (the Signature Match Percentile, SMP). This measure provides a principled way to leverage alternate methods for measuring individual members within a complex ecosystem, such as quantitative culture or use of a different sequencing methodology, to validate findings identified by the originally used method. SMP values $>50\%$ indicate specific matches between the culture and sequence-based signatures. In contrast, SMP values $<50\%$ indicate relatively non-specific matches between the culture and sequence-based signatures, raising the possibility that the observed correspondence may be due to chance alone, and thus may be less reflective of true *in vivo* behaviors.

The three *Lactobacillus* species exhibited specific matches between culture and sequence-derived signatures for each organism in ileum and cecum (Table 3). In both sources of data, *L. johnsonii* exhibited a *decrease* during acute infection (79th-centile SMP in ileum and 80th-centile SMP in cecum; Fig. 6A), *L. murinus* exhibited an *increase* during acute infection (94th-centile SMP in ileum and cecum; Fig. 6B), and *L. reuteri* exhibited a *decrease* in acute infection (91th-centile SMP in ileum and 80th-centile SMP in cecum; Fig. 6C). The high SMP values indicate that the identified signatures are in close agreement between the two data sources, suggesting that each *Lactobacillus* species may play different functional roles within the ileal and cecal ecosystems.

MC-TIMME detects differing signatures among low abundance commensal species phylogenetically related to *C. rodentium*

Enterobacter hormachei and *Proteus vulgaris*, both members of family Enterobacteriaceae, normally attain their highest biomass in the

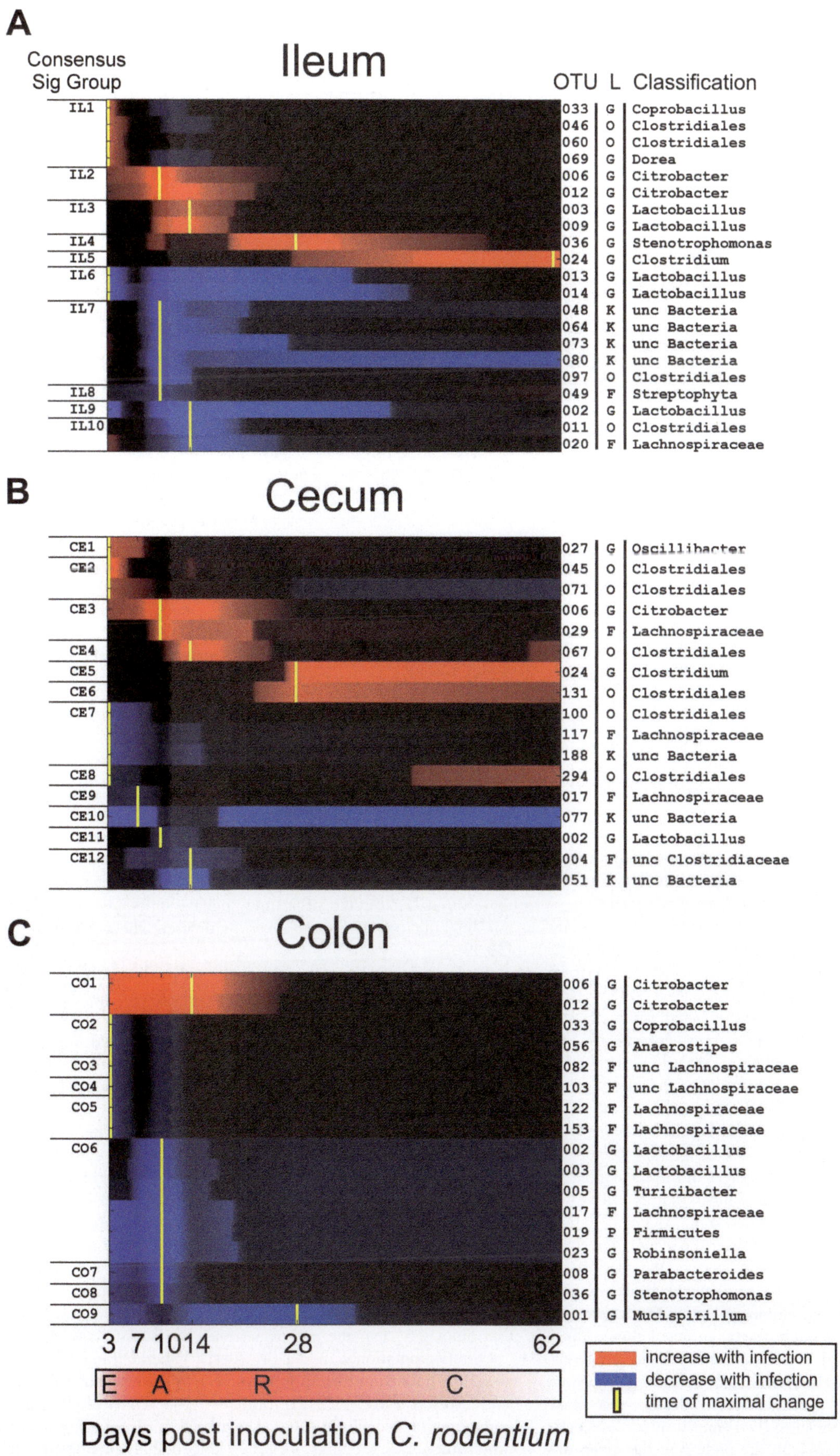
A
Ileum
Consensus
Sig Group
OTU L Classification
IL1
IL2
IL3
IL4
IL5
IL6
IL7
IL8
IL9
IL10
033 G Coprobacillus
046 O Clostridiales
060 O Clostridiales
069 G Dorea
006 G Citrobacter
012 G Citrobacter
003 G Lactobacillus
009 G Lactobacillus
036 G Stenotrophomonas
024 G Clostridium
013 G Lactobacillus
014 G Lactobacillus
048 K unc Bacteria
064 K unc Bacteria
073 K unc Bacteria
080 K unc Bacteria
097 O Clostridiales
049 F Streptophyta
002 G Lactobacillus
011 O Clostridiales
020 F Lachnospiraceae
B
Cecum
CE1
CE2
CE3
CE4
CE5
CE6
CE7
CE8
CE9
CE10
CE11
CE12
027 G Oscillibacter
045 O Clostridiales
071 O Clostridiales
006 G Citrobacter
029 F Lachnospiraceae
067 O Clostridiales
024 G Clostridium
131 O Clostridiales
100 O Clostridiales
117 F Lachnospiraceae
188 K unc Bacteria
294 O Clostridiales
017 F Lachnospiraceae
077 K unc Bacteria
002 G Lactobacillus
004 F unc Clostridiaceae
051 K unc Bacteria
C
Colon
CO1
CO2
CO3
CO4
CO5
CO6
CO7
CO8
CO9
006 G Citrobacter
012 G Citrobacter
033 G Coprobacillus
056 G Anaerostipes
082 F unc Lachnospiraceae
103 F unc Lachnospiraceae
122 F Lachnospiraceae
153 F Lachnospiraceae
002 G Lactobacillus
003 G Lactobacillus
005 G Turicibacter
017 F Lachnospiraceae
019 P Firmicutes
023 G Robinsoniella
008 G Parabacteroides
036 G Stenotrophomonas
001 G Mucispirillum
3 7 10 14 28 62
E A R C
Days post inoculation C. rodentium
increase with infection
decrease with infection
time of maximal change

Figure 4. Time-maps reveal tissue-specific cascading and coordinate changes among commensal taxa in response to infection. MC-TIMME detected 45 Operational Taxonomic Units (OTUs) that change over time between infected and uninfected mice in at least one intestinal site, and assigned these OTUs to Consensus Signature Groups. Time-maps use CSG information to visualize interrelations among the dynamic responses of OTUs within intestinal ecosystems. X-axis indicates days post-challenge with *C. rodentium*. Rows in each panel depict OTUs with detectable changes in infected mice relative to uninfected controls. Red indicates an increase in the signature for infected mice relative to baseline, and blue indicates a corresponding decrease. Intensity of colors corresponds to magnitude of change of signatures; regions in which changes were not detected are attenuated in color. Yellow marks indicate the time of maximal increase or decrease relative to the uninfected baseline controls for the CSG. Note that magnitudes of changes are not directly comparable across OTUs in this visualization. Phases of infection are E = early, A = acute, R = recovery, C = convalescence. Vertical axis on the right-hand side indicates the OTU#, level of taxonomic assignment by the Ribosomal Database Project (RDP) classifier (L = level; K = kingdom; P = phylum; O = order; F = family; G = genus), and RDP taxonomic assignment.

mouse cecum where they comprise a minor (10^3–10^5 CFU/g), but consistently present, population of the ecosystem involved in rodent hindgut metabolism [34]. With 2000–3000 sequencing reads/sample these organism are effectively undetectable. Thus, we used culture-based methods and input these counts into MC-TIMME to gain an understanding of the dynamics of these organisms during the *C. rodentium* infection. *E. hormachei* showed a detectable increase then decrease that correlated with the signature of *C. rodentium* (Fig 7A, C, E). In contrast, *P. vulgaris* demonstrated a detectable decrease in ileum and distal colon (Fig 7B, 7F), though negligible effects in cecum (Fig 7D).

Discussion

Our study demonstrates the power of using an integrated experimental and computational approach to analyze complex microbial ecosystems over time. Utilization of a defined experimental system enabled collection of high-quality samples from multiple intestinal sites, and reduced confounding factors though use of age-matched controls and biological replicates. Our computational model was designed to effectively leverage sparse and noisy time-series of counts produced from high-throughput sequencing and quantitative culture data. These approaches enabled robust detection of temporal changes in the behavior of the microbiota at multiple levels of resolution. Furthermore, this approach characterized reproducible changes in predominant organisms to the level of culturable isolates, illustrating the utility of combining molecular, microbiologic and computational methods for study of dynamically changing microbial ecosystems.

Many of the time-dependent microbial signatures discovered by our analyses are associated with key events in the host response to *C. rodentium* infection. For example, Consensus Signature Group (CSG) CO9, represented by the genus *Mucispirillum* (OTU#1) in colon, showed a prolonged decrease in infected mice. *In vivo*, *Mucispirillum* inhabits the mucus layer over colonocytes. *C. rodentium*'s adherence to colonic epithelium actively destroys this microenvironment [35–37]. Full regeneration of the mucus layer occurs some time after the pathogen's clearance, providing a possible explanation for the observed delay in *Mucispirillum*'s recolonization of distal colon. This CSG could thus provide a marker for health of the surface mucus layer in distal colon, with potential application to other models of inflammatory colitis. As another example, MC-TIMME identified multiple CSGs exhibiting maximal decreases in infected mice over the second and third weeks post-challenge. This timeframe covers the period of maximal pathogen burden in the distal colon, the height of host symptoms, and the early development of pathogen-specific adaptive immune responses. Interestingly, some of these CSGs were detected in ileum and cecum, sites in which free-living populations of the pathogen reside at considerably lower biomasses than populations in the colon. These microbial signatures, which occur proximal to the primary site of infection, could reflect direct interactions with the pathogen, but given the lower biomasses of pathogen at these sites, may well reflect effects of systemic host responses elicited in response to the colitis. Finally, MC-TIMME detected CSGs exhibiting prolonged increases after clearance of the pathogen, such as CSGs represented by the genus *Clostridium* (OTU#24) in ileum or cecum. During the recovery and convalescent phases, the host resolves local inflammatory responses and actively repairs damaged tissues, a process that takes weeks to complete. The finding that some organisms attain higher biomass during this phase suggests that these taxa may promote or otherwise benefit from changes in intestinal microenvironments that occur during tissue repair and resolution of inflammation, a pattern warranting further evaluation in other models of colitis.

Consensus Signature Groups provide a useful method to compress the potentially enormous variety of dynamic behaviors seen across taxa into smaller sets of interpretable patterns. Many CSGs contained phylogenetically diverse taxa, identifying groups of organisms that may share functional traits needed to compete with the pathogen or otherwise take advantage of microenvironments changing in concert with evolving host responses. These CSGs identify candidate taxa and species to validate in further directed experimental systems. Conversely, CSG analysis also found closely related taxa with dramatically different behaviors during infection, such as members of the families Clostridiaceae and Lactobacillaceae in cecum and ileum. In particular, MC-TIMME used sequencing data to discover three *Lactobacillus* species with different dynamic behaviors, which were subsequently validated with quantitative culture data. Of note, prior studies in *C. rodentium* and other mouse models of colitis have suggested that these three *Lactobacillus* species may behave differently in specific intestinal microenvironments, exerting different effects on the host [38,39].

Analyses on species detectable only with culture also demonstrated significant effects of *C. rodentium* infection on the dynamics of related *Enterobacteriacaea*, namely mouse commensal species of *P. vulgaris* and *E. hormachei*. In the cecum of infected mice, *E. hormachei's* trajectory generally paralleled that of *C. rodentium*. *Enterobacter* species are the most similar phylogenetically and metabolically to *Citrobacter* species [40]. These findings would thus suggest that, rather than directly competing for nutrients, *C. rodentium*'s introduction into the ecosystem results in conditions also favorable for *E. hormachei*, with negligible or negative impact upon *P. vulgaris*. The mechanism(s) underlying these interactions are likely complex, but subject to experimental analysis in *in vitro* co-culture systems and specific association studies in germfree mice, and may include direct interactions among species, as well as host-elicited alterations to the gut luminal environment, stimulated by active infection and necessary immune and epithelial responses that ultimately clear the pathogen from enteric environments.

Our computational model provides a general framework for analyzing data from longitudinal studies of the microbiota. In this study, we used MC-TIMME to analyze 16S rDNA phylotyping and quantitative culture data. The model is also applicable to other time-series of counts, including longitudinal sequence-based

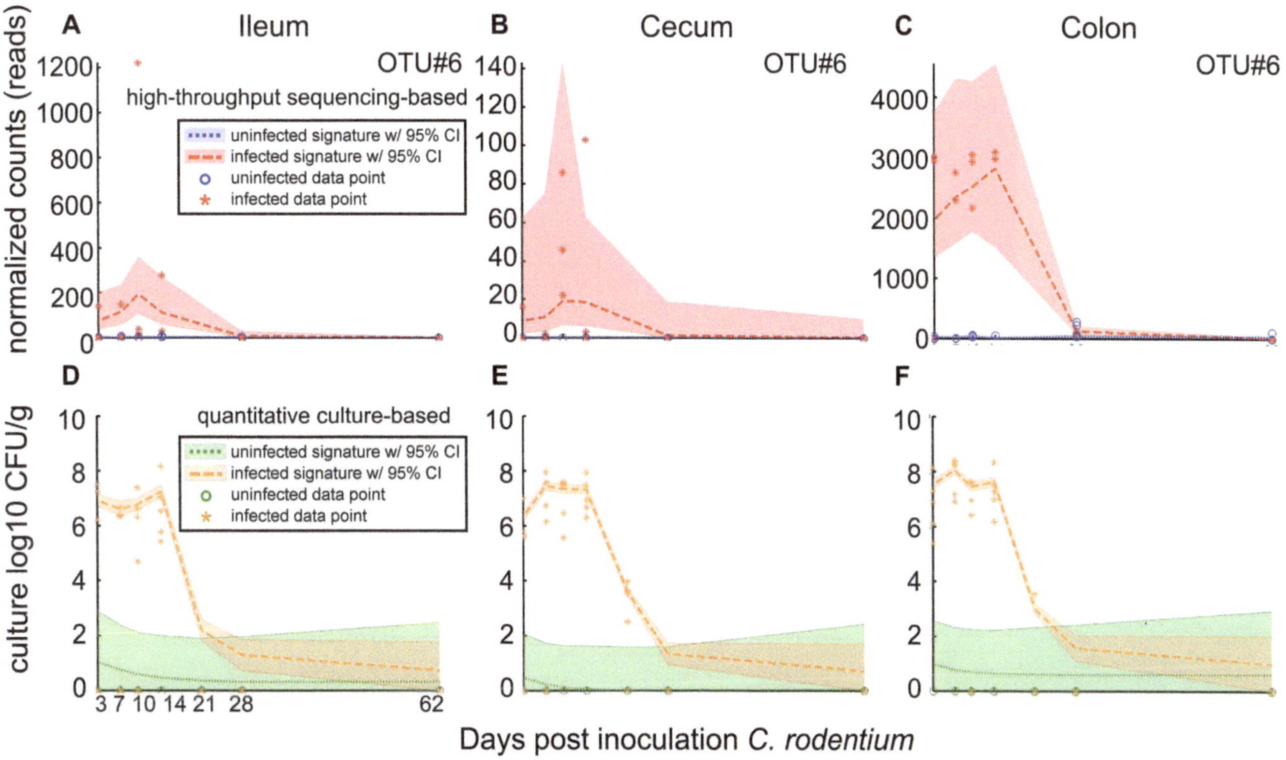

Figure 5. Sequence and culture-based signatures for *Citrobacter rodentium* in all tissues. Horizontal axis indicates days post-inoculation with *C. rodentium*. Dashed lines indicate the inferred median signature shape for each trajectory. Shaded regions indicate the 95% credible interval. **(A, B, C)** Signatures derived from sequencing data for the predominant *C. rodentium* Operational Taxonomic Unit (OTU) in ileum, cecum, and colon. Vertical axis indicates the number of normalized sequencing counts. **(D, E, F)** Signatures derived from culture-based data for *C. rodentium* in ileum, cecum and colon. Vertical axis indicates $\log_{10}$ Colony Forming Units (CFUs) per gram of input tissue. Of note, although *C. rodentium* was not cultured from uninfected mice, the estimated upper bound of the 95% credible interval for the culture counts in uninfected and infected mice trends to $1X10^2$ CFU/g, which was the threshold of detection when using MacConkey agar for selective culture (Table 3).

metagenomic and transcriptomic data. The model's ability to automatically discover and compress a potentially large range of dynamic patterns into smaller and refined sets holds value for extracting relevant signatures from rich time-series datasets. Further, because MC-TIMME uses a fully specified probabilistic model, additional covariates may be readily incorporated into the analysis, which will be essential for supporting future studies linking dynamic changes in the human microbiota to patient phenotypes and outcomes from disease.

Methods

C. rodentium infection and tissue harvest

This study was carried out in strict accordance with the recommendations in the Guide for the Care and Use of Laboratory Animals of the National Institutes of Health. The protocol was approved by the Institutional Animal Care and Use Committee (IACUC) for Brigham & Women's Hospital (Permit Number: A-3431-01). All efforts were made to minimize suffering. Male C57BL/6 mice at 3 weeks of age were purchased from Taconic Farms (Taconic, NY) and maintained in-house for 2 weeks prior to challenge at 5 weeks of age with 5 X 10^7 CFU/mouse of strain DBS100 of *C. rodentium* as described [23,41]. Prior to inoculation, bedding was mixed and distributed among cages twice a week to limit development of varied flora across mice. Post-inoculation, longitudinal tissue samples were harvested at the same time each day. For tissue collection, mice were placed in pre-sterilized plastic containers for up to 30 minutes to allow clearance of fecal pellets from the distal colon through normal defecation. Mice were anesthetized and sacrificed by overdose with volatile isoflurane (Vedco Inc., St. Joseph, MO). The abdomen was sprayed with 70% ethanol to wet the fur. To prevent cross-contamination of intestinal contents across samples harvested from gut tissues, three sets of sterile, UV-irradiated tools were used on each mouse to open skin, the abdominal cavity, and then to remove the digestive tract *en bloc*. Individually sterilized razor blades were used on each tissue location to remove 1.0 cm segments of ileum, 2–5 cm proximal to the ileo-cecal valve, 0.5 cm of cecum taken 0.5 cm proximal from the cecal tip, and terminal 3 cm of distal colon, starting 0.5 cm from the anal canal. Segments of distal colon were harvested that lacked fecal pellets, to insure detection of resident colonic flora.

Ileum, cecum and distal colon (devoid of fecal pellets) from infected mice and uninfected controls were harvested at days 3, 7, 10, 14, 21 and 28 post-inoculation for 16S rDNA gene sequencing and quantitative culture of the pathogen and predominant commensals. Samples to be subjected to 16S rDNA gene sequencing were snap frozen on liquid nitrogen and stored at −80°C until processed for DNA. Samples to be used for culturing were placed in 1.0 mL of pre-reduced phosphate buffered saline (PBS) containing 40 mM L-cysteine-HCl (Sigma Chemical, St. Louis, MO). Remaining segments were placed in 10% zinc-buffered formalin for fixation and paraffin-embedding to evaluate tissue histopathology.

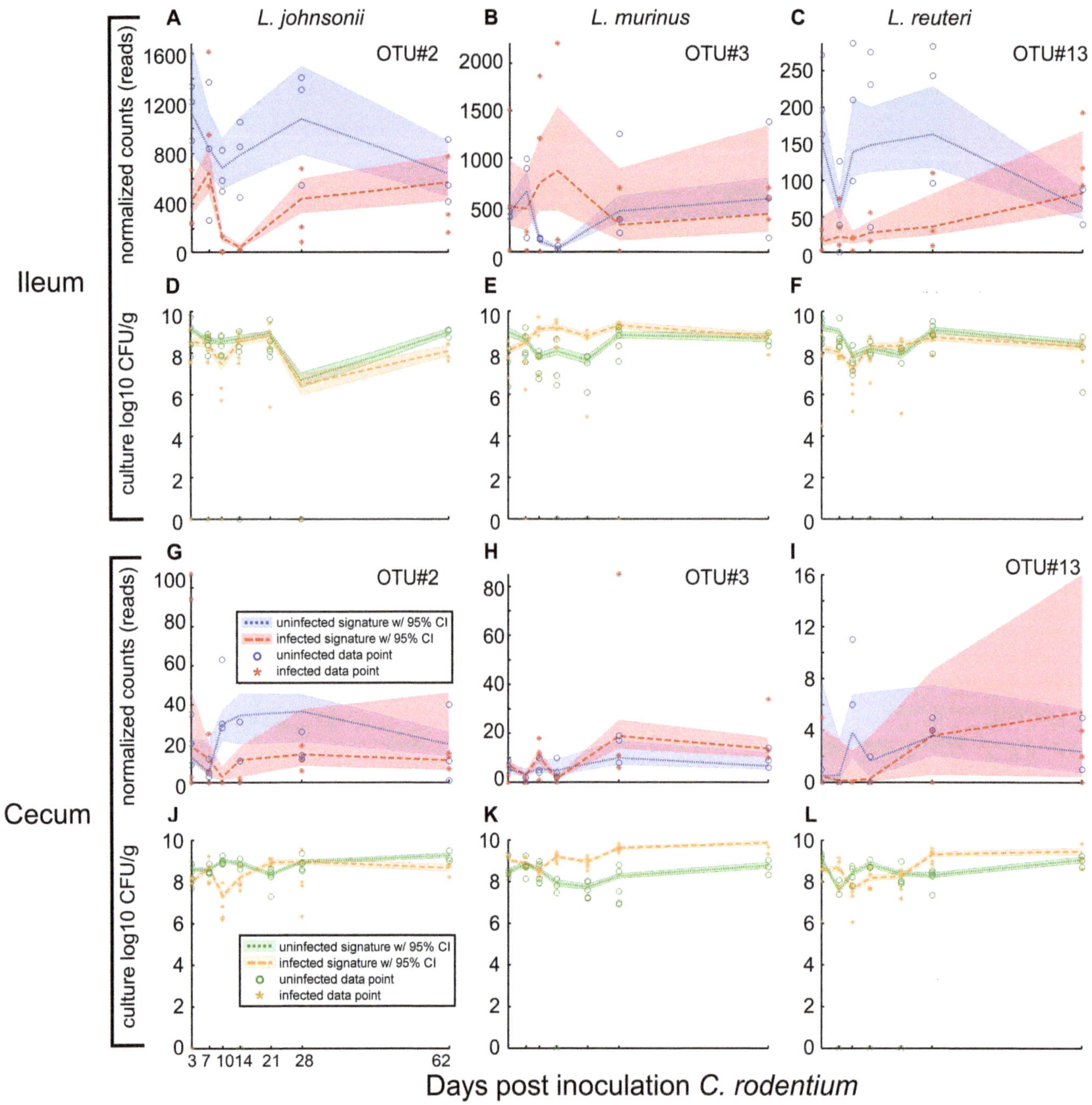

Figure 6. Sequence and culture-based signatures for three *Lactobacillus* species in ileum and cecum. Horizontal axis indicates days post-inoculation with *Citrobacter rodentium*. For sequence-based signatures, vertical axis indicates the number of normalized sequencing counts for the Operational Taxonomic Unit (OTU). For culture-based signatures, vertical axis indicates $\log_{10}$ Colony Forming Units (CFUs) per gram of input tissue. Dashed lines indicate the inferred median signature shape for each trajectory. Shaded regions indicate the 95% credible interval. **(A, G)** Sequence and **(D, J)** culture-based signatures for *Lactobaillus johnsonii*. **(B, H)** Sequence and **(E, K)** culture-based signatures for *Lactobacillus murinus*. **(C, I)** Sequence and **(F, L)** culture-based signatures for *Lactobacillus reuteri*.

Quantitative aerobic and anaerobic culture

Preliminary cultures, conducted prior to infection, were used to identify dominant aerotolerant and obligately anaerobic commensals (data not shown). In particular, efforts were directed at 1) commensal species demonstrating counts of 1 X 10^8 CFU/gram or higher in cecum, and preferably in ileum or distal colon, with the intent of providing "reference" species that could be detected both by culture and pyrosequencing, and 2) members of the Enterobacteriaceae to ascertain if the 16S rDNA V1 and V2 gene regions could discriminate enteric commensals from *C. rodentium*. From this initial screen, species of *Lactobacillus* and were found to be the most prevalent, culturable commensals with biomass > 10^8 CFU/g of tissue. Though members of the *Bacteroidales* and *Clostridiales* were also cultured, these species did not exceed this threshold.

Samples were weighed prior to homogenization in an anaerobic Coy chamber with serial dilution and plating to the media described in Table 2. Quantitative culture was used to obtain

Table 3. Correlation of Sequence and Culture-based Signatures.

Tissue	Organism	Uncentered correlation of culture vs. sequence signatures	Percentile of correlation
Ileum	*C. rodentium*	0.82	98%
	L. johnsonii	0.59	79%
	L. murinus	0.66	94%
	L. reuteri	0.69	91%
Cecum	*C. rodentium*	0.82	98%
	L. johnsonii	0.63	80%
	L. murinus	0.76	94%
	L. reuteri	0.50	80%
Colon	*C. rodentium*	0.89	>99%

species-level counts for the pathogen, *Citrobacter rodentium*, and for the commensals listed in Table 2. With the exception of MacConkey agar with 10 μg/mL of Tetracycline (Sigma Chemical, St. Louis, MO) all agar media listed in Table 2 were commercially purchased from Remel (Lenexa, KS). Aerobic incubation was in a 5% CO_2 humidified incubator at 37°C. Colony types were enumerated at 24 and 48 hours of incubation. Anaerobic incubation was conducted in a Coy chamber with atmosphere of 10% carbon dioxide, 10% hydrogen and 80% nitrogen at 37°C. Plates were incubated for a minimum of 72 hours after which colony types were enumerated and described. Representative isolates from each tissue sample were re-streaked to anaerobic Brucells modified broth (BMB) agar, Gram stained, and aerotolerance tests were performed to confirm obligate anaerobes versus aerotolerant species.

Citrobacter rodentium was identified by growth on MacConkey agar as dark pink, lactose-positive, indole-negative colonies, producing Gram-negative rods by morphology. Representative isolates were typed by API-20E panels (Biomérieux, Durham, NC) and full 16S rDNA sequencing.

Enterobacter hormachei was identified as opaque, light pink, lactose-fermenting, Gram-negative rods on MacConkey agar, and typed by API-20E panel and full 16S rDNA sequence. This species produced larger colonies than *C. rodentium* at 24 hours of growth and thus was distinguishable within dense growth of the pathogen. Putative isolates of *E. hormachei* from acutely infected mice were selected and re-streaked to fresh MAC plates to verify the species.

Lactobacillus johnsonii was identified by growth on CNA agar, producing slender, elongated and non-sporulating Gram-positive rods that were catalase negative. Isolates were susceptible to vancomycin. Full 16S rDNA sequence was used to speciate representative isolates from each time point.

Lactobacillus murinus was identified by growth on CNA and BKV agar (vancomycin-resistant) that produced non-sporulating Gram-positive rods that were catalase negative. Isolates grew on BKV containing vancomycin. Full 16S rDNA sequence was used to speciate representative isolates from each time point.

Lactobacillus reuteri was identified by preferential growth under anaerobic conditions on BMB and CNA agar. Cell morphology by Gram stain showed shorter Gram-positive, non-spore forming rods that were catalase-negative. Full 16S rDNA sequence was used to speciate representative isolates from each time point.

Proteus vulgaris was the only organism to grow on MacConkey agar + 10 μl/mL tetracycline where it produced pale, lactose-negative colonies. Identification was confirmed by API-20E typing of representative isolates from each time point, and by full 16S rDNA sequence.

After final counts of each species at the highest dilution at which it was detected, counts were entered into a spreadsheet containing the starting tissue mass and dilution factor to obtain $\log_{10}$ of colony forming units per gram of input tissue mass ($\log_{10}$ CFU/g).

454 Pyrosequencing of 16S rDNA gene signatures

DNA was extracted from mouse fecal pellets using the MoBio Fecal DNA extraction kit per the manufacturer's specifications (MoBio Laboratories Inc, Carlsbad, CA). Fragments of the 16S rRNA gene, spanning the V1 and V2 hypervariable regions were PCR amplified from each tissue sample using sample-specific barcodes adapted to universal 16S rDNA primers 27F (5′-AGAGTTTGATCMTGGCTCAG-3′) and 338R (5′-GACTCC-TACGGGAGGCWGCAG-3′). DNA was sequenced using a Genome Sequencer FLX and GS-LR70 kit (Roche Applied Sciences, Indiannapolis, IN) at Duke University's IGSP Sequencing Core Facility (Durham, NC). Raw sequence datasets have been deposited at NCBI's Short Read Archive (SRA) under BioProjectID PRJNA202962.

Data preprocessing

Mothur v.1.14.0 was used to preprocess sequences, and to construct and taxonomically classify OTUs [4]. The pipeline used was as follows:

1. Barcodes were trimmed and sequences were filtered based on quality scores with the following parameters used: window average quality score $\geq$ 35, window size = 50, no ambiguous bases, homopolymer length $\leq$ 8, primer differences $\leq$ 2, barcode differences $\leq$ 1, and length between 200 bp and 300 bp.
2. Sequences were then chopped to 200 bp and aligned against the SILVA compatible database provided with mothur. Sequences that started before the 2.5-percentile or ended after the 97.5-percentile in the alignment were filtered out.
3. The ChimeraSlayer and preclustering algorithms implemented in mothur were run.

Table 4 indicates the number of reads after each filtering step.

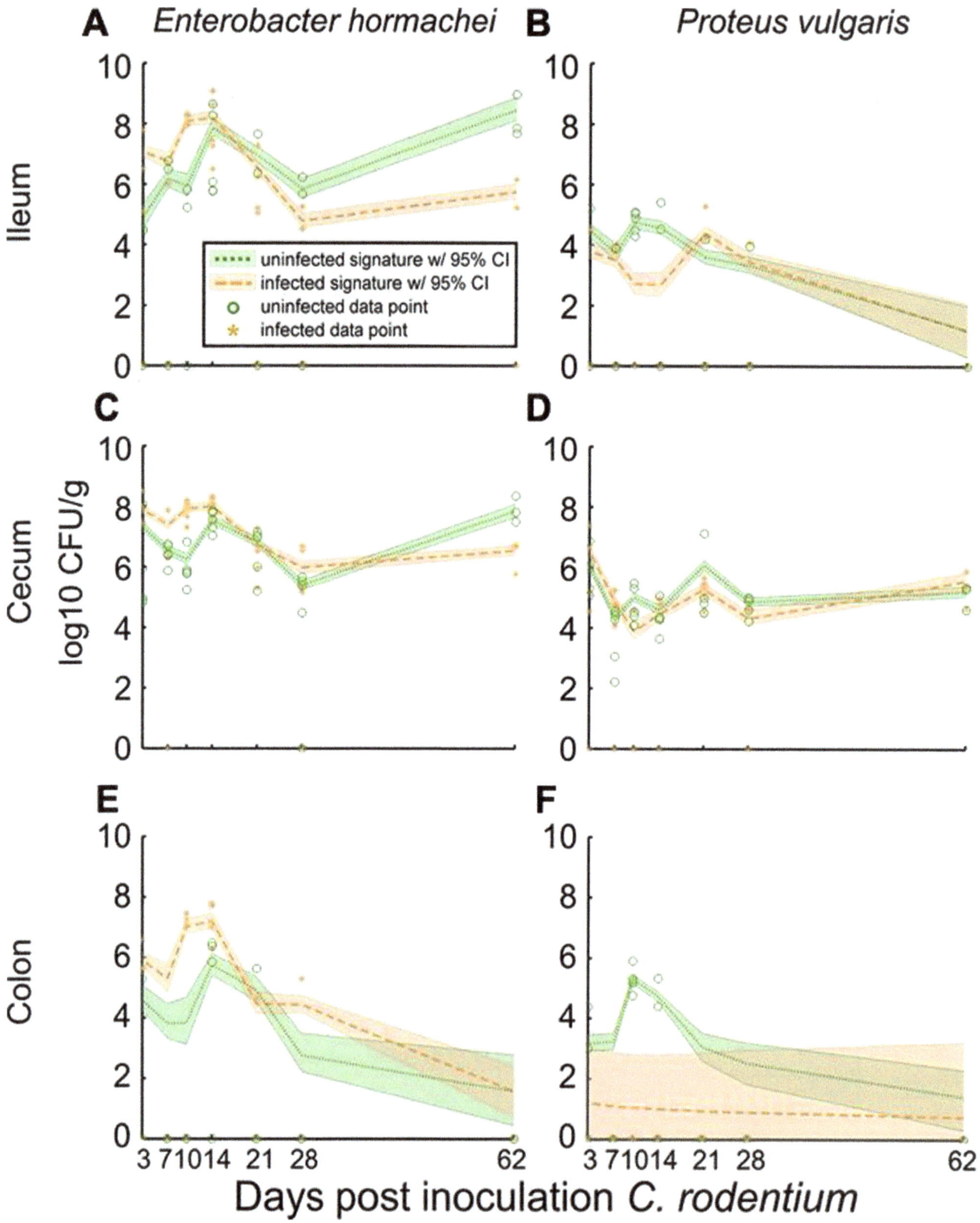

Figure 7. Culture-based signatures for low abundance commensal species phylogenetically related to *Citrobacter rodentium* in ileum, cecum, and colon. Horizontal axis indicates days post-inoculation with *C. rodentium*. Vertical axis indicates log_{10} Colony Forming Units (CFUs) per gram of input tissue. Dashed lines indicate the inferred median signature shape for each trajectory. Shaded regions indicate the 95% credible interval. **(A, C, E)** Signatures for *Enterobacter hormachei*. **(B, D, F)** Signatures for *Proteus vulgaris*.

Table 4. Filtering steps in bioinformatics pipeline and remaining sequencing reads.

Filtering step	Sequences remaining after filter
None	678,247
QC scores, ambiguous bases, homopolymers, min/max length, match barcodes	300,462
Alignment to 16S rDNA reference database	294,996
ChimeraSlayer	274,882

Sequences were assigned to OTUs using the furthest neighbor method and a threshold of 97% similarity. Distances between sequences were calculated in mothur, and sequences were assigned to OTUs using the furthest neighbor method with a threshold of 97% similarity. **OTU numbers were assigned by the mothur pipeline, which orders OTUs in descending order by the sum of the sequencing reads across all samples assigned to the OTU**. The consensus taxonomic classification for each OTU was determined using the naïve Bayes classifier [33] implemented in mothur against the RDP database (bootstrap cutoff of 60%).

After filtering OTUs that failed to have ≥5 reads in ≥3 samples, 210 OTUs were available for input to MC-TIMME. Relationships among the 210 OTUs were visualized by building a tree of the most abundant sequence from each OTU and rendering the results using the Dendroscope software [42]. The tree was constructed using mothur, by aligning sequences against the included ARB SILVA reference database, calculating distances between sequences using the dist.seqs command (default options), and then building a neighbor joining tree with the clearcut command (default options).

Sequencing count normalization

To make sequence counts for OTUs comparable across samples, we used a nonparametric regression method, Locally Weighted Scatterplot Smoothing (LOWESS). This technique allows for nonlinear normalization, which has been shown to be important for data in which error characteristics differ substantially at the extreme ranges of values such as with DNA microarray data [43]. This error characteristic is evident in our data, as shown in Figure S1. The LOWESS normalization on our data compensates for differing numbers of sequencing reads in samples, with allowance for a nonlinear relationship for samples with very low or high numbers of reads.

To perform the LOWESS regression, all OTU counts across all samples were included in the regression. The independent variable for the regression was the relative abundance of the OTU in the sample, and the dependent variable was the counts for the OTU in the sample. We used the MATLAB 2010b (MathWorks, Natick, MA) malowess function with the robust regression option and with second degree curves to perform the analysis. Counts were then rescaled and rounded based on the LOWESS output. These normalized counts were then used as input for the MC-TIMME algorithm (Dataset S1).

Microbial Counts Trajectories Infinite Mixture Model Engine (MC-TIMME)

MC-TIMME is based on Dirichlet Process or Infinite Mixture Models, a class of nonparametric Bayesian models in which data is assumed to be generated from a weighted mixture of a potentially unlimited number of components [29]. Mixture components correspond to prototype signatures, with component k defined by a set of signature shape variables δ_{kt} and a variable controlling data variance ε_k. Observed count data is modeled using a Generalized Linear Model (GLM) with a Negative Binomial Distribution (NDB) data model with mean dependent on δ_{kt} and variance controlled by ε_k. The model also specifies prior probability distributions on model parameters controlled by hyperparameters. We extended the original MC-TIMME model to use a continuous-time model of dynamics based on a Gaussian random with time-scaled variance (Protocol S1). We derived an efficient Markov Chain Monte Carlo (MCMC) approximate inference algorithm with Gibbs sampling steps for assignments of time-series of counts for OTUs to prototype signatures, and specialized Metropolis-Hastings (MH) steps for trajectory variable and hyperparameter updates. We used 25,000 MCMC iterations for burn-in and then an additional 75,000 iterations to estimate the posterior distribution.

Detecting signature changes in response to infection

Our model is fully Bayesian and thus does not use p-values. We instead directly estimate probabilities of signatures differing (Protocol S1). We used the following criteria to detect changes in OTU signatures in response to introduction of the pathogen: a) the posterior probability of infected and uninfected samples sharing a prototype signature was <5%, and b) the 95% credible intervals for infected and uninfected signatures did not overlap for at least two time-points. Using these criteria, we estimated the false rate of detection at <5% across our dataset.

Correlation analyses of sequence and culture-based trajectories:

To assess the similarity of culture and sequence-derived signatures, we performed the following analysis. For the culture-derived signatures, we computed the difference between signatures in the infected and uninfected states for each organism in each tissue site. We performed analogous computations for all sequence-derived OTU signatures (including for OTUs for which there was no corresponding culture data). Then, for each differenced culture-derived signature, we computed its uncentered correlation coefficient [44] against all the differenced sequence-derived signatures. This analysis provided an indication of how similar or different a given culture signature was from all OTU signatures analyzed in the longitudinal dataset, including the OTU signature associated with the given isolate cultured. We then computed the percentage of these correlations that were less than the actual correlation between the culture-derived differenced signature and the corresponding sequence-derived differenced signature. For the computation of this percentage, we excluded OTUs classified to the same genus as the cultured organism being evaluated.

Supporting Information

Figure S1 Locally Weighted Scatterplot Smoothing (LOWESS) normalization of sequence count data. The LOWESS non-parametric regression method was applied to sequencing data to normalize counts obtained across samples. Data (blue plus signs) are plotted with the relative abundance as the independent variable and sequence counts as the dependent variable. Each data point represents an Operational Taxonomic Unit (OUT) from a single sample (a particular time-point, tissue and replicate). The red line indicates the best linear fit to the data. Deviations from the linear fit are evident at the lower and upper range of data values. The black dots represent the fitted values from the LOWESS regression. The LOWESS estimated values were rounded down and used as the effective number of reads for each OTU.

Protocol S1 Detailed description of the MC-TIMME algorithm and associated computational methods.

Table S1 Sample key.

Table S2 Sample barcode key for samples submitted for 16S rRNA gene sequencing.

Dataset S1 LOWESS normalized counts for Operational Taxonomic Units from 16S rDNA phylotyping data.

Dataset S2 Colony Forming Units per gram of input tissue for cultured organisms.

Acknowledgments

The authors would like to thank the Duke University Sequencing Core for Roche 454 sequencing services and Laurie Comstock, Wendy Garrett, David Gifford, Curtis Huttenhower, Neil Herring and Wayne Lencer for critical reading of the manuscript.

Author Contributions

Conceived and designed the experiments: CB GG GR CC AH AO LB. Performed the experiments: CB GG GR MD AD QL VB VY AH AO CC LB. Analyzed the data: CB GG GR MD AD VB AH AO CC LB. Contributed reagents/materials/analysis tools: GG AH AO CC LB. Wrote the paper: CB GG AH AO LB.

References

1. Metzker ML (2010) Sequencing technologies - the next generation. Nat Rev Genet 11: 31–46.
2. Caporaso JG, Kuczynski J, Stombaugh J, Bittinger K, Bushman FD, et al. (2010) QIIME allows analysis of high-throughput community sequencing data. Nat Methods 7: 335–336.
3. Hamady M, Lozupone C, Knight R (2010) Fast UniFrac: facilitating high-throughput phylogenetic analyses of microbial communities including analysis of pyrosequencing and PhyloChip data. ISME J 4: 17–27.
4. Schloss PD, Westcott SL, Ryabin T, Hall JR, Hartmann M, et al. (2009) Introducing mothur: open-source, platform-independent, community-supported software for describing and comparing microbial communities. Appl Environ Microbiol 75: 7537–7541.
5. Arumugam M, Raes J, Pelletier E, Le Paslier D, Yamada T, et al. (2011) Enterotypes of the human gut microbiome. Nature 473: 174–180.
6. Group NHW, Peterson J, Garges S, Giovanni M, McInnes P, et al. (2009) The NIH Human Microbiome Project. Genome Res 19: 2317–2323.
7. Human Microbiome Jumpstart Reference Strains C, Nelson KE, Weinstock GM, Highlander SK, Worley KC, et al. (2010) A catalog of reference genomes from the human microbiome. Science 328: 994–999.
8. Manichanh C, Rigottier-Gois L, Bonnaud E, Gloux K, Pelletier E, et al. (2006) Reduced diversity of faecal microbiota in Crohn's disease revealed by a metagenomic approach. Gut 55: 205–211.
9. Round JL, Mazmanian SK (2009) The gut microbiota shapes intestinal immune responses during health and disease. Nat Rev Immunol 9: 313–323.
10. Turnbaugh PJ, Gordon JI (2009) The core gut microbiome, energy balance and obesity. J Physiol 587: 4153–4158.
11. Turnbaugh PJ, Ridaura VK, Faith JJ, Rey FE, Knight R, et al. (2009) The effect of diet on the human gut microbiome: a metagenomic analysis in humanized gnotobiotic mice. Sci Transl Med 1: 6ra14.
12. Dethlefsen L, Relman DA (2011) Incomplete recovery and individualized responses of the human distal gut microbiota to repeated antibiotic perturbation. Proc Natl Acad Sci U S A 108 Suppl 1: 4554–4561.
13. Ravel J, Gajer P, Abdo Z, Schneider GM, Koenig SS, et al. (2011) Vaginal microbiome of reproductive-age women. Proc Natl Acad Sci U S A 108 Suppl 1: 4680–4687.
14. Borenshtein D, McBee ME, Schauer DB (2008) Utility of the Citrobacter rodentium infection model in laboratory mice. Curr Opin Gastroenterol 24: 32–37.
15. Mundy R, MacDonald TT, Dougan G, Frankel G, Wiles S (2005) Citrobacter rodentium of mice and man. Cell Microbiol 7: 1697–1706.
16. Ghosh S, Dai C, Brown K, Rajendiran E, Makarenko S, et al. (2011) Colonic microbiota alters host susceptibility to infectious colitis by modulating inflammation, redox status, and ion transporter gene expression. Am J Physiol Gastrointest Liver Physiol 301: G39–49.
17. Hoffmann C, Hill DA, Minkah N, Kirn T, Troy A, et al. (2009) Community-wide response of the gut microbiota to enteropathogenic Citrobacter rodentium infection revealed by deep sequencing. Infect Immun 77: 4668–4678.
18. Lupp C, Robertson ML, Wickham ME, Sekirov I, Champion OL, et al. (2007) Host-mediated inflammation disrupts the intestinal microbiota and promotes the overgrowth of Enterobacteriaceae. Cell Host Microbe 2: 119–129.
19. Wlodarska M, Willing B, Keeney KM, Menendez A, Bergstrom KS, et al. (2011) Antibiotic treatment alters the colonic mucus layer and predisposes the host to exacerbated Citrobacter rodentium-induced colitis. Infect Immun 79: 1536–1545.
20. Wiles S, Pickard KM, Peng K, MacDonald TT, Frankel G (2006) In vivo bioluminescence imaging of the murine pathogen Citrobacter rodentium. Infect Immun 74: 5391–5396.
21. Bishop AL, Wiles S, Dougan G, Frankel G (2007) Cell attachment properties and infectivity of host-adapted and environmentally adapted Citrobacter rodentium. Microbes Infect 9: 1316–1324.
22. Mangan PR, Harrington LE, O'Quinn DB, Helms WS, Bullard DC, et al. (2006) Transforming growth factor-beta induces development of the T(H)17 lineage. Nature 441: 231–234.
23. Bry L, Brigl M, Brenner MB (2006) CD4+-T-cell effector functions and costimulatory requirements essential for surviving mucosal infection with Citrobacter rodentium. Infect Immun 74: 673–681.
24. White JR, Nagarajan N, Pop M (2009) Statistical methods for detecting differentially abundant features in clinical metagenomic samples. PLoS Comput Biol 5: e1000352.
25. Bar-Joseph Z (2004) Analyzing time series gene expression data. Bioinformatics 20: 2493–2503.
26. Bar-Joseph Z, Gerber G, Simon I, Gifford DK, Jaakkola TS (2003) Comparing the continuous representation of time-series expression profiles to identify differentially expressed genes. Proc Natl Acad Sci U S A 100: 10146–10151.
27. Ernst J, Nau GJ, Bar-Joseph Z (2005) Clustering short time series gene expression data. Bioinformatics 21 Suppl 1: i159–168.
28. Ramoni MF, Sebastiani P, Kohane IS (2002) Cluster analysis of gene expression dynamics. Proc Natl Acad Sci U S A 99: 9121–9126.
29. Gerber GK, Onderdonk AB, Bry L (2012) Inferring dynamic signatures of microbes in complex host ecosystems. PLoS Comput Biol 8: e1002624.
30. Good IJ (1953) The Population Frequencies of Species and the Estimation of Population Parameters. Biometrika 40: 237–264.
31. Chao A, Shen T (2003) Nonparametric estimation of Shannon's index of diversity when there are unseen species in sample. Environmental and Ecological Statistics 10: 429–443.
32. Cole JR, Chai B, Farris RJ, Wang Q, Kulam SA, et al. (2005) The Ribosomal Database Project (RDP-II): sequences and tools for high-throughput rRNA analysis. Nucleic Acids Res 33: D294–296.
33. Wang Q, Garrity GM, Tiedje JM, Cole JR (2007) Naive Bayesian classifier for rapid assignment of rRNA sequences into the new bacterial taxonomy. Appl Environ Microbiol 73: 5261–5267.
34. Hintz HF, Schryver HF, Stevens CE (1978) Digestion and absorption in the hindgut of nonruminant herbivores. Journal of Animal Science 46: 1803–1807.
35. Bergstrom KS, Kissoon-Singh V, Gibson DL, Ma C, Montero M, et al. (2010) Muc2 protects against lethal infectious colitis by disassociating pathogenic and commensal bacteria from the colonic mucosa. PLoS Pathog 6: e1000902.
36. Linden SK, Florin TH, McGuckin MA (2008) Mucin dynamics in intestinal bacterial infection. PLoS One 3: e3952.
37. Robertson BR, O'Rourke JL, Neilan BA, Vandamme P, On SL, et al. (2005) Mucispirillum schaedleri gen. nov., sp. nov., a spiral-shaped bacterium

colonizing the mucus layer of the gastrointestinal tract of laboratory rodents. Int J Syst Evol Microbiol 55: 1199–1204.
38. Chen CC, Louie S, Shi HN, Walker WA (2005) Preinoculation with the probiotic Lactobacillus acidophilus early in life effectively inhibits murine Citrobacter rodentium colitis. Pediatr Res 58: 1185–1191.
39. Johnson-Henry KC, Nadjafi M, Avitzur Y, Mitchell DJ, Ngan BY, et al. (2005) Amelioration of the effects of Citrobacter rodentium infection in mice by pretreatment with probiotics. J Infect Dis 191: 2106–2117.
40. Murray PR, Holmes B, Aucken HM (2010) Citrobacter, Enterobacter, Klebsiella, Plesiomonas, Serratia, and Other Members of the Enterobacteriaceae. Topley and Wilson's Microbiology and Microbial Infections. Hoboken: John Wiley & Sons, Ltd.
41. Bry L, Brenner MB (2004) Critical role of T cell-dependent serum antibody, but not the gut-associated lymphoid tissue, for surviving acute mucosal infection with Citrobacter rodentium, an attaching and effacing pathogen. J Immunol 172: 433–441.
42. Huson DH, Richter DC, Rausch C, Dezulian T, Franz M, et al. (2007) Dendroscope: An interactive viewer for large phylogenetic trees. BMC Bioinformatics 8: 460.
43. Yang YH, Dudoit S, Luu P, Lin DM, Peng V, et al. (2002) Normalization for cDNA microarray data: a robust composite method addressing single and multiple slide systematic variation. Nucleic acids research 30: e15.
44. Rodgers JL, Nicewander WA (1988) Thirteen Ways to Look at the Correlation Coefficient. The American Statistician 42: 59–66.

2

Soil Bacterial Communities Respond to Mowing and Nutrient Addition in a Steppe Ecosystem

Ximei Zhang[1], Quansheng Chen[2], Xingguo Han[1,2]*

1 State Key Laboratory of Forest and Soil Ecology, Institute of Applied Ecology, Chinese Academy of Sciences, Shenyang, China, **2** State Key Laboratory of Vegetation and Environmental Change, Institute of Botany, Chinese Academy of Sciences, Beijing, China

Abstract

In many grassland ecosystems, nitrogen (N) and phosphorus (P) are added to improve plant productivity, and the aboveground plant biomass is mowed and stored as hay for the bullamacow. Nutrient addition and mowing affect the biodiversity and ecosystem functioning, and most of the previous studies have primarily focused on their effects on macro-organisms, neglecting the responses of soil microbial communities. In this study, we examined the changes in three community attributes (abundance, richness, and composition) of the entire bacterial kingdom and 16 dominant bacterial phyla/classes in response to mowing, N addition, P addition, and their combinations, by conducting a 5-year experiment in a steppe ecosystem in Inner Mongolia, China. Overall, N addition had a greater effect than mowing and P addition on most of these bacterial groups, as indicated by changes in the abundance, richness and composition in response to these treatments. N addition affected these soil bacterial groups primarily through reducing soil pH and increasing available N content. Meanwhile, the 16 bacterial phyla/classes responded differentially to these experimental treatments, with Acidobacteria, Acidimicrobidae, Deltaproteobacteria, and Gammaproteobacteria being the most sensitive. The changes in the abundance, richness, and composition of various bacterial groups could imply some potential shift in their ecosystem functions. Furthermore, the important role of decreased soil pH caused by N addition in affecting soil bacterial communities suggests the importance of restoring acidified soil to maintain soil bacterial diversity.

Editor: Stefan Bereswill, Charité-University Medicine Berlin, Germany

Funding: This work was supported by the National Natural Science Foundation (30830026) and Postdoctoral Science Foundation (2012T50158, 2011M500440) of China. The funders had no role in study design, data collection and analysis, decision to publish, or preparation of the manuscript.

Competing Interests: The authors have declared that no competing interests exist.

* E-mail: xghan@ibcas.ac.cn

Introduction

With the growth of human population and improvement in the living standard, there is an urgent demand for the development of livestock husbandry. The traditional strategy used for the development of graziery is to increase the bullamacow number; however, this has led to the degeneration of grassland ecosystems [1–4]. To reconcile the conflict between graziery development and ecosystem conservation, it is suggested that specific methods (such as addition of nitrogen (N) and phosphorus (P) and mowing) be adopted to improve forage grass productivity in a small area of grassland and leave a large part of the grassland area for maintaining its other ecosystem functions [5,6]. Mowing and nutrient addition have profound influences on the biodiversity and ecosystem functioning, and previous studies have mainly focused on their influence on higher organisms [5–7]. Although soil microbial communities are among the most abundant and diverse groups of organisms on Earth and are responsible for numerous key ecosystem processes [8–10], their response to these treatments and, especially, their combinations, has not been comprehensively explored.

Because anthropogenic N deposition is one of the major environmental changes that have greatly altered the processes, functions and services of various terrestrial ecosystems [11], its influence on soil microbial communities and their associated functions has received much attention in the past few decades. Actually, only the response of some microbial functional groups (e.g. the ammonia-oxidizing bacteria and denitrifying bacteria) or the entire soil microbial (or bacterial) community as a whole was often investigated [12,13]. In particular, the changes in some community attributes (e.g. abundance, diversity and composition), enzyme activities and functional indices (e.g. N cycling rates) were often detected, and these changes were even found to be interrelated sometimes [14]. For example, it was found that in temperate hardwood and pine forests, chronic N deposition reduced the bacterial/fungal biomass ratio and the activity of a fungal lignin-degrading enzyme, and it might be these reductions that caused the decreased litter decomposition rates and altered N cycling rates [15]. Soil microbial communities comprise diverse taxonomic groups (e.g. different phyla/classes) with different phylogenetic histories, physiological traits and ecological functions; however, the response of various taxonomic groups to the environmental changes has not been investigated comprehensively due to the technological limitation [16]. For example, the traditional Sanger DNA sequencing technology was too labor-intensive for the simultaneous measurement of community structure of various bacterial phyla/classes. Fortunately, the current metagenomic approaches that exploit next-generation sequencing technologies (such as the pyrosequencing) can acquire millions or even billions of DNA sequences in only a run, thus enabling the investigation of the diversity and composition of various bacterial phyla/classes rapidly, accurately and simultaneously [17,18].

Carbon (C), N and P are the most basic compositional elements of an organism. Logically, mowing removes much plant biomass from the ecosystem, and the main effect of mowing on soil microbial communities is the reduction in the amount of C resource (or energy resource) supplied to soil microorganisms [19,20]. In contrast, N and P addition increases the content of soil nutrients supplied to the soil microbes. Thus, these human activities will likely affect the abundance, richness, and composition of soil microbial communities and various microbial taxonomic groups may respond differently. Meanwhile, a relatively stable C/N/P ratio exists in the elemental composition of a given soil microbial taxonomic group [21,22]. Because mowing, N and P addition will change the stoichiometric ratios (including C/N, C/P and N/P) of these elements in the soil, these changes will also affect soil microbial communities. In addition, these human activities may even have interactive effects on various microbial communities. For example, P addition was found to only affect the abundance of soil ammonia-oxidizing bacteria in the presence of added N, and not in the absence of N addition [23]. Furthermore, besides the contents of soil C, N and P, these human activities may change other soil physicochemical factors, which will also lead to the alteration in soil microbial communities. For example, it was found that N addition not only leads to increased soil total N content, but also changes in the NH_4^+-N content, NO_3^--N content and pH, as well as changes in the heterogeneity of the four indices [24]. And N addition also leads to the changes in the biomass, diversity and composition of the plant communities. All these changes may further lead to the alteration in the soil microbial abundance, diversity and composition [24]. It is worth noting that soil pH has been found to be the most important ecological factor determining soil bacterial community structure [18,25], suggesting that the changes in soil pH may be the primary route through which N addition alters the bacterial community. Overall, there is an urgent need to study the effects of mowing, N addition, P addition and their combinations on soil microbial communities and to elucidate the underlying mechanisms.

The semiarid temperate steppe in northern China is an important part of the Eurasian grassland biome. Adoption of treatments such as mowing and nutrient addition has been suggested to improve plant productivity in this area [5,6]. To comprehensively examine the response of soil microbial communities to these managements in the steppe ecosystem, a field manipulative experiment of mowing, N addition, and P addition with eight treatments (control, mowing, N addition, P addition, mowing and N addition, mowing and P addition, N and P addition, and simultaneous mowing, N addition and P addition) was conducted from 2005. The pyrosequence technology targeting bacterial 16S rRNA gene was used to measure the structure of soil bacterial communities [17]. The specific questions of this study were as follows: (1) Whether and how the abundance, richness, and composition of the entire soil bacterial kingdom and the dominant phyla/classes get affected by mowing, N addition, P addition, and their combinations? (2) What are the ecological mechanisms of these treatments altering soil microbial communities?

Materials and Methods

Study site and experimental design

This study was part of a long-term experiment conducted at the Duolun Restoration Ecology Station of Institute of Botany, Chinese Academy of Sciences, approximately 30 km from Duolun County (42°02′N, 116°17′E), Inner Mongolia Autonomous Region of China. Our field studies did not involve endangered or protected species, so no specific permissions were required for the location/activity. The experimental site was a typical temperate zone characterized by a semiarid continental monsoon climate. Mean annual temperature was 2.1°C with monthly mean temperature ranging from 18.9°C in July to −17.5°C in January. Mean annual precipitation was about 385.5 mm with 80% precipitation occurred from June to September. Soil was chestnut soil (Chinese classification), corresponding to Calcis-orthic Aridisol in the US Soil Taxonomy classification, with sand, silt, and clay being 62.7%, 20.3%, and 17.0%, respectively [26,27]. Mean soil bulk density was 1.31 g/cm^3. This temperate steppe was dominated by perennials, including *Stipa krylovii*, *Artemisia frigida*, *Potentilla acaulis*, *Cleistogenes squarrosa*, *Allium bidentatum*, and *Agropyron cristatum*.

This experiment commenced from 2005. The effects of mowing, N, P, and their combinations were investigated using a nested design with mowing as the primary factor and nutrient addition as the secondary factor. There were four replicates for each of the eight treatments. Within a 199 m×265 m area, eight 60 m×92 m primary plots were set-up with a 5-m-wide buffer zone among plots. Four primary plots were randomly assigned to the mowing treatment and the other four were controls. Aboveground plants were mowed on 20 August every year, leaving only 10 cm of stubble. Within each of the eight primary plots, four 28 m×44 m secondary plots were set-up, with a 1-m buffer zone between them. Each of the four secondary plots was randomly assigned to N addition, P addition, simultaneous N and P addition, and control treatments. N was added in the form of urea in 2005 and NH_4NO_3 in 2006–2010 at a rate of 10 g N m^{-2} y^{-1}. P was added in the form of calcium superphosphate at a rate of 5 g PO_4^{-3} m^{-2} y^{-1}. Both N and P were added on a rainy day in the middle of July every year.

Sampling and measurement of bacterial abundance and community structure

Soil samples were taken on 22 August of 2010. Four soil cores (10 cm deep, 3.5 cm diameter) were collected from each secondary plot at random and thoroughly mixed. Soil total C (TC) content was quantified with the potassium dichromate–vitriol oxidization method [28]. Soil total N (TN) content was measured using an Alpkem autoanalyzer (Kjektec System 1026 Distilling Unit, Sweden) according to the Kjeldahl acid–digestion method [28]. Soil total P concentrations (TP; % of dry mass) were measured by the ammonium molybdate method after persulfate oxidation [28]. Soil available N content was determined on a FIAstar 5000 Analyzer (Foss Tecator, Denmark) after extraction of fresh soil with 1 mol/L KCl. Soil pH was measured in 1:2.5 (W/V) suspensions of soil in distilled water. Soil water content was determined as the weight loss after drying for 24 h at 105°C. For each soil sample, we extracted DNA from 0.5 g of mixed soil using the Fast DNA SPIN kit for soil according to the manufacturer's instructions (Qbiogene, Carlsbad, CA, USA); however, we used 350 µL instead of 50 µL DNA elution solution to elute the DNA in the tenth step of the procedure. Purity and quantity of DNA were checked using a Nanodrop® ND-1000 UV–vis Spectrophotometer (Thermo, USA). The DNA solution was then stored at −20°C until analysis.

To estimate the abundance of the entire bacterial community, the content of bacterial 16S rRNA gene was measured using real-time PCR according to a procedure similar to that described by Fierer et al. [16]. The standard template was a plasmid containing a copy of the *Escherichia coli* 16S rRNA gene, and the standard curve was generated using a 10-fold serial dilution of the template across five orders of magnitudes ($2.73\times10^4 \sim 2.73\times10^8$ copies). The 20 μL PCR reaction mixtures contained 10 μL SYBR Premix (2×) (TaKaRa Biotechnology Co., Ltd., China), 0.4 μL each of 10 μmol/L forward and reverse primers (Eub338: 5′-ACT CCT ACG GGA GGC AGC AG-3′; Eub518: 5′-ATT ACC GCG GCT GCT GG-3′), 0.4 μL Rox II, 2 μL BSA (10 mg/mL), and 5.8 μL sterile and DNA-free water. The amount of standard and soil DNA samples added per reaction was 1.0 μL (1.2–5.1 ng). The reaction was conducted with a Roche LightCycler™ Real-time PCR system using the following program: 95°C for 1 min followed by 40 cycles of 95°C for 5 sec, 55°C for 15 sec and 72°C for 15 sec. The melting curve analysis was performed to confirm PCR product specificity after amplification by measuring fluorescence continuously as the temperature increased from 65°C to 95°C. Gel electrophoresis analyses were also conducted to confirm that the amplified products were of the appropriate size. The equation $\text{Eff} = [10^{(-1/\text{slope})} - 1]$ was used to calculated the amplification efficiencies and resulted in the value of 97%. The bacterial 16S rRNA gene copy number was calculated using a regression equation that related the cycle threshold (Ct) value to the known number of copies in the standards. Three no-template controls were run for each quantitative PCR assay. For each soil sample, the qPCR reactions were repeated three times. We added BSA to the PCR reaction mixtures to reduce the inhibitory effects of co-extracted polyphenolic compounds in the soil. Additionally, three rounds of PCR were conducted after adding known amounts of standard plasmid with the soil DNA extract to estimate the possible inhibitory effects of co-extracted polyphenolic compounds. The inhibitory effects were found to be negligible.

The method of 454 pyrosequences was used to measure the bacterial community structure of each soil sample. The primers 27F (5′-AGA GTT TGA TCC TGG CTC AG-3′) and 338R (5′-TGC TGC CTC CCG TAG GAG T-3′) were used to amplify the fragment of 16S rRNA gene. To measure all 32 samples in a run, a unique 10-mer tag for each sample was added to the 5′-end of the primer 338R [17]. Each 20 μl PCR mixture contained 4 μl FastPfu Buffer (5×; Transgen), 2 μl of 2.5 mM dNTPs, 0.4 μl of each primer (5 μM), 0.8 μl of DNA template, and 0.4 μl of FastPfu Polymerase (Transgen). The PCR protocol was as follows: 95°C for 2 min (denature); 25 cycles of 95°C for 30 sec (denature), 55°C for 30 sec (anneal), 72°C for 30 sec (elongate); and 72°C for 5 min (elongate). Three replicates of PCR were performed for each sample, after which the products were combined and purified by agarose gel electrophoresis, recovered, and quantified with Pico-Green using a TBS-380 Mini-Fluorometer. Equal molar concentrations of PCR products for each sample were then pooled and sequenced in a Roche 454 Genome Sequencer FLX Titanium system at Shanghai Majorbio Bio-pharm Technology Co., Ltd. The sequence reads for all samples have been deposited in the National Center for Biotechnology Information Sequence Reads Archive (accession no. SRA057669).

Data analysis

The pyrosequence reads were analyzed mainly with the Mothur software (Version 1.19) [29]. Briefly, the raw reads were first assigned to samples according to their tags and then the standard primers and barcodes were trimmed off, after which reads with length less than 150 bp or with ambiguous characters were removed. The chimeric sequences were also excluded by the chimera.uchime command with default parameters. While the existence of pyronoise may lead to the overestimation of bacterial OTU richness, the removal of pyronoise may remove some actual but rare sequences and further result in the underestimation of OTU richness. Therefore, the step of pyronoise removal was not adopted in this study, and the influence of pyronoise was taken as a systematic error. The V3 region of 16S rRNA gene of the remaining reads were aligned to the Silva database (Version 106) to determine their taxonomic classification and non-bacterial reads were further removed [30]. To minimize the influence of unequal sampling on the following calculated indices, we randomly selected 3,478 reads for each sample. All these sequences (3,478×32) were clustered into OTUs (operational taxonomic units) with larger than 97% similarity, and the consensus taxonomy for an OTU was determined using classify.otu command with confidence threshold 80%. The OTU number of 3,478 reads for each sample was used to represent the OTU richness of the entire bacterial community. For each pair of samples, the Bray-Curtis distance basing on OTU abundance was calculated to represent the compositional variation of the entire bacterial community [31]. Briefly, we first calculated the difference in the number of reads for each OTU within this pair of samples, and then calculated the sum of the absolute values of these differences for all OTUs. Finally, we divided the sum by 6,956 (3,478×2) to represent the Bray-Curtis distance.

For each of the 16 dominant bacterial phyla/classes, (its sequence number in all 3,478 sequences)/3,478 was calculated to represent its relative abundance. To compare the OTU richness of each phylum/class among different samples, we used a different calculation approach from that for the entire bacterial community, because there was different number of sequences for the same phylum/class among different samples. For each phylum/class, we randomly selected a certain number of representative sequences from a sample and calculated the OTU number represented by these sequences, repeated this process 1,000 times and used the mean OTU number to represent its richness. For each of the 16 phyla/classes, we also calculated the Bray-Curtis distance basing on the abundance of OTUs to represent the compositional variation between each pair of samples [31].

Three-way analysis of variance (ANOVA) was used to determine the main and interactive effects of mowing, N addition and P addition on the nine soil physicochemical indices, including soil TC content, TN content, TP content, the three stoichiometric characteristics (TC/TN, TC/TP, and TN/TP), available N content, pH and water content. Three-way ANOVA was also used to determine the main and interactive effects of mowing, N addition, and P addition on the abundance and OTU richness of the entire bacterial community as well as the relative abundance and OTU richness of each of the 16 dominant phyla/classes. To visualize the compositional variation of the entire bacterial community among different treatments, we used non-metric multidimensional scaling (NMDS) plots with the software of Primer (PRIMER 5 for Windows). Permutational multivariate analysis of variance (PERMANOVA) was further used to reveal the effects of experimental treatments on the composition of the entire bacterial community and that of each of the 16 phyla/classes [32]. Stepwise regression analysis was used to identify the factor that could effectively explain the changes in the abundance, richness, and composition from the aforementioned nine potential soil physicochemical indices. In particular, principal coordinate analyses were first used to determine the difference in community composition among different samples, and the first principal coordinates of the 17 bacterial groups explained 5.01–12.9% of community compositional variation. The first principal coordinate was subsequently used to represent the dependent variable in the stepwise regression analyses. Before regressions, all the data were tested for normal distribution. Collinearity was detected by calculating the condition index for each explanatory variable and it was less than 50 for each variable, suggesting autocorrelation did not occur. Three-way ANOVA and stepwise regression analyses were conducted with SPSS software (SPSS 13.0 for WINDOWS).

Results

Soil physicochemical indices

The effects of these experimental treatments on some of the nine soil physicochemical indices have been described previously [23]. Briefly, for all treatments of mowing, N addition, P addition, and their combinations, only N addition significantly affected the three indices of soil TC content, available N content and pH ($P<0.05$; Table S1). Only P addition significantly altered the three indices of soil TP content, TC/TP ratio and TN/TP ratio; while none of the treatments changed the remaining three indices of soil TN content, TC/TN ratio, and water content.

Bacterial abundance or relative abundance

Among the treatments of mowing, N addition, and P addition, only N addition significantly decreased the abundance of the entire soil bacterial community ($P<0.05$; Fig. 1a). Furthermore, the effects of the treatments on the relative abundances of nine dominant bacterial phyla (Acidobacteria, Actinobacteria, Bacteroidetes, Chloroflexi, Firmicutes, Gemmatimonadetes, Nitrospirae, Planctomycetes, and Proteobacteria) and seven dominant classes (Acidimicrobidae, Actinobacteridae, Rubrobacteridae, Alphaproteobacteria, Betaproteobacteria, Deltaproteobacteria, and Gammaproteobacteria) were also examined. N addition significantly decreased the relative abundances of Acidobacteria, Acidimicrobidae, and Deltaproteobacteria, and increased those of Actinobacteria, Firmicutes, Gemmatimonadetes, Actinobacteridae, and Gammaproteobacteria (Fig. 2). Mowing and N addition had an interactive effect on the relative abundance of Firmicutes (Fig. 2e); in particular, although N addition increased the relative abundance of Firmicutes, the extent of increase was less following simultaneous mowing and N addition. Furthermore, mowing, N addition, and P addition also had an interactive effect on the relative abundance of Deltaproteobacteria (Fig. 2o).

OTU richness

Among the treatments of mowing, N addition, and P addition, only N addition significantly decreased the OTU richness of the entire soil bacterial community ($P<0.05$; Fig. 1b). Furthermore, N addition also significantly decreased the OTU richness of Acidobacteria, Chloroflexi, Proteobacteria, Rubrobacteridae, and Deltaproteobacteria, and increased that of Gammaproteobacteria (Fig. 3). On the other hand, P addition significantly decreased the OTU richness of Proteobacteria and Alphaproteobacteria (Fig. 3i, m). N addition and P addition had an interactive effect on the OTU richness of Bacteroidetes (Fig. 3c); in particular, N addition alone and P addition alone had little effect on its OTU richness, but the simultaneous addition of N and P decreased it. Mowing and N addition had an interactive effect on the OTU richness of Nitrospirae (Fig. 3g), whereas mowing, N addition, and P addition had an interactive effect on the OTU richness of Actinobacteridae (Fig. 3j).

Bacterial composition

Among the treatments of mowing, N addition, and P addition, only N addition significantly altered the composition of the entire soil bacterial communities ($P<0.05$; Fig. 1c). As shown in the NMDS plots in Fig. 1c, the bacterial communities in all the N addition treatments (including N addition, N and P addition, mowing and N addition, and simultaneous mowing, N addition and P addition) shifted relative to those in all the treatments without N addition (including control, P addition, mowing, and mowing and P addition). Furthermore, N addition also significantly altered the composition of each of the 16 bacterial phyla/classes ($P<0.05$; Table 1). Mowing, N addition, and P addition had an interactive effect on Nitrospirae composition, whereas mowing and P addition had an interactive effect on Planctomycetes composition (Table 1).

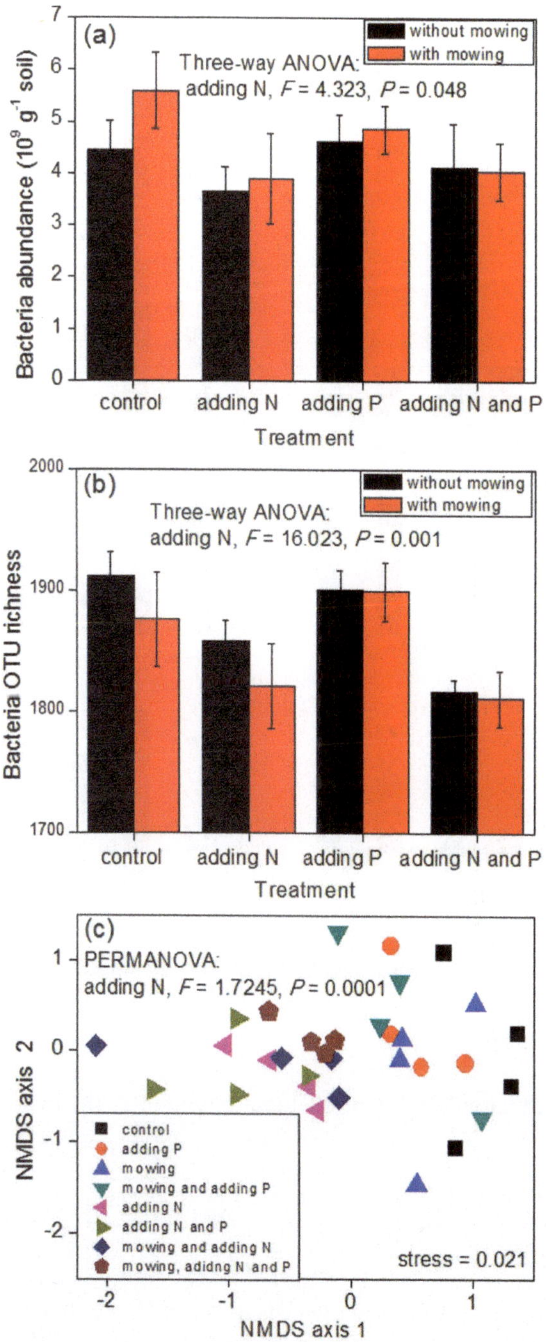

Figure 1. Effects of experimental treatments on the abundance, richness, and composition of the entire bacterial community. For clarity, only the significant statistical results ($P<0.05$) are shown in the figure. In (a) and (b), the bars represent one standard error ($n=4$).

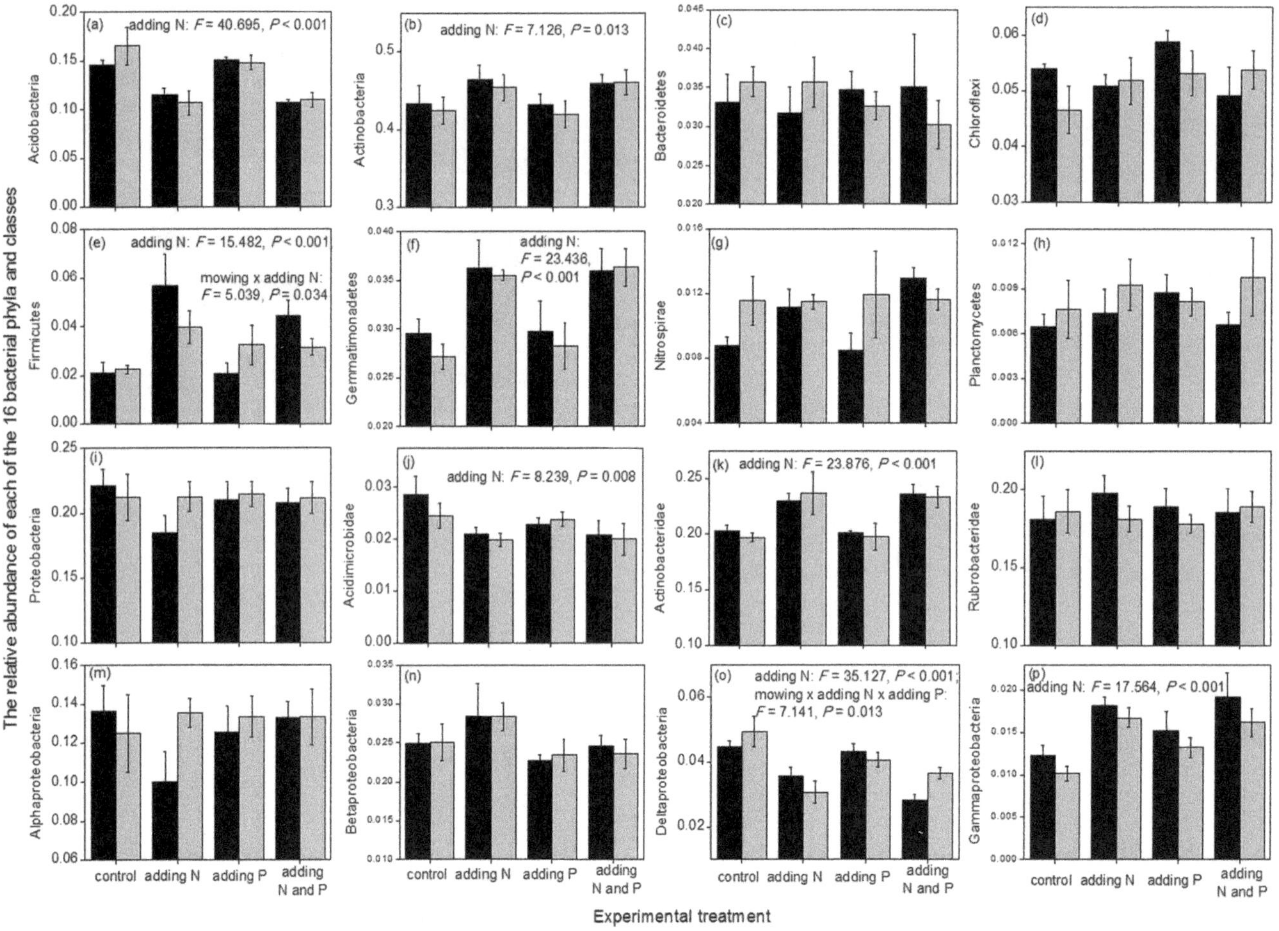

Figure 2. Effects of experimental treatments on the relative abundances of 16 dominant bacterial phyla/classes. Three-way ANOVA was used to test the effect of experimental treatments. For clarity, only the significant statistical results ($P<0.05$) are shown in the figure. The bars represent one standard error ($n=4$). The black and gray columns represent the treatments without and with mowing, respectively.

Variables correlated with bacterial abundance, richness, and composition

Among all the 17 bacterial groups (including the entire bacterial kingdom and 16 phyla/classes), stepwise regression analyses revealed that the abundance or relative abundances of 12 groups were explained by the nine measured soil physicochemical factors (Table 2). In particular, the relative abundances of four groups (Bacteria, Firmicutes, Gemmatimonadetes, and Gammaproteobacteria) were correlated with soil pH alone, those of two groups (Actinobacteria and Chloroflexi) correlated with soil available N content alone, and those of two groups (Acidobacteria and Deltaproteobacteria) correlated with both soil pH and available N content. In addition, the relative abundance of Nitrospirae was correlated with TN/TP ratio, that of Acidimicrobidae correlated with both soil pH and TN/TP ratio, that of Actinobacteria correlated with both available N content and TC/TP ratio, and that of Bacteroidetes with soil pH, available N content and water content (Table 2).

Among all the 17 groups, the OTU richness of 11 groups was explained by the soil physicochemical factors (Table 2). Specifically, the richness of four groups (Chloroflexi, Proteobacteria, Alphaproteobacteria, and Gammaproteobacteria) were correlated with soil pH alone, those of three groups (Bacteria, Acidobacteria, and Deltaproteobacteria) correlated with soil available N content alone, and that of one group (Planctomycetes) with both soil pH and available N content. In addition, the richness of Firmicutes, Bacteroidetes, and Gemmatimonadetes was correlated with soil N content, P content, and soil TC/TN ratio, respectively (Table 2).

Among all the 17 bacterial groups, the compositions of 16 groups were explained by the soil physicochemical factors (Table 2). In particular, the compositions of four groups (Firmicutes, Nitrospirae, Deltaproteobacteria, and Gammaproteobacteria) were correlated with soil pH alone, that of one group (Rubrobacteridae) correlated with soil available N content alone, and those of five group (Bacteria, Acidobacteria, Bacteroidetes, Proteobacteria, and Alphaproteobacteria) with both soil pH and available N content. In addition, the composition of Chloroflexi was correlated with soil C/P ratio, that of Actinobacteria correlated with both soil available N content and P content, that of Gemmatimonadetes correlated with both soil P content and water content, that of Acidimicrobidae correlated with both soil available N content and N/P ratio, that of Actinobacteria correlated with both soil available N content and C/P ratio, and that of Betaproteobacteria with all of soil pH, available N content and N/P ratio (Table 2).

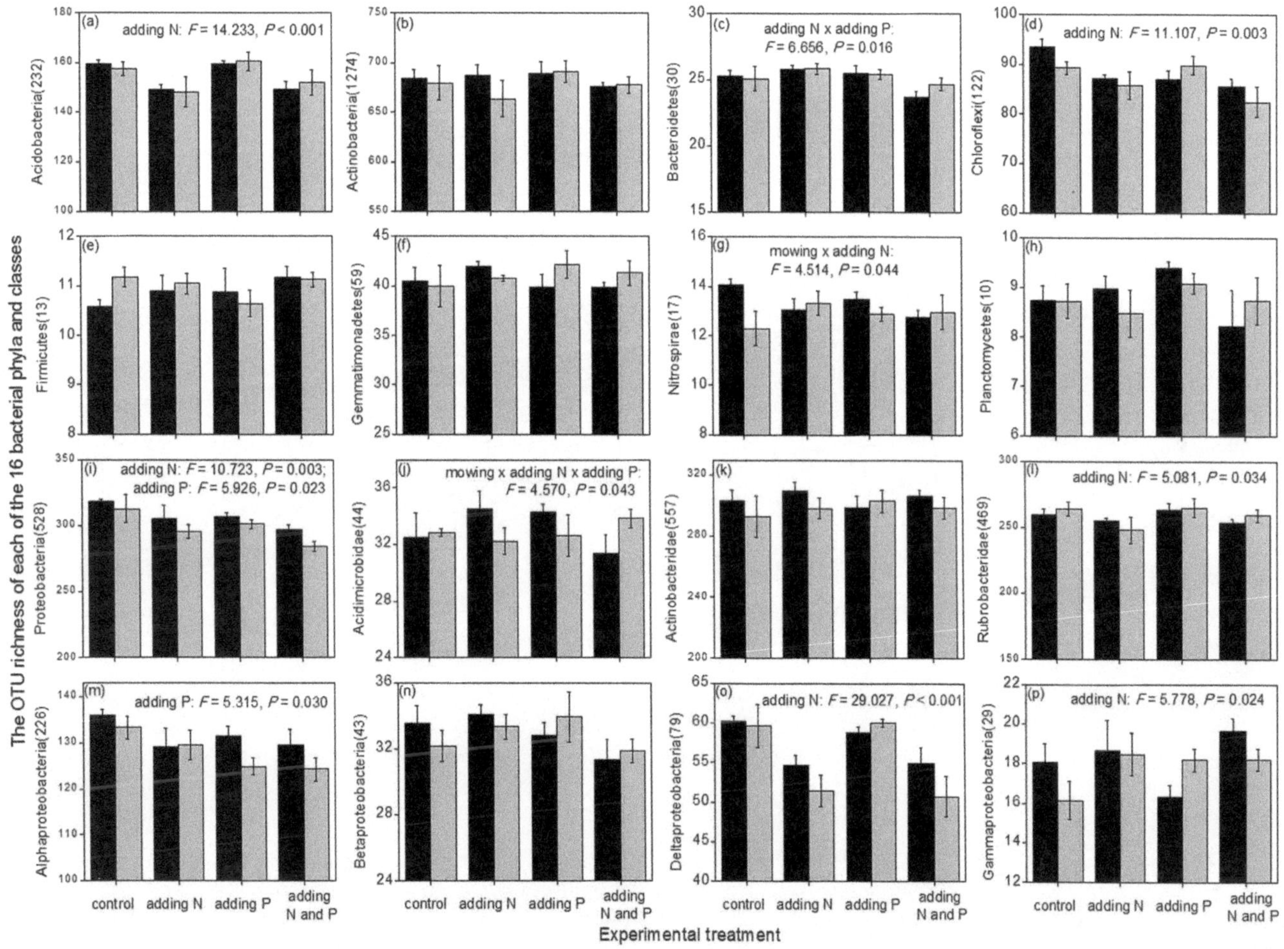

Figure 3. Effects of experimental treatments on the OTU richness of 16 dominant bacterial phyla/classes. Three-way ANOVA was used to test the effect of experimental treatments. For clarity, only the significant statistical results ($P<0.05$) are shown in the figure. The bars represent one standard error (n = 4). The black and gray columns represent the treatments without and with mowing, respectively. The number in the brackets following the phylum/class name (e.g., 232 in Acidobacteria(232) in Fig. 3a) represents the sampled sequence number from which OTU richness was calculated.

Discussion

Four types of evidence revealed that N addition had a much greater influence on the soil bacterial communities than mowing and P addition. First, with respect to the entire bacterial kingdom, only N addition significantly ($P<0.05$) affected its abundance, richness, and composition, and mowing and P addition had non-significant effects (Fig. 1). Second, with regard to the relative abundances of the 16 bacterial phyla/classes, N addition significantly affected eight groups, whereas mowing affected only two groups (Firmicutes and Deltaproteobacteria) and P addition affected only one group (Deltaproteobacteria; Fig. 2). Third, regarding the OTU richness of the 16 bacterial phyla/classes, N addition significantly affected nine groups, whereas mowing affected only two groups (Nitrospirae and Acidimicrobidae) and P addition affected only four groups (Bacteroidetes, Proteobacteria, Acidimicrobidae, and Alphaproteobacteria; Fig. 3). Finally, with respect to the composition of the 16 bacterial phyla/classes, N addition significantly affected all these groups, whereas mowing as well as P addition, respectively, affected only two groups (Nitrospirae and Planctomycetes) (Table 1).

Mowing had limited influence on soil bacterial communities possibly because it was carried out at the end of the growth season. In other words, the plants would have already supplied much organic C resource (or energy resource) to the soil microorganisms before the mowing time. However, if mowing is performed earlier, then the influence might be much greater. Nevertheless, this hypothesis needs to be tested in future studies. P addition had limited influence on soil bacterial communities, implying that P was not a key limiting element for soil microorganisms in this steppe ecosystem [33].

These experimental treatments affected various soil physicochemical factors (Table S1), which might further lead to changes in the abundance, richness and composition of soil microbial communities. For example, N addition increased soil available N content from 12.912 to 31.369 mg/kg soil (independent of other treatments), which is favorable to soil bacterial communities because N is a limiting element in this steppe ecosystem [33]. However, N addition decreased soil pH from 7.094 to 6.501 (independent of other treatments), which is unfavorable to most of the soil microbial communities because they are most adapted to neutral environments. Stepwise regression analysis revealed the mechanism of these experimental treatments altering soil bacterial communities. For example, the first principal coordinate of the composition of the entire soil bacterial kingdom showed significant

Table 1. The effects of mowing (M), nitrogen addition (N), phosphorus addition (P) and their combination on the composition of the 16 dominant bacterial phyla/classes revealed by PERMANOVA.

Bacterial groups	M		N		P		M×N		M×P		N×P		M×N×P	
	F	*P*	*F*	*P*	*F*	*P*	*F*	*P*	*F*	*P*	*F*	*P*	*F*	*P*
Acidobacteria	1.024	0.343	2.188	0.0001*	1.008	0.401	0.961	0.645	0.903	0.879	0.941	0.732	0.937	0.762
Actinobacteria	0.970	0.706	1.433	0.0001*	1.058	0.125	0.972	0.691	1.007	0.407	0.966	0.744	0.986	0.584
Bacteroidetes	1.126	0.135	2.269	0.0001*	1.075	0.222	1.052	0.284	1.107	0.167	0.863	0.896	1.042	0.313
Chloroflexi	1.076	0.165	1.224	0.0063*	1.045	0.277	0.974	0.614	1.108	0.093	1.133	0.052	1.021	0.382
Firmicutes	1.246	0.102	2.010	0.0008*	0.936	0.597	1.299	0.072	0.941	0.578	1.019	0.394	0.836	0.825
Gemmatimonadetes	1.074	0.252	1.556	0.0001*	0.903	0.790	0.822	0.948	1.087	0.216	0.989	0.511	0.982	0.536
Nitrospirae	1.259	0.117	1.738	0.0018*	0.981	0.510	1.260	0.118	1.313	0.079	0.786	0.851	1.533	0.016*
Planctomycetes	0.956	0.650	1.236	0.0193*	1.166	0.067	1.075	0.230	1.194	0.044*	1.070	0.252	0.908	0.805
Proteobacteria	1.012	0.375	2.014	0.0001*	1.077	0.156	0.911	0.885	1.045	0.243	0.910	0.897	1.025	0.320
Acidimicrobidae	1.110	0.174	1.552	0.0001*	0.934	0.718	0.917	0.770	0.934	0.715	0.995	0.514	0.864	0.893
Actinobacteridae	1.024	0.339	1.598	0.0001*	1.101	0.085	1.028	0.317	1.033	0.289	0.938	0.807	0.996	0.490
Rubrobacteridae	0.904	0.941	1.233	0.0004*	1.023	0.346	0.942	0.822	0.931	0.863	1.012	0.417	0.962	0.721
Alphaproteobacteria	1.010	0.402	2.068	0.0001*	1.062	0.248	0.850	0.936	1.057	0.269	0.842	0.952	1.010	0.412
Betaproteobacteria	1.129	0.175	2.283	0.0001*	1.045	0.340	0.874	0.817	1.021	0.405	0.966	0.572	1.009	0.438
Deltaproteobacteria	0.942	0.720	1.674	0.0001*	1.056	0.277	1.017	0.410	0.977	0.579	0.986	0.544	1.065	0.238
Gammaproteobacteria	0.938	0.612	2.307	0.0001*	1.051	0.366	0.956	0.565	1.021	0.422	1.124	0.234	0.744	0.929

The composition of each bacterial phylum/class means the relative abundance of each OTU within this phylum/class. See the effect of these treatments on the composition of the entire bacterial kingdom in Fig. 1c. * denotes $P<0.05$.

correlations with both soil pH and available N content (Table 2). Because soil bacterial composition was demonstrated to be altered significantly only by the treatment of N addition (Fig. 1c), these results suggest that soil pH and available N content are the primary drivers of the change. Overall, there were a total of 39 bacterial groups' attributes correlated with the nine measured soil physicochemical indices (Table 2). In particular, soil pH and available N content were correlated with 23 and 20 bacterial groups' attributes, respectively. All the three stoichiometric indices (TC/TN, TC/TP, and TN/TP) were correlated with eight groups' attributes. Soil TN, TP, and water contents were correlated with one, three, and two groups' attributes, respectively (Table 2). These results suggested that soil pH and available N content affected more bacterial phyla/classes than other measured soil indices.

Terrestrial ecosystems are generally N-limited, so it is traditionally thought that N addition/deposition would favor soil microbial communities by increasing nutrient content [34–36]. Although the influence of N addition/deposition on other physicochemical indices (e.g., soil pH and trace elements) has also been widely demonstrated, their contribution to the soil bacterial changes in response to N addition/deposition has not received enough attention [24,34]. Here our results showed the important role of the decreased soil pH. It is likely because that the intracellular pH of most bacterial species is usually within one pH unit of neutral [25] and the decreased soil pH become a serious selective pressure for them. Actually, a recent study found that the decrease in soil pH caused by N addition was the factor most closely related with the alteration in soil microbial community composition as well as the decline in microbial respiration [37]. Together, these results suggest that some specific strategies (e.g., adding calcium oxide) should be considered to restore soil pH and to maintain soil microbial diversity and their ecosystem functions.

The 16 bacterial phyla/classes examined responded to the experimental treatments with different sensitivities. Acidobacteria, Acidimicrobidae, Deltaproteobacteria, and Gammaproteobacteria were the most sensitive groups, because they exhibited a significant response to the treatments with regard to all the three community attributes of relative abundance, OTU richness, and composition ($P<0.05$; Figs. 2 and 3; Table 1). In contrast, Planctomycetes and Betaproteobacteria were the least sensitive groups, because they responded significantly to the treatments only with respect to the community composition. The remaining 10 groups, including Actinobacteria, Bacteroidetes, Chloroflexi, Firmicutes, Gemmatimonadetes, Nitrospirae, Proteobacteria, Actinobacteridae, Rubrobacteridae, and Alphaproteobacteria, had middling sensitivity, because they responded with respect to two community attributes (Figs. 2 and 3; Table 1).

Soil bacterial communities drive many types of ecosystem functions, and the changes in their abundance, richness, and composition can suggest potential functional shifts. For example, a large proportion of bacteria in the phylum Chloroflexi can acquire energy and fix CO_2 through photosynthesis [38–40]; thus, the decreased OTU richness of Chloroflexi following N addition implies reduction in their C sink function (Fig. 3d). Furthermore, a large proportion of bacteria in the phylum Nitrospirae can transform nitrite into nitrate [41,42]; thus, the decreased OTU richness of Nitrospirae following mowing (Fig. 3g) suggests that their N-cycling function may be reduced. However, there were no differences in the OTU richness of Nitrospirae following N addition and simultaneous treatment of mowing and N addition, implying that this function was not affected. Many plant pathogenic bacteria belong to the phylum Firmicutes [43]. Although N addition promoted the growth of plant community in the steppe ecosystem [5,6], it also led to an increase in the relative abundance of Firmicutes (Fig. 2e), implying that the health of the plant community may be threatened. In contrast, mowing decreased plant biomass, and simultaneous mowing and N addition decreased the relative abundance of Firmicutes than N

Table 2. Variables responsible for the changes in abundance, richness, and composition of various bacterial groups.

Index	Bacterial group	Result	r^2	*P*	effective factors
	Bacteria	$y = -4.336*10^9+1.285*10^9pH$	0.132	0.041	N
	Acidobacteria	y = −0.064−0.001Nav+0.032pH	0.610	<0.001	N
	Actinobacteria	y = 0.414+0.001Nav	0.229	0.006	N
	Bacteroidetes	y = −0.001−0.302water+0.00032Nav+0.007pH	0.463	0.001	
	Chloroflexi	y = 0.057−0.00023Nav	0.171	0.019	
abundance	Firmicutes	y = 0.210−0.026pH	0.302	0.001	N; M×N
	Gemmatimonadetes	y = 0.097−0.010pH	0.647	<0.001	N
	Nitrospirae	y = 0.017−0.001N/P	0.196	0.011	
	Acidimicrobidae	y = −0.021+0.008pH−0.001N/P	0.391	0.001	N
	Actinobacteria	y = 0.152+0.001Nav+0.002C/P	0.608	<0.001	N
	Deltaproteobacteria	y = −0.018+0.009pH−0.00025Nav	0.459	<0.001	N; M×N×P
	Gammaproteobacteria	y = 0.053−0.006pH	0.237	0.005	N
	Bacteria	y = 1913.940−2.348Nav	0.257	0.003	N
	Acidobacteria	y = 163.918−0.428Nav	0.452	<0.001	N
	Bacteroidetes	y = 28.067−7.826P	0.251	0.003	N×P
	Chloroflexi	y = 47.073+5.959pH	0.212	0.008	N
	Firmicutes	y = 13.186−0.899N	0.147	0.030	
richness	Gemmatimonadetes	y = 35.152+1.967C/N	0.199	0.010	
	Planctomycetes	y = 15.814−0.046Nav−0.883pH	0.387	0.001	
	Proteobacteria	y = 161.767+20.720pH	0.254	0.003	N; P
	Alphaproteobacteria	y = 83.713+6.770pH	0.163	0.022	P
	Deltaproteobacteria	y = 61.650−0.241Nav	0.400	<0.001	N
	Gammaproteobacteria	y = 34.431−2.421pH	0.198	0.011	N
	Bacteria	y = 0.944+0.005Nav−0.157pH	0.794	<0.001	N
	Acidobacteria	y = −1.247+0.201pH−0.006Nav	0.719	<0.001	N
	Actinobacteria	y = −0.026−0.007Nav+0.464P	0.594	<0.001	N
	Bacteroidetes	y = −1.424−0.007Nav+0.232pH	0.664	<0.001	N
	Chloroflexi	y = −0.271+0.014C/P	0.295	0.001	N
	Firmicutes	y = 1.957−0.288pH	0.333	0.001	N
	Gemmatimonadetes	y = 0.649−6.484water−0.647P	0.299	0.006	N
composition	Nitrospirae	y = 1.239−0.182pH	0.141	0.034	N; M×N×P
	Proteobacteria	y = 1.048+0.006Nav−0.174pH	0.810	<0.001	N
	Acidimicrobidae	y = −0.553+0.007Nav+0.058N/P	0.489	<0.001	N
	Actinobacteria	y = −0.276+0.006Nav+0.007C/P	0.564	<0.001	N
	Rubrobacteridae	y = 0.086−0.004Nav	0.222	0.007	N
	Alphaproteobacteria	y = 0.888+0.006Nav−0.151pH	0.756	<0.001	N
	Betaproteobacteria	y = 1.025+0.008Nav−0.201pH+0.023N/P	0.801	<0.001	N
	Deltaproteobacteria	y = 1.716−0.252pH	0.374	<0.001	N
	Gammaproteobacteria	y = −1.807+0.266pH	0.382	<0.001	N

Effective factors represent the factors with significant effects on soil bacterial communities. See the details in Figs 1–3 and Table 1. In this part of effective factors, M, N, and P represent mowing, N addition, and P addition, respectively. '×' represents the interaction among different treatments. In the result part, pH, N, P, water, Nav, C/N, C/P, N/P represent soil pH, N content, P content, water content, available N content, C/N ratio, C/P ratio and N/P ratio, respectively. There were 32 samples for the regressions.

addition alone (Fig. 2e), implying a lower threat to plant health. Nevertheless, as soil bacterial communities are very complex, knowledge about the ecosystem functions of most of the bacterial phyla/classes is still very limited [44], not to mention the functional shifts under the treatments of mowing and nutrient addition.

Author Contributions

Conceived and designed the experiments: XZ XH. Performed the experiments: XZ QC. Analyzed the data: XZ. Wrote the paper: XZ.

References

1. Chen Q, Hooper DU, Lin S (2011) Shifts in species composition constrain restoration of overgrazed grassland using nitrogen fertilization in Inner Mongolian steppe, China. PLoS ONE 6: e16909.
2. Schönbach P, Wan HW, Gierus M, Bai YF, Müller K, et al. (2011) Grassland responses to grazing: effects of grazing intensity and management system in an Inner Mongolian steppe ecosystem. Plant and Soil 340: 103–115.
3. Liu YY, Evans JP, McCabe MF, de Jeu RAM, van Dijk AIJM, et al. (2013) Changing climate and overgrazing are decimating Mongolian steppes. PLoS ONE 8: e57599.
4. Cease AJ, Elser JJ, Ford CF, Hao S, Kang L, et al. (2012) Livestock overgrazing promotes locust outbreaks by lowering plant nitrogen content. Science 335: 467–469.
5. Bai YF, Wu JG, Clark CM, Naeem S, Pan QM, et al. (2010) Tradeoffs and thresholds in the effects of nitrogen addition on biodiversity and ecosystem functioning: Evidence from Inner Mongolia grasslands. Global Change Biology 16: 358–372.
6. Yang H, Jiang L, Li L, Li A, Wu M, et al. (2012) Diversity-dependent stability under mowing and nutrient addition: evidence from a seven-year grassland experiment. Ecology Letters 15: 619–626.
7. Zhang XM, Liu W, Bai YF, Zhang GM, Han XG (2011) Nitrogen deposition mediates the effects and importance of chance in changing biodiversity. Molecular Ecology 20: 429–438.
8. Torsvik V, Øvreås L, Thingstad TF (2002) Prokaryotic diversity–magnitude, dynamics, and controlling factors. Science 296: 1064–1066.
9. Fierer N, Strickland MS, Liptzin D, Bradford MA, Cleveland CC (2009) Global patterns in belowground communities. Ecology Letters 12: 1238–1249.
10. Zhou JZ, Xue K, Xie JP, Deng Y, Wu LY, et al. (2012) Microbial mediation of carbon-cycling feedbacks to climate warming. Nature Climate Change 2: 106–110.
11. Sala OE, Chapin III FS, Armesto JJ, Berlow E, Bloomfield J, et al. (2000) Global biodiversity scenarios for the year 2100. Science 287: 1170–1174.
12. Deiglmayr K, Philippot L, Kandeler E (2006) Functional stability of the nitrate-reducing community in grassland soils towards high nitrate supply. Soil Biology and Biochemistry 38: 2980–2984.
13. Kandeler E, Brune T, Enowashu E, Dörr N, Guggenberger G, et al. (2009) Response of total and nitrate-dissimilating bacteria to reduced N deposition in a spruce forest soil profile. FEMS Microbiology Ecology 67: 444–454.
14. Cusack DF, Silver WL, Torn MS, Burton SD, Firestone MK (2011) Changes in microbial community characteristics and soil organic matter with nitrogen additions in two tropical forests. Ecology 92: 621–632.
15. Frey SD, Knorr M, Parrent JL, Simpson RT (2004) Chronic nitrogen enrichment affects the structure and function of the soil microbial community in temperate hardwood and pine forests. Forest Ecology and Management 196: 159–171.
16. Fierer N, Jackson JA, Vilgalys R, Jackson RB (2005) Assessment of soil microbial community structure by use of taxon-specific quantitative PCR assays. Applied and Environmental Microbiology 71: 4117–4120.
17. Hamady M, Walker J, Harris J, Gold N, Knight R (2008) Error-correcting barcoded primers for pyrosequencing hundreds of samples in multiplex. Nature Methods 5: 235–237.
18. Rousk J, Bååth E, Brookes PC, Lauber CL, Lozupone C, et al. (2010) Soil bacterial and fungal communities across a pH gradient in an arable soil. The ISME Journal 4: 1340–1351.
19. Kowalchuk GA, Buma DS, de Boer W, Klinkhamer PG, van Veen JA (2002) Effects of above-ground plant species composition and diversity on the diversity of soil-borne microorganisms. Antonie Van Leeuwenhoek 81: 509–520.
20. Berg G, Smalla K (2009) Plant species and soil type cooperatively shape the structure and function of microbial communities in the rhizosphere. FEMS Microbiology Ecology 68: 1–13.
21. Chrzanowski TH, Kyle M, Elser JJ, Sterner RW (1996) Element ratios and growth dynamics of bacteria in an oligotrophic Canadian shield lake. Aquatic Microbial Ecology 11: 119–125.
22. Makino W, Cotner JB, Sterner RW, Elser JJ (2003) Are bacteria more like animals than plants? Growth rate and resource dependence of bacterial C:N:P stoichiometry. Functional Ecology 17: 121–130.
23. Zhang XM, Liu W, Zhang GM, Chen QS, Huang JH, et al. (2013) Differential sensitivities of various soil nitrogen-cycling functional groups under multi-factorial global changes. PLoS ONE 10.1371/journal.pone.0076500.
24. Zhang XM, Han XG (2012) Nitrogen deposition alters soil chemical properties and bacterial communities in the Inner Mongolia grassland. Journal of Environmental Science 24: 1483–1491.
25. Fierer N, Jackson RB (2006) The diversity and biogeography of soil bacterial communities. Proceedings of the National Academy of Sciences USA 103: 626–631.
26. Wang JW, Cai WQ (1988) Studies on genesis, types and characteristics of the soils of the Xilin River Basin. Inner Mongolia grassland ecosystem research station. In: Research on Grassland Ecosystem (No. 3). Beijing: Science Press. 23–83. (in Chinese with English abstract)
27. Bai YF, Wu JG, Clark CM, Naeem S, Pan QM, et al. (2010) Tradeoffs and thresholds in the effects of nitrogen addition on biodiversity and ecosystem functioning: evidence from inner Mongolia Grasslands. Global Change Biology 16: 358–372.
28. Bao SD (2000) Soil and agricultural chemistry analysis, 3rd edn. China Agriculture Press, Beijing.
29. Schloss PD, Westcott SL, Ryabin T, Hall JR, Hartmann M, et al. (2009) Introducing mothur: open-source, platform-independent, community-supported software for describing and comparing microbial communities. Applied and Environmental Microbiology 75: 7537–7541.
30. Pruesse E, Quast C, Knittel K, Fuchs BM, Ludwig W, et al. (2007) SILVA: a comprehensive online resource for quality checked and aligned ribosomal RNA sequence data compatible with ARB. Nucleic Acids Research 35: 7188–7196.
31. Bray JR, Curtis JT (1957) An ordination of the upland forest communities of Southern Wisconsin. Ecological Monograph 27: 325–349.
32. Anderson MJ (2005) PERMANOVA: a FORTRAN computer program for permutational multivariate analysis of variance. Department of Statistics, University of Auckland, New Zealand.
33. Zhang LX, Bai YF, Han XG (2004) Differential responses of N: P stoichiometry of Leymus chinensis and Carex korshinskyi to N additions in a steppe ecosystem in Nei Mongol. Acta Botanica Sinica 46: 259–270.
34. Lovett GM, Goodale CL (2011) A new conceptual model of nitrogen saturation based on experimental nitrogen addition to an oak forest. Ecosystems 14: 615–631.
35. Ramirez KS, Lauber CL, Knight R, Bradford MA, Fierer N (2010) Consistent effects of nitrogen fertilization on soil bacterial communities in contrasting systems. Ecology 91: 3463–3470.
36. Fierer N, Lauber CL, Ramirez KS, Zaneveld J, Bradford MA, et al. (2011) Comparative metagenomic, phylogenetic and physiological analyses of soil microbial communities across nitrogen gradients. The ISME Journal 6: 1007–1017.
37. Wei CZ, Yu Q, Bai E, Lü XT, Li Q, et al. (2013) Nitrogen deposition weakens plant–microbe interactions in grassland ecosystems. Global Change Biology doi: 10.1111/gcb.12348
38. Klappenbach JA, Pierson BK (2004) Phylogenetic and physiological characterization of a filamentous anoxygenic photoautotrophic bacterium 'Candidatus Chlorothrix halophila'. Gen. Nov., sp. Nov., recovered from hypersaline microbial mats. Archives of Microbiology 181: 17–25.
39. Hanada S, Hiraishi A, Shimada K, Matsuura K (1995) Chloroflexus aggregans sp. nov., a Filamentous Phototrophic Bacterium Which Forms Dense Cell Aggregates by Active Gliding Movement. International Journal of Systematic Bacteriology 45: 676–681.
40. Pierson BK, Castenholz RW (1974) A phototrophic gliding filamentous bacterium of hot springs, Chloroflexus aurantiacus, gen. And sp. Nov. Archives of Microbiology 100: 5–24.
41. Watson SW, Bock E, Valois FW, Waterbury JB, Schlosser U (1994) Nitrospira marina gen. nov. sp. nov.: a chemolithotrophic nitrite-oxidizing bacterium. Archives of Microbiology 144: 62–9.
42. Ehrich S, Behrens D, Lebedeva E, Ludwig W, Bock E (1995) A new obligately chemolithoautotrophic, nitrite-oxidizing bacterium, Nitrospira moscoviensis sp. nov. and its phylogenetic relationship. Archives of Microbiology 164: 16–23.
43. Hogenhout SA, Loria R (2008) Virulence mechanisms of Gram-positive plant pathogenic Bacteria. Current Opinion in Plant Biology 11: 449–456.
44. Quaiser A, Ochsenreiter T, Lanz C, Schuster SC, Treusch AH, et al. (2003) Acidobacteria form a coherent but highly diverse group within the bacterial domain: evidence from environmental genomics. Molecular Microbiology 50: 563–75.

3

Exploring the Dynamic Core Microbiome of Plaque Microbiota during Head-and-Neck Radiotherapy Using Pyrosequencing

Yue-jian Hu[1,3]⁹, Zi-yang Shao[2]⁹, Qian Wang[1], Yun-tao Jiang[1], Rui Ma[1], Zi-sheng Tang[1], Zheng Liu[1], Jing-ping Liang[1]*, Zheng-wei Huang[1]*

1 Department of Endodontics, Ninth People's Hospital, Shanghai Jiao Tong University School of Medicine, Shanghai Key Laboratory of Stomatology, Shanghai, China, **2** Division of Radiation Oncology, Department of Oral and Maxillofacial Surgery, Ninth People's Hospital, Shanghai Jiao Tong University School of Medicine, Shanghai, China, **3** Department of Endodontics, The Affiliated Stomatology Hospital of Tongji University, Tongji University School of Stomatology, Shanghai, China

Abstract

Radiotherapy is the primary treatment modality used for patients with head-and-neck cancers, but inevitably causes microorganism-related oral complications. This study aims to explore the dynamic core microbiome of oral microbiota in supragingival plaque during the course of head-and-neck radiotherapy. Eight subjects aged 26 to 70 were recruited. Dental plaque samples were collected (over seven sampling time points for each patient) before and during radiotherapy. The V1–V3 hypervariable regions of bacterial 16S rRNA genes were amplified, and the high-throughput pyrosequencing was performed. A total of 140 genera belonging to 13 phyla were found. Four phyla (Actinobacteria, Bacteroidetes, Firmicutes, and Proteobacteria) and 11 genera (*Streptococcus, Actinomyces, Veillonella, Capnocytophaga, Derxia, Neisseria, Rothia, Prevotella, Granulicatella, Luteococcus,* and *Gemella*) were found in all subjects, supporting the concept of a core microbiome. Temporal variation of these major cores in relative abundance were observed, as well as a negative correlation between the number of OTUs and radiation dose. Moreover, an optimized conceptual framework was proposed for defining a dynamic core microbiome in extreme conditions such as radiotherapy. This study presents a theoretical foundation for exploring a core microbiome of communities from time series data, and may help predict community responses to perturbation as caused by exposure to ionizing radiation.

Editor: Ramy K. Aziz, Cairo University, Egypt

Funding: This work was supported by a grant from the National Natural Science Foundation (NSFC) No. 81070826/30872886 and partly sponsored by Shanghai Rising-Star Program No. 12QH1401400. Patient enrollment was funded by the Shanghai Jiaotong University grant (YG2011MS67). The funders had no role in study design, data collection and analysis, decision to publish, or preparation of the manuscript.

Competing Interests: The authors have declared that no competing interests exist.

* E-mail: huangzhengwei@shsmu.edu.cn (ZWH); liangjp123@yahoo.com.cn (JPL)

⁹ These authors contributed equally to this work.

Introduction

While radiation therapy is a mainstay of treatment for head-and-neck cancers, it is often accompanied by significant side effects. Acute and chronic oral complications that may occur in patients receiving radiotherapy include mucositis, candidiasis, and radiation caries [1,2]. These side effects probably stem from radiation-induced hyposalivation and consequent shifts in microbial population selected toward a pathogenic flora [3–5]. It is widely accepted that the composition of the commensal oral flora is controlled by complex interactions among the oral microorganisms themselves, the host tissues, and the mechanical flushing action and antimicrobial activity of saliva [6]. If one of these factors is greatly altered, changes might be expected to occur in the oral microflora, and disease will manifest itself.

Recent studies of oral microbiota using next generation high-throughput sequencing indicated that there was a core microbiome among oral microbiota phylotypes across communities from different individuals [7]. Taxa found in samples of all time points of all subjects are designated as common [8]. Generally, the term "core microbiome" is defined as the set of members shared among microbial assemblages associated with a particular habitat (e.g. the oral cavity), and is present in all or the vast majority of humans [9]. It has been proposed that the core microbiomes (or common taxa, or shared operational taxonomic units) are likely critical to the ecosystem function of the community within the oral cavity and play a key role in health maintenance. Thus, a better understanding of the composition and variation of these commonly occurring microorganisms is essential to guide the manipulation of communities to achieve desired outcomes and predict community responses to perturbation as caused by exposure to ionizing radiation [10].

Dental plaque harbors a highly diverse resident community of microorganisms. Varying degrees of overlap of plaque microbiota of healthy individuals have been revealed by recent studies of oral microbiome, but little characteristic information of core microbiome in the diseased condition is available so far. Here we hypothesized that radiotherapy of the head and neck induces shifts in the core microbiome of oral microbial communities. Thus, in this study, the high-throughput sequencing technique (pyrosequencing) was used to define the core microbiome of plaque microbiota of irradiated patients compared with the compositional profiles before radiotherapy.

Materials and Methods

Ethics Statement

This study was approved by the ethics committee of Shanghai Jiao Tong University and conducted according to the principles expressed in the Declaration of Helsinki. Written informed consent was obtained from all participating patients.

Enrollment of Subjects

Potential study subjects were identified from a group of patients who were scheduled to receive radiation therapy. Inclusion/exclusion criteria are listed in Table 1. After we had obtained informed consent, the subject population comprised eight individuals who had head-and-neck cancers. Their ages ranged from 26 to 70 years. An oral health examination was performed before radiation therapy. If necessary, carious lesions were restored, endodontic treatment performed, and doubtful teeth extracted. The patients were given oral hygiene instruction, but no special fluoride remedy was added.

Radiation Therapy Protocols

Patients were placed in the supine position on a commercial thermoplastic head-neck mask attached to a carbon-fiber laminate base plate. The transverse images represented 2.5-mm thick slices from 5 cm superior to the skull base to the clavicle heads. The acquired images were transferred directly to a XiO treatment planning system (Computerized Medical Systems, St. Louis, MO, USA). Radiotherapy is concerned with the delivery of the correct radiation dose to the tumour mass. The gray (Gy) is the standard unit of absorbed ionizing-radiation dose expressed in terms of absorbed energy per unit mass of tissue, equivalent to one joule per kilogram (1 J/kg). The primary field was irradiated through lateral parallel-opposed portals with 6-MV photons, 2.0 Gy/30 fractions. Each patient received 10 Gy per week for 6 weeks, with cumulative dose of 60 Gy. The parotid and submandibular glands were directly adjacent to the target volume and could not be spared.

Microbial Sampling

For each of the eight subjects, before and during radiotherapy, microbial samples were collected at 7-day intervals using the method mentioned in the *Manual of Procedures for Human Microbiome Project* (http://hmpdacc.org/tools_protocols/tools_protocols.php) with minor modifications. Briefly, a sterile Gracey curette was used to collect a pooled supragingival plaque sample from the buccogingival surfaces of the maxillary first molar after the site had been isolated with cotton rolls and dried. The collected plaque sample was released from the curette by agitation in 300 μL of TE buffer (10 mM Tris-Cl [pH 7.5] and 1 mM EDTA). The microbial samples were immediately transported on ice to the laboratory for further DNA extraction and pyrosequencing analysis. All samples were collected at seven time points within 7 weeks. The samples collected at the time point PT (prior to treatment, no dose received) was used as a control group. The following 6-week treatment period included 10 Gy (the first week of radiotherapy), 20 Gy (second week), 30 Gy (third week), 40 Gy (fourth week), 50 Gy (fifth week), 60 Gy (sixth week, the end of radiotherapy).

DNA Extraction and Pyrosequencing Analysis

The plaque samples were lysed in a Mini-Beadbeater-16 (Biospec Products, Bartlesville, OK, USA) according to the manufacturer's instructions. The total genomic DNA was obtained from the lysate using a Bacterial Genomic DNA Extraction Kit (QIAGEN, Valencia, USA). All DNA was stored at −20°C before further analysis. PCR amplification of the 16S rDNA hypervariable V1–V3 region [8] was carried out using the forward primer 8F (5′-AGAGTTTGATCCTGGCTCAG-3′) and reverse primer 533R (5′-TTACCGCGGCTGCTGGCAC-3′). Unique barcode sequences were incorporated into the primers so that sequences

Table 1. Admission criteria.

Inclusion criteria
a. Presence of caries-free maxillary first molar in the oral cavity and absence of periodontal disease
b. Able to continue current diet and regimen of oral care for duration of the study
c. No anticipated chemotherapy during the course of the study
d. Life expectancy of at least 2 months
e. Eighteen years of age or older
f. Written informed consent
Exclusion criteria
a. Untreated cavitated carious lesions or oral abscesses
b. Periodontal pockets ≥4 mm
c. Clinically meaningful halitosis as determined by organoleptic assessment of an experienced clinician
d. Previous diagnosis of Sjögren's syndrome or any disease characterized by xerostomia
e. Receiving antibiotics during therapy or within 3 months before the study
f. Major salivary glands involved in the surgery region
g. Previous head and neck irradiation
h. Unable to maintain oral hygiene during the study

from different samples can be identified. The PCR program used was as follows: 2 minutes of initial denaturation at 95°C, followed by 20 cycles of denaturation (95°C for 30 seconds), annealing (55°C for 30 seconds) and extension (72°C for 30 seconds) with a final extension of 5 minutes at 72°C. Amplicon pyrosequencing was performed with standard Roche 454 GS-FLX protocols [11]. Raw sequences were processed in a data curation pipeline implemented in the program MOTHUR (version 1.23.1; http://www.mothur.org/). Sequences that were less than 200 bp, contained ambiguous bases or homopolymeric stretches, had a low quality score (<25), or checked as chimeric artifact were discarded after removing the primer sequences and 8-bp barcode. The qualified sequences were submitted to the SILVA database (SILVA 106; http://www.arb-silva.de) for taxonomic analysis. MOTHUR was applied to generate the operational taxonomic units (OTUs) and OTU rarefaction curves. The correlation analyses between the number of OTUs and radiation dose were performed by SAS statistical analysis system (version 8.02; SAS Institute Inc., USA).

Results

Overall Sequence Data

Of 189,305 sequences obtained, 147,232 passed quality control, representing about 78% of the total number of sequences obtained. The average number of sequences per sample at each time point ranged from 1,963 to 3,031 (Table 2). The richness of bacterial communities within supragingival dental plaque at each time point was estimated by rarefaction curves (Figure S1). The curve representing the control group (PT) presented the steepest slope compared with that of the following process of treatment (Figure 1A). Given that the number of sequences obtained from a sample or group had a strong correlation with the number of observed OTUs, comparison of communities from different time point should be made using an equal number of sequences [12]. For a given number of sequences sampled (e.g., 5,000 sequences, 10,000 sequences, or 15,000 sequences; Figure 1B), the control group (PT) had the largest number of OTUs compared with the other time points during radiotherapy (10 Gy–60 Gy). With the radiation dose increasing over time, the distribution of rarefaction curves presented a certain characteristic (Figure 1A). Fewer OTUs were found in the later period (40 Gy, 50 Gy, 60 Gy) than in the early stage (10 Gy, 20 Gy, 30 Gy), and there was a negative correlation between the number of OTUs and dosage ($P<0.01$; Figure 1B). The number of OTUs found across all time points ranged from 165 to 263 by clustering at the 10% dissimilarity level, from 372 to 630 at the 5% dissimilarity level. When a more stringent cut-off was used, more OTUs were formed across all time points, ranging from 580 to 1,038 at the 3% dissimilarity level (Table 2).

Composition and Variation of the Bacterial Community

After eliminating unidentified sequences, 13 phyla were found in the oral microbiota at seven time points, which were dominated by eight major phyla (Figure 2). The predominant phyla were Actinobacteria, Bacteroidetes, Firmicutes, Fusobacteria, Proteobacteria, Spirochaetes, Synergistetes, and candidate division TM7. Among them, Synergistetes were identified in six of our subjects, though their richness and relative abundance were very low. The top 4 phyla in the control group (PT) included, in order of prevalence at the time point PT, Proteobacteria (30.38% of the sequences taxonomically assigned at the phylum level), Firmicutes (29.77%), Bacteroidetes (23.52%) and Actinobacteria (9.29%). By contrast, the four most abundant phyla during radiotherapy (10 Gy–60 Gy) were, in order of average prevalence across six time points, Firmicutes (ranging from 25.13% to 48.09% during the course of radiotherapy), Actinobacteria (10.59%–34.7%), Proteobacteria (13.46%–28.24%) and Bacteroidetes (5.6%–20.84%) (Figure 2). The above mentioned four phyla (Actinobacteria, Bacteroidetes, Firmicutes, and Proteobacteria) were found at all time points (before and during treatment) in all subjects and comprised the core microbiome (95% of all sequences). Fusobacteria, Spirochaetes, Synergistetes, and Candidate division TM7 were identified at all time points of some subjects but absent at all time points of other subjects and were designated as subject-specific taxa. The other phyla, such as Chloroflexi, Tenericutes, and Candidate division SR1, which had the lowest relative abundance in sequences, were not found in some time points possibly because their numbers in each sample were below the detection limit of the assay. Other rare phyla like Cyanobacteria were most likely transient "contaminants" of the mouth, gaining access to the oral cavity through food intake or exposure to airborne pollen [8].

At the genus level, sequences from samples of seven time points represented 140 different genera after eliminating the unclassified sequences. The top 10 taxa that constituted roughly 80% of the sequences in relative abundance belonged to *Streptococcus, Veillonella, Actinomyces, Capnocytophaga, Derxia, Neisseria, Rothia, Prevotella, Lactobacillus,* and *Leptotrichia*. Among these, six genera (*Streptococcus, Actinomyces, Capnocytophaga, Neisseria, Granulicatella,* and *Gemella*) were found in the control group (PT) in all subjects and were designated as common at the time point PT. By contrast, only two genera (*Streptococcus* and *Actinomyces*) were present across the time points from 10 Gy to 60 Gy in all subjects. These were designated as "common taxa" during radiotherapy, but were fewer than that in the control group or in previous reports on oral microbial communities of healthy individuals [7,8]. The relative abundance of *Streptococcus* fluctuated between 21.33% (20 Gy) and 3.2% (50 Gy), and that of *Actinomyces* remained stable (about 4.48% to 4.85%) in the early stage of the time points (10 Gy, 20 Gy, 30 Gy), but rose to 23.32% at the 50 Gy time point. In addition to the two genera mentioned above, several genera including *Veillonella, Capnocytophaga, Derxia, Neisseria, Rothia, Prevotella, Granulicatella, Luteococcus,* and *Gemella* could be identified in all subjects during radiotherapy (10 Gy–60 Gy) but were absent at some time points of each subject. These nine genera could be recognized as "potential common taxa" in this study (discussed below). The most abundant genus at the time point PT was *Neisseria* (15.54%). However, at the time points 10 Gy and 20 Gy, the most abundant genus was *Streptococcus* (12.34% and 21.33%, respectively). At the time points 30 Gy and 40 Gy, *Veillonella* predominated (13.77% and 20.98%, respectively), even though this genus was not found in some samples. At the time points 50 Gy and 60 Gy, *Actinomyces* ranked first in relative abundance (23.32% and 20.38%, respectively; Figure 3). The major genera varied significantly in relative abundance across different time points. About 2.7% (4,038) of all sequences could not be identified at the genus level and were classified at the higher taxonomic levels (family, order, class, and phylum).

At the species level, defined OTUs on the basis of 3% dissimilarity, the number of phylotypes at seven time points ranged from 580 to 1,038, while a more conservative approach–5% dissimilarity level among OTUs–yielded 372 to 630 phylotypes. Individual species or strains were not described because of the relatively short reads and the low taxonomic resolution below the genus level [13,14]. The rarefaction curves were generated at the 3% dissimilarity level in this study (Figure 1). Generally, the cut-off of 3% dissimilarity was used in species level

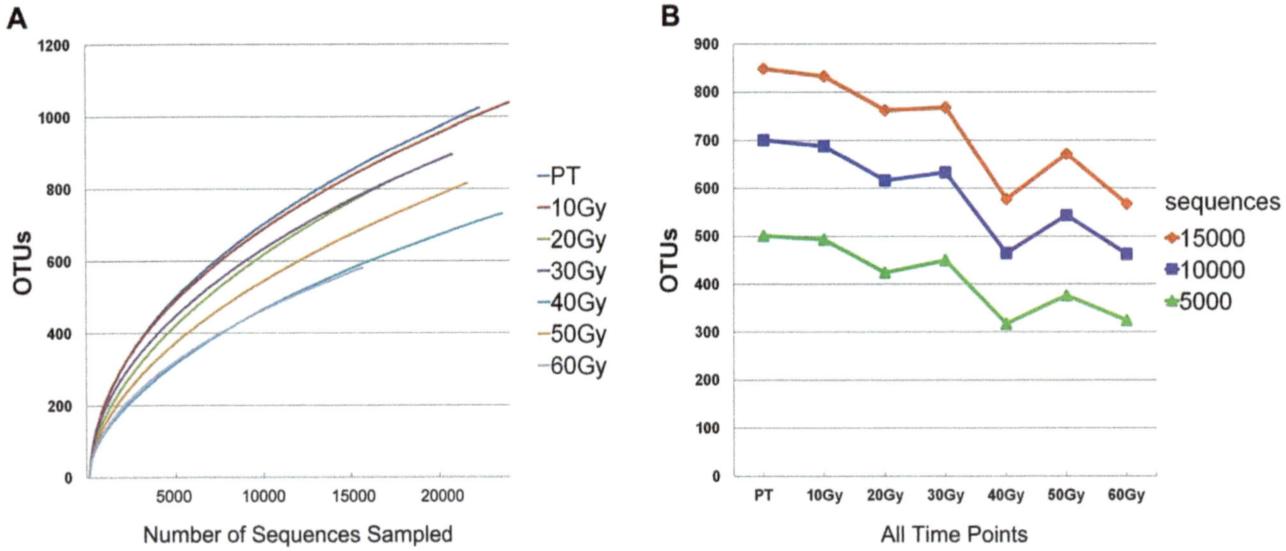

Figure 1. Characterization of all time points rarefaction curves and the correlation analyses between the number of OTUs and radiation dose. (A) Rarefaction curves were used to estimate richness among the seven time points (0.03 dissimilarity level). (B) For a given number of sequences sampled (e.g., 5,000 sequences, 10,000 sequences, or 15,000 sequences), there was a negative correlation between the number of OTUs and dosage ($P<0.01$).

analyses, which facilitates a direct comparison with other studies [7,13,15].

Discussion

Radiation therapy is the most widely used approach to treat head-and-neck cancers. This technique involves three coplanar isocentric fields designed to adequately cover the target volumes, with the major salivary glands being inevitably exposed to the radiation field [16,17]. The high radiosensitivity of these glands causes the salivary secretion rate to continuously decrease and the microflora composition of the saliva to change. Saliva is important for a variety of functions in the oral cavity. It contains buffering systems that neutralize the acids formed during bacterial carbohydrate metabolism. Salivary components such as lysozyme, histatins and lactoferrin provide an antimicrobial function crucial to prevent progression of oral diseases [18]. As a result of radiation-induced hyposalivation, an imbalance in the oral microbial ecosystem is inevitable. In addition, the microbial changes may result from a direct or indirect effect of ionizing radiation [19,20], lifestyle, dietary habit, medical history, or other factors that vary between individuals. In order to eliminate adverse effect brought by inter-individual variations, a self-comparative analysis was proposed to reduce selection biases and achieve more reliable results. Thus, in this study, the samples collected before treatment was used as a control group and compared with the following treatment procedures [14].

One fundamental question raised by sponsored human microbiome projects (HMPs) is whether there is an identifiable "core microbiome" of shared organisms, genes or functional capabilities found in a given body habitat (e.g., the oral cavity) of all or the vast majority of humans [21]. Several models of "core microbiome" have been presented in the past few years [10,22], but the most typical approach used to identify a core was to report the number of phylotypes (or OTUs) found across samples from a similar habitat and find the overlap of different communities based on a presence/absence data set. For instance, if one OTU was present in all or the vast majority of subjects, this OTU should be considered as a member of the "core" [9], whether it had high or low abundance. The most stringent definition of a core microbiome would require its presence in all subjects (100%) sampled [23]. Zaura et al. obtained the first insight into the oral core

Table 2. Comparison of number of OTUs at the 3% (0.03), 5% (0.05), and 10% (0.10) dissimilarity level.

Time point	Number of sequences	Average number of sequences per sample (Median, SD)	0.03	0.05	0.10
PT	22580	2823 (2825, 138.8)	1024	617	263
10 Gy	24246	3031 (3051, 148.1)	1038	630	257
20 Gy	18010	2251 (2189, 180.2)	836	524	232
30 Gy	20940	2618 (2655, 144.7)	895	578	247
40 Gy	23865	2983 (2934.5, 183.7)	731	436	182
50 Gy	21887	2736 (2789, 143.3)	815	513	221
60 Gy	15704	1963 (1964, 101.4)	580	372	165

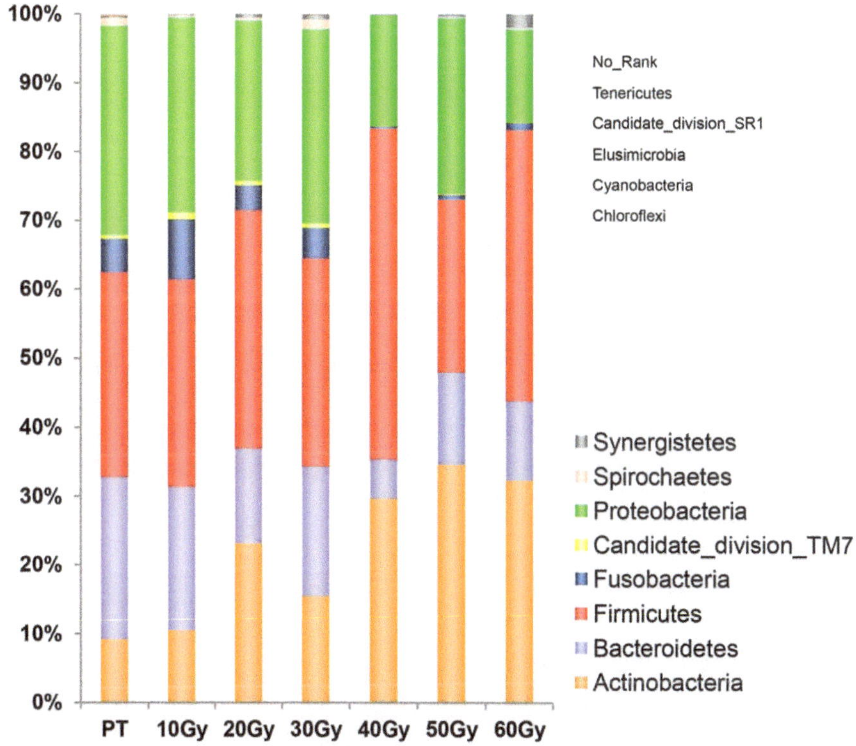

Figure 2. Relative abundance of predominant phyla at seven time points. Members of Actinobacteria, Bacteroidetes, Firmicutes, and Proteobacteria were found at all time points in all subjects and comprised the core microbiome. Phyla having the lowest relative abundance cannot be identified in histogram, and they were listed in small font.

microbiome and found the overlap of oral microbial communities from a small population of 3 individuals [7]. Huse et al. explored the core microbiome with a larger sample size [23], but they defined "core" as those shared amongst 95% or more of the subjects and didn't incorporate time series data (samples obtained at different time points). A recent study presented a more systematic conceptual framework for identifying core microbiomes [10] in which the core microbiome from time series data was comprised of OTUs that were consistently observed across all time points and shared by all subjects (OTU e in Table 3). However, this framework is only suitable for the analyses of communities in the stable or healthy status with little perturbation. Some potential cores may be overlooked under extreme conditions such as radiotherapy or other chemical and physical irritants. A typical example in this study was the assemblage of above mentioned nine genera which could be identified in all subjects but not at all time points during treatment. Actually, the consistent presence of these genera over time was due to computational analysis of pooled data. These genera, defined as core microbiomes in previous studies in healthy individuals [7,8], were probably potential common taxa and their presences and absences at some time points might be due to radiation-induced changes in the oral ecosystem. Thus, we recommend that taxa detected in all subjects but absent at some time points under extreme conditions (OTU f-i in Table 3) should be considered as a dynamic core microbiome when dealing with the data of time series (suppose that methodological errors are not involved). Briefly, we define the dynamic "core" during radiotherapy as those that are present in all subjects but not necessarily persistent over time (Table S1). Moreover, the lowest depth of sampling in the present study was about 1,963 sequences/sample (60 Gy) which was possibly insufficient to reveal common taxa when present at extremely low abundance. Therefore, our findings on core microbiomes should be considered conservative. It was also possible that sampling of the buccogingival surface of the maxillary first molar as representative of the dental plaque microbiome could not reflect the entire oral microecosystem, because the microbiota might vary depending on the intraoral sites from which the plaque samples were obtained. For example, a higher diversity was likely to be found in the samples taken at the approximal surfaces and the lingual surfaces of the front teeth [7], which could lead to a more complex core microbiome. More efforts are needed to achieve a better understanding of the core microbiome of oral microbial communities.

The major challenge facing the various HMPs is how to best relate community composition and microbial changes to physiological impacts. The data in our study showed that the above-mentioned 11 genera (*Streptococcus, Actinomyces, Veillonella, Capnocytophaga, Derxia, Neisseria, Rothia, Prevotella, Granulicatella, Luteococcus* and *Gemella*), which were designated as a core microbiome in this study, varied in relative abundance during the course of radiotherapy (illustrated in detail in Figure 3). *Granulicatella*, for example, is very fastidious and difficult to cultivate. Our study indicates that this genus was dramatically reduced in dental plaque following radiation. It is therefore tempting to speculate that this genus is unlikely to contribute to the pathogenesis of post-radiation diseases such as caries. Meanwhile, a notable fluctuation (e.g., *Streptococcus*) was observed during radiotherapy. *Streptococcus* was one of the predominant genera in oral cavity and consisted of a large numbers of cariogenic and non-cariogenic species, including *S. sobrinus, S. mutans, S. oralis, S. mitis,* and *S. pneumonia,* etc. However, the current technology is generally much more

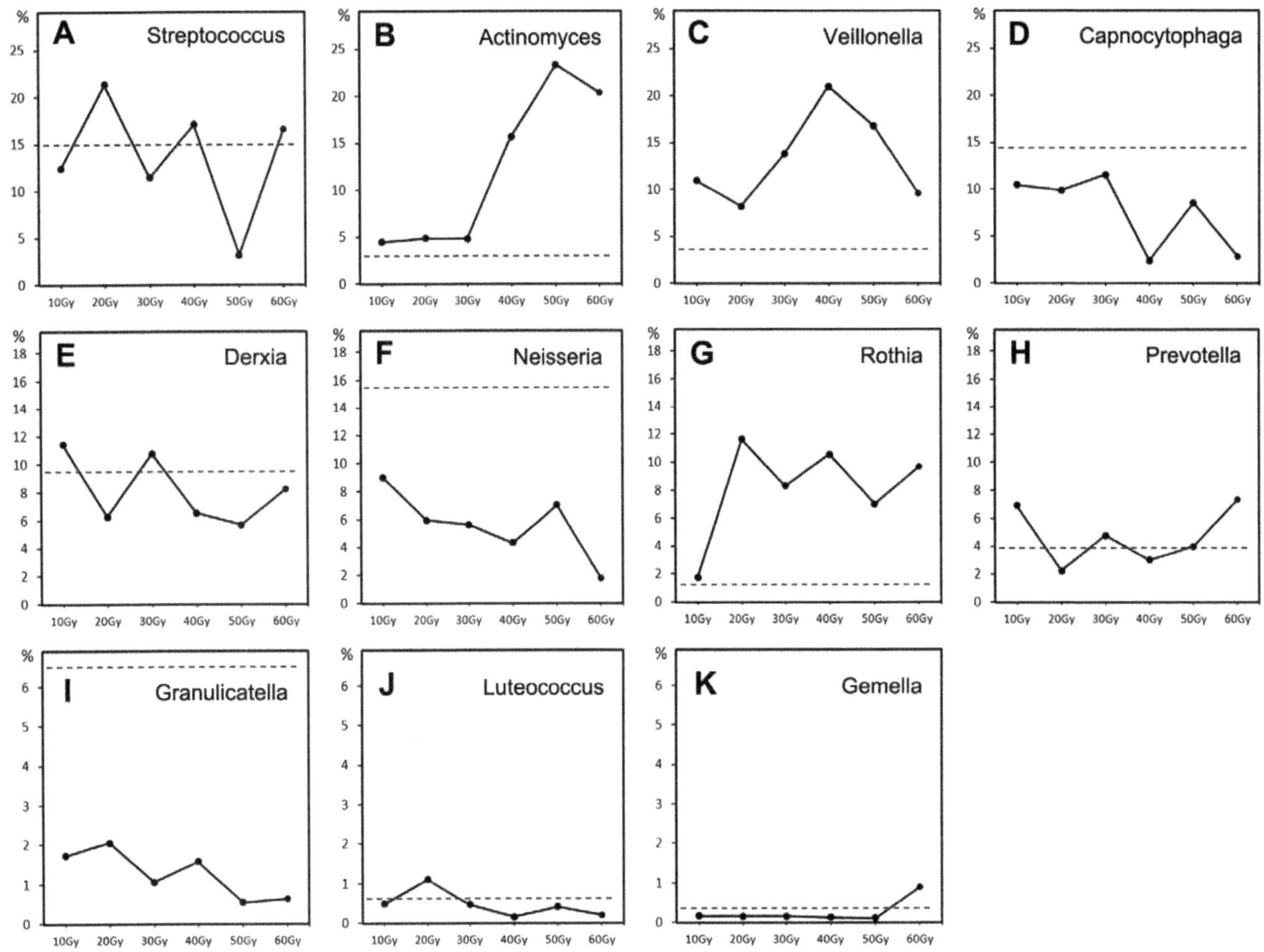

Figure 3. Temporal variation in relative abundance of 11 genera found in all subjects (two common taxa and nine "potential common taxa"). These 11 genera constituted 75.27% of the total sequences. The abscissas and ordinates represent six time points and relative abundance of each genus respectively (different in ordinate scale). The levels of relative abundance of the control group (prior to treatment) are indicated by dotted lines.

Table 3. The conceptual framework of a dynamic core microbiome in extreme conditions.

	Subject A			**Subject B**				
	time 1	**time 2**	**time 3**	**time 1**	**time 2**	**time 3**		
OTU a	•	•	•				**subject-specific taxa**	
OTU b				•	•	•		
OTU c	•	•					**potential subject-specific taxa**	
OTU d				•		•		
OTU e	•	•	•	•	•	•	**common taxa**	**core microbiome**
OTU f	•	•	•		•	•	**potential common taxa**	
OTU g	•	•		•	•			
OTU h		•	•			•		
OTU i	•			•				

OTU a-i represent the units of interest defined at a given taxonomic level. Subject A and B represent time series communities to be compared using the OTU table. The presences of OTUs are represented by the symbols "•". The OTUs present at all time points of all subjects are designated as common taxa (OTU e). Those present in all subjects but not at all time points are designated as potential common taxa (OTU f-i). Both taxa (common taxa and potential common taxa) can be recognized as core microbiomes from time series data.

effective in the identification of higher level taxonomic assignments such as phyla, classes, orders, families, and genera [14]. Thus it was difficult to relate the general fluctuation of *Streptococcus* to dental caries. Additional efforts are needed to identify actual species or strains. Furthermore, a change in number of OTUs during radiotherapy was also observed. The negative correlation between the number of OTUs and dosage indicated that there might be a general decrease in number of species as the radiation dose was increased. It is known that oral microbial communities can remain relatively stable over time in healthy individuals [8,24]. Our results show that dramatic physical and chemical fluctuations (as caused by administration of ionizing radiation) can impose significant changes in the composition of the oral microbiome.

Conclusion

In summary, the core microbiomes of plaque microbiota in patients receiving head-and-neck radiotherapy were explored by high-throughput pyrosequencing. Among all 13 phyla and 140 genera, 4 phyla and 11 genera were found during radiotherapy in all subjects, supporting the concept of a core microbiome. A negative correlation between the number of OTUs and radiation dose was also observed. Moreover, we proposed an optimized conceptual framework for defining a dynamic core microbiome in extreme conditions such as radiotherapy. This study provides researchers with valuable information about the profiles of oral microbial communities during radiotherapy, and presents a theoretical foundation for exploring a core microbiome of communities from time series data.

Acknowledgments

We express our sincere appreciation to all the patients who took part in this study. We thank Frank A. Scannapieco for advices.

Author Contributions

Conceived and designed the experiments: ZWH JPL. Performed the experiments: ZWH YJH ZYS QW. Analyzed the data: ZWH YJH ZYS. Contributed reagents/materials/analysis tools: ZYS YTJ RM ZST ZL. Wrote the paper: ZWH YJH.

References

1. Farah CS, Hong S, Wanasaengsakul S, Elahi S, Pang G, et al. (2001) Irradiation-induced oral candidiasis in an experimental murine model. Oral Microbiol Immunol 16: 358–363.
2. de Castro GJ, Federico MH (2006) Evaluation, prevention and management of radiotherapy-induced xerostomia in head and neck cancer patients. Curr Opin Oncol 18: 266–270.
3. Alvarado Y, Bellm LA, Giles FJ (2002) Oral mucositis: time for more studies. Hematology 7: 281–289.
4. Nederfors T (2000) Xerostomia and hyposalivation. Adv Dent Res 14: 48–56.
5. Tong HC, Gao XJ, Dong XZ (2003) Non-mutans streptococci in patients receiving radiotherapy in the head and neck area. Caries Res 37: 261–266.
6. MacFarlane TW, Mason DK (1974) Changes in the oral flora in Sjögren's syndrome. J Clin Pathol 27: 416–419.
7. Zaura E, Keijser BJ, Huse SM, Crielaard W (2009) Defining the healthy "core microbiome" of oral microbial communities. BMC Microbiol 9: 259.
8. Lazarevic V, Whiteson K, Hernandez D, Francois P, Schrenzel J (2010) Study of inter- and intra-individual variations in the salivary microbiota. BMC Genomics 11: 523.
9. Turnbaugh PJ, Ley RE, Hamady M, Fraser-Liggett CM, Knight R, et al. (2007) The Human Microbiome Project. Nature 449: 804–810.
10. Shade A, Handelsman J (2012) Beyond the Venn diagram: the hunt for a core microbiome. Environ Microbiol 14: 4–12.
11. Margulies M, Egholm M, Altman WE, Attiya S, Bader JS, et al. (2005) Genome sequencing in open microfabricated high density picolitre reactors. Nature 437: 376–380.
12. Schloss PD, Gevers D, Westcott SL (2011) Reducing the effects of PCR amplification and sequencing artifacts on 16S rRNA-based studies. PLoS One 6: e27310.
13. Keijser BJ, Zaura E, Huse SM, van der Vossen JM, Schuren FH, et al. (2008) Pyrosequencing analysis of the oral microflora of healthy adults. J Dent Res 87: 1016–1020.
14. Siqueira JJ, Fouad AF, Rocas IN (2012) Pyrosequencing as a tool for better understanding of human microbiomes. J Oral Microbiol 4: 10743. doi: 10.3402/jom.v4i0.10743.
15. Ling Z, Kong J, Jia P, Wei C, Wang Y, et al. (2010) Analysis of oral microbiota in children with dental caries by PCR-DGGE and barcoded pyrosequencing. Microb Ecol 60: 677–690.
16. Schuurhuis JM, Stokman MA, Roodenburg JL, Reintsema H, Langendijk JA, et al. (2011) Efficacy of routine pre-radiation dental screening and dental follow-up in head and neck oncology patients on intermediate and late radiation effects. A retrospective evaluation. Radiother Oncol 101: 403–409.
17. Shao ZY, Tang ZS, Yan C, Jiang YT, Ma R, et al. (2011) Effects of intensity-modulated radiotherapy on human oral microflora. J Radiat Res 52: 834–839.
18. Almståhl A, Wikström M (1999) Oral microflora in subjects with reduced salivary secretion. J Dent Res 78: 1410–1416.
19. Alamri OD, Cundy AB, Di Y, Jha AN, Rotchell JM (2012) Ionizing radiation-induced DNA damage response identified in marine mussels, Mytilus sp. Environ Pollut 168: 107–112.
20. Shih A, Miaskowski C, Dodd MJ, Stotts NA, MacPhail L (2003) Mechanisms for radiation-induced oral mucositis and the consequences. Cancer Nurs 26: 222–229.
21. Turnbaugh PJ, Hamady M, Yatsunenko T, Cantarel BL, Duncan A, et al. (2009) A core gut microbiome in obese and lean twins. Nature 457: 480–484.
22. Hamady M, Knight R (2009) Microbial community profiling for human microbiome projects: Tools, techniques, and challenges. Genome Res 19: 1141–1152.
23. Huse SM, Ye Y, Zhou Y, Fodor AA (2012) A core human microbiome as viewed through 16S rRNA sequence clusters. PLoS One 7: e34242.
24. Costello EK, Lauber CL, Hamady M, Fierer N, Gordon JI, et al. (2009) Bacterial community variation in human body habitats across space and time. Science 326: 1694–1697.

4

Long-Term Temporal Analysis of the Human Fecal Microbiota Revealed a Stable Core of Dominant Bacterial Species

Inés Martínez, Catherine E. Muller, Jens Walter*

Department of Food Science and Technology, University of Nebraska, Lincoln, Nebraska, United States of America

Abstract

Next-generation sequencing has greatly contributed to an improved ecological understanding of the human gut microbiota. Nevertheless, questions remain regarding the characteristics of this ecosystem and the ecological processes that shape it, and controversy has arisen regarding the stability of the bacterial populations and the existence of a temporal core. In this study, we have characterized the fecal microbial communities of three human individuals over a one-year period by 454 pyrosequencing of 16S rRNA tags in order to investigate the temporal characteristics of the bacterial communities. The findings revealed a temporal core of 33 to 40 species-level Operational Taxonomic Units (OTUs) within subjects. Although these OTUs accounted only for around 12% of the total OTUs detected, they added up to >75% of the total sequences obtained for each individual. In order to determine the capacity of the sequencing and bioinformatic approaches applied during this study to accurately determine the proportion of a core microbiota, we analyzed the fecal microbiota of nine mice with a defined three-member community. This experiment revealed that the sequencing approach inflated the amount of rare OTUs, which introduced a significant degree of artificial variation across samples, and hence reduced the apparent fraction of shared OTUs. However, when assessing the data quantitatively by focusing on dominant lineages, the sequencing approaches deliver an accurate representation of the community. In conclusion, this study revealed that the human fecal microbiota is dominated by around 40 species that maintain persistent populations over the duration of one year. The findings allow conclusions about the ecological factors that shape the community and support the concept of a homeostatic ecosystem controlled largely by deterministic processes. Our analysis of a three-member community revealed that methodological artifacts of OTU-based approaches complicate core calculations, and these limitations have to be considered in the interpretation of microbiome studies.

Editor: Jack Anthony Gilbert, Argonne National Laboratory, United States of America

Funding: The authors have no support or funding to report.

Competing Interests: The authors have declared that no competing interests exist.

* E-mail: jwalter2@unl.edu

Introduction

The gastrointestinal tract of humans is colonized by a complex microbial community dominated by bacteria referred to as the gut microbiota. The bacterial cells in the gastrointestinal tract outnumber somatic cells by at least an order of magnitude, so it is not surprising that the gut microbiota is of profound importance for human health and physiology [1]. Alterations of the gut microbiota have been linked to several chronic immunological and metabolic diseases in humans (including obesity, heart disease, colon cancer, and a variety of inflammatory conditions), and in several animal models, aberrations in gut microbiota composition play a causative role in the development of these pathologies [2,3,4,5]. The association of the gut microbiota with human disease opens avenues for the development of therapies that aim to restore the ecosystem, but their implementation requires a mechanistic understanding about the ecological principles that shape and regulate microbial communities [6].

It is becoming increasingly clear that our understanding of the human microbiome will require the application of ecological theory, and the development of concepts that apply specifically to host associated microbial communities [7,8]. To be successful, this will first require a thorough characterization of the communities in terms of temporal and spatial diversity in different environmental contexts. Despite several decades of research, conflicting perspectives remain regarding the nature of the human gut microbiota, with older concepts being challenged in light of new evidence [9]. In a classic review, Dwayne Savage considered the assembly of the gut microbiota a predictable niche-driven process that ultimately results in the establishment of a climax community with a high degree of temporal stability [10]. In his model, which is essentially the classic deterministic ecological perspective [11], niches characterized by nutrients, environmental filters, and the principle of competitive exclusion determine species membership, abundance, distributions and diversity, and are ultimately stably occupied by the best adapted competitor. Savage referred to these community members as 'autochthonous', and they were not only expected to maintain stable populations in normal adults but were also supposed to be always detected in individuals of the host species [12]. This classic deterministic view has undeniable success in providing explanations for some ecological characteristics of gut microbiomes, such as the presence of specific communities with particular traits in different intestinal compartments and the occurrence of colonization resistance [13,14]. In addition, some aspects of Savage's early concepts were more recently supported by findings indicating that individual members can be detected in a majority of individuals [15,16,17], that mammals assemble gut microbial communities whose composition is phylogenetically conserved [18], that gut microbes can show specific adaptations to the niche environment in particular hosts [19], and that whole communities show a remarkable stability [20,21,22,23,24].

However, new concepts in community ecology and findings obtained through next-generation sequencing techniques have challenged the conventional concepts of host-associated microbial communities [9]. The profiling of ribosomal RNA (rRNA) sequences, which exceeds previous techniques in terms of phylogenetic resolution and dynamic range, has suggested that the human fecal microbiota is highly individualized at the species level, even among monozygotic twins [25]. The factors driving the substantial inter-subject variation observed in humans have not been determined yet, but modern concepts of community ecology presume that, apart from deterministic factors, historic and neutral processes can contribute significantly to ß-diversity [11]. The debate as to what extent gut microbiomes are shaped by deterministic, neutral, and historic processes falls within the larger historic controversy about the nature of communities, and an assessment of their relative importance is essential for our ability to explain and predict community structure and dynamics and ultimately develop models and ecological theories. In this respect, it is important to point out that communities shaped largely by either historic or neutral principles would have fundamentally different characteristics. A community whose assembly is largely impacted by historic processes would be niche-centered, although niches would not be determined solely by environmental filters but also physicochemical changes of the habitat caused by community members, species interactions, and adaptive processes during assembly [26]. The niche-environment would be influenced by the historic trajectory of the assembly process, which is inherently stochastic and could account for ß-diversity [14]. Despite the importance of non-deterministic factors, specialization for habitats would still play a central role in determining diversity and species abundance, and in the assembled community, niches would be stably occupied by what is the best adapted organism. In contrast, the neutral model presumes that communities are composed of species that are ecologically equivalent, and composition at a local scale and ß-diversity arise solely by dispersal limitation, disturbance, and other random processes [27]. Consequently, communities are open to additional colonists, are continuously changing, and have non-equilibrium assemblages of species which show functional redundancies [11]. Given the different characteristics of communities shaped by deterministic and neutral factors, temporal patterns of populations can be interpreted to deduce the ecological principles by which they are governed.

Recently, Caporaso and co-workers have provided the first long-term (two subjects, daily samples for 15 months) analysis of the bacterial communities at different body sites (feces, mouth, skin) employing Illumina sequencing and 454 pyrosequencing [28]. This work exceeded previous studies in sequencing depth, duration, and taxonomic resolution, and it revealed a pronounced temporal variability in an individual's fecal microbiota. Only a minor fraction of bacterial taxa were detected to be persistent, and the authors concluded that there is little evidence for a core temporal fecal microbiome. The findings of this study contradicted previous studies that were more short-term and relied on non-sequencing based methods with lower taxonomic resolution [20,21,22,23,24,29], and suggested that the human gut microbiome might be significantly less stable than previously acknowledged. To gain an independent perspective on the long-term temporal dynamics of the human gut microbiome, we characterized the fecal bacterial communities of three human individuals over the duration of one year to the species level. This was achieved by analyzing sequence data sets generated with 454 pyrosequencing of 16S rRNA tags during two consecutive nutritional studies [30,31]. In agreement with Caporaso and colleagues, we found only a small proportion of species-level OTUs that persisted over a one-year period. However, by employing a quantitative approach, we arrived at different conclusions in terms of the long-term temporal characteristics of the human microbiota, and a temporal core that dominated the fecal microbiota was identified. To evaluate the ability of the OTU-selection approach used in this study to accurately determine the proportion of a core microbiota, we characterized the fecal microbiota of nine mice with a defined three-member community. This experiment revealed that the sequencing approach inflated diversity measures and reduced the apparent fraction of shared OTUs, and these methodological artifacts were considered in the interpretation of the human data.

Results

General Characteristics of the Fecal Microbiota in Three Human Individuals

To investigate the temporal patterns of the human fecal microbiota ove[r a 56-week period, we characterized the bacterial populations in three healthy subjects that participated in two dietary trials that tested the impact of resistant starches (types 2 and 4) and GOS on the bacterial populations [31,32]. The sequencing approach used resulted in sequences of 500 bp spanning the entire V1-V3 region of the 16S rRNA gene, allowing clear assignments to the species level in most cases. 33 fecal samples from subjects 1 and 2, and 32 samples from subject 3, were analyzed (see Figure 1 for an overview of the study design). All three subjects remained healthy throughout the entire study period with the exception of minor infections. Subject 1 underwent a 1-week antibiotic treatment during the 24-week period in which no sampling took place (week 19, corresponding to the 2nd week of the non-sampling period). Subject 3 did a deliberate lifestyle change between the dietary trials that included an increase in exercise, healthier nutrition, and moderate weight loss (7 kg).

As shown in many previous studies, the subjects' microbiota were conserved at the phylum level and dominated by Firmicutes and Bacteroidetes phyla, and to a lesser degree Actinobacteria, Verrucomicrobia and Proteobacteria. At lower taxonomic scales, the microbiota was individualized and clustered by subject (Figure S1). The lifestyle changes adopted by subject 3 were reflected in the separate clustering of the fecal microbial community among samples taken in the two trial periods (but were still distinct from those of the other subjects), and was associated with a pronounced decrease in the Firmicutes:Bacteroidetes ratio (from 20±10 in the first study period, to 2±1 in the second trial). Equivalent changes of this ratio have been linked to weight loss in a previous study [33], but the mechanisms that cause these shifts are not yet understood.

Temporal Dynamics of the Human Fecal Microbiota at Different Taxonomic Scales

Sequence proportions determined by pyrosequencing were used to characterize the temporal dynamics of the fecal microbiota within individuals. In accordance with previous studies [24,28], the temporal stability of the fecal microbiota is dependent on the

	TRIAL 1 (Resistant starches type 2 and 4 + native starch) Weekly sampling																		TRIAL 2 (Dose-response feeding of GOS) Weekly sampling															
	Baseline		Treatment 1			Washout		Treatment 2			Washout		Treatment 3			Washout		24 week interval (no sampling)	Baseline		0 g GOS			2.5 g GOS			5.0 g GOS			10 g GOS			Washout	
Week	1	2	3	4	5	6	7	8	9	10	11	12	13	14	15	16	17		41	42	43	44	45	46	47	48	49	50	51	52	53	54	55	56
Sample	1	2	3	4	5	6	7	8	9	10	11	12	13	14	15	16	17		18	19	20	21	22	23	24	25	26	27	28	29	30	31	32	33

Figure 1. Experimental design. 33 (two subjects) and 32 (one subject) samples were collected in a 56-week time period from human subjects that participated in two dietary trials testing the effect of resistant starches and GOS on the fecal microbiota [31,32]. Fecal samples were collected weekly throughout both trials (17 in Trial 1 and 16 in trial 2), and the two trials were interspaced by a 24-week period without sampling.

taxonomic scale, decreasing from the phylum to the species level (Figure S2). To determine the temporal core at the species level, we identified the OTUs that were detected in at least 80% of the samples within an individual. The criteria for identifying core members are not standardized, and the criteria used in our study differed from that used by Caporaso and colleagues [28], who defined persistent core members as those that were observed across all sampling events, ignoring single, isolated zero-counts. However, this standard might be too stringent as it would select against core members that become temporarily undetectable when dropping below the detection threshold (which accounts for around 10^7 cells/gram with the sequencing depth obtained in our study). Given the dynamic nature of the gut microbiota and the limited analysis depth, we considered OTU detected in >80% of the samples as members of the temporal core.

Only a small proportion of the total species-level OTUs were determined to make up an individual's temporal core (Figure 2). Out of the total of 411±119 OTUs detected per subject, 12% ±3% of the OTUs were detected as stable. Overall, 69 OTUs were determined to be persistent in at least one subject (39, 40 and 33 in the individual subjects), out of which 16 were shared among the 3 subjects (Table 1). It is important to point out that the majority of these OTUs were detected throughout the entire period. Only 6 OTUs became undetectable in 3 consecutive weeks (1 in subject 1, 2 in subject 2, and 3 in subject 3) and only one OTU in subject 2 was undetected in 4 consecutive weeks.

The Human Fecal Microbiota is Dominated by a Small Number of Persistent Species

The small proportion of OTUs that were stably detected over a 56-week time period could give the impression of a negligible temporal core within the human fecal microbiota. However, when the abundance of these OTUs is considered (calculated from the proportion of sequences represented by each OTU), the analysis revealed that the majority of the human fecal microbiota is stably maintained (Figures 2), as 75%, 81%, and 79% of the total sequences from the individual subjects belonged to stable OTUs (Figure 2A to C). Moreover, even if only OTUs detected in 100% of the samples within a subject are considered (18, 19, and 16 OTUs for subjects 1, 2, and 3, respectively), they composed 62% ±4% of the total microbiota, on average. In addition, the 16 stable OTUs that were shared among the three subjects comprised an average of 47% of the total reads per individual. Thus, the proportion of the inter-individual core microbiota in our study is comparable to that identified in a previous human study [17].

The 69 species that remained stable in at least one of the subjects (Table 1) belonged to the four phyla Firmicutes (48 OTUs), Bacteroidetes (14 OTUs), Actinobacteria (6 OTUs), and Verrumicrobia (1 OTU). Out of the 16 stable OTUs that were detected in all 3 subjects, twelve showed >97% sequence similarity to described species: *Bifidobacterium longum*, *Bacteroides vulgatus*, *Bacteroides putredinis*, *Faecalibacterium prausnitzii* (3 OTUs, related to strains SL3/3, KLE1255, and A2-165), *Ruminococcus obeum*, *Blautia wexlerae*, *Dorea formicigenerans*, *Eubacterium ramulus*, *Eubacterium rectale*, and *Coprococcus comes*. Another 11 OTUs were stably detected in 2 of the 3 subjects, and included phylotypes closely related to *Bifidobacterium adolescentis*, *Collinsella aerofaciens*, *Bacteroides thetaiotaomicron*, *Bacteroides uniformis*, *Odoribacter splanchnicus*, *Parabacteroides distansonis*, *Streptococcus salivarius*, *Roseburia inulinivorans*, as well as one additional lineage within each the Ruminococcaceae, Clostridiales and Erysipelotrichaceae (Table 1). Out of these 11 OTUs, 7 were also detected in the third subject, but in less than 80% of the samples. Finally, 42 OTUs were determined to be persistent only in one individual, although 32 of these OTUs were transiently detected in at least one other subject (Table 1). Firmicutes comprised most of the transient members of the fecal microbiota, whereas Actinobacteria, Bacteroidetes and Verrucomicrobia populations remained more stably associated to the gut ecosystem (Figure 2D to F).

Resilience of Core Members to Short-term Dietary Modulations

Diet is a major factor in shaping the structure of the gut microbiota [34], and long-term dietary preferences have been linked to consistent differences in the microbial community structure among humans [35]. As previously described, the dietary modulations used in the two trials (RS2, RS4, native starch, and GOS) had individualized and reversible effects on the fecal microbiota of the subjects [30,31,32]. Interestingly, the most substantial diet-driven changes detected concerned species of the temporal core, namely *Bifidobacterium adolescentis* (induced through RS2, RS4, and GOS) and *Parabacteroides distasonis* (induced through RS4). As shown in Figure S3, the diet-induced changes in the abundance of these taxa was individualized, and tightly linked to the dietary modulation, with populations returning to baseline levels after treatment cessation. These findings provide evidence that shifts within core members can be induced though diet but populations show resilience.

Characterization of the Temporal Fluctuations in Relative Abundance of Core Members

Virtually all members of the human fecal microbiota, including members of the temporal core, showed temporal fluctuations in relative abundance (Figure 3 and S4–S7). Normal variations in the individual's diet are one likely cause of these patterns. Accordingly, the taxa with the highest fluctuations belong to species that have been identified to respond to dietary compounds or possess enzymatic capabilities to degrade dietary carbohydrates. Examples are *Bacteroides xylanisolvens* (utilizes xylan), *Blautia wexlerae* (enriched by whole grains [36]), *Clostridium clostridiiformes* (enriched by RS4 [31]), *Eubacterium rectale* and *Ruminococcus bromii* (enriched by RS2 and RS3 [31,37]), *Faecalibacterium prausnitzii* (enriched through the prebiotics inulin, oligofructose, GOS [34]), and several *Roseburia* species (enriched by whole grains [36]).

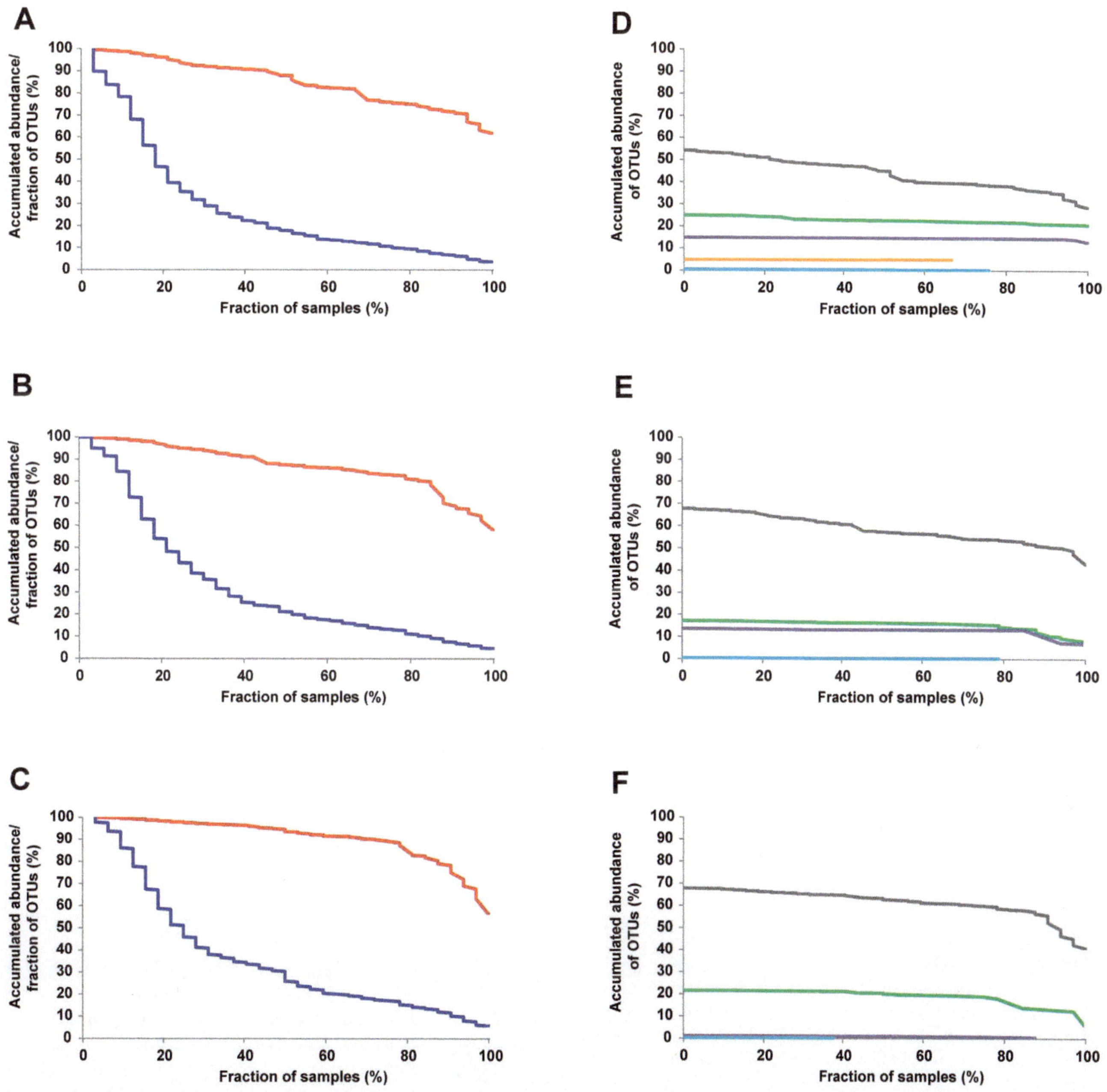

Figure 2. Individual fecal temporal core. Fraction of species-level OTUs shared across samples of the total OTUs within these samples (blue line) and the proportion that these OTUs represent of the total sequences obtained (red line) for subjects 1, 2, and 3 (A,B,C). Proportion of total sequence reads obtained of OTUs shared across sampled within the different phyla (Firmicutes: grey, Bacteroidetes: green, Actinobacteria: purple, Verrucomicrobia: orange, Proteobacteria: light blue) (D,E,F).

Evidence for Events of Invasion and Extinction

Although the findings obtained during this study suggest that most of the dominant bacterial species in the human gut are persistent over a one-year time span, the analysis revealed several patterns that are indicative of invasion and extinction events. Such events were indicated by stable populations of OTUs over longer timespans that abruptly appeared or disappeared. For example, such patterns suggested the extinction of OTU 36 (*Akkermansia muciniphila*) in subject 1, and the invasion of the same species in subject 2 (Figures 3 and S8). In subject 1, several phylotypes among the Firmicutes (OTUs 4262, 4257, 4256) became extinct between the two nutritional trials (Figures 3 and S8). This "mass extinction" event coincided with a short-term clarithromycin treatment. Interestingly, all four OTUs are closely related to members of the fecal microbiota (*Clostridium clostidiiforme*, *Eubacterium ramulus*, *Eubacterium rectale*, *Eubacterium ventriosum* and *Roseburia cecicola)* that have been shown to be affected by long-term clarithromycin therapy [38].

Some population dynamics are suggestive of niche-related processes, potentially associated with cooperation or competition. For example, several distinct OTUs belonging to the Bacteroidetes emerge simultaneously after week 49 in subject 1 (OTUs 354, 479 and 513) (Figure 3), probably associated with the availability of a new niche or a syntrophic relationship between these species. In

Table 1. Abundance (%) of bacterial OTUs that comprise the temporal core within the fecal microbiota of three humans over a one-year period.

OTU #	Phylum	Closest related type strain* (average percent similarity)	Subject 1 Mean ± SD (percent of samples in which OTU was detected)	Subject 2 Mean ± SD (percent of samples in which OTU was detected)	Subject 3 Mean ± SD (percent of samples in which OTU was detected)
2	Actinobacteria	*Bifidobacterium adolescentis* (100)	8.67±7.15 (100)	5.80±11.47 (88)	ND
5	Actinobacteria	*Bifidobacterium adolescentis* (96)	3.91±3.32 (100)	0.04±0.08 (27)	ND
358	Actinobacteria	*Bifidobacterium pseudocatenulatum* (99)	0.13±0.26 (42)	3.68±2.89 (100)	ND
11	Actinobacteria	*Bifidobacterium longum* (99)	1.19±0.92 (97)	0.44±0.47 (94)	0.28±0.42 (84)
928	Actinobacteria	*Adlercreutzia equilifaciens* (98)	ND	ND	0.39±0.51 (91)
40	Actinobacteria	*Collinsella aerofaciens* (98)	0.50±0.43 (97)	3.15±5.20 (100)	0.34±0.57 (47)
16/548	Bacteroidetes	*Bacteroides thetaiotaomicrom* (98)	1.78±0.81 (100)	0.40±0.31 (91)	0.09±0.14 (50)
20	Bacteroidetes	*Bacteroides xylanisolvens* (99)	4.39±4.49 (100)	0.32±0.38 (79)	ND
273	Bacteroidetes	*Bacteroides intestinalis* (98)	ND	0.81±0.61 (97)	ND
18	Bacteroidetes	*Bacteroides uniformis* (99)	3.37±1.78 (100)	1.29±1.14 (100)	0.57±0.70 (66)
1	Bacteroidetes	*Bacteroides vulgatus* (99)	8.98±5.09 (100)	5.15±3.98 (100)	7.11±8.24 (97)
512	Bacteroidetes	*Bacteroides dorei* (100)	0.10±0.22 (21)	0.56±0.38 (97)	0.80±1.19 (75)
269	Bacteroidetes	*Bacteroides ovatus* (97)	0.15±0.20 (67)	0.25±0.33 (82)	0.06±0.09 (50)
118	Bacteroidetes	*Odoribacter splanchnicus* (99)	0.17±0.14 (88)	0.36±0.33 (82)	0.08±0.15 (44)
226	Bacteroidetes	*Parabacteroides distasonis* (97)	0.35±0.61 (85)	0.14±0.18 (70)	1.43±2.88 (85)
513	Bacteroidetes	*Alistipes shahii* (100)	0.03±0.07 (21)	0.22±0.18 (91)	0.14±0.18 (53)
235	Bacteroidetes	*Alistipes finegoldii* (100)	0.42±0.49 (88)	ND	0.07±0.10 (50)
19	Bacteroidetes	*Bacteroides putredinis* (99)	2.14±1.16 (100)	1.62±1.33 (100)	5.29±3.80 (100)
479	Bacteroidetes	*Bacteroides intestinihominis* (99)	0.14±0.30 (24)	0.56±0.47 (97)	ND
4400/4617	Bacteroidetes	Bacteroidales	ND	2.83±2.92 (91)	ND
114	Firmicutes	*Flavonifactor*	0.11±0.13 (70)	0.04±0.05 (52)	0.40±0.35 (91)
12556	Firmicutes	Ruminococcaceae bacterium D16 (97)**	ND	ND	0.44±0.39 (91)
359	Firmicutes	*Oscillibacter* spp.	0.09±0.09 (67)	0.16±0.15 (82)	0.15±0.16 (66)
5715	Firmicutes	Ruminococcaceae	0.02±0.04 (15)	0.03±0.07 (27)	0.50±0.54 (94)
32	Firmicutes	*Butyricicoccus* spp.	0.21±0.19 (91)	0.03±0.06 (39)	ND
2817	Firmicutes	Clostridiales	ND	0.55±0.74 (85)	ND
5522	Firmicutes	*Ruminococcus bromii* (99)	0.80±1.92 (21)	1.02±3.95 (18)	4.90±5.32 (100)
4705	Firmicutes	Ruminococcaceae	0.09±0.25 (21)	0.41±0.46 (85)	0.16±0.28 (50)
4256	Firmicutes	Ruminococcaceae	1.48±3.07 (45)	4.56±3.73 (100)	3.74±3.29 (100)
161/4277	Firmicutes	Clostridiales	0.12±0.12 (85)	0.22±0.18 (88)	ND
2308	Firmicutes	Clostridiales	0.14±0.24 (58)	1.33±0.90 (94)	ND
29	Firmicutes	*Faecalibacterium prausnitzii* SL3/3 (99)**	1.36±1.17 (100)	2.01±1.30 (100)	3.05±2.22 (100)
35	Firmicutes	*Faecalibacterium prausnitzii* KLE-1255 (98)**	0.85±0.77 (97)	0.03±0.04 (45)	0.03±0.07 (22)
6	Firmicutes	*Faecalibacterium prausnitzii* KLE-1255 (97)**	2.72±2.20 (100)	1.51±1.46 (100)	1.60±1.33 (97)
17	Firmicutes	*Faecalibacterium prausnitzii* A2-165 (97)**	4.68±2.23 (100)	3.56±1.67 (100)	2.65±3.10 (100)
31/136	Firmicutes	*Dialister* spp.	0.41±0.29 (91)	ND	ND
288	Firmicutes	*Dialister invisus* (100)	ND	ND	3.07±2.68 (91)
66	Firmicutes	*Streptococcus salivarius* (98)	0.73±0.66 (94)	0.27±0.45 (64)	0.64±0.87 (94)
21	Firmicutes	*Streptococcus parasanguinis* (99)	0.22±0.36 (88)	0.02±0.05 (24)	0.04±0.07 (34)
22	Firmicutes	*Clostridium spiroforme* (99)	0.18±0.20 (73)	ND	0.17±0.20 (81)
53	Firmicutes	Erysipelotrichaceae	0.43±0.33 (88)	ND	ND
9864	Firmicutes	Erysipelotrichaceae	ND	2.45±1.97 (100)	1.21±1.35 (100)
9902	Firmicutes	Erysipelotrichaceae	ND	0.14±0.11 (81)	ND

Table 1. Cont.

OTU #	Phylum	Closest related type strain* (average percent similarity)	Subject 1 Mean ± SD (percent of samples in which OTU was detected)	Subject 2 Mean ± SD (percent of samples in which OTU was detected)	Subject 3 Mean ± SD (percent of samples in which OTU was detected)
48/23/365/27/262/10	Firmicutes	*Ruminococcus obeum* (98)	3.13±1.56 (100)	7.46±3.74 (100)	5.73±3.53 (100)
3/7	Firmicutes	*Blautia wexlerae* (99)	6.62±5.24 (100)	1.87±1.95 (100)	6.02±4.24 (100)
361	Firmicutes	Lachnospiraceae	0.12±0.12 (82)	0.11±0.11 (70)	0.12±0.15 (69)
9873	Firmicutes	*Dorea formicigenerans* (97)	ND	0.05±0.07 (48)	0.42±0.49 (97)
51	Firmicutes	*Dorea formicigenerans* (99)	0.79±0.43 (100)	0.62±0.49 (100)	2.62±3.07 (100)
76/4328	Firmicutes	*Dorea* spp.	0.23±0.21 (85)	0.51±0.25 (100)	1.04±0.81 (100)
78	Firmicutes	*Ruminococcus torques* (97)	0.04±0.13 (27)	0.04±0.13 (27)	0.89±1.20 (91)
9942	Firmicutes	*Ruminococcus torques* (99)	ND	0.19±0.17 (88)	0.08±0.13 (50)
24/511/403	Firmicutes	*Coprococcus comes* (98)	0.57±0.51 (97)	3.01±2.01 (100)	3.86±1.99 (100)
12	Firmicutes	*Ruminococcus gnavus* (99)	0.36±0.29 (94)	0.07±0.19 (36)	0.14±0.15 (72)
26	Firmicutes	*Eubacterium ramulus* (99)	0.64±0.34 (97)	0.28±0.29 (91)	0.82±1.04 (91)
42	Firmicutes	*Roseburia inulinivorans* (100)	0.26±0.27 (91)	0.17±0.37 (55)	1.44±1.55 (91)
120	Firmicutes	Lachnospiraceae	0.36±0.25 (97)	0.04±0.05 (52)	0.09±0.11 (53)
14/1107	Firmicutes	Lachnospiraceae	2.53±1.64 (100)	4.61±2.32 (100)	1.38±1.15 (94)
63/370	Firmicutes	*Clostridium clostridiiformes* (96)	1.24±1.58 (100)	4.42±3.18 (100)	2.34±2.70 (100)
9865	Firmicutes	*Coprococcus euctatus* (99)	ND	3.79±2.91 (100)	ND
3927	Firmicutes	*Coprococcus catus* GD/7 (98)**	0.08±0.10 (48)	0.15±0.13 (91)	ND
30	Firmicutes	Lachnospiraceae	0.21±0.27 (82)	0.03±0.11 (18)	0.01±0.03 (25)
9	Firmicutes	Lachnospiraceae	0.83±0.93 (82)	0.04±0.08 (36)	0.13±0.15 (56)
4264	Firmicutes	*Eubacterium eligens* (100)	0.06±0.16 (41)	0.30±0.25 (91)	0.12±0.11 (63)
4	Firmicutes	*Eubacterium rectale* (97)	6.49±5.44 (100)	9.07±9.32 (100)	4.06±3.53 (100)
4262/2326	Firmicutes	*Roseburia intestinalis* (98)	2.26±2.72 (55)	0.56±1.36 (27)	4.26±3.40 (100)
45	Firmicutes	Lachnospiraceae	1.03±0.67 (100)	0.29±0.36 (70)	0.19±0.37 (69)
15	Firmicutes	*Clostridium* sp. SS2/1 (98)**	1.50±1.19 (100)	0.24±0.26 (85)	0.80±1.09 (100)
5348	Firmicutes	*Eubacterium ventriosum* (97)	0.06±0.16 (18)	0.26±0.22 (79)	0.41±0.48 (88)
36	Verrucomi crobia	*Akkermansia muciniphila* (98)	4.70±7.15 (67)	0.30±0.94 (36)	5.82±8.92 (100)
		Sum	**85.17%**	**84.44%**	**82.19%**

Mean and standard deviations were calculated with all samples across the study period.
*Similarity was assessed by alignment of the sequences with ClustalW (p-distances).
**Not the type strain.

addition, within the Ruminococcaceae family, a direct succession of closely related OTUs was observed in subject 1 (OTU 4321 followed by 5219 and then 5522) and subject 2 (OTU 5522 followed by 5218 followed by 10354), suggesting events of invasion coupled with competitive exclusion of closely related taxa (Figure 3, Figure S8).

Pyrosequencing Overestimates Diversity in a Defined Three-member Community, Artificially Augmenting β-diversity

As described previously [28], this study revealed that most OTUs detected by sequencing in human fecal samples are not stably detected over time. These OTUs could constitute species that are transient. However, limitations of sequence surveys, such as sequencing errors, chimeras, and over-splitting of OTUs, can result in artifacts that can confound the interpretation of community data, especially in terms of diversity determination [39,40,41] and the definition of a core [42]. To assess the capacity of the OTU-based sequencing approach used in our study to investigate measures of bacterial community diversity, we characterized the fecal microbiota in nine gnotobiotic mice with a simple defined microbiota composed of *Bacteroides thetaiotaomicron*, *Bifidobacterium adolescentis*, and *Escherichia coli*. Taxonomic classification of the sequences down to the genus level with the Classifier (RDP) tool identified 3 genera as being the only members of the fecal microbiota in the animals (Figure S9), with 19.7% of the sequences accounting for the genus *Bifidobacterium*, 20.3% for *Escherichia*, and 58.7% for *Bacteroides*. This showed that the Classifier algorithm (which performs taxonomic assignments of individual reads and is not reliant on multiple sequence alignments) is able to correctly assign most sequences to the bacterial genera that were present in these mice.

However, the OTU picking led to a substantial overestimation of diversity, as 237 OTUs were detected in the data set (an average of 42 per animal). Only 9 OTUs were detected in all 9 animals (Figure S9), which accounts for only 3.8% of the total OTUs. Therefore, the approach led to artificial measures of both α- and

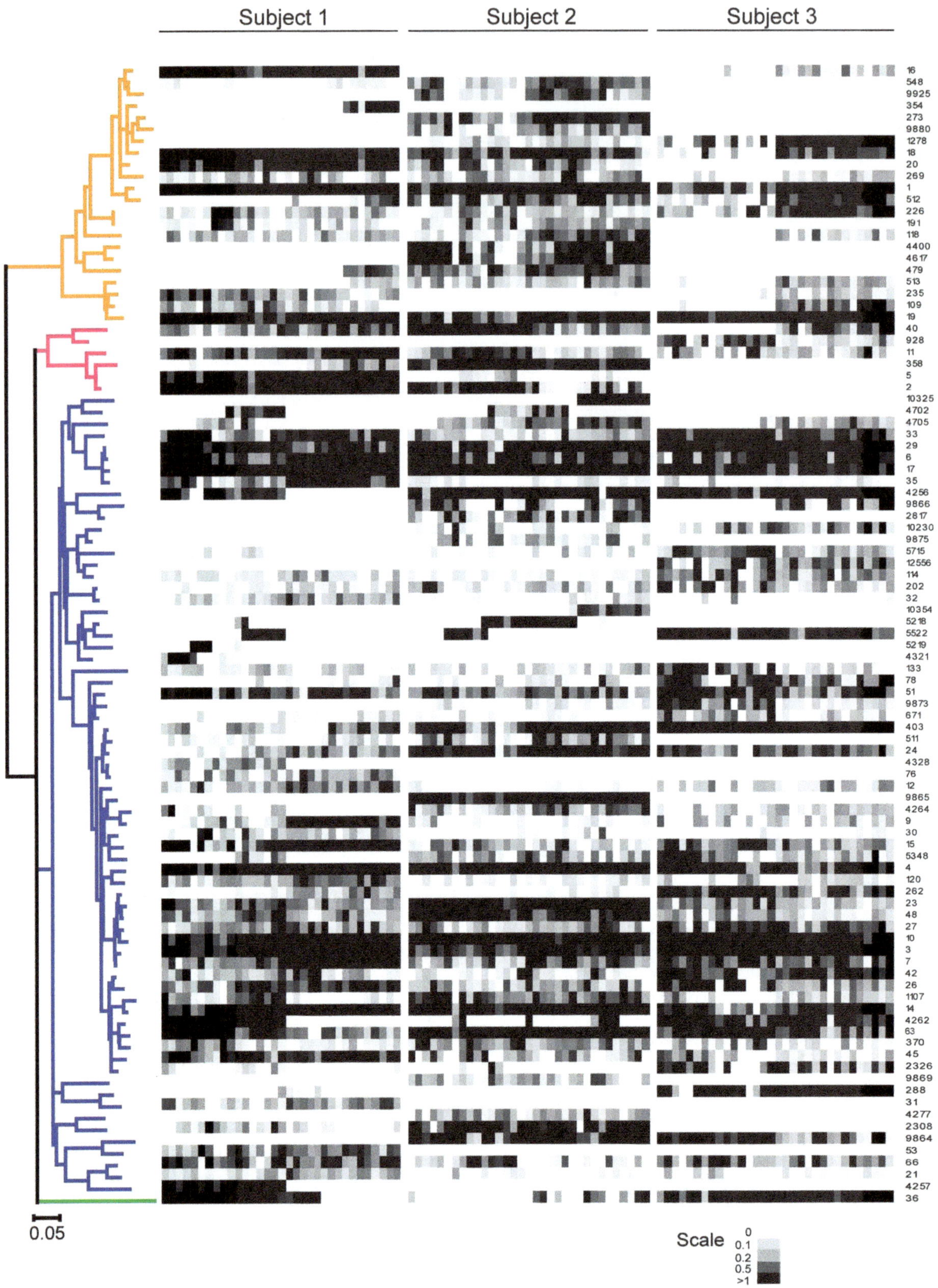

Figure 3. Abundance of the dominant bacterial OTUs in the fecal samples of subjects throughout the study. All OTUs that had an average proportion >0.2% were included. Representative sequences of OTUs were used to construct the Maximum Likelihood phylogenetic tree, which is colored-coded according to phylum: Bacteroidetes (orange), Actinobacteria (magenta), Firmicutes (blue), Verrucomicrobia (green). The abundance of each phylotype is indicated by grayscale. The OTU numbers are presented to the right of the heatmap.

β-diversity, giving the false impression that the microbiota across these animals is highly variable and that the fraction of core species is very small. However, it is important to point out that the 9 OTUs detected in all samples represented virtually the entire bacterial population (96.6% of the sequences) (Figure S9). Importantly, the three most abundant OTUs accounted for the three individual species present in the mice, and comprised 79% of the total sequences, 17% for *Bifidobacterium adolescentis*, 17% for *Escherichia coli*, and 45% for *Bacteroides thetaiotaomicron*. In conclusion, this validation experiment showed that OTU picking in combination with sequencing errors overestimates the number of OTUs and β-diversity among samples.

Discussion

The diversity and temporal dynamics are important characteristics of a bacterial community are a reflection of the ecological processes that shape the ecosystem. The human gut microbiota has historically been considered stable [10,23,29], but recent work using next-generation sequencing of 16S rRNA tags has challenged older findings and suggested a pronounced intra- and inter-subject variability [25,28]. The new insight perpetuated the notion that the human gut microbiota is composed of hundreds of species that show little conservation, not only across individuals but also over time [6]. In a recent review, Fierer and coworkers argued that "it is apparent that the adult microbiome is in a constant state of flux as microbial community composition on and in an individual varies substantially over time" [43]. This perspective, which is essentially contrary to the traditional concepts in gut microbial ecology proposed by Savage 35 years ago [10], would have major conceptual implications on how we view the gut microbial ecosystem, as it proposes an important role of neutral processes in shaping the bacterial communities [44]. If correct, it would have repercussions that go beyond ecology and would impact how we conceive host-microbiota symbiosis and the microbiome's role in health and disease. If host-associated microbiomes were mainly governed by neutral processes, it would be extremely difficult for the host to select and maintain beneficial symbionts over ecological and evolutionary time frames. The development of a beneficial relationship would be extremely unlikely, as mutualism is favored by the selection of true mutualists and their stable maintenance over evolutionary time [45]. Accordingly, a recent modeling approach predicted that in a neutral host without control over the bacterial symbionts, mutualism would be intrinsically fragile [46].

The temporal analysis of the fecal microbiota of three humans that we report here is more supportive of the traditional perspective on gut microbial ecology [10], as we were able to detect a stable temporal core comprising around 40 OTUs per human individual over a one-year period. In accordance to the findings by Caporaso et al. [28], this temporal core represents a very small fraction of the total OTUs. However, it is important to emphasize that it constitutes, quantitatively, the majority (>75%) of the microbial community. Therefore, our findings indicate that the human fecal microbiome is dominated by species that form stable populations over longer periods of time. This is consistent with the continuous detection of strains in human fecal samples [47,48,49,50] and the stability of the microbiota described with alternative fingerprinting techniques [20,21,22,23,24,29]. In addition, it appears that a stable microbiota is a general feature in mammals, as wild chimpanzees and lab mice also harbor stable microbial communities in their gut [51,52].

Although this study revealed a temporal core, it is likely that the sequencing techniques used underestimate the relative proportion of this core. First, amplicon sequencing in combination with the bioinformatic approaches substantially overestimates diversity due to sequencing artifacts and limitations in data processing [42]. Most importantly, all current time- and processing-efficient cluster algorithms, including the one used in this study, lead to an over-splitting of OTUs (often 10–100 fold) [41]. This generates an artificial increase in α-diversity, and when different samples are compared, β-diversity, which was clearly demonstrated in our experiments with triple-associated mice (Figure S9). That methodological artifacts cause an artificial increase of OTUs and a concomitant reduction in the apparent fraction that are shared has been recognized by other researcher [42], but this notion has not been considered in studies that assessed inter- and intrapersonal microbiome variation [25,28], and it is likely that these parameters have been over-estimated. Second, it has been shown that the size of the core microbiota is dependent on the depth of the analysis, and it is likely that the core, both within and between individuals, has been underestimated due to undersampling [53]. Accordingly, most of the dominant species detected in our study were persistent, while taxa present in lower abundance were generally detected in a smaller number of samples and were, therefore, not considered to be core members. However, given that most of these species were identified throughout the study period, it is likely that they represent stable members of the community occasionally falling below the detection threshold due to normal temporal fluctuations. However, despite the limitations of the sequencing approaches used in this study, a stable temporal core that dominated the fecal microbiome in humans was detected. In addition, the findings obtained in our mouse experiment indicate that a quantitative approach that focuses on abundance of OTUs can provide a more accurate representation of bacterial populations, and thus allows a better interpretation of community characteristics.

The temporal characteristics of the human fecal microbiota revealed during this study support the concept of an ecosystem that operates, to a large degree, in a state of homeostasis and shows resilience to perturbations caused by diet and lifestyle changes. Perturbations appear to mainly alter the relative proportions of the individual's bacterial populations but do not cause extensive changes in its membership, as suggested by Rajilic-Stojanovic and co-workers [29]. Therefore, the gut microbiota appears to function as an ecological unit whose composition and local diversity is largely determined by niche-driven processes, a view that is also supported by theoretical model calculations [54,55]. The significance of niches provides an explanation for ecosystem characteristics such as resilience and colonization resistance. Diet can alter the niche landscape through the provision of novel nutrients, leading to fluctuations in microbiome structure that are however reversible upon cessation of the dietary stimulus [30,31]. The use of antibiotics and invasions by better adapted microbes have the potential to remove members from the community, whose loss might alter the niche environment and may lead to more global and permanent changes [56]. The seemingly conflicting characteristics of the human gut microbiota, intra-individual stability despite substantial inter-individual diversity (even in mono-zygotic twins), can not only be explained by differences in host genotype, diet, and health, but also by a historical perspective of community assembly [14]. This view emphasizes the importance of historical patterns of dispersal and acquisition for the composition of the emerging community. These patterns are inherently stochastic, influencing colonization order and adaptive processes during assembly, thereby impacting the physicochemical properties of the niche-environment and the outcome of the assembly process. Inter-individual diversity could therefore be due to current and past personalized environmental differences (diet, antibiotics, exposure, host physiology, genetics and immunity, age, and metabolic state).

However, although our study confirmed that the gut microbiota is individualized, it also revealed a substantial inter-individual overlap, in terms of total sequences, between the microbiomes of the three subjects. The 16 stable OTUs present in the three individuals comprised, quantitatively, almost 50% of the sequences obtained from the individual subjects. Importantly, most of the stable species detected in our study in residents of the USA were also identified as core members in individuals residing in Europe [15,50], and 27 of the species detected recently by Schloissnig and co-workers as dominant member of the human microbiome were also among the stable core in our study. Our findings suggested that the majority of the human fecal microbiota is composed of only around 69 species (Table 1), which is in accordance to recent findings obtained with whole metagenome sequencing that revealed that the gut microbiome of 207 individuals is composed of 66 bacterial species that account for 99% of the mapped reads [50]. These findings support the idea of a dominant core within human subjects [16], and these species could be considered autochthonous members of the human microbiota. In light of the novel sequence data, the requirements for autochthony proposed by Savage are probably too strict in that they call for members to be dominant throughout the entire life span of a host and present in all individuals of the host species [10]. It is unlikely that many lineages fulfill these requirements (also due to the impact of environment, diet, age, and health status on gut microbiota composition). However, the available data still support the concept of a human gut microbiota that is dominated by species that occupy long-term niches and are likely to share an evolutionary relationship with humans. Although the relative proportions of these lineages fluctuate and are susceptible to environmental cues and host physiology, they are still inherently human, are likely to play an important ecological role, and potentially support human health.

It is increasingly recognized that it will require the application of ecological theory to refine our understanding of the human microbiome, explain and predict community characteristics, and to inform strategies to successfully reshape the microbiota [6,7,8]. This study allowed inferences about the ecological principles that govern the human gut microbiota as it revealed that the human fecal microbiota is dominated by species that are temporarily stable and that overlap in individuals from the US and Europe. The temporal and spatial variation and dynamic nature of the gut microbiota revealed by next-generation sequencing has made some scientists focus on the plasticity of the gut microbiota and the lack of core species [6,28,43]. However, our evaluation of OTU-based sequencing approaches in gnotobiotic mice strongly suggested that much of this variation and the small apparent fraction of core species might be caused by methodological artifacts. We argue that the data obtained in this study, and most of the data that is now available (including metagenomic datasets [50]), is in favor of a stable human microbiota whose dominant members are, although in varying proportions, shared among a majority of humans in the western world. This view stresses the importance of niche-driven processes in shaping diversity and host control over the community [57], which would allow for a targeted selection of beneficial microbial communities during an individual's life-span and over evolutionary times. Clearly, the development of ecological concepts that apply to the gut microbiota will require improved approaches for the characterization of microbial ecosystems that accurately determine diversity, and their careful interpretation, and efforts remain necessary to improve current tools.

Materials and Methods

The human trials that are part of this study were approved by the Institutional Review Board of the University of Nebraska (IRB Approval Numbers: 2008038840EP and 2009019551EP), and written informed consent has been obtained from all subjects.

Triple-associated B6 mice were maintained at the University of Nebraska Gnotobiotic Mouse Facility and all experiments were performed with approval of the Institutional Animal Care and Use Committee (Project ID 731).

Study Subjects and Experimental Design

Fecal samples of three individuals who had participated in two dietary trials conducted by our group [30,31,32] were included in this study. At the moment of the first sample collection subject 1 (female) was 25 years old, subject 2 (male) was 26 years old, and subject 3 (male) 23 years old. A total of 33 samples for subjects 1 and 2, and 32 samples for subject 3 were obtained throughout a 56-week time period. The time schedule of sample collection is shown in Figure 1. The first study was a randomized, double-blind, placebo-controlled, cross-over trial that lasted 17 weeks, in which crackers containing resistant starches types 2 or 4 (RS2 and RS4, respectively), or native starch, were consumed by the subjects. Weekly samples across the 17-week period were collected, including 2 samples at baseline, 3 samples per treatment (9 total) and 6 samples corresponding to 2-week washout periods between treatments and at the end of the final treatment [31]. 17 samples were thus collected for subjects 1 and 2, and 16 samples for subject 3 (one sample during a wash-out period was not collected). After a 24-week interval without sampling, a second trial was conducted in which weekly fecal samples were obtained throughout 16 weeks. The three subjects underwent 3-week treatments of increasing doses of galactooligosaccharides (GOS) (0 g, 2.5 g, 5 g, and 10 g), with additional 2-week baseline and washout sampling points at the beginning and end of the trial respectively [32].

The three subjects participating in these studies had no history of chronic gastrointestinal diseases, abnormal gastrointestinal symptoms, and did not take antibiotics 3 months prior to the beginning of each sampling period, or throughout the two sampling periods. Subject 1 received a 1-week antibiotic (clarithromycin) treatment during the 24-week period in which no sampling took place (week 19, corresponding to the 2nd week of the non-sampling period).

Characterization of the Fecal Microbiota by 454 Pyrosequencing

Both studies used the same protocols for sample collection, DNA extraction, and fecal microbiota characterization by pyrosequencing [31]. Briefly, the fecal bacterial community was characterized by massive parallel sequencing with the Roche Genome Sequencer GS-FLX using the Titanium platform (454 Life Sciences). The V1-V3 region of the 16S rRNA gene was initially amplified using a mixture (4:1) of the forward primers B-8FM (5′- *CCTATCCCCTGTGTGCCTTGGCAGTCTCAG*AGAGTTTGATCMTGGCTCAG–3′) and B-8FMBifido (5′-*CCTATCCCCTGTGTGCCTTGGCAGTCTCAG*AGGGTTCGATTCTGGCTCAG–3′), and the primer A-518R as the reverse primer (5′- *CCATCTCATCCCTGCGTGTCTCCGACTCAG*BBBBBBBBATTACCGCGGCTGCTGG–3′) (Martínez et al., 2010). The Titanium adaptor sequences (A and B) are shown in italics, and a unique 8-base barcode (BBBBBBBB) in the reverse primer was used to tag individual samples within the same run. Sequencing was performed following the manufacturer's protocol.

Quality filtering and demultiplexing of the resulting sequence set was performed with the QIIME pipeline [58]. Sequences that were shorter than 300 or longer than 520 bases, contained one or more ambiguous nucleotides, had at least one mismatch to the primer or barcode, showed an average quality score below 25, or contained homopolymer runs over 6 bases, were removed. Chimeras were identified with the Blast Fragments Algorithm implemented in QIIME and removed. A total of 303509 sequences were obtained after quality controls and used for further analyses (108090, 107505, and 87914 for Subjects 1, 2 and 3, respectively). The sequences used for analysis can be found in the MG-RAST database, with the following accession numbers: 4521358.3, 4521394.3, 4521382.3, 4521416.3, 4521406.3, 4521437.3, 4521428.3, 4521415.3, 4521404.3, 4521432.3, 4521423.3, 4521414.3, 4521403.3, 4521395.3, 4521383.3, 4521373.3, 4521363.3, 4521393.3, 4521381.3, 4521356.3, 4521368.3, 4521377.3, 4521388.3, 4521398.3, 4521409.3, 4521419.3, 4521429.3, 4521400.3, 4521410.3, 4521424.3, 4521434.3, 4521402.3, 4521413.3, 4521418.3, 4521425.3, 4521436.3, 4521361.3, 4521370.3, 4521379.3, 4521389.3, 4521362.3, 4521372.3, 4521355.3, 4521366.3, 4521374.3, 4521385.3, 4521401.3, 4521412.3, 4521420.3, 4521430.3, 4521397.3, 4521408.3, 4521435.3, 4521427.3, 4521417.3, 4521407.3, 4521392.3, 4521378.3, 4521371.3, 4521359.3, 4521396.3, 4521384.3, 4521365.3, 4521353.3, 4521387.3, 4521375.3, 4521426.3, 4521391.3, 4521405.3, 4521422.3, 4521433.3, 4521411.3, 4521399.3, 4521431.3, 4521421.3, 4521364.3, 4521354.3, 4521386.3, 4521376.3, 4521367.3, 4521357.3, 4521390.3, 4521380.3, 4521369.3, 4521360.3.

Taxonomic assignments of sequences down to the genus level were made using the Classifier tool from the Ribosomal Database Project (RDP) [59]. Sequences were also designated to Taxonomic Operational Units (OTUs) with 97% similarity cutoff, using the Mothur v. 1.26.0 sequence analysis pipeline [60]. Clustering was performed with sequences from all three subjects in one single alignment. Very low abundance OTUs (5 sequences or less per subject) were removed from further analyses. The abundances of the phylotypes were computed as percent proportions based on the total number of sequences in each sample. OTU picking was also performed in the QIIME pipeline with the uclust algorithm (and default parameters), however, this procedure determined around double the amount of OTUs with QIIME (977+/−238 OTUs) when compared to MOTHUR (411+/−119 OTUs). Given that OTU picking algorithms have been shown to over-split OTUs [41], we used the amount of OTUs obtained with the two methods as a criterion to select the methodology and used the approach that resulted in the lowest amount of OTUs (MOTHUR). However, a visual inspection of the OTUs that represented the same phylotype showed that they followed the same temporal dynamics independent on the method used to generate the clustering (data not shown).

Identification of the Stable Core Microbiota

OTUs detected in at least 80% of the samples within a subject were considered persistent members of the gut microbiota for that individual. Representative sequences of core OTUs were taxonomically classified at the phylum level with the Classifier tool. Next, pairwise alignments were performed for sequences within each phylum with ClustalW and default parameters in MEGA 5.05 [61]. Distance matrices were further constructed based on these alignments, and sequences with >97% similarity were combined within one OTU.

Triple-associated Gnotobiotic Mice

We developed gnotobiotic mice from germ-free mice housed in the Gnotobiotic Mouse Facility at the University of Nebraska. *E. coli* MG1655 was streaked on MacConkey (BD) agar and grown aerobically overnight at 37°C. *Bifidobacterium adolescentis* BD1 and *Bacteroides thetaiotaomicron* VPI-5482 (ATCC-29148) were grown anaerobically on Rogosa SL (BD) (96 hours) and Bile Esculin agar (BD) (48 hours), respectively, at 37°C. We picked colonies from *E. coli*, *B. adolescentis*, and *B. thetaiotaomicron* and grew them in LB (BD) media, MRS (BD) with 5% L-cysteine, and TYG medium, respectively. *E. coli* was grown aerobically overnight (14 hours) at 37°C, while the other two cultures were grown anaerobically for 48 hours at 37°C. Cultures were washed once with PBS and mixed in volumetrically equal proportions immediately before inoculating germ-free C57BL/6 mice by allowing mice to drink the bacterial mixture and pouring it on their fur. Colonized mice were maintained and bred under gnotobiotic conditions.

The presence of the three bacterial species was confirmed by Denaturing Gradient Gel Electrophoresis (DGGE) (data not shown) and selective culture. DGGE also confirmed the absence of other bacterial species from the community. Selective culture was performed from fecal samples on Rogosa SL for *B. adolescentis*, Brain Heart Infusion (BD) with 10% sheep blood and 0.2 mg/mL gentamicin for *B. thetaiotaomicron*, and MacConkey for *E. coli*. Rogosa SL and Brain Heart Infusion plates incubated anaerobically at 37°C for 96 hours and 48 hours, respectively, before enumeration. MacConkey agar plates incubated aerobically at 37°C for 24 hours before enumeration. DNA from 1–3 fecal pellets per individual mouse was extracted following a conventional phenol-chloroform protocol with enzymatic and mechanic cell lysis [62]. Sequencing of the bacterial community, and quality processing of the sequences was performed as described above, using 1,000 randomly selected sequences per animal. Quality-controlled sequences were taxonomically characterized with the Classifier tool and OTU picking was done with MOTHUR and QIIME, as described above. QIIME resulted in less total OTUs then MOTHUR (185 versus 237), but produced more average OTUs per animal (51 versus 42). As MOTHUR was used to analyze the human data, we used the OTUs obtained with this software for the diversity analysis.

Supporting Information

Figure S1 Principal-coordinates plots of the beta-diversity measurements based on unweighted UniFrac distances among samples. Samples were color-coded by subject (A) and by subject and trial period (B).

Figure S2 Presence-absence patterns of dominant bacterial taxa in the fecal samples of human subjects over the entire study period. Sequences were taxonomically classified (Classifier, RDP) and the presence (red) and absence (white) patterns of the most dominant bacterial taxa are presented for each sample at the phylum, order, family and genus levels. Samples are grouped by subject and presented in chronological order.

Figure S3 Resilience of core members to dietary perturbations. Abundance of *Bifidobacterium adolescentis* in fecal samples of (A) subject 1 (showing an increase in abundance due to the intake of both resistant starches and GOS) and (B) subject 2 (showing increase only with resistant starches) and (C) *Parabacter-*

oides distasonis, which was only significantly increased in subject 1 during consumption of resistant starch 4.

Figure S4 Temporal dynamics of core taxa within the Actinobacteria phylum. Abundances of the phylotypes identified as persistent (present in >80% of the samples) within subjects are presented in their phylogenetic context. A representative sequence of each OTU and the closest related type-strains were used to build trees with the neighbor-joining algorithm (1000 bootstrap replicates).

Figure S5 Temporal dynamics of core taxa within the Bacteroidetes phylum. Abundances of the phylotypes identified as persistent (present in >80% of the samples) within subjects are presented in their phylogenetic context.

Figure S6 Temporal dynamics of core taxa within the Firmicutes phylum (Clostridia cluster XIV). Abundances of the phylotypes identified as persistent (present in >80% of the samples) within subjects are presented in their phylogenetic context.

Figure S7 Temporal dynamics of core taxa within the Firmicutes phylum (Clostridia clusters non-XIV). Abundances of the phylotypes identified as persistent (present in >80% of the samples) within subjects are presented in their phylogenetic context. Phylogenetic trees were constructed as described in Suppl. Fig. 4.

Figure S8 Invasion and extinction events within the fecal bacterial community. The abundance of OTUs that displayed temporal dynamics that suggested events of invasion or extinction are shown. See text for details.

Figure S9 Characterization of the fecal microbiota of triple-associated gnotobiotic mice. Genus level classification (Classifer, RDP) of sequences obtained for each fecal sample of mice colonized by with *Bifidobacterium adolescentis* BD-1, *Escherichia coli* MG1655, and *Bacteroides thetaiotaomicron* VPI-5482 (A). Accumulated fraction of OTUs of total OTUs (blue line) and total sequences (red line) shared across all nine samples (B). The expected fraction and abundance represented by the broken black graph.

Author Contributions

Conceived and designed the experiments: IM CEM JW. Performed the experiments: IM CEM. Analyzed the data: IM CEM JW. Wrote the paper: IM CEM JW.

References

1. Neish AS (2009) Microbes in gastrointestinal health and disease. Gastroenterology 136: 65–80.
2. Wang Z, Klipfell E, Bennett BJ, Koeth R, Levison BS, et al. (2011) Gut flora metabolism of phosphatidylcholine promotes cardiovascular disease. Nature 472: 57–63.
3. Henao-Mejia J, Elinav E, Jin C, Hao L, Mehal WZ, et al. (2012) Inflammasome-mediated dysbiosis regulates progression of NAFLD and obesity. Nature 482: 179–185.
4. Vijay-Kumar M, Aitken JD, Carvalho FA, Cullender TC, Mwangi S, et al. (2010) Metabolic syndrome and altered gut microbiota in mice lacking Toll-like receptor 5. Science 328: 228–231.
5. Devkota S, Wang Y, Musch MW, Leone V, Fehlner-Peach H, et al. (2012) Dietary-fat-induced taurocholic acid promotes pathobiont expansion and colitis in Il10−/− mice. Nature 487: 104–108.
6. Lozupone CA, Stombaugh JI, Gordon JI, Jansson JK, Knight R (2012) Diversity, stability and resilience of the human gut microbiota. Nature 489: 220–230.
7. Costello EK, Stagaman K, Dethlefsen L, Bohannan BJ, Relman DA (2012) The application of ecological theory toward an understanding of the human microbiome. Science 336: 1255–1262.
8. Robinson CJ, Bohannan BJ, Young VB (2010) From structure to function: the ecology of host-associated microbial communities. Microbiol Mol Biol Rev 74: 453–476.
9. Dethlefsen L, Eckburg PB, Bik EM, Relman DA (2006) Assembly of the human intestinal microbiota. Trends Ecol Evol 21: 517–523.
10. Savage DC (1977) Microbial ecology of the gastrointestinal tract. Annu Rev Microbiol 31: 107–133.
11. Cavender-Bares J, Kozak KH, Fine PV, Kembel SW (2009) The merging of community ecology and phylogenetic biology. Ecol Lett 12: 693–715.
12. Savage DC (1972) Associations and physiological interactions of indigenous microorganisms and gastrointestinal epithelia. Am J Clin Nutr 25: 1372–1379.
13. Stecher B, Hardt WD (2011) Mechanisms controlling pathogen colonization of the gut. Curr Opin Microbiol 14: 82–91.
14. Walter J, Ley R (2011) The human gut microbiome: ecology and recent evolutionary changes. Annu Rev Microbiol 65: 411–429.
15. Qin J, Li R, Raes J, Arumugam M, Burgdorf KS, et al. (2010) A human gut microbial gene catalogue established by metagenomic sequencing. Nature 464: 59–65.
16. Tap J, Mondot S, Levenez F, Pelletier E, Caron C, et al. (2009) Towards the human intestinal microbiota phylogenetic core. Environ Microbiol 11: 2574–2584.
17. Willing BP, Dicksved J, Halfvarson J, Andersson AF, Lucio M, et al. (2010) A pyrosequencing study in twins shows that gastrointestinal microbial profiles vary with inflammatory bowel disease phenotypes. Gastroenterology 139: 1844–1854 e1841.
18. Ochman H, Worobey M, Kuo CH, Ndjango JB, Peeters M, et al. (2010) Evolutionary relationships of wild hominids recapitulated by gut microbial communities. PLoS Biol 8: e1000546.
19. Frese SA, Benson AK, Tannock GW, Loach DM, Kim J, et al. (2011) The Evolution of Host Specialization in the Vertebrate Gut Symbiont Lactobacillus reuteri. PLoS Genet 7: e1001314.
20. Scanlan PD, Shanahan F, O'Mahony C, Marchesi JR (2006) Culture-independent analyses of temporal variation of the dominant fecal microbiota and targeted bacterial subgroups in Crohn's disease. J Clin Microbiol 44: 3980–3988.
21. Tannock GW, Munro K, Harmsen HJ, Welling GW, Smart J, et al. (2000) Analysis of the fecal microflora of human subjects consuming a probiotic product containing Lactobacillus rhamnosus DR20. Appl Environ Microbiol 66: 2578–2588.
22. Vanhoutte T, Huys G, De Brandt E, Swings J (2004) Temporal stability analysis of the microbiota in human feces by denaturing gradient gel electrophoresis using universal and group-specific 16S rRNA gene primers. FEMS Microbiol Ecol 48.
23. Zoetendal EG, Akkermans AD, De Vos WM (1998) Temperature gradient gel electrophoresis analysis of 16S rRNA from human fecal samples reveals stable and host-specific communities of active bacteria. Appl Environ Microbiol 64: 3854–3859.
24. Jalanka-Tuovinen J, Salonen A, Nikkila J, Immonen O, Kekkonen R, et al. (2011) Intestinal microbiota in healthy adults: temporal analysis reveals individual and common core and relation to intestinal symptoms. PLoS One 6: e23035.
25. Turnbaugh PJ, Hamady M, Yatsunenko T, Cantarel BL, Duncan A, et al. (2009) A core gut microbiome in obese and lean twins. Nature 457: 480–484.
26. Emerson BC, Gillespie RG (2008) Phylogenetic analysis of community assembly and structure over space and time. Trends Ecol Evol 23: 619–630.
27. Hubbell SP (2006) Neutral theory and the evolution of ecological equivalence. Ecology 87: 1387–1398.
28. Caporaso JG, Lauber CL, Costello EK, Berg-Lyons D, Gonzalez A, et al. (2011) Moving pictures of the human microbiome. Genome Biol 12: R50.
29. Rajilic-Stojanovic M, Heilig HG, Tims S, Zoetendal EG, de Vos WM (2012) Long-term monitoring of the human intestinal microbiota composition. Environ Microbiol: Epub ahead of print.
30. Davis LM, Martinez I, Walter J, Goin C, Hutkins RW (2011) Barcoded pyrosequencing reveals that consumption of galactooligosaccharides results in a highly specific bifidogenic response in humans. PLoS One 6: e25200.
31. Martinez I, Kim J, Duffy PR, Schlegel VL, Walter J (2010) Resistant starches types 2 and 4 have differential effects on the composition of the fecal microbiota in human subjects. PLoS One 5: e15046.

32. Davis LM, Martinez I, Walter J, Hutkins R (2010) A dose dependent impact of prebiotic galactooligosaccharides on the intestinal microbiota of healthy adults. Int J Food Microbiol 144: 285–292.
33. Ley RE, Turnbaugh PJ, Klein S, Gordon JI (2006) Microbial ecology: human gut microbes associated with obesity. Nature 444: 1022–1023.
34. Flint HJ, Scott KP, Louis P, Duncan SH (2012) The role of the gut microbiota in nutrition and health. Nat Rev Gastroenterol Hepatol 9: 577–589.
35. Wu GD, Chen J, Hoffmann C, Bittinger K, Chen YY, et al. (2011) Linking long-term dietary patterns with gut microbial enterotypes. Science 334: 105–108.
36. Martinez I, Lattimer JM, Hubach KL, Case JA, Yang J, et al. (2013) Gut microbiome composition is linked to whole grain-induced immunological improvements. ISME J 7: 269–280.
37. Walker AW, Ince J, Duncan SH, Webster LM, Holtrop G, et al. (2011) Dominant and diet-responsive groups of bacteria within the human colonic microbiota. ISME J 5: 220–230.
38. Jernberg C, Lofmark S, Edlund C, Jansson JK (2010) Long-term impacts of antibiotic exposure on the human intestinal microbiota. Microbiology 156: 3216–3223.
39. Schloss PD, Gevers D, Westcott SL (2011) Reducing the effects of PCR amplification and sequencing artifacts on 16S rRNA-based studies. PLoS One 6: e27310.
40. Kunin V, Engelbrektson A, Ochman H, Hugenholtz P (2010) Wrinkles in the rare biosphere: pyrosequencing errors can lead to artificial inflation of diversity estimates. Environ Microbiol 12: 118–123.
41. Cheng L, Walker AW, Corander J (2012) Bayesian estimation of bacterial community composition from 454 sequencing data. Nucleic Acids Res 40: 5240–5249.
42. Ursell LK, Metcalf JL, Parfrey LW, Knight R (2012) Defining the human microbiome. Nutr Rev 70 Suppl 1: S38–44.
43. Fierer N, Ferrenberg S, Flores GE, Gonzalez A, Kueneman J, et al. (2012) From Animalcules to an Ecosystem: Application of ecological concepts to the human microbiome. Annu Rev Ecol Evol Syst 43: 137–155.
44. Gordon JI, Klaenhammer TR (2011) A rendezvous with our microbes. Proc Natl Acad Sci U S A 108 Suppl 1: 4513–4515.
45. Walter J, Britton RA, Roos S (2011) Host-microbial symbiosis in the vertebrate gastrointestinal tract and the Lactobacillus reuteri paradigm. Proc Natl Acad Sci U S A 108 Suppl 1: 4645–4652.
46. Schluter J, Foster KR (2012) The evolution of mutualism in gut microbiota via host epithelial selection. PLoS Biol 10: e1001424.
47. Gossling J, Slack JM (1974) Predominant gram-positive bacteria in human feces: numbers, variety, and persistence. Infect Immun 9: 719–729.
48. Johnson JL (1980) Specific strains of Bacteroides species in human fecal flora as measured by deoxyribonucleic acid homology. Appl Environ Microbiol 39: 407–413.
49. McCartney AL, Wenzhi W, Tannock GW (1996) Molecular analysis of the composition of the bifidobacterial and lactobacillus microflora of humans. Appl Environ Microbiol 62: 4608–4613.
50. Schloissnig S, Arumugam M, Sunagawa S, Mitreva M, Tap J, et al. (2013) Genomic variation landscape of the human gut microbiome. Nature 493: 45–50.
51. Degnan PH, Pusey AE, Lonsdorf EV, Goodall J, Wroblewski EE, et al. (2012) Factors associated with the diversification of the gut microbial communities within chimpanzees from Gombe National Park. Proc Natl Acad Sci U S A 109: 13034–13039.
52. Schloss PD, Schubert AM, Zackular JP, Iverson KD, Young VB, et al. (2012) Stabilization of the murine gut microbiome following weaning. Gut Microbes 3: 383–393.
53. de Vos WM, de Vos EA (2012) Role of the intestinal microbiome in health and disease: from correlation to causation. Nutr Rev 70 Suppl 1: S45–56.
54. Jeraldo P, Sipos M, Chia N, Brulc JM, Dhillon AS, et al. (2012) Quantification of the relative roles of niche and neutral processes in structuring gastrointestinal microbiomes. Proc Natl Acad Sci U S A 109: 9692–9698.
55. Sloan WT, Lunn M, Woodcock S, Head IM, Nee S, et al. (2006) Quantifying the roles of immigration and chance in shaping prokaryote community structure. Environ Microbiol 8: 732–740.
56. Dethlefsen L, Huse S, Sogin ML, Relman DA (2008) The pervasive effects of an antibiotic on the human gut microbiota, as revealed by deep 16S rRNA sequencing. PLoS Biol 6: e280.
57. McFall-Ngai M (2007) Adaptive immunity: care for the community. Nature 445: 153.
58. Caporaso JG, Kuczynski J, Stombaugh J, Bittinger K, Bushman FD, et al. (2010) QIIME allows analysis of high-throughput community sequencing data. Nat Methods 7: 335–336.
59. Cole JR, Wang Q, Cardenas E, Fish J, Chai B, et al. (2009) The Ribosomal Database Project: improved alignments and new tools for rRNA analysis. Nucleic Acids Res 37: D141–145.
60. Schloss PD, Westcott SL, Ryabin T, Hall JR, Hartmann M, et al. (2009) Introducing mothur: open-source, platform-independent, community-supported software for describing and comparing microbial communities. Appl Environ Microbiol 75: 7537–7541.
61. Tamura K, Peterson D, Peterson N, Stecher G, Nei M, et al. (2011) MEGA5: molecular evolutionary genetics analysis using maximum likelihood, evolutionary distance, and maximum parsimony methods. Mol Biol Evol 28: 2731–2739.
62. Martinez I, Wallace G, Zhang C, Legge R, Benson AK, et al. (2009) Diet-induced metabolic improvements in a hamster model of hypercholesterolemia are strongly linked to alterations of the gut microbiota. Appl Environ Microbiol 75: 4175–4184.

5

Metabolic Proximity in the Order of Colonization of a Microbial Community

Varun Mazumdar[1], Salomon Amar[1,2], Daniel Segrè[1,3]*

1 Bioinformatics Program, Boston University, Boston, Massachusetts, United States of America, **2** Center for Anti-Inflammatory Therapeutics; Boston University Goldman School of Dental Medicine, Boston, Massachusetts, United States of America, **3** Department of Biology and Department of Biomedical Engineering, Boston University, Boston, Massachusetts, United States of America

Abstract

Microbial biofilms are often composed of multiple bacterial species that accumulate by adhering to a surface and to each other. Biofilms can be resistant to antibiotics and physical stresses, posing unresolved challenges in the fight against infectious diseases. It has been suggested that early colonizers of certain biofilms could cause local environmental changes, favoring the aggregation of subsequent organisms. Here we ask whether the enzyme content of different microbes in a well-characterized dental biofilm can be used to predict their order of colonization. We define a metabolic distance between different species, based on the overlap in their enzyme content. We next use this metric to quantify the average metabolic distance between neighboring organisms in the biofilm. We find that this distance is significantly smaller than the one observed for a random choice of prokaryotes, probably reflecting the environmental constraints on metabolic function of the community. More surprisingly, this metabolic metric is able to discriminate between observed and randomized orders of colonization of the biofilm, with the observed orders displaying smaller metabolic distance than randomized ones. By complementing these results with the analysis of individual vs. joint metabolic networks, we find that the tendency towards minimal metabolic distance may be counter-balanced by a propensity to pair organisms with maximal joint potential for synergistic interactions. The trade-off between these two tendencies may create a "sweet spot" of optimal inter-organism distance, with possible broad implications for our understanding of microbial community organization.

Editor: Jens Kreth, University of Oklahoma Health Sciences Center, United States of America

Funding: This research was partially funded by the United States Department of Energy, grant DE-SC0004962, and by NIH grants R01DE15345, R01HL6801 and R0115989. The funders had no role in study design, data collection and analysis, decision to publish, or preparation of the manuscript.

Competing Interests: The authors have declared that no competing interests exist.

* E-mail: dsegre@bu.edu

Introduction

In many natural environments, bacteria and other microorganisms are part of spatially structured ecosystems, and engage in complex interactions, involving the exchange of nutrients and chemical signals [1,2]. Such communities provide their members with protection from environmental perturbations, and allow for effective utilization of available resources. Modifications of the environment, such as a change in diet in a human host, can cause shifts in the composition of a microbial community. In turn, the collective metabolic activity of the community itself can substantially modify the environment, and set the stage for transitions between health and disease states. While 16S rRNA studies [3] and metagenomic DNA sequencing [4] have been very helpful in providing provide global snapshots of the composition and biological functions of a community, a big gap still exists in our understanding of the forces that shape specific interactions between different organisms. Systems biology approaches have started to provide valuable insight into the metabolic basis of interactions between different species in elementary [5] and complex [6] microbial ecosystems. However, a lot is still unknown on how properties of individual species give rise to global ecosystem organization.

Here, we address this problem by presenting an intermediate-scale approach to elucidate the role of metabolism in determining the order of colonization in a microbial community. Our approach captures the complexity of how eleven species spatially organize in a biofilm, by using a mathematical description of metabolism that lies in between the detailed quantitative power of stoichiometric models [7], and the coarse enrichment analyses typically obtained from metagenomic studies [4,8–11]. We focus specifically on one of the most intensively characterized biofilm systems, the dental biofilm, which plays a crucial role in tooth and gum diseases. Oral pathogens such as *Porphyromonas gingivalis* are also known to be able to enter the blood stream, possibly causing cardiovascular disease. The collection of bacteria present on the surface of teeth, anchored to the salivary pellicle, has a specific spatial structure, which has been mapped and investigated in detail [12–14]. The biofilm structure is made up of a number of different species, which aggregate together over time to form a complex structure. The aggregation process is not random [15]; instead it appears to be a repeatable sequential process mediated by bacterial adhesins that allow organisms to aggregate to surfaces and other bacteria by binding to specific receptor moieties (Fig. 1A). The initial colonizers are capable of binding to salivary pellicle receptors, and the subsequent organisms proceed to bind to the initial colonizers [12,14,16]. Late colonizers, such as *P. gingivalis,* which has been linked to periodontal disease, are found in the final layer of this complex structure [17]. Some steps of the colonization process can lead to mutual exclusion between closely related

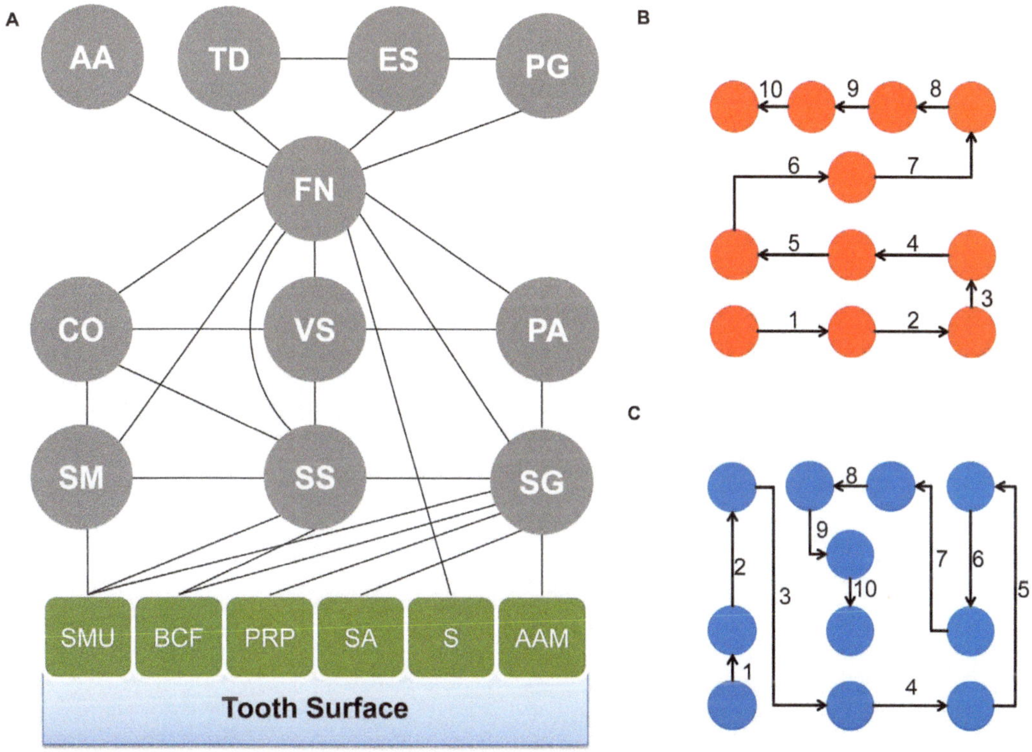

Figure 1. A simplified model of dental biofilm. (A) Rectangular nodes represent components of the salivary pellicle while circular nodes represent organisms in the biofilm. Lines represent known interactions (often mediated by adhesin molecules) between different components of the biofilm. The organism abbreviations are as follows: Layer 1: SM-*Streptococcus mitis*, SS-*Streptococcus sanguinis* and SG-*Streptococcus gordonii*. Layer 2: CO-*Capnocytophaga ochracea*,VS-*Veionella* (represented by *Veillonella parvula*), and PA-*Propionibacterium acnes*. Layer 3: FN-*Fusobacterium nucleatum*. Layer 4: AA-*Aggregatibacter actinomycetemcomitans*, TD-*Treponema denticola*, ES-*Eubacterium* (represented by *Eubacterium eligens*), and PG-*Porphyromonas gingivalis*. The salivary receptors have the following abbreviations: SMU-Sialylated mucins, BCF-Bacterial cell fragment, PRP-Proline rich protein, SA-Salivary agglutinin, S- Statherin and AAM-Alpha amylase. (B) A schematic representation of one of the many possible step-wise orders of colonization that conforms to the layered organization inferred from the literature, i.e. is such that the path that walks through the different species is monotonically departing from the salivary pellicle upwards. In our calculations of inter-species metabolic distances, we average the distances between any two species connected by a segment. This calculation is performed for all paths that reflect the order of colonization, giving rise to the distributions shown in Fig. 2. (C) A schematic representation of one of the many possible randomized orders of colonization that do not follow the order of the literature-derived layers; for the 11 organisms present in the biofilm there are 11! possible permutations.

species (e.g. streptococci), leading to drastically different macroscopic disease-related outcomes [18].

In this work we test the hypothesis that metabolism is a predictor, and potentially a major driving force, of the order of colonization in the oral biofilm. Specifically, based on the individual inter-species interactions mapped by Kolenbrander ([12][19], Fig. 1) we quantify the overlap in metabolic functions between adjacent organisms, and compare the distribution of such overlaps to the one obtained for randomized biofilms or for random assemblages of bacteria. We find that the real biofilm is characterized by a significantly larger overlap in metabolic functions between adjacent species, relative to randomized biofilm compositions and structures. Specific metabolic pathways can be associated with the different layers, providing a snapshot of the gradient of metabolic requirements across the biofilm. The observed tendency towards maximal metabolic overlap is likely counteracted by an opposite trend driven by the synergistic advantage of combining the metabolic capabilities of sufficiently different species. In all, these findings suggest that an optimal tradeoff between resource sharing and functional synergy may constitute a fundamental property of structured microbial communities. Our approach, much more detailed than broad functional enrichment studies, but much less demanding than stoichiometric flux balance models, should be broadly applicable to other microbial ecosystems, where spatial or temporal order matters.

Results

Metabolic Proximity among Different Layers in the Oral Biofilm

We ask whether the order of colonization in the human dental biofilm may reflect a quantitative principle of microbial ecosystem organization. While the structure of the biofilm from the Kolenbrander model (Fig. 1A, and [12]) reflects known mutual binding between adjacent species, we develop our analysis based on the premise that such binding effects reflect fundamental adaptations to environmental gradients and mutual metabolic exchange between species. Hence, we analyze the biofilm structure in terms of mutual distances between adjacent organisms. 16S rRNA-based distances are standard practice when estimating the similarity of organisms without dealing with the complexities of whole genome alignment [3,10]. The assumption that 16S rRNA is conserved allows investigators to ascertain evolutionary relationships between organisms. Here, however, we wish to utilize a metabolic, rather than an evolutionary distance [17] (See

Methods). Such a metabolic distance will provide a method to gauge the difference in biochemical functions between different species. It is important to note that a metabolic metric can be used to compare biochemical abilities of different organisms without constructing full-fledged genome-scale stoichiometric models, such as the ones built for several microbial species [20–26], including the oral pathogen *P. gingivalis* [17]. We expect that organisms with similar enzyme profiles (based on the above metric) will have a comparable ability to utilize and process metabolites from their environment. The specific metabolic distance we use in this work is a standard metric (Jaccard's distance, J) gauging the degree of dissimilarity between the sets of enzymes present in the two organisms (see Methods).

As a first step towards ascertaining the validity of this premise, one can test whether the 11 organisms that are in the dental biofilm are on average closer to each other than 11 randomly chosen prokaryotes from the KEGG database [27–30]. To this end, 1000 random groups of 11 organisms were chosen from the list of KEGG prokaryotes and the sum of pairwise metabolic distances were calculated for each permutation (11 factorial potential orders) for a given random group. The average of all permutation scores was calculated for each random group. We found that the groups of 11 organisms belonging to the oral biofilm model tend to have a smaller average of pairwise metabolic distances (J) than 1000 randomly selected groups of prokaryotes (Fig. 2). This is not unexpected, as the organisms inhabit the same niche in the human body and hence must have some underlying metabolic similarity to cope with the common environment. It is interesting to note that the mean distance value between organisms of the dental biofilm is far from being close to a global minimum, when compared to its position in the distribution. This could be due to the fact that organisms with similar metabolic requirements will compete for the same environmental niche and probably would not be found in close proximity to each other within a complex multi-species biofilm.

The next question we ask is whether a special pattern of inter-species metabolic distances can be observed between adjacent organisms in the layered structure of the oral biofilm. In Fig. 1A we present a simplified version of the model presented in [12]. The simplified model contains only organisms whose genome has been sequenced, and whose annotation is available in KEGG. We translate the Kolenbrander map into a set of possible orders of colonization by assuming that an organism can join the biofilm only if it can bind to an organism that that is already present in the biofilm, or to an environmental anchor point. In this way, we determine 864 orders in which the bacteria may join, consistently with the reduced Kolenbrander model (Fig. 1B). The 864 orders ($3! \times 3! \times 4!$) come from all permutations that allow organisms to be placed in their correct layers. Conversely, there would be 39,916,800 (i.e. 11!) orders which disregard the network of experimentally known interactions that constitute the layered model (Fig. 1A). For each given order of colonization, we compute the average Jaccard's distance between metabolic compositions of adjacent organisms ($\langle J \rangle$, see Methods). Fig. 2 shows that the distribution of average distances for the orders of colonization compatible with the Kolenbrander model is markedly shifted to the left relative to the background distribution of random orders ($P<2\cdot10^{-7}$). This implies that spatially adjacent organisms in the biofilm have a larger number of common metabolic enzymes relative to randomly chosen pairs. In order to ascertain the overall robustness of this result we repeated the analysis upon different types of perturbations. In particular, we tried to omit from the calculation the *Streptococci* species, which are phylogenetically and metabolically very close to each other, and might therefore bias the result towards high significance. Despite removing these organisms, we still found a statistically significant p-value ($P<1.2\cdot10^{-4}$). Furthermore we verified that the results are not too sensitive to removal of specific enzymes. In fact, we found that the result is still significant when up to 60% of the enzymes used in the analysis are removed (Fig. S1). The "minimal metabolic distance" criterion apparently satisfied by the non-random orders of colonization may be indicative of the way dental biofilm is thought to form. Each group of organisms creates a micro-environment that is conducive for the next set of organisms. Each new organism joining the biofilm can take advantage of the micro-environment generated by organisms that precede it, provided that the inter-species metabolic distance is small enough to make this build-up favorable.

The above analysis demonstrates that adjacent species in the correct order of colonization tend to minimize their mutual metabolic enzyme distance. This finding does not rule out the possibility that similar effects might be observable based on alternative metrics that take into account other (non-metabolic) gene categories. In other words, is the order of colonization pattern reported predominantly a metabolic effect, or does it reflect a general inter-species distance? To address this question, we performed the same analysis shown in Fig. 2 for a large set of non-metabolic genes (see Methods). As shown in Fig. 3A, the correct and random order distributions are still significantly different ($P<0.0023$), but their separation is much less dramatic than what is found with metabolic enzymes ($P<2\cdot10^{-7}$). Given this result, we can infer that metabolism is one of the most important factors in oral biofilm organization, more so than other classes of genes as a whole.

Finally, to confirm that the pattern observed is not strongly dependent on the metric used, we verified that an alternative, widely used metric can discriminate between correct vs. randomized order of colonization. In particular, we calculated the amount of Shannon information that each organisms adds to the system relative to the previous organism in the order of colonization (see Methods). As with the distance calculation, the overall score is the sum of the pairwise added information values for a given order of colonization. Using a 2-sample Komogrov-Smirnov test [31], we found that literature-informed orders are significantly ($P<4.5\cdot10^{-6}$) smaller than randomized distributions (Fig. 3B). An interesting aspect of the information content metric is that, in contrast to the metabolic distance defined above, it can also capture directionality, i.e. discriminate between a colonization order that starts with the *Streptococci* layer (the first, pellicle-bound layer of the biofilm, Fig. 1A) and one that ends with *Streptococci* layer.

If, as suggested by the above results, metabolism is a fundamental determinant of the order of colonization, we would expect to be able to find that specific metabolic pathways can be associated with different layers of the biofilm. Indeed, by performing a GSEA (Gene Set Enrichment Analysis) for metabolic functions (see Methods and Dataset S1), we found that gradients of metabolic functionalities span the different layers (Fig. 4). The pathways that displayed significant enrichment are: arginine and proline metabolism, biosynthesis of alkaloids, carbon fixation, glyoxylate and dicarboxylate metabolism, glycine, serine and threonine metabolism, nitrogen metabolism, porphyrin and chlorophyll metabolism, pentose and glucuronate interconversions, propanoate metabolism, pyruvate metabolism, terpenoid backbone biosynthesis, and tricarboxylic acid cycle. Enzymes related to carbohydrate and proline metabolism are enriched at the initial colonizer stage, possibly allowing for utilization of available carbohydrates as a source of energy that is present in the

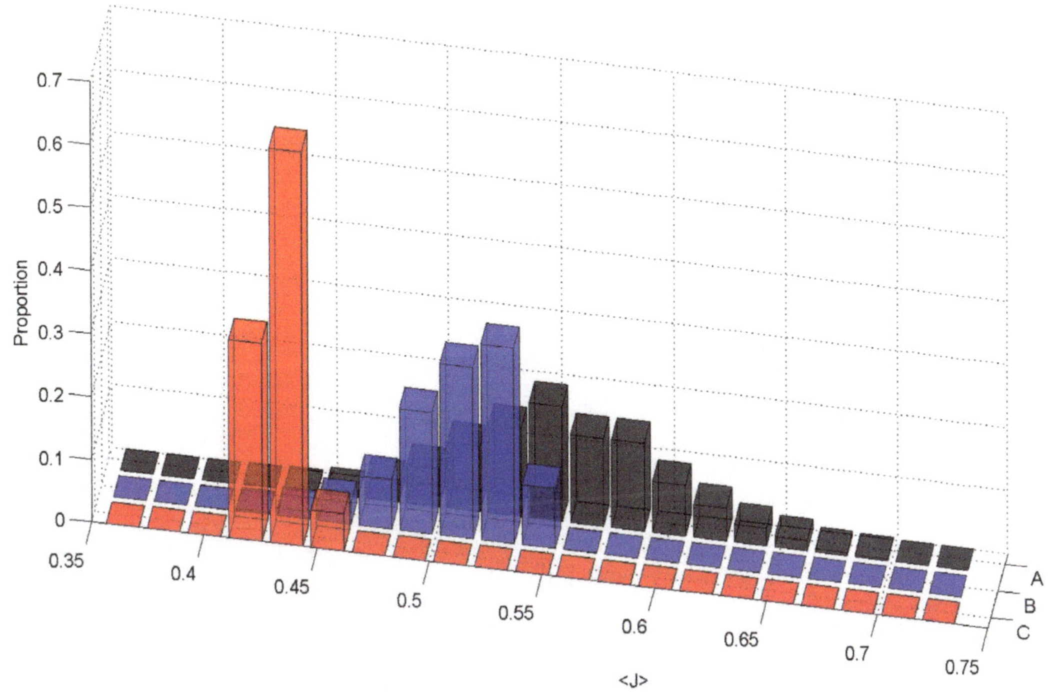

Figure 2. Metabolic distance distributions for correct and randomized orders of colonization. Distributions of average pairwise Jaccard's distance are compared across different computational realizations of the 11-species biofilm. In particular, we show in red (C) the distribution of average pairwise distances between the 11 organisms for all paths that reflect the layered structure of the Kolenbrander map (see Fig. 1B). In blue (B) we show the distribution obtained for all possible random orders that do not necessarily reflect the layered order of colonization (e.g. path shown in Fig. 1C). The last distribution (grey, A) is obtained from choosing in random order 11 random prokaryotes from the KEGG database.

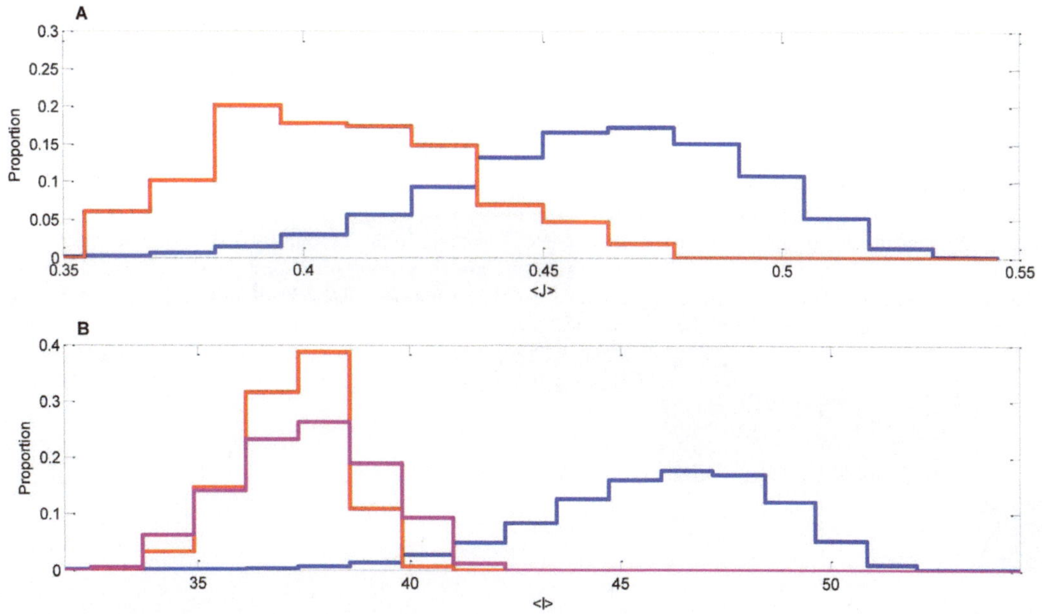

Figure 3. Distributions of alternative metrics for correct and randomized orders of colonization. (A) Similar to what shown in Fig. 1B and Fig. 2, we computed inter-species distance between organisms along paths that respect (red) or do not respect (blue) the layered order of colonization of the Kolenbrander map. Here, however, as opposed to Fig. 2, we compute the Jaccard distance between two species based on their profiles of non-enzyme genes (as identifiable through KEGG KO numbers). (B) The correct and incorrect orders of colonization are compared based on an information metric, rather than on the Jaccard distance. In walking along a colonization order path from one organism to the next, we compute (in Nats) the amount of information added due to the presence of previously absent enzymes. The added information for each pair of adjacent organisms is summed to form the added information score, along paths that respect (red) or do not respect (blue) the layered order of colonization. The purple distribution is obtained by computing the added information scores for orders of colonization that reflect the layered structure, but walk through it in reverse order (i.e. from the outer layer downwards towards the salivary pellicle).

saliva [32]. The second biofilm layer contains both propionate and TCA pathways. Both require, as input, compounds such as lactate, a byproduct of carbohydrate metabolism which in turn is converted into cytotoxic byproducts [33]. Additionally, butyrate is known to affect gingival epithelial cells inducing apoptosis in sufficient concentrations [34]. Apoptosis of tissues provides organisms with a highly enriched food source [35]. Fumarase and succinate dehydrogenase, are both enriched in this layer of the biofilm. Both enzymes provide a pathway for proline metabolism byproducts to be funneled into energy production which takes advantage of proline catabolism enrichment in the previous biofilm layer. The third layer of the biofilm is enriched for porphyrin metabolism, a pathway that is essential to *Porphyromonas gingivalis*. The organism has a well-known requirement for heme and has been implicated in a number of disease processes such as chronic periodontitis [36]. It is however not capable of producing heme for itself, and must therefore scavenge it from the environment. Correspondingly, we found enrichment for enzymes related to heme production, in particular along a pathway that converts L-glutamate to 5-aminolevulinate, a precursor of heme. In the fourth layer of the colonization process, we find enrichment for nitrogen-related and TCA-cycle genes. This specific combination could reflect amino acids from tissue degradation being shunted into cellular metabolism via entry points within the TCA cycle.

Determination of Optimal Metabolic Overlap Using Flux Modes

The above analysis of metabolic distances between adjacent organisms in the oral biofilm demonstrates that the correct order of colonization displays close to minimal average metabolic distance relative to randomized orders. Another way of formulating this principle is that, within a biofilm, organisms next to each other will be as close as possible in terms of their metabolic enzyme content. Taken to the extreme, this principle would suggest that biofilms may be preferably composed of rather similar and uniform species (compatibly with the biofilm size and environmental gradients). This is likely not the case, both because organisms metabolically too close to each other may engage in fierce competition for survival [37], and because there may be a physiological advantage to pairing organisms that are neither too close nor too distant from each other.

To formulate a specific hypothesis about this last scenario, we evaluated the metabolic potential of conjoined metabolic networks as a function of their metabolic distance, using elementary flux modes. Elementary flux modes analysis identifies all minimal non-zero flows through a metabolic network [38]. At a first approximation, the number of elementary flux modes can be thought of as an estimate of the size of the space of possible paths through a metabolic network. Here, we sought to estimate the increase in the number of such paths for a pair of interacting metabolic networks (e.g. two bacterial species), relative to the metabolic capabilities of isolated networks, as a function of the similarity between the two networks. The intuition is that when two networks are very similar to each other, there is little added benefit in combining them with each other. At the opposite extreme, if two networks are too different from each other they will "speak different metabolic languages" and barely be able to build significant synergistic pathways. In between these two extremes, there may be an inter-species metabolic distance that provides a maximal synergistic benefit.

Indeed, upon computing a metabolic synergy score for randomly generated pairs of metabolic networks with a given Jaccard's distance between each other (see Methods), we found that the mean score displays a distinct profile as a function of inter-network metabolic distance (Fig. 5). In particular, there is a peak at a Jaccard's coefficient of 0.2. This means that having 33.3% reaction overlap between the component networks is optimal, in the sense that it generates the maximum number of useable balanced pathways (elementary flux modes). The range of Jaccard's coefficients (i.e. metabolic similarity) within which synergistic interaction is expected extends up to approximately $Jc = 0.55$ (i.e. $J = 0.45$). This means that metabolic networks with a distance below $J = 0.45$ will have little potential for increased biochemical capabilities through metabolic cross-talk. Interestingly, this Jaccard's distance is very close to the average of the

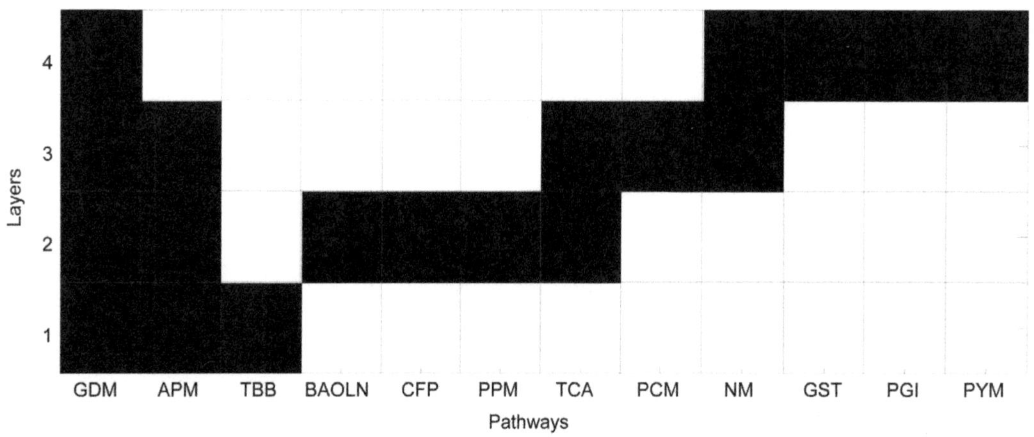

Figure 4. Metabolic pathway enrichment across layers. Based on the enzyme content of the different species found in different layers of the biofilm (with layers labeled from 1 to 4, see Fig. 1), one can estimate whether any given layer is enriched for specific metabolic functions. Enzyme and pathway enrichments for each layer are computed based on a standard GSEA algorithm. Black boxes in the pathways-by-layers matrix denote enrichment of a particular KEGG pathway in a given layer. The pathway abbreviations are as follows: APM-Arginine and proline metabolism, BAOLN-Biosynthesis of alkaloids derived from ornithine lysine and nicotinic acid, CFP-Carbon fixation pathways in prokaryotes, GDM-Glyoxylate and dicarboxylate metabolism, GST-Glycine, serine and threonine metabolism, NM-Nitrogen metabolism, PCM-Porphyrin and chlorophyll metabolism, PGI-Pentose and glucuronate interconversions, PPM-Propanoate metabolism, PYM-Pyruvate metabolism, TBB-Terpenoid backbone biosynthesis, and TCA-Tricarboxylic acid cycle.

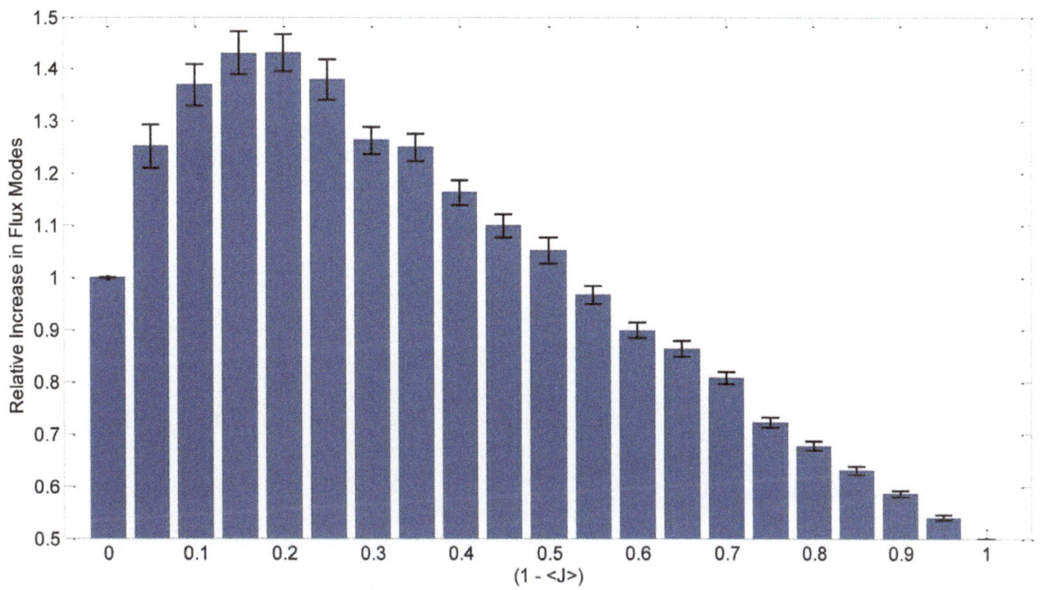

Figure 5. Expected synergy between metabolic networks as a function of metabolic distance. The synergy is computed as the count of elementary flux modes (pathways) that are feasible for a metabolic network that is the union of two networks with a given Jaccard's distance from each other, normalized to the count of elementary flux modes of the constituent networks. The count of elementary flux modes can be thought of as an estimate of the number of distinct metabolic tasks that the network can perform, i.e. its versatility. Hence, the graph shows how the versatility of two conjoined networks relative to the constituent networks is maximal for an intermediate Jaccard's distance between such networks. 100 random paired networks were generated for each of several possible Jaccard's distances. Bar heights reflect the average normalized increase in the number of elementary flux modes, whereas error bars represent the standard error of the mean.

distance among adjacent species in oral biofilm (Fig. 2). A possible interpretation of this result is that the players in the community encounter a tradeoff between maximizing their metabolic overlap, and still not losing the benefit of possible synergistic interactions (i.e. going *beyond* $J = 0.45$, i.e. $J_C = 0.55$).

Discussion

We have addressed the question of whether the spatio-temporal organization of a biofilm can be understood in terms of the differential metabolic properties of individual organisms relative to their neighboring organisms. We implemented a simplified model of the spatial organization of the biofilm, based on experimental evidence of individual pairwise interactions between species. This abstraction of the colonization process enabled us to discretize the order of succession, and systematically investigate all potential permutations of organisms in a linear fashion. A more complex model could more realistically capture inter-species dynamics in physical space [15], without the limitations imposed by taking into account only pairwise interactions and a step-by-step "walk" through the different layers of the biofilm. However, our simplified approach overcomes in an effective way the combinatorial complexity of multi-species networks, and takes advantage of the available pairwise interaction data. We found that metabolic similarity is a highly informative indicator of vicinity in the biofilm. The metabolic structure of the biofilm is reflected in the existence of multiple layers enriched for specific biochemical pathways. This structure lends credence to the idea that each layer contributes to a gradient of metabolic properties, causing environmental modifications that pave the way for subsequent layers of bacteria. We cannot exclude the possibility that the observed effect might be just a result of metabolically similar organisms adapting to environmentally present gradients (e.g. abundance of oxygen). However, the fact that next-to-minimal metabolic distance is significantly associated with binding between organisms in the biofilm suggests this metabolic similarity is truly reflective of particular inter-species interactions. The current work is limited to the eleven organisms of the Kolenbrander map whose annotated sequence was publically available (in KEGG) at the time of the analysis. Future extensions could include additional organisms, benefit from improved genome annotation approaches [39] and gradually move to more mechanistic models of microbe-microbe interactions, such as ecosystem-level flux balance models [40]. In addition, while the current evidence we use for a putative order of colonization is based on a collection of multiple *in vitro* individual pairwise interactions, more comprehensive *in vivo* measurements [19] could in the future be used as a more accurate baseline for testing hypotheses.

Our analysis focused on a specific microbial community in which the order of colonization is manifested both in the chronological sequence of events leading to the full biofilm, as well as in the final spatial architecture of the biofilm itself. We envisage that this analysis could be extended to other microbial ecosystem with a similar spatio-temporal organization, such as biofilms on catheters and medical instruments [41–44] or microbial mats in hot springs and desert environments [45,46]. However, the approach we proposed is not limited to communities with a well-defined or known spatial structure, and could be extended to analyze purely temporal orders of colonization in microbial ecosystems whose biomass is found largely in a planktonic phase, or whose detailed spatial structure is not easily observable (e.g., in the gut microbiome [10,47]). Metagenomic sequencing projects frequently produce 16S rRNA population composition data, which in a longitudinal study provides us with changes in population composition over time. Combining population data with enzymatic profiles from KEGG would make it possible to test whether metabolic proximity is significantly predictive of temporal species-to-species shifts in an ecosystem.

An interesting outcome of our analysis is the hypothesis that multiple counteracting forces may ultimately determine, at the evolutionary scale, an optimal steady state genomic and spatial configuration of different species in a biofilm. Close metabolic proximity seems to be one desirable criterion for spatial vicinity, motivated by uniformity of environmental conditions, and by multiple chances for metabolic cross-feeding. At the same time, organisms which are metabolically too close to each other would likely compete for common metabolic resources. In addition, as we found in Fig. 5, they would have minimal chance for true synergism, i.e. for the capacity to contribute novel metabolic capabilities to the group as a whole. The specific Jaccard's distances at which the optimum occurs are likely dependent on the specific topology of the underlying reaction networks utilized, and may not be universal (see Methods). However, in our synergy calculations based on elementary flux modes, we retain a similar degree distribution and reaction topology as would be seen in the bacterial species of the oral biofilm. It will be interesting to explore the possibility that the general shape of the curve observed in Fig. 5 could be derived analytically.

Finally, our finding poses an interesting evolutionary chicken and egg dilemma: did the observed metabolic proximity pattern precede or follow the emergence of specific binding affinities between receptors and ligands across species? On one hand, energy and food-related requirements may be hypothesized to dictate the emergence of a biofilm structure. Subsequent adaptations could have optimized inter-species binding interactions to facilitate the formation of an efficient nutrient and energy flow. Conversely, we cannot rule out the alternative possibility that metabolic proximity may have arisen between organisms with a tendency to bind to each other, e.g. through horizontal gene transfer, or by forcing each other to face specific selective pressures. This may be an interesting challenge for future research, in which the experimental investigation of evolving symbiotic system [48] could be complemented by computational studies of evolutionary rates in genomic sequences [49–51].

Methods

Parsing an Experimental Map of the Oral Biofilm Structure

The present analysis uses the biofilm organization map that is presented in [12], which is based on the collection of several individual experimental papers. This model (to which we will refer as the Kolenbrander model) contains 22 organisms, 11 of which were sequenced and annotated [27–30] at the time of our analysis. The 11 organisms for which data is available are listed in Fig. 1A. The map also includes known connections between some organisms and the salivary pellicle, from which we assume that the biofilm starts developing. In the Kolenbrander model, reproduced in simplified form in Fig. 1A, nodes correspond to species (with the exception of *Veionella* and *Eubacterium*, which were reported only at the genus level). Organisms for which there was no KEGG data were omitted from our analysis. An edge between two nodes in the network of Fig. 1A denotes a documented capacity of the two corresponding biofilm constituents to bind to each other.

Calculation of Enzyme Based Distances

The KEGG database contains information that describes the number and type of enzymes present in an organism's genome [27–30]. Each enzymatic function is associated with an Enzyme Commission (EC) number [52]. In this case, we are not interested in the abundance of any particular enzyme or in its substrate/product stoichiometry, but simply in the presence or absence of such enzyme in a given organism's genome. Given this information, a binary vector $\boldsymbol{S}^{(A)}$ can be defined to describe the enzyme composition for an organism A, with component $S^{(A)}_i = 1$ if enzyme i is present in organism A, and $S^{(A)}_j = 0$ otherwise. We then evaluate the difference between the metabolic profiles of two organisms by computing the Jaccard's distance $J(A,B)$ between $\boldsymbol{S}^{(A)}$ and $\boldsymbol{S}^{(B)}$, defined as follows:

$$J(A,B) = 1 - \frac{|\boldsymbol{S}^{(A)} \cap \boldsymbol{S}^{(B)}|}{|\boldsymbol{S}^{(A)} \cup \boldsymbol{S}^{(B)}|}$$

If A and B have the same metabolic enzymes, then $J = 0$. If they have no enzyme in common, it is $J = 1$. This metric will also be used to quantify metabolic similarity between species, in the form of the Jaccard's coefficient ($J_c = 1 - J$) [53].

Calculation of Non-metabolic Distances

In addition to enzyme content, the KEGG database includes data on the presence of different categories of non-metabolic genes. Using this data we can generate binary vectors $\boldsymbol{S}^{(A)}$ and $\boldsymbol{S}^{(B)}$, that represent the non-metabolic gene content for organisms A and B, just as was done for metabolic distance. We then evaluate the difference between the non-metabolic profiles of two organisms by computing their Jaccard's distance $J(A,B)$.

Calculation of Added Information

All organisms in the community contribute to an overall super-set of enzymes that represents the metabolic potential of the community. In examining the gradual build-up of the oral biofilm, we can ask how much novelty is introduced by each new organism joining an increasingly complex ecosystem. This can be achieved by calculating the amount of metabolic information added to the current super-set upon introducing a new organism to the biofilm. If we call N_i the number of organisms in which enzyme i is present ($N_i \in \{1, 2, \ldots, N_{organisms}\}$, with $N_{organisms} = 11$), then the probability to find a given enzymatic function i in the whole biofilm is $P_i = N_i / N_{organisms}$. For an ordered pair of organisms (A,B), we can identify the set $K(A,B)$ of all EC numbers k such that $S^{(A)}_k = 0$ and $S^{(B)}_k = 1$. The information added when organism B is added to organism A is then computed as the Shannon information content of all enzymes that are currently added and were not present in the prior organisms, i.e.:

$$\Delta I(A,B) = -\sum_{k \in K(A,B)} P_k \ln P_k$$

$\Delta I(A,B)$ represents the amount of Shannon information (relative to the overall abundance of EC numbers in the entire biofilm), added by an organism B upon colonization on top of an organism A. Note that ΔI is not symmetric, i.e., in general, $\Delta I(A,B) \neq \Delta I(B,A)$. Hence, this added information metric allows us to distinguish between orders of colonization that would be indistinguishable using the Jaccard's metric J defined above.

Layer Specific Pathway Enrichment

To determine possible layer-specific metabolic pathway enrichment in the oral biofilm, we first calculate the proportion of organisms in a given layer that contain a given enzyme. This calculation uses the same organism-specific binary vector $\boldsymbol{S}^{(A)}$ defined above. If the set of organisms in biofilm layer x is defined as L_x, then a biofilm layer profile ($\boldsymbol{B}^{(x)}$), describing the occurrence of each enzyme in any given organism in layer x can be calculated as follows:

$$\boldsymbol{B}^{(x)} = \frac{\sum_{i \in L_X} \boldsymbol{S}^{(i)}}{|L_x|}$$

These profiles can be used to estimate the enrichment of the different layers for specific metabolic processes. This is achieved by using gene set enrichment analysis (GSEA). In addition to the $\boldsymbol{B}^{(x)}$ profiles for the four different layers (Fig. 1A), the GSEA algorithm utilizes a binary mapping matrix $K_{(i,j)}$, where $K_{(i,j)} = 1$ if enzyme i is present in pathway j. This matrix maps enzymes to corresponding KEGG metabolic pathways. We next look for specific enzyme and pathway enrichment in a given layer relative to other layers within the dental biofilm. The enrichment calculation is performed using a standard GSEA application [54]. Pathways with a nominal p-value of 0.05 or less and an FDR of less than 0.25 were chosen (with FDR accounting for multiple testing biases).

Construction of Randomized Paired Networks

In order to estimate the metabolic benefit derived from the cooperation of two species, we perform an analysis of elementary flux modes in appropriately modified versions of the *E. coli* FBA model [21]. In particular, we used the *E. coli* FBA model (cytoplasm reactions only) as an initial main network (of size R_{TOT}) to generate random metabolic networks with a degree distribution and topology similar to that of real metabolic networks. Random networks are generated in pairs, with a specified metabolic similarity (Jaccard's coefficient, J_c) between them. The algorithmic pipeline to generate such pairs of networks proceeds as follows:

(i) Out of the main source network of size R_{TOT}, we choose a subnetwork (the "source network") of size $R_s = P \times R_{TOT}$, where P is a given percent coverage, chosen in order to guarantee tractability of the elementary flux modes calculations. The standard value used throughout this work is $P = 0.8$.

(ii) Given the desired degree of metabolic overlap (J_c) between the two networks, and the size of the source network (R_s), we compute the size R_{int} of the set N^{int} of reactions that should be in common between the two networks. This number is simply

$$Rint = Jc \times Rs$$

(iii) We select two random sets (N^1, N^2) of non-overlapping reactions from the source network. Each of these reaction sets is chosen to have size

$$Rn = (Rs - Rint)/2$$

(iv) We use the sets of reactions N^{int}, N^1 and N^2 to build reaction sets for the two desired randomized networks, and for their union. These reaction sets for individual networks ($C^{(1)}$, $C^{(2)}$) and for their joint combination ($C^{(1,2)}$) are defined as follows:

$$C^{(1)} = N^1 \cup N^{int}$$

$$C^{(2)} = N^2 \cup N^{int}$$

$$C^{(1,2)} = N^1 \cup N^2 \cup N^{int}$$

(v) The reaction sets $C^{(1)}$ $C^{(2)}$ and $C^{(1,2)}$ are mapped to stoichiometric matrices M^1 and M^2 and $M^{(1,2)}$ respectively. When generating the combined stoichiometric matrix $M^{(1,2)}$, we need to make sure that no additional overlap (in addition to the chosen N^{int} reactions) is introduced between the two random networks. This is achieved by ascribing new names (i.e. new stoichiometric matrix rows) to the metabolites contributed by N^1 and N^2 which are not already present in N^{int}.

Computation of Degree of Metabolic Synergy Using Elementary Flux Modes

For the stoichiometric matrices described in the previous paragraph, the number of elementary flux modes (EFM) is calculated using the *efmtool* software [38]. For each stoichiometric matrix passed to it, *efmtool* returns an EFM matrix describing the various elementary modes through the network. The number of columns of the matrix corresponds to the number of EFMs possible for the given network. A normalized score estimating the increase in the number of EFMs obtained upon conjoining two networks can be computed as follows:

$$\Delta EFM = \frac{EFM(1,2)}{EFM(1) + EFM(2)}$$

Where $EFM(1,2)$ is the number of elementary flux modes from the joint network as defined by $M^{(1,2)}$, while $EFM(1)$ and $EFM(2)$ represent the numbers of elementary flux modes generated by the constituent randomized stoichiometric matrices M^1 and M^2 respectively. ΔEFM represents the increase in the number of flux modes, relative to the constituent networks.

Supporting Information

Figure S1 Sensitivity analysis of our metabolic approach for recapitulating the order of colonization, upon gradual removal of information. The distributions of pairwise metabolic distances for correct (literature-informed) and randomized orders of colonization are plotted for different percentages of enzymes removed from the dataset. Between 20 and 80 percent of enzymes were removed.

Acknowledgments

We are grateful to members of the Segrè lab for helpful feedback and discussions.

Author Contributions

Conceived and designed the experiments: VM SA DS. Performed the experiments: VM. Analyzed the data: VM DS. Wrote the paper: VM DS.

References

1. Stoodley P, Sauer K, Davies DG, Costerton JW (2002) Biofilms as complex differentiated communities. Annual review of microbiology 56: 187–209.
2. Donlan RM (2002) Biofilms: microbial life on surfaces. Emerging infectious diseases 8: 881–890.
3. Dewhirst FE, Chen T, Izard J, Paster BJ, Tanner ACR, et al. (2010) The Human Oral Microbiome. Journal of Bacteriology 192: 5002–5017.
4. Turnbaugh PJ, Ridaura VK, Faith JJ, Rey FE, Gordon JI (2010) Metagenomic Analysis in Humanized Gnotobiotic Mice. 1: 1–19.
5. Klitgord N, Segre D (2010) Environments that Induce Synthetic Microbial Ecosystems. PLoS Computational Biology 6: e1001002.
6. Borenstein E, Kupiec M, Feldman MW, Ruppin E (2008) Large-scale reconstruction and phylogenetic analysis of metabolic environments. Proceedings of the National Academy of Sciences 105: 14482–14487.
7. Kauffman KJ, Prakash P, Edwards JS (2003) Advances in flux balance analysis. Curr Opin Biotechnol 14: 491–496.
8. Belda-Ferre P, Alcaraz LD, Cabrera-Rubio R, Romero H, Simón-Soro A, et al. (2011) The Oral Metagenome in Health and Disease. The ISME journal: 1–11.
9. Diaz-Torres ML, Villedieu A, Hunt N, McNab R, Spratt DA, et al. (2006) Determining the antibiotic resistance potential of the indigenous oral microbiota of humans using a metagenomic approach. FEMS microbiology letters 258: 257–262.
10. Gill SR, Pop M, Deboy RT, Eckburg PB, Turnbaugh PJ, et al. (2006) Metagenomic Analysis of the Human Distal Gut Microbiome. Science 312: 1355–1359.
11. Turnbaugh P, Ley R, Hamady M, Fraser-Liggett C, Knight R, et al. (2007) The Human Microbiome Project. Nature 449: 804–810.
12. Kolenbrander PE, Palmer RJ, Periasamy S, Jakubovics NS (2010) Oral Multispecies Biofilm Development and the Key Role of Cell-cell Distance. Nature reviews Microbiology 8: 471–480.
13. Rickard AH, Gilbert P, High NJ, Kolenbrander PE, Handley PS (2003) Bacterial coaggregation: an integral process in the development of multi-species biofilms. Trends in Microbiology 11: 94–100.
14. Kolenbrander PE, Andersen RN, Blehert DS, Egland PG, Foster JS, et al. (2002) Communication among oral bacteria. Microbiology and molecular biology reviews 66: 486–505.
15. Valm AM, Welch JLM, Rieken CW, Hasegawa Y, Sogin ML, et al. (2011) Systems-level analysis of microbial community organization through combinatorial labeling and spectral imaging. Proceedings of the National Academy of Sciences 108: 4152–4157.
16. Kolenbrander PE, Andersen RN (1986) Multigeneric aggregations among oral bacteria: a network of independent cell-to-cell interactions. Journal of bacteriology 168: 851–859.
17. Mazumdar V, Snitkin ES, Amar S, Segrè D (2009) Metabolic Network Model of a Human Oral Pathogen. Journal of Bacteriology 191: 74–90.
18. Li L, Guo L, Lux R, Eckert R, Yarbrough D, et al. (2010) Targeted antimicrobial therapy against Streptococcus mutans establishes protective non-cariogenic oral biofilms and reduces subsequent infection. International journal of oral science 2: 66.
19. Palmer Jr RJ, Wu R, Gordon S, Bloomquist CG, Liljemark WF, et al. (2001) [27] Retrieval of biofilms from the oral cavity. Methods in enzymology 337: 393–403.
20. Becker SA, Palsson BO (2005) Genome-scale Reconstruction of the Metabolic Network in *Staphylococcus aureus* N315: an Initial Draft to the Two-dimensional Annotation. BMC Microbiology 5: 8.
21. Feist AM, Henry CS, Reed JL, Krummenacker M, Joyce AR, et al. (2007) A Genome-scale Metabolic Reconstruction for *Escherichia coli* K-12 MG1655 that Accounts for 1260 ORFs and Thermodynamic Information. Molecular Systems Biology 3: 121.
22. Forster J, Famili I, Fu P, Palsson BO, Nielsen J (2003) Genome-scale reconstruction of the *Saccharomyces cerevisiae* metabolic network. Genome Res 13: 244–253.
23. Nogales J, Palsson B, Thiele I (2008) A Genome-scale Metabolic Reconstruction of Pseudomonas putida KT2440: iJN746 as a Cell Factory. BMC Systems Biology 2: 79.
24. Oliveira AP, Nielsen J, Forster J (2005) Modeling *Lactococcus lactis* Using a Genome-scale Flux Model. BMC Microbiology 5: 39.
25. Senger RS, Papoutsakis ET (2008) Genome-scale Model for *Clostridium acetobutylicum*: Part I. Metabolic Network Resolution and Analysis. Biotechnology and Bioengineering: 1036–1052.
26. Thiele I, Vo TD, Price ND, Palsson BO (2005) Expanded metabolic reconstruction of *Helicobacter pylori* (iIT341 GSM/GPR): an in silico genome-scale characterization of single- and double-deletion mutants. Journal of bacteriology 187: 5818–5830.
27. Kanehisa M, Goto S, Furumichi M, Tanabe M, Hirakawa M (2010) KEGG for Representation and Analysis of Molecular Networks Involving Diseases and Drugs. Nucleic acids research 38: D355–360.
28. Kanehisa M, Goto S, Hattori M, Aoki-Kinoshita KF, Itoh M, et al. (2006) From Genomics to Chemical Genomics: New Developments in KEGG. Nucleic Acids Research 34: D354–357.
29. Kanehisa M, Goto S (2000) KEGG: kyoto encyclopedia of genes and genomes. Nucleic Acids Res 28: 27–30.
30. Ogata H, Goto S, Sato K, Fujibuchi W, Bono H, et al. (1999) KEGG: Kyoto Encyclopedia of Genes and Genomes. Nucleic Acids Res 27: 29–34.
31. Massey Jr FJ (1951) The Kolmogorov-Smirnov test for goodness of fit. Journal of the American Statistical Association: 68–78.
32. Lendenmann U, Grogan J, Oppenheim F (2000) Saliva and dental pellicle-a review. Advances in Dental Research 14: 22–28.
33. Singer RE, Buckner BA (1981) Butyrate and propionate: important components of toxic dental plaque extracts. Infection and Immunity 32: 458–463.
34. Tsuda H, Ochiai K, Suzuki N, Otsuka K (2010) Butyrate, a Bacterial Metabolite, Induces Apoptosis and Autophagic Cell Death in Gingival Epithelial cells. Journal of Periodontal Research 45: 626–634.
35. Abe N, Kadowaki T, Okamoto K, Nakayama K, Ohishi M, et al. (1998) Biochemical and functional properties of lysine-specific cysteine proteinase (Lys-gingipain) as a virulence factor of *Porphyromonas gingivalis* in periodontal disease. Journal of biochemistry 123: 305–312.
36. Olczak T, Simpson W, Liu X, Genco CA (2005) Iron and Heme Utilization in *Porphyromonas gingivalis*. FEMS Microbiology Reviews 29: 119–144.
37. Ley RE, Peterson DA, Gordon JI (2006) Ecological and evolutionary forces shaping microbial diversity in the human intestine. Cell 124: 837–848.
38. Terzer M, Stelling J (2008) Large-scale Computation of Elementary Flux Modes with Bit Pattern Trees. Bioinformatics 24: 2229–2235.
39. Roberts RJ, Chang YC, Hu Z, Rachlin JN, Anton BP, et al. (2011) COMBREX: a project to accelerate the functional annotation of prokaryotic genomes. Nucleic acids research 39: D11–D14.
40. Klitgord N, Segrè D (2010) Environments that induce synthetic microbial ecosystems. PLoS computational biology 6: e1001002.
41. Fu W, Forster T, Mayer O, Curtin JJ, Lehman SM, et al. (2010) Bacteriophage cocktail for the prevention of biofilm formation by Pseudomonas aeruginosa on catheters in an in vitro model system. Antimicrobial agents and chemotherapy 54: 397–404.
42. Hall-Stoodley L, Costerton JW, Stoodley P (2004) Bacterial biofilms: from the natural environment to infectious diseases. Nature Reviews Microbiology 2: 95–108.
43. Hawser SP, Douglas LJ (1994) Biofilm formation by Candida species on the surface of catheter materials in vitro. Infection and Immunity 62: 915–921.
44. Nickel J, Ruseska I, Wright J, Costerton J (1985) Tobramycin resistance of *Pseudomonas aeruginosa* cells growing as a biofilm on urinary catheter material. Antimicrobial agents and chemotherapy 27: 619–624.
45. Canfield DE, Des Marais DJ (1991) Aerobic sulfate reduction in microbial mats. Science (New York, NY) 251: 1471.
46. Ward DM, Ferris MJ, Nold SC, Bateson MM (1998) A natural view of microbial biodiversity within hot spring cyanobacterial mat communities. Microbiology and Molecular Biology Reviews 62: 1353–1370.
47. Arumugam M, Raes J, Pelletier E, Le Paslier D, Yamada T, et al. (2011) Enterotypes of the human gut microbiome. Nature 473: 174–180.
48. Xu J, Mahowald MA, Ley RE, Lozupone CA, Hamady M, et al. (2007) Evolution of symbiotic bacteria in the distal human intestine. PLoS biology 5: e156.
49. Barrick JE, Yu DS, Yoon SH, Jeong H, Oh TK, et al. (2009) Genome evolution and adaptation in a long-term experiment with *Escherichia coli*. Nature 461: 1243–1247.
50. Boussau B, Karlberg EO, Frank AC, Legault BA, Andersson SGE (2004) Computational inference of scenarios for α-proteobacterial genome evolution. Proceedings of the National Academy of Sciences of the United States of America 101: 9722–9727.
51. Ochman H, Elwyn S, Moran NA (1999) Calibrating bacterial evolution. Proceedings of the National Academy of Sciences 96: 12638–12643.
52. Webb EC (1992) Enzyme nomenclature 1992. Recommendations of the Nomenclature Committee of the International Union of Biochemistry and Molecular Biology on the Nomenclature and Classification of Enzymes: Academic Press.
53. Jaccard P (1908) Nouvelles recherches sur la distribution florale.
54. Subramanian A, Tamayo P, Mootha VK, Mukherjee S, Ebert BL, et al. (2005) Gene Set Enrichment Analysis: a Knowledge-based Approach for Interpreting Genome-wide Expression Profiles. Proceedings of the National Academy of Sciences of the United States of America 102: 15545–15550.

6

Quantification of Carbon and Phosphorus Co-Limitation in Bacterioplankton: New Insights on an Old Topic

Irene Dorado-García[1]*, Juan Manuel Medina-Sánchez[2], Guillermo Herrera[1], Marco J. Cabrerizo[1], Presentación Carrillo[1]

1 Instituto del Agua, Universidad de Granada, Granada, Spain, **2** Departamento de Ecología, Facultad de Ciencias, Universidad de Granada, Granada, Spain

Abstract

Because the nature of the main resource that limits bacterioplankton (e.g. organic carbon [C] or phosphorus [P]) has biogeochemical implications concerning organic C accumulation in freshwater ecosystems, empirical knowledge is needed concerning how bacteria respond to these two resources, available alone or together. We performed field experiments of resource manipulation (2×2 factorial design, with the addition of C, P, or both combined) in two Mediterranean freshwater ecosystems with contrasting trophic states (oligotrophy vs. eutrophy) and trophic natures (autotrophy vs. heterotrophy, measured as gross primary production:respiration ratio). Overall, the two resources synergistically co-limited bacterioplankton, i.e. the magnitude of the response of bacterial production and abundance to the two resources combined was higher than the additive response in both ecosystems. However, bacteria also responded positively to single P and C additions in the eutrophic ecosystem, but not to single C in the oligotrophic one, consistent with the value of the ratio between bacterial C demand and algal C supply. Accordingly, the trophic nature rather than the trophic state of the ecosystems proves to be a key feature determining the expected types of resource co-limitation of bacteria, as summarized in a proposed theoretical framework. The actual types of co-limitation shifted over time and partially deviated (a lesser degree of synergism) from the theoretical expectations, particularly in the eutrophic ecosystem. These deviations may be explained by extrinsic ecological forces to physiological limitations of bacteria, such as predation, whose role in our experiments is supported by the relationship between the dynamics of bacteria and bacterivores tested by SEMs (structural equation models). Our study, in line with the increasingly recognized role of freshwater ecosystems in the global C cycle, suggests that further attention should be focussed on the biotic interactions that modulate resource co-limitation of bacteria.

Editor: Rodolfo Paranhos, Instituto de Biologia, Brazil

Funding: This research was supported by Junta de Andalucía (Excelencia P09-RNM-5376 to JMMS) and the Spanish Ministry Ciencia e Innovación (CGL2011-23681 to PC). The funders had no role in study design, data collection and analysis, decision to publish, or preparation of the manuscript.

Competing Interests: The authors have declared that no competing interests exist.

* E-mail: idorado@ugr.es

Introduction

In aquatic ecosystems, bacterioplankton is regulated by different factors [1], including: abiotic such as temperature [2–4] or inorganic and organic nutrient sources [5,6]; and biotic, such as predation [7–9]. In turn, bacterioplankton regulate such ecological processes as water quality, atmospheric composition, nutrient cycling [10], and the breakdown of organic matter [11]. Specifically, heterotrophic bacteria remineralize nutrients [1,12,13] via general or specialized biogeochemical pathways and, in turn, transfer them to high trophic levels through the microbial loop [14,15]. In this regard, bacteria convert dissolved organic carbon (DOC) to biomass throughout bacterial production (BP) and/or oxidize it to CO_2 through bacterial respiration (BR) [16].

Depending on relative availability of organic carbon (C) and nutrients (e.g. P), bacteria can be C or P limited. The nature of the resource that limits bacteria has biogeochemical implications concerning organic C accumulation in freshwater ecosystems. Thus, when bacteria are limited by C having enough inorganic nutrients available, bacteria consume labile C for growth, while semi-labile C tends to accumulate in the surrounding waters [12,17]. Contrarily, if bacteria are limited by inorganic nutrients, e.g. P, the labile share of the C pool also accumulates in the ecosystem. This connects with the "malfunctioning microbial loop" hypothesis of [18], proposing an accumulation of degradable C because food-web interactions restrict bacterial growth. Controversy persists concerning resource limitation of bacteria in aquatic ecosystems. On the one hand, bacteria are presumed to be P limited in oligotrophic lakes [19,20] and/or in those with high DOC:P ratios [21], despite that bacteria have a greater affinity for P than do phytoplankton in P-poor aquatic ecosystems [18,22,23]. On the other hand, some studies have shown that bacteria are C limited, preferring autochthonous C provided by algae [24–26], even though they could use other C resources such allochthonous DOC [24,27], old DOC (i.e. not recently produced [25,28]) or semi-labile DOC. Several studies show, however, a prevalence of mineral-nutrient and organic carbon co-limitation in aquatic microbial communities, deduced from stronger responses to combined resources than to single-nutrient additions [13,26].

Most studies using the nutrient manipulation of bacteria in freshwater ecosystems (increasing C and P) have been conducted in oligotrophic ecosystems because their nutrient-limited condition makes them sensitive to nutrient inputs, and they can act as early-warning ecosystems for eutrophication due to human activities [20,26]. Thus, studies performed in freshwater ecosystems involve arctic and alpine lakes [20,29], tropical regions [26,30,31] or humic lakes [13], whereas inland waters of Mediterranean region have received less attention. Noticeably, the Mediterranean ecoregion shows a broad intersystem trophic gradient for inland waters, not only throughout the entire ecoregion, but also locally [32]. This has been promoted by traits such as scarceness of vegetation and developed soil on watersheds [33], the high frequency of extreme weather events [34,35], and millennia of land use (i.e. farming and cattle raising). Moreover, those ecosystems tend to be more autotrophic with greater eutrophy, as has been reported based on estimated BR:PP ratio values [32], probably related to the restricted terrestrial C input [36]. Likewise, the Mediterranean freshwater ecosystems undergo intense and irregular Saharan dust loads that constitute notable inputs of P [37–39] and of soil-derived organic components to these water bodies [39,40]. These resource inputs could alter the microbial community and its functioning [38,41] and hence biogeochemical cycles in which they are involved [10,42].

It has been proposed that bacterial P limitation would be more frequent towards oligotrophy, C limitation towards eutrophy, and combined C and P limitation in a wide variety of eutrophic-oligotrophic ecosystems [21]. This trend is partially supported by an across-system study [43] where the effect of P additions diminished towards eutrophy, while the effect of added organic carbon did not correlate with the trophic state. Controversy remains, however, because other studies (e.g. [23]) have found that bacterial production was not stimulated by mineral nutrient (P and N) addition in oligotrophy. Regarding the possible relationship between resource co-limitation and the trophic state, [32] found that P and C co-limited bacteria towards the more eutrophic systems.

Several studies focussing on the co-limitation in primary producers have tried to categorize the kinds of co-limitation based on several mathematical approaches [44–47]. While co-limitation has been generally defined as the higher response to combined nutrient additions than to single ones, some studies (e.g. [44–47]) have distinguished different kinds of co-limitation. These have been generically based on whether responses to joint factors were higher (synergistic or super-additive co-limitation), equal (additive co-limitation) or lower than the sum of single responses (antagonistic or sub-additive co-limitation). However, these mathematical approaches have not been applied to studies examining co-limitation of bacteria. A unified classification of these kinds and their terminology is still lacking and, moreover, there is scant theory to predict the prevailing kinds of co-limitation of bacteria expected in ecosystems with different trophic states.

Few studies dealing with nutrient limitation of bacteria measure both structural (e.g. biomass) and functional (BP and BR) response variables. In fact, hardly any studies have measured BR in this context due to the difficulties of separating the bacterial fraction from the rest of planktonic community [13]. Therefore, BR has often been estimated from BP values using empirical models (e.g. [32,48]), despite of troubles posed by the universal application of these estimated measurements.

Because the nature of the resource (C, P) that limits bacteria has biogeochemical implications concerning organic C accumulation in freshwater ecosystems, it is important to increase the knowledge concerning how bacteria respond to C and/or P. Accordingly, we used an experimental approach to evaluate whether the simultaneous addition of C and P (co-) limited bacteria by measuring their responses to C and P addition alone or in combination in two contrasting ecosystems (in terms of the trophic state) in the Mediterranean region. The two ecosystems chosen were a eutrophic reservoir (Cubillas, Granada, Spain) and an oligotrophic high-mountain lake (La Caldera, Sierra Nevada National Park, Spain), two ecosystems that provide major services to surrounding human populations.

Based on patterns of resource limitation reported formerly by our research team [32] for the Mediterranean region, and the positive correlation between bacterial abundance and DOC concentration found in alpine lakes [39], we hypothesise that the bacterial community would be C limited in the oligotrophic high-mountain ecosystem, but limited mainly by P in the eutrophic ecosystem. In addition, we expect that, in both study ecosystems, bacterial responses to joint C and P additions would be more intense (synergistic), according to the generalized resource co-limitation found in aquatic ecosystems, at least for primary producers [46,47,49]. In this regard, the use of a modified index based on [47] enabled us to quantify the type of resource co-limitation in bacterioplankton when simultaneous resources, organic carbon (energy substrate) and phosphorus (mineral nutrient) were added.

Materials and Methods

Study Area

Nutrient-manipulation experiments were conducted in two Mediterranean ecosystems with contrasting trophic states (Cubillas and La Caldera) located in southern Spain. Cubillas reservoir, of medium size, has a capacity of 21×10^6 m^3 and a surface area of 2.0 km^2 [40]. The water column is characterized by low Secchi depth (<0.5 m), and high values of chemical and biological variables (total nitrogen [TN] >2000 mg m^{-3}; total P [TP] > 25 mg m^{-3}; chlorophyll *a* [Chl a] >30 mg m^{-3}; [40]) inherent of eutrophic trophic state [50]. Cubillas reservoir was also selected because it provides cultural ecosystem services to the surrounding urban populations.

On the other hand, La Caldera is a fishless oligotrophic high-mountain lake with a total surface area of 0.02 Km^2 located in Sierra Nevada National Park (southern Spain) at an elevation of 3050 m.a.s.l. The water is highly transparent (Secchi's visibility for the entire water column) and registers low values of chemical and biological variables (TP<5 mg m^{-3}; Chl a <2 mg m^{-3}; DOC ~0.5 mg C L^{-1}; [51,52]), inherent of oligotrophic trophic state [50].

Research permits for this study were provided by Agencia de Medio Ambiente y Agua (Consejería de Medio Ambiente y Ordenación del Territorio, Junta de Andalucía, Spain) for Cubillas, and by Sierra Nevada Parque Nacional (Spain) for La Caldera.

Experimental Design

A 2×2 factorial nutrient-manipulation design was implemented for 15 days in Cubillas reservoir (15 to 29 May 2012) and 19 days in La Caldera lake (31 August to 18 September 2011). The experimental setup consisted of transparent semi-spherical polyethylene bags (microcosms, 0.58 m diameter) filled with 100 L of water accurately measured with a flowmeter. The volume of microcosms was considered to be sufficient to minimize the bottle effect over the experimental incubation periods [53–55]. The experiments had four treatments per triplicate (control, addition of P [P treatment], of C [C treatment] and of C plus P [CP treatment]). The water used to fill in the microcosms was taken from upper layers of each water body (from 0.2 m to 0.5 m depth) with water pump, sieved through a 45-µm mesh to remove large zooplankton and mixed before and after nutrient additions. The microcosms were fixed to a PVC-tube structure and incubated *in situ* in the upper layers (from surface to ~0.5 m depth). The C and P were added once at the beginning of the experiment with concentrations of 30 μgP L^{-1} and 290.2 μgC L^{-1}, added as Na_2HPO_4 and sucrose, respectively. The concentrations were selected to ensure nutrient availability throughout the experiments, maintaining the N:P molar ratio of 30 in P-added treatments (i.e. P and CP) and C:P molar ratio of 300 in C-added treatments (i.e. C and CP). Thus, to maintain the N:P ratio, we added 406.4 μgN L^{-1} as NH_4NO_3. Nutrients were prepared by dissolving and thoroughly mixing the nutrients in double-distilled water before the addition.

Sampling

Before the sampling, the bags (microcosms) were homogenized by gentle beating the exterior by hand. Afterwards, water was slowly pumped from each microcosm with a small water pump through a plastic tube to fill 1.5 L acid-cleaned bottles, which were transported to the laboratory for analysis.

The sampling schedule was designed to monitor the changes in the response variables to nutrient additions in three different experimental periods, defined as early, middle, and late. Early period means sampling on each of the first three days after additions in Cubillas and on each of the first four days in La Caldera; middle and late periods mean sampling, respectively, on the 5th and the 14th days in Cubillas, and the 14th and 18th days in La Caldera after additions. These periods (and the response variables measured in each) were previously fitted for each ecosystem (time as adjusted 'within-subjects' factor) to account for the differences in the generation times of organisms between the ecosystems, based on water temperature and plankton dynamics [56,57]. These periods allowed us to appreciate both the rapid and delayed responses of the dependent variables to nutrient additions (as fixed 'between-subjects' factors).

Physico-chemical Analyses

Temperature was measured every 0.5 m depth through the water column with YSI MPS-556 multi-parametric sensor (YSI Incorporated, OH, USA). Samples for TP and total dissolved P (TDP) were persulphate-digested and analysed as soluble reactive P applying the acid molybdate technique [58]. Samples for TN and total dissolved nitrogen (TDN) were persulphate-digested and measured as nitrate [58].

Samples for DOC were filtered by through a pre-combusted 0.7-μm Whatman GF/F glass-fiber filters, then acidified with HCl (final pH<2) and stored in darkness at 4°C until analysed [51]. DOC concentrations were measured with the high-temperature catalytic oxidation method in a Shimadzu TOC analyser (Model 5000) equipment [59].

Functional Variables

Bacterial production. Bacterial production (BP) was measured following the ^{3}H-thymidine method [60] modified by [61]. Briefly, to each vial (2 mL micro-centrifuge tubes) containing 1.5 mL of sample (3 replicates and 2 blanks for each experimental treatment), ^{3}H-thymidine (S.A. = 48–50 Ci $mmol^{-1}$, Perkin Elmer) was added to a final saturating concentration (12 nM in La Caldera; 20 nM in Cubillas). Vials were incubated at *in situ* temperature for 60 min in darkness. Extraction was carried out with 5% (final concentration) cold trichloroacetic acid (TCA). The tubes were centrifuged at 16,000 g, rinsed twice with 5% TCA, and measured in a scintillation counter equipped with autocalibration (Beckman LS 6000 TA). In all the calculations, data were corrected by blanks (bacteria were killed with 5% TCA before addition of the radiotracer). To convert incorporated tracers to BP expressed in C terms, we applied the conversion factors 1×10^{18} cells mol^{-1} of thymidine [62], and 2×10^{-14} g C per cell [63].

Primary production. Primary production (PP) was measured following the ^{14}C method [64]. Briefly, each 50-mL quartz flask (three clear and one dark from each microcosm), added with 0.37 MBq of $NaH^{14}CO_3$ (Specific Activity: 310.8 MBq $mmol^{-1}$, DHI Water and Environment, Germany) was incubated in the upper layers (0.5 m depth) for 4 h symmetrically distributed at noon. PP was measured as total organic carbon, and algal excretion of organic carbon (EOC) as the organic measured in the <1 μm fraction, following the laboratory procedure of [65]. Gross primary production (GPP) was calculated as the sum of PP, standardized for 14.5 h of daytime, and planktonic community respiration (R, see below), standardized for 24 h (entire day), and thus these ratios were assumed to be constant throughout the day.

Community and bacterial respiration. Respiration of planktonic community (<45 μm fraction; R) and bacterial community (<0.7 μm fraction, filtrate of samples through glass-fiber filters Whatman GF/F; BR) were estimated from oxygen depletion measured with sensor-spot optodes (SP-PSt3-NAU-D5-YOP and Fibox3; PreSens GmbH, Germany). Briefly, optodes followed a two-point calibration (0 and 100% oxygen saturation), 0% point calibration was performed adding sodium sulphite (Na_2SO_3) to a final concentration exceeding 0.1 mg mL^{-1} and 100% by putting wet cotton wool into the closed flask to ensure a 100% O_2-saturated water-vapour air. Samples for each fraction and experimental treatment were transferred to 25-mL quartz flasks with a sensor spot attached inside and closed with glass stoppers. Respiration rates were calculated from least-squares regressions after confirming that oxygen fitted a linear model during the first 24 h after sampling.

Bacterial carbon demand and availability of algal C supply. The bacterial carbon demand (BCD) was estimated as the sum of BP and BR [13,66]. Because of the uncertainty of algal C supply during the night time both BCD and EOC were standardized for 14.5 h of day time, as in [32]. We calculated the BCD:EOC ratio to quantify the proportion of bacterial C demand met by algae-released C, and whether it underwent late period variations compared to initial conditions. The use of EOC in this ratio is based on the bacterial preference for autochthonous C [24] and experimental findings of bacterial dependence on organic C freshly released by algae in an Iberian lake [32,67].

Biological Variables

Bacterial abundance. Bacterial abundance (BA) was determined by flow cytometry technique (FACSCanto II, Becton Dickinson Biosciences, Oxford, UK) fixing 1.5 mL of sampling water with particle-free 20% (w/v) paraformaldehyde, 1% final concentration followed by liquid nitrogen frozen and stored at −80°C [68,69]. Before being analysed, the samples were thawed and stained with Syber Green I DNA (Sigma-Aldrich) 1:5000 final dilution of initial stock [69,70]. Yellow-green 1-μm beads at a standard concentration (Fluoresbrite Microparticles, Polysciences, Warrington, PA, USA) were added in order to determine absolute cell concentrations [69,71].

Algae and ciliate abundance and biomass. Autotrophic picoplankton (<2 μm) was checked in samples fixed with 1% paraformaldehyde, and immediately filtered through black 0.2-μm pore-size Nuclepore filters, through fluorescence microscopy with an inverted microscope (Axio Observer.A1 model, Carl Zeiss Microscopy GmbH, Germany). Phytoplankton and ciliates were preserved in glass bottles with alkaline Lugol's solutions until analysed. Water samples (50 mL) were allowed to settle in Utermöhl chambers for 24 h and cells were counted at 400x and 1000x following [72]. At least 400 individuals of the most abundant species were counted [73], and 20 individuals per species were measured in each treatment. Biomass was estimated by approximating cell volume to their geometric shape [74,75] and to transform it to C units following suitable conversion factors [76,77].

Data Analysis

For each ecosystem, the nutrient-addition effects over time were tested by two-way repeated-measures ANOVA (RM-ANOVA). Homoscedasticity (by Cochran's and Levene's tests), and correlation between means and standard deviations were checked for each data group in order to verify the assumptions required by RM-ANOVA. The interaction between time and resources (C, P) was tested using Pillai, Hotelling and Roy's multivariate tests that avoid the problem of assuming compound symmetry and sphericity. When the interactive effect was significant, Fisher's LSD *post hoc* test was applied to detect statistical significance between treatments. The effect size of nutrient additions was calculated as the quotient between the mean value of each resource-added treatment and the mean value of the control, for each response variable.

To investigate the evolution of co-limitation on bacteria over time, we estimated the type of co-limitation for each period by applying the zero-centred interaction ratio response index ("IRR") reported by [47], but incorporating its error term (calculated by the general equation of propagation of errors) and considering zero as the threshold to distinguish among super-additive (IRR ± error>0), additive (IRR ± error = 0), and sub-additive positive responses.

$$IRR = ((CP + Control) - (C + P))/Control$$

where CP, C, P, and Control represent the mean value of the response variable for each treatment.

Subsequently, the diversity of kinds of co-limitation provided by the IRR index was "summarized" in a unifying classification according to [47] and following more simplifying criteria equivalent to those reported by [46], despite that we discarded the direct use of their interaction effect index because of log-transformation biases, as discussed by [47]. We selected these two classifications because they were the only ones that enabled us to quantify the type of co- limitation, applied to bacterioplankton, and determined by the IRR values. Thus, IRR>0 indicates that the response to the simultaneous addition of both resources is higher than that of a single resource, considered as synergistic co-limitation by [46], while subdivided into three subtypes by [47] (Figure S1). IRR = 0 indicate: i) no response, when no effects of single resources are found, ii) single limitation, when only one of the single resources has a significant effect, and iii) additive (*sensu* [46]) or Independent Co-Limitation Additive (*sensu* [47]), when single resources show a significant effect, but the sum of both effects is lower than the effect of the two simultaneous resources. Finally, IRR<0 indicates that the additive effect is higher than the effect of combined resources (i.e. antagonistic responses, *sensu* [46]) or no interaction *sensu* [47]).

This unifying classification is summarized in Figure S1 as well as the abbreviations used hereafter in the text that correspond to the types of co-limitation derived only from the positive or null responses to resource additions depicted in Figure S1.

We used a SEM (structural equation model) to test whether the dynamics of inorganic and organic resources (i.e. P and C) and of potential predators (mixotrophs and ciliates) influenced BA dynamics. Our model proposed that BA is regulated by bottom-up (resources) and top-down (potential predators) controls.

All variables were assessed for normality prior to statistical analyses. The GLS→ML method was used to estimate standardized path coefficients in our model. The degree of fit of the model to the observed data was tested by χ^2. Non-significant χ^2 indicates that the model could be accepted. Additionally, the degree of fit was supplemented with other goodness-of-fit indices such as the Bentler–Bonnet Normed FitIndex (NFI) and the Goodness-of-Fit Index (GFI), as recommended by [78–80].

For all statistical tests, we assumed p<0.05 as a threshold level of acceptance, and STATISTICA 7 software (StatSoftInc, 2005) was used.

Results

Starting Conditions of the Experiments

Water temperature differed between the two ecosystems studied (Table 1). This was >22.5°C in the upper layers, declining gradually to 17.5°C at the bottom (6.5 m) with no stratification pattern, in Cubillas. By contrast, water temperature reached only 14°C in La Caldera, with a homogeneous vertical profile to the maximum depth (10 m).

Chemical variables, such as DOC, TP, and TN showed values between 5- and 9-fold higher in Cubillas than in La Caldera, thus representing the contrasting trophic states of the two ecosystems. This difference was further indicated by water transparency (Secchi disk depth was 18-fold higher in La Caldera) and the biotic variables, which showed higher values in Cubillas than in La Caldera (Chl *a*, 3.6-fold; algal biomass, 44-fold; BA, 10.3-fold; BP, 16.1-fold; R [<45 µm fraction], 3.8-fold; BR [<1.2 µm fraction], 8.9-fold; Table 1). Therefore, hereafter La Caldera will be designated as the oligotrophic ecosystem, and Cubillas as the eutrophic ecosystem.

Another important differentiating feature between the two ecosystems was the nanoplanktonic community composition. Thus, in the eutrophic ecosystem, the algal community was dominated by potential mixotrophic algae (up to 92% of the algal biomass) and ciliate abundance reached 27 cells mL^{-1}. By contrast, in the oligotrophic ecosystem, strict autotrophic algae dominated the algal community (Chlorophyceae, up to 100%) and ciliates were not detected (Table 1).

Contrary to our expectations, the eutrophic ecosystem showed a heterotrophic nature (GPP:R<1, PP:R<1, PP:BP = 2.20) while the oligotrophic one showed the opposite scenario (autotrophic nature; GPP:R>1, PP:R>1, PP:BP = 34.47).

Resource Variation Over the Experiments

In both ecosystems, TDP followed a decreasing trend over time (RM-ANOVA, effect time p<0.05, Table S1) with a steeper decline in the P-added treatments (Figure 1A, B) while concentration of TDN did not change (RM-ANOVA, time effect p-value>0.05). By contrast, DOC showed a different pattern in each ecosystem, as it did not vary over time in the eutrophic ecosystem (RM-ANOVA, time effect p-value>0.05, Figure 1), but decreased in CP treatment over both the middle and late periods, reaching values similar to those of control in the oligotrophic ecosystem (Figure 1C, D; Table S1).

Bacterial Production Response to Resource Addition

In the eutrophic ecosystem, BP responded positively to the P and C treatments over the early period. However, these responses vanished over the middle period to reappear over the late period

Table 1. Mean values of the main physical, chemical and biological variables measured under initial conditions of the experiment.

Variable	Eutrophic ecosystem	Oligotrophic ecosystem
Temperature (°C)	>22.5	14
Secchi disk (m)	0.5	9
TP (μgP L^{-1})	15.13±0.13	2.18±1.11
TDP (μgP L^{-1})	3.73±0.33	2.12±0.70
TN (mgN L^{-1})	2.93±0.01	0.32±0.04
TDN (mgN L^{-1})	2.75±0.01	0.28±0.02
DOC (mgC L^{-1})	3.50±0.98	0.59±0.14
BP (μg C L^{-1} h^{-1})	1.13±0.17	0.07±0.01
R (μg C L^{-1} h^{-1})	6.02±2.19	0.89±0.10
BR (μg C L^{-1} h^{-1})	2.19±1.57	0.33±0.09
BA (cells mL^{-1})	$3.6\times10^6\pm5.3\times10^5$	$3.5\times10^5\pm2.4\times10^4$
Chl a (μg L^{-1})	9.0±0.1	2.5±0.14
AB (μg C L^{-1})	824±275	18.7±3.2
CB (μg C L^{-1})	2±3	0.0±0.0

Abbreviations: Temperature (average water column); TP = total phosphorus; TDP = total dissolved phosphorus; TN = total nitrogen; TDN = total dissolved nitrogen; DOC = dissolved organic carbon; BP = bacterial production; R = planktonic respiration; BR = bacterial respiration; BA = bacterial abundance; Chl a = chlorophyll a; AB = algae biomass; CB = ciliate biomass.

(Figure 2A; Tables 2 and S2). Notably, BP strongly responded to CP treatment over the early and middle periods, although the magnitude of this positive response (effect size) diminished after 24 h and decreased further over the late period, when BP equalled the control or each single addition treatment (Figure 2A; Table 2). As a result, the kind of resource co-limitation for BP changed within the early period from additive (ICLA) to synergistic (SSL-P), which persisted over the middle period, but became antagonistic (ICLSbA) over the late period (Figure 2A; Table 3). In La Caldera, BP positively responded to P treatment after 48 h until the end of the experiment (Figure 2B; Table 4) but not to C treatment at any time (Figure 2B, Table 4). BP strongly responded to CP treatment after 48 h until the end of the experiment, with the strongest positive response at middle period (Figure 2B; Tables 4 and S2). As a result, the kind of resource co-limitation for BP was synergistic during that period.

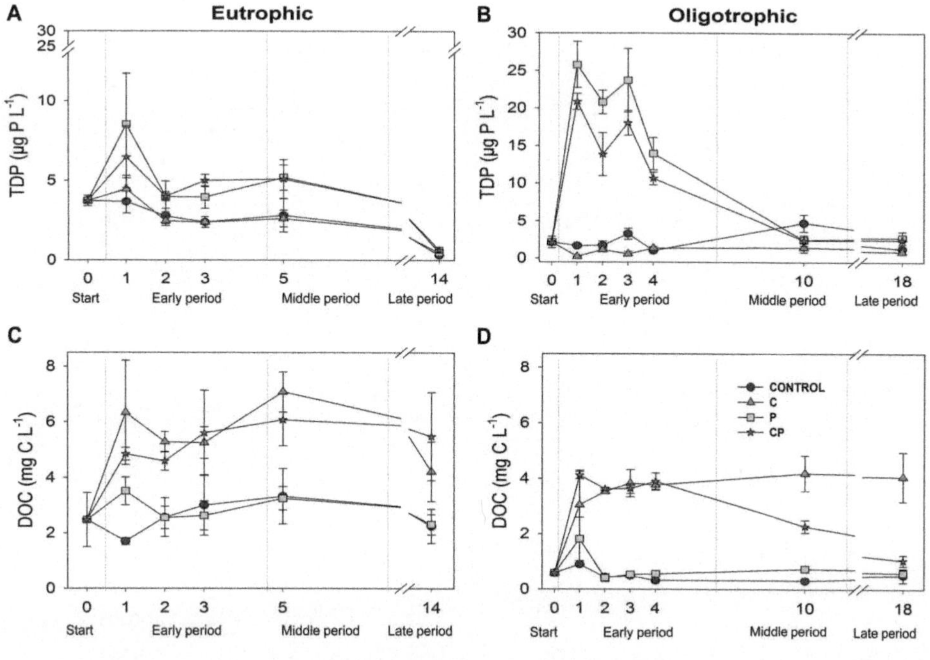

Figure 1. Experimental dynamics of DOC and TDP in both study ecosystems. (A) TDP in the eutrophic ecosystem (Cubillas); (B) TDP in the oligotrophic ecosystem (La Caldera); (C) DOC in the eutrophic ecosystem (Cubillas reservoir); (D) DOC in the oligotrophic ecosystem (La Caldera lake). Symbols represent mean values and error bars represent standard deviations. DOC = dissolved organic carbon. TDP = total dissolved phosphorus.

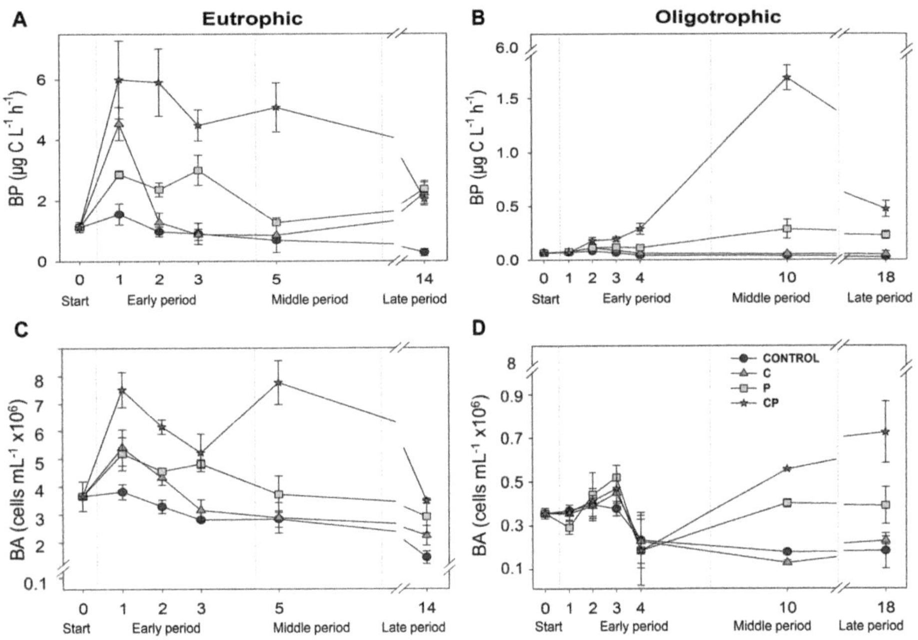

Figure 2. Temporal responses of bacterial production and abundance to experimental treatments in the two ecosystems. Left panels show the bacterial production (A) and abundance (C) responses in the eutrophic ecosystem, while the right panels show the same response variables (B and D, respectively) for the oligotrophic ecosystem. Symbols represent mean values and error bars represent standard deviations.

The Amount of Algal C Supply that Met Bacterial C Demand

In the eutrophic ecosystem, the BCD:EOC ratio yielded values >1 in all treatments (i.e. up to 16.2), suggesting that bacterial demands exceeded the supply of photosynthetic C. By contrast, in the oligotrophic ecosystem, the BCD:EOC ratio was <1 (up to 0.7) in all treatments, suggesting the opposite scenario. In each ecosystem, the values of the ratio were similar to those found for the respective starting conditions (Table S3).

Response of the Microbial Plankton Structure to Resource Addition

In the eutrophic ecosystem, BA positively responded to all treatments with resource additions over the early period, although

Table 2. Effect size and p-value (Fisher's *post hoc* test) for BP and BA for the eutrophic ecosystem.

			BP		BA	
Eutrophic ecosystem						
			Effect size	**p-value**	**Effect size**	**p-value**
C	Early period	24 h	2.90	**<0.01**	1.42	**<0.01**
		48 h	1.32	0.39	1.31	**<0.01**
		72 h	0.97	0.94	1.12	0.41
	Middle period		1.24	0.64	1.02	0.90
	Late period		8.08	**<0.01**	1.54	0.08
P	Early period	24 h	1.83	**<0.01**	1.36	**<0.01**
		48 h	2.43	**<0.01**	1.39	**<0.01**
		72 h	3.31	**<0.01**	1.72	**<0.01**
	Middle period		1.85	0.12	1.32	**<0.01**
	Late period		8.57	**<0.01**	2.01	**<0.01**
CP	Early period	24 h	3.83	**<0.01**	1.97	**<0.01**
		48 h	6.06	**<0.01**	1.87	**<0.01**
		72 h	4.94	**<0.01**	1.87	**<0.01**
	Middle period		7.37	**<0.01**	2.74	**<0.01**
	Late period		7.35	**<0.01**	2.41	**<0.01**

Abbreviations: BP = bacterial production; BA = bacterial abundance.

Table 3. IRR values and their correspondence with type of co-limitation for BP and BA for eutrophic ecosystem.

			BP			BA		
Eutrophic ecosystem								
			IRR	**Type**	**Type**	**IRR**	**Type**	**Type**
CP	Early period	24 h	0.10±0.92	AD	ICLA	0.19±0.29	AD	ICLA
		48 h	3.30±1.26	SC	SSL(P)	0.17±0.13	SC	ICLSpA
		72 h	1.66±0.85	SC	SSL(P)	0.02±0.28	n.d.	SL(P)
	Middle period		5.28±1.52	SC	SCL	1.41±0.42	SC	SSL(P)
	Late period		−8.30±4.29	AA	ICLSbA	−0.15±0.51	n.d.	SL(P)
Oligotrophic ecosystem								
CP			**IRR**	**Type**	**Type**	**IRR**	**Type**	**Type**
	Early period	24 h	−0.16±0.29	n.d.	NR	0.23±0.12	n.d.	NR
		48 h	0.43±0.47	n.d.	NR	−0.06±0.26	n.d.	NR
		72 h	0.73±0.39	SC	SSL(P)	−0.33±0.31	AC	ASL(P)
		96 h	3.56±1.45	SC	SSL(P)	0.02±0.97	n.d.	NR
	Middle period		35.39±3.78	SC	SSL(P)	1.15±0.12	SC	SSL(P)
	Late period		8.73±4.01	SC	SSL(P)	1.15±0.12	SC	SSL(P)

IRR = Interaction Ratio Response index ± error term and its correspondence with the type of co-limitation according to a modified classification based on [46] (Type) and [47] (Type) for each variable response. BP = bacterial production and BA = bacterial abundance. Abbreviations: AD = additive; SC = synergistic co-limitation; AC = antagonism co-limitation; AA = absolute antagonism; NR = no response; ICLA = independent co-limitation additive; ICLSpA = independent co-limitation super-additive; SL(P) = single limitation (P); SSL(P) = *synergistic* serial limitation (P); ICLSbA = independent co-limitation sub-additive; ASL(P) = *antagonism* serial limitation (P); n.d. = not defined.

Table 4. Effect size and p-value (Fisher's *post hoc* test) for BP and BA for oligotrophic ecosystem.

			BP		BA	
Oligotrophic ecosystem						
			Effect size	**p-value**	**Effect size**	**p-value**
C	Early period	24 h	1.05	0.90	0.96	0.79
		48 h	1.38	0.29	0.98	0.90
		72 h	1.36	0.41	1.19	0.21
		96 h	1.44	0.51	0.98	0.91
	Middle period		1.42	0.57	0.73	0.21
	Late period		2.10	0.38	1.25	0.40
P	Early period	24 h	1.07	0.84	0.79	0.18
		48 h	1.38	0.30	1.12	0.41
		72 h	1.80	0.09	1.38	**<0.05**
		96 h	2.63	**<0.05**	0.80	0.40
	Middle period		7.32	**<0.01**	2.25	**<0.01**
	Late period		9.30	**<0.01**	2.13	**<0.01**
CP	Early period	24 h	0.96	0.91	0.98	0.86
		48 h	2.19	**<0.01**	1.04	0.77
		72 h	2.88	**<0.01**	1.24	0.12
		96 h	6.63	**<0.01**	0.79	0.39
	Middle period		43.12	**<0.01**	3.13	**<0.01**
	Late period		19.13	**<0.01**	3.95	**<0.01**

Abbreviations: BP = bacterial production; BA = bacterial abundance.

the magnitude of the response (effect size) to CP treatment varied over the early period; consequently, the kind of co-limitation shifted from additive (ICLA) to synergistic (ICLSpA). Over the middle period, BA responded slightly to the P and strongly to CP treatments (Figure 2C; Tables 2 and 3), generating a synergistic co-limitation (SSL-P). Over the late period, the response of BP to CP treatment decreased and, therefore, co-limitation was replaced by single limitation by P (Table 3).

In the oligotrophic ecosystem, BA followed a completely different dynamic, due to the lack of response to resource addition until the middle period. BA positively responded to P and CP treatments from the middle to late period (Figure 2D; Tables 3 and 4), which generated a synergistic co-limitation (SSL-P).

Figure 3 depicts the dynamics of nanoplanktonic community (autotrophic picoplankton was not detected) and its response to resource additions over the middle and late period in each ecosystem. In the eutrophic ecosystem, algal biomass showed the highest values over the late period within each treatment. In contrast to initial conditions, strict autotrophic algae dominated algal biomass over the middle and late periods, regardless of the resource addition (up to 91% of total algal biomass), although mixotrophic algae developed only slightly over the late period in all treatments (reaching up to 8.8% of algal biomass; Figure 3A). Ciliates developed in all treatments, except in control (ciliate biomass was 86.7-fold in C, 50-fold in P, and 23-fold in CP treatment) over the middle period. However, over the long period, ciliates vanished in the C treatment or sharply diminished in the P treatment (coinciding with the development of mixotrophic algae), but only slightly decreased in the CP treatment. By contrast, in the oligotrophic ecosystem, ciliates were not detected, and strict autotrophic algae became the dominant group (Figure 3B) in all experimental treatments until the middle period. Only over the late period and under P-added conditions (P and CP treatments) did mixotrophs strongly develop, reaching up to 93.5% (P treatment) or 84% (CP treatment) of algal biomass. As expected, algal biomass strongly increased under P-added conditions at the middle period (up to 155.48 $\mu g\ C\ L^{-1}$ in P treatment), particularly in autotrophic algae (i.e. 7.56-fold in P and 7.60-fold in CP). However, over the late period, algal biomass decreased in all treatments.

SEM Analysis

The SEM provided a good fit with the observed data for each ecosystem, as indicated by the non-significant χ^2 (p-value>0.1), and by goodness-of-fit indices (NFI and GFI>0.9 in each case; Figure 4). In the eutrophic ecosystem, SEM showed significant and positive standardized path coefficients for TDP and DOC regarding their effects on BA (0.598 and 0.580 respectively; p-values<0.000) while non-significant coefficients were found regarding their direct effects on potential predators (p-values> 0.05). Besides, the effect of the predators to BA was negative (-0.343; p-value<0.01). In the oligotrophic ecosystem, both resources were positively correlated (p-value>0.05), and therefore this correlation was included in the SEM, which was simpler than that of the eutrophic ecosystem. In this case, the resources affected BA but did not affect the mixotrophs. Standardized path coefficients for TDP and DOC regarding their effects on BA were positive and significant or marginally significant (TDP: 0.625, p-value<0.001; DOC: 0.460, p-value<0.05). Contrarily to the eutrophic ecosystem, the standardized path coefficient between mixotrophs and BA was positive and significant (0.716, p-value< 0.001).

Discussion

This study contributes new insights on resource co-limitation of bacterioplankton by two different resources, i.e. organic carbon (energetic resource) and phosphorus (mineral nutrient) in freshwater ecosystems, particularly in a largely unstudied ecoregion. Thus, through a modified index (adapted to bacterioplankton) based on that reported by [47], we quantified the type of resource co-limitation, and its dynamics over time. In addition, the trophic-nature gradient (i.e. autotrophy-heterotrophy axis) proves to be a key feature determining the expected types of resource co-limitation of bacteria, which are summarized in a proposed theoretical framework, based on the types found in the literature for primary producers [45–47,49,81]. On this basis, the actual types of resource co-limitation, and their dynamics, highlight the major role exerted by ecological interactions, mainly predation, altering the expected responses of bacterioplankton to limiting resources.

The Resource (C vs. P) that Mainly Limited Bacterioplankton

Based on the IRR index (which measures the relative magnitude of the bacterial responses to joint C and P addition compared to single-resource addition), our experimental test of resource limitation consistently showed that both resources (C and P) co-limited bacterioplankton in two ecosystems with contrasting trophic states. In this regard, the SEM for each ecosystem showed a positive path-analysis effect of both resource dynamics on bacterioplankton (BA). Therefore, our results extend to bacterioplankton the prevalence of resource co-limitation found for primary producers [45,47] and agree with the current criticism regarding the broad relevance of Liebig's law of the minimum for more complex organizational levels than individual organisms [81,82].

Through an analysis of the bacterial response to single-resource additions, we distinguish which of them mainly limited bacteria in each ecosystem. Overall, our results regarding BP and BA responses contradicted our initial hypothesis and the conclusions of [32] that C (not P) was the main limiting resource for bacteria in oligotrophic ecosystems. That is, in our study, bacteria did not respond to C but did to P treatments in the oligotrophic ecosystem. Nevertheless, our results agree with the conclusions of [32] stating that mainly P and secondarily C were co-limiting resources under eutrophic conditions because, in the eutrophic ecosystem of our study, bacteria responded to P and C treatments. Notably, the response to the P treatment in most cases had a greater magnitude (effect size) than in C treatment. These interpretations of resource limitation are also supported by the values shown by the BCD:EOC ratio, a proxy for bacterial limitation of autochthonous C. This proxy measures the ability of algal C (EOC) to meet bacterial demands for C, based on the reported bacterial preference for this C source [25,26,32]. Thus, the BCD:EOC ratio indicates that bacteria were limited by autochthonous C in the eutrophic ecosystem (BCD:EOC>1), but not in the oligotrophic lake (BCD:EOC<1) at starting conditions and over the late period.

The discrepancies regarding the expected bacterial limitations can be explained by the trophic nature (i.e. autotrophy-heterotrophy) of the ecosystems studied. Thus, the eutrophic ecosystem was heterotrophic, and the BCD:EOC ratio indicated that the EOC was not sufficient to support bacterial demands; by contrast, the oligotrophic lake was autotrophic, and the BCD:EOC ratio showed the opposite scenario of (autochthonous) C sufficiency. In this regard, the trophic nature of the two ecosystems, measured as

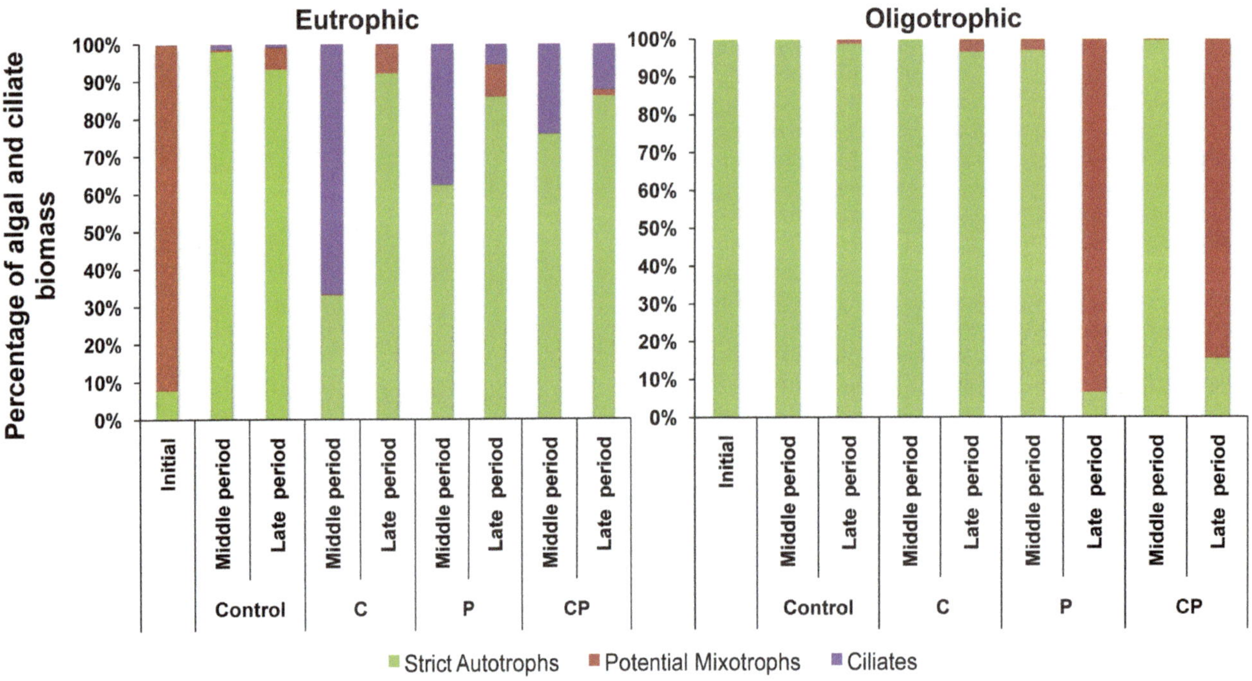

Figure 3. Experimental dynamics of algal and ciliate biomass in the two ecosystems. (A) eutrophic ecosystem; (B) oligotrophic ecosystem. Columns represent the percentage of mean biomass values of the different planktonic functional groups (strict autotrophs, potential mixotrophs and ciliate) over the experiment.

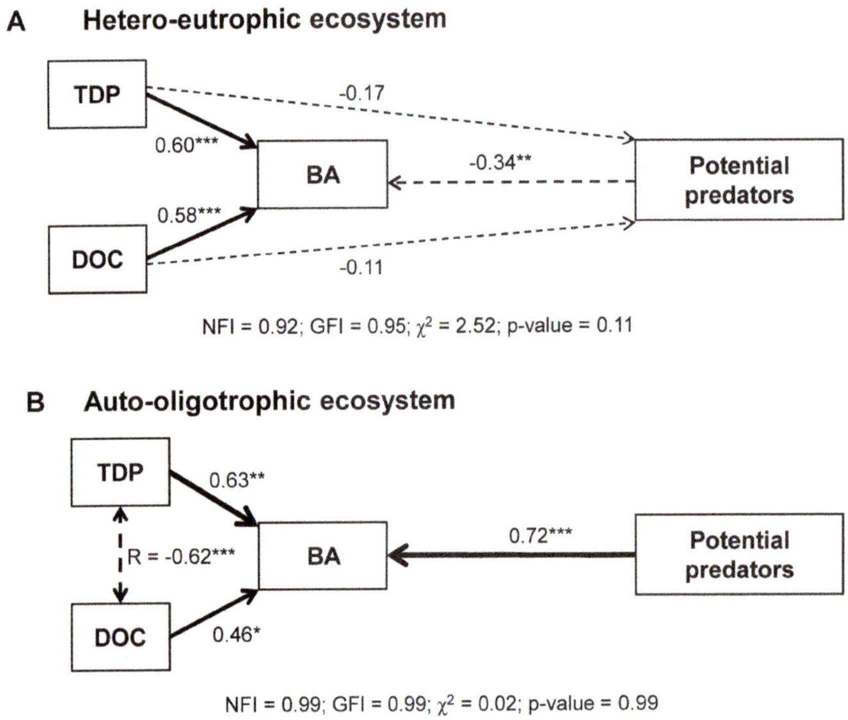

Figure 4. Path diagram representing the relationships between the dynamics of resources, bacterial abundance, and potential predators (mixotrophs and ciliates) for each ecosystem. One-headed arrows depict causal relationships whereas two-headed arrows depict correlations. Positive effects are indicated by solid lines and negative effects by dashed lines. Arrow widths are proportional to path coefficients. Numbers near the paths indicate standardized path coefficients. Significance path p-values are denoted with *p-value<0.05, **p-value<0.01, ***p-value<0.001, while non-significant paths are denoted with the superscript 'n.s.'. Fit statistics (Goodness of fit index, GFI; normal fit index, NGI; χ^2; p-value).

GPP:R or PP:R ratios, deviated from the expected pattern for the Mediterranean inland waters with respect to the trophic nature linked to the trophic gradient (i.e. increasing autotrophy towards eutrophy, [32]). It is important to point out that the trophic nature of the ecosystems studied was appropriately estimated by the quantification of respiration in the two major fractions of the planktonic community (i.e. community and picoplankton) and the lack of autotrophic picoplankton in both ecosystems. Therefore, we consider that our estimation of the trophic nature of the two ecosystems through the GPP:R ratio, from the direct measurements of respiration of both fractions, is more realistic than if it had been estimated from PP:BP or BR:BP ratios with BR determined from empirical models, as in [32]. In this sense, [83] underline the importance of the evaluation of both primary production and key heterotrophic processes (e.g. respiration) necessary to define autotrophic and heterotrophic states that expand the concept of the trophic state in aquatic ecosystems. In this regard, the correspondence between (autochthonous) C sufficiency for bacteria in the autotrophic ecosystem and the opposite scenario in the heterotrophic ecosystem in our study lead us to conclude that the trophic-nature gradient, rather than trophic-state gradient (i.e. oligotrophy-eutrophy axis, based in TP), can have a major role determining the resource-limitation patterns. Consequently, based on the trophic-nature gradient, we propose a theoretical framework (Figure 5) to distinguish among the main types of co-limitation expected for bacteria and described in the literature for primary producers [45–47]. In addition, we propose a unifying classification from the diversity of types of co-limitation reported in the literature, as summarized in Figure S1. In this context, our findings disagree with the conclusions of [32] because the relation between the trophic-nature and trophic-state gradients (i.e. more autotrophy with eutrophy) does not necessarily hold in the Mediterranean ecoregion, as our more accurate results regarding autotrophy or heterotrophy reveals. Additionally, our experimental findings confirm the intersystem variability existing in the Mediterranean ecoregion, even at local scale (e.g. oligotrophic-autotrophic vs. eutrophic-heterotrophic ecosystems).

The Potential Effect of Ecological Interactions on the type of Co-limitation Over Time

With respect to the fastest and most responsive bacterial variable to resource addition (i.e. BP), our results fit the expected theoretical types of co-limitation depicted in Figure 5, although with a lesser degree of synergism. Thus, in the eutrophic-heterotrophic ecosystem, the maximum BP response to CP treatment (over the early period) generated an additive co-limitation (CP>C>P>Control; ICLA), partially deviating from the theoretical expectations of synergistic co-limitation (CP>>C>P>Control; ICLSpA; Figures 5 and S1). Plausible mechanisms explaining these deviations are ecological interactions that also regulate the bacterioplankton, such as predation, the role of which as an external force to physiological limitations in promoting antagonistic-like co-limitations was also discussed by [46]. In our study, this interpretation is supported by the negative path-analysis effect that the dynamics of mixotrophic algae and ciliates exert on the structural bacterioplankton variable (BA). Moreover, this interpretation also agrees with the delay in the transference of the predominant type of co-limitation from function (BP, cell division) to structure (BA); thus SSL-P emerged for BP over the early period, but was delayed for BA until the middle period. The uncoupling between structural and functional variables has widely been described in the literature as evidence of top-down regulation (from [7]), also shown through empirical models (see [8]) describing how regulation of bacteria can shift over time, with a relative stronger regulation by predation at early stages. In this view, the role of nutrient regeneration mediated by grazers can be a major mechanism promoting the maintenance of predator-prey system [8,84], which accounted for the bacterial responses found over middle and late periods. Thus, it is remarkable that the type of synergistic co-limitation found for BP in the eutrophic-heterotrophic ecosystem shifted from SSL-P (CP>P>C = Control) to SCL (CP>>P = C = Control, Figure S1) over the middle period, and even that synergistic co-limitation became absolute antagonistic co-limitation (ICLSbA) over the late period, an infrequent type of co-limitation strongly influenced by extrinsic forces [46].

In the oligotrophic-autotrophic ecosystem, the fact that the synergistic co-limitation for both BP and BA was delayed until the middle period, when TDP was already exhausted, may be due to the lower water temperature instead of to a severe top-down control. It is known that low temperature (<15°C) can limit bacterial productivity [85] and diminish the positive response of bacteria to limiting nutrient [86,87], while mixotrophs and ciliates (the main predators on bacteria in this ecosystem, [41,52]) were minor nanoplanktonic components from initial conditions to middle period. In this respect, the SEM shows a positive path-analysis effect of mixotrophic algal dynamics on BA, supporting the idea that the predatory control was negligible. This fact could promote or, at least not hinder, the coupling of synergistic co-limitation (SSL-P) for BA and BP over the middle period.

In conclusion, our work shows the predominance of carbon and phosphorus co-limitation of heterotrophic bacteria, and that the types of co-limitation shifted over time, partially deviating from the theoretical expectations. This deviation may be due to modulation by extrinsic ecological forces, such as predatory control. Our results indicate the difficulty of accurate predicting which resource mainly limits organisms based only on their physiological limitations. Our approach helps explain which type of resource co-limitation actually occurs in ecosystems with contrasting trophic states, underlining the preponderant role of the trophic nature over the trophic state.

Ecological Implications

Based on our results, the nature of the resource that mainly limits bacteria depending on the trophic nature of their host ecosystems is an important trait that may determine whether freshwaters act as a net C source or sink. Thus, eutrophic-heterotrophic ecosystems in which heterotrophic processes predominate (GPP<R) and with bacteria capable of consuming C (alone or combined with P) may act as net C sources through bacterial respiration. By contrast, oligotrophic ecosystems where autotrophic processes predominate (GPP>R) and with bacteria unable to respond to C alone, may act as a net C sink. Moreover, in the oligotrophic-autotrophic ecosystems, inputs of organic carbon and inorganic nutrients would not alter their behaviour as net C sinks, because their autotrophic feature (GPP>R) would remain unchanged even under these resource-enriched conditions. Therefore, in these autotrophic ecosystems, the contribution of bacteria in C accumulation may be lower than in the eutrophic-heterotrophic ecosystems, due to the high GPP:R ratio, showing the major contribution of primary producers. Our work is in line with the increasingly recognized role of freshwater ecosystems in the global C cycle [88], and highlights the need for more attention to be placed on the heterotrophic microbial compartment because of their key role in freshwaters, particularly in the heterotrophic ecosystems.

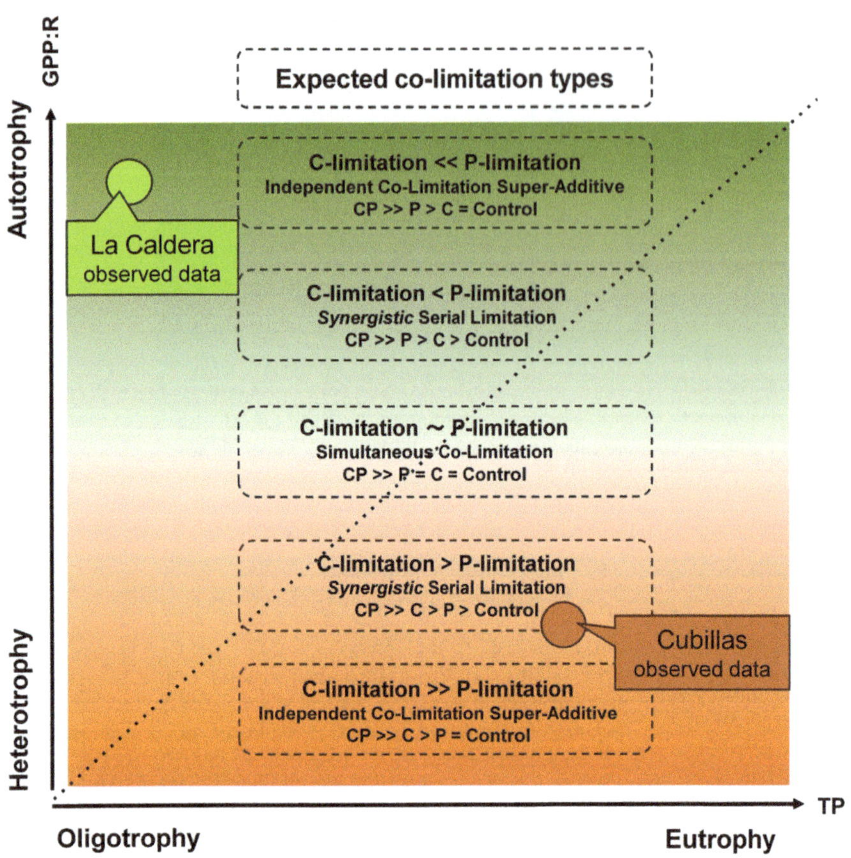

Figure 5. Theoretical diagram of expected types of C and P co-limitation of bacteria based on the trophic-nature gradient. The diagram represents the different types of expected co-limitation according to the trophic-nature gradient (Y-axis), indicated by the colour gradient from dark brown (maximum heterotrophy) to dark green (maximum autotrophy). The diagonal dashed line arrow crossing the diagram represents the trophic-state gradient (X-axis) in the sense of more autotrophy more eutrophy, as found in [32], which does not reflect the major role of trophic-nature gradient found in the present study. The squares with dashed lines include: (i) the single resource expected to be more limiting, according the trophic nature axis; (ii) the name of the expected co-limitation type *sensu* [47], and its meaning as effects of the treatments. The circles represent the position found for the ecosystems according to both axes; brown circle is the hetero-eutrophic ecosystem (Cubillas), green circle is the auto-oligotrophic ecosystem (La Caldera).

Remarkably, the prevalence of C and P co-limitation of bacteria in both types of ecosystems indicates that bacterioplankton is highly responsive to human increases in mineral nutrients and organic matter. This increment, therefore, may alter not only bacterial structure and function [89], but also scale to ecosystem functioning [90,91], ultimately deteriorating the ecosystem services that they provide [10,42].

Supporting Information

Figure S1 Flow chart of logical tests used to categorize the types of resource co-limitation for bacteria. The proposed types from the positive responses to both resources are depicted according to a modified classification based on [47] (black abbreviations) and [46] (blue text). See text for more details. Y or N correspond to "yes" or "no" (logical true or false). Synergistic co-limitation (SC) indicates that the response to the CP treatment is greater than the sum of the response to C alone and P alone; additive co-limitation (AD) indicates that the response to CP is equal to the sum of the response to C alone and P alone; antagonistic co-limitation (AC) indicates that the response to CP is greater than that of either C or P alone, but not their sum; absolute antagonism (AA) indicates that the CP response is less than that of either C or P alone. ICLSpA = Independent Co-Limitation Supper-Additive; *S*SL = *Synergistic* Serial Limitation; SCL = Simultaneous Co-Limitation; ICLA = Independent Co-Limitation Additive; SL = Single Limitation; NR = No Response; ICLSbA = Independent Co-Limitation Sub-Additive; *A*SL = *Antagonistic* Serial Limitation; NI = Negative Interaction.

Table S1 Results from two-way RM-ANOVA and multivariate tests of Pillai, Hotelling, and Roy for dissolved organic carbon, total dissolved phosphorus and total dissolved nitrogen for both aquatic ecosystems. F values with their corresponding degrees of freedom and significance levels (p) are shown for each resource treatment and resource treatment × time in each variable. DOC = dissolved organic carbon; TDP = total dissolved phosphorus; TDN = total dissolved nitrogen.

Table S2 Results from two-way RM-ANOVA and multivariate tests of Pillai, Hotelling, and Roy of bacterial production and bacterial abundance for both aquatic ecosystems. F values with their corresponding degrees of freedom and significance levels (p) are shown for each resource treatment and resource treatment × time in each variable. BP = bacterial production; BA = bacterial abundance.

Table S3 Results from two-way ANOVA of the bacterial carbon demands: excretion of organic carbon ratio for both ecosystems. F values with their corresponding degrees of freedom and significance levels (p) are shown for each resource treatment and resource treatment × time in the bacterial carbon demand: excretion of organic carbon (BCD:EOC).

Acknowledgments

The authors are grateful to the staff of Cubillas reservoir for their invaluable help during the sample period and thank to Joaquina María García, Silvia Rosillo and Sergio Velasco for their assistance with the analysis of water samples. The authors appreciate the suggestions made by two anonymous reviewers that greatly improved this manuscript. Additionally, the authors also thank David Nesbitt for English writing assistance.

Author Contributions

Conceived and designed the experiments: IDG JMMS PC. Performed the experiments: IDG JMMS GH MJC PC. Analyzed the data: IDG GH MJC. Contributed reagents/materials/analysis tools: JMMS PC. Wrote the paper: IDG JMMS.

References

1. Pomeroy LR, leB Williams PJ, Azam F, Hobbie JE (2007) The Microbial Loop. Oceanography 20: 28–33.
2. Pomeroy LR, Wiebe WJ (2001) Temperature and substrates as interactive limiting factors for marine heterotrophic bacteria. Aquat Microb Ecol 23: 187–204.
3. Berggren M, Laudon H, Jonsson A, Jansson M (2010) Nutrient Constraints on Metabolism Affect the Temperature Regulation of Aquatic Bacterial Growth Efficiency. Microb Ecol 60: 894–902. doi:10.1007/s00248-010-9751-1.
4. Degerman R, Dinasquet J, Riemann L, Sjöstedt de Luna S, Andersson A (2013) Effect of resource availability on bacterial community responses to increased temperature. Aquat Microb Ecol 68: 131–142. doi:10.3354/ame01609.
5. Hoikkala L, Aarnos H, Lignell R (2009) Changes in Nutrient and Carbon Availability and Temperature as Factors Controlling Bacterial Growth in the Northern Baltic Sea. Estuaries Coasts 32: 720–733. doi:10.1007/s12237-009-9154-z.
6. Solic M, Krstulovic N, Vilibic I, Bojanic N, Kuspipilic G, et al. (2009) Variability in the bottom-up and top-down controls of bacteria on trophic and temporal scales in the middle Adriatic Sea. Aquat Microb Ecol 58: 15–29. doi:10.3354/ame01342.
7. Billen G, Servais P, Becquevort S (1990) Dynamics of bacterioplankton in oligotrophic and eutrophic aquatic environments: bottom-up or top-down control? Hydrobiologia 207: 37–42.
8. Pernthaler J (2005) Predation on prokaryotes in the water column and its ecological implications. Nat Rev Micro 3: 537–546. doi:10.1038/nrmicro1180.
9. Herndl GJ, Agogué H, Baltar F, Reinthaler T, Sintes E, et al. (2008) Regulation of aquatic microbial processes: the "microbial loop" of the sunlit surface waters and the dark ocean dissected. Aquat Microb Ecol 53: 59–68. doi:10.3354/ame01225.
10. Ducklow H (2008) Microbial services: challenges for microbial ecologists in a changing world. Aquat Microb Ecol 53: 13–19. doi:10.3354/ame01220.
11. Bell T, Newman JA, Silverman BW, Turner SL, Lilley AK (2005) The contribution of species richness and composition to bacterial services. Nature 436: 1157–1160. doi:10.1038/nature03891.
12. Carlson CA, Giovannoni SJ, Hansell DA, Goldberg SJ, Parsons R, et al. (2002) Effect of nutrient amendments on bacterioplankton production, community structure, and DOC utilization in the northwestern Sargasso Sea. Aquat Microb Ecol 30: 19–36.
13. Vidal LO, Granéli W, Daniel CB, Heiberg L, Roland F (2011) Carbon and phosphorus regulating bacterial metabolism in oligotrophic boreal lakes. J Plankton Res 33: 1747–1756. doi:10.1093/plankt/fbr059.
14. Pomeroy LR (1974) The ocean's food web, a changing paradigm. BioSciences: 499–504.
15. Azam F, Fenchel T, Field JG, Gray JS, Meyer-Reil LA, et al. (1983) The Ecological Role of Water-Column Microbes in the Sea. Mar Ecol Prog Ser 10: 257–263.
16. Farjalla VF, Amado AM, Suhett AL, Meirelles-Pereira F (2009) DOC removal paradigms in highly humic aquatic ecosystems. Environ Sci Pollut Res 16: 531–538. doi:10.1007/s11356-009-0165-x.
17. Vadstein O, Olsen LM, Busch A, Andersen T, Reinertsen HR (2003) Is phosphorus limitation of planktonic heterotrophic bacteria and accumulation of degradable DOC a normal phenomenon in phosphorus-limited systems? A microcosm study. FEMS Microbiol Ecol 46: 307–316. doi:10.1016/S0168-6496(03)00195-8.
18. Thingstad TF, Hagström Å, Rassoulzadegan F (1997) Accumulation of degradable DOC in surface waters: Is it caused by a malfunctioning microbial loop? Limnol Oceanogr: 398–404.
19. Cotner JB, Biddanda BA (2002) Small Players, Large Role: Microbial Influence on Biogeochemical Processes in Pelagic Aquatic Ecosystems. Ecosystems 5: 105–121. doi:10.1007/s10021-001-0059-3.
20. Bertoni R, Callieri C, Balseiro E, Modenutti B (2008) Susceptibility of bacterioplankton to nutrient enrichment of oligotrophic and ultraoligotrophic lake waters. J Limnol 67: 120–127.
21. Jansson M, Bergström A-K, Lymer D, Vrede K, Karlsson J (2006) Bacterioplankton Growth and Nutrient Use Efficiencies Under Variable Organic Carbon and Inorganic Phosphorus Ratios. Microb Ecol 52: 358–364. doi:10.1007/s00248-006-9013-4.
22. Vadstein O (2000) Heterotrophic, planktonic bacteria and cycling of phosphorus: Phosphorus requirements, competitive ability and food web interactions. Adv Microb Ecol: 115–168.
23. Stets EG, Cotner JB (2008) The influence of dissolved organic carbon on bacterial phosphorus uptake and bacteria-phytoplankton dynamics in two Minnesota lakes. Limnol Oceanogr: 137–147.
24. Kritzberg ES, Cole JJ, Pace MM, Granéli W (2005) Does autochthonous primary production drive variability in bacterial metabolism and growth efficiency in lakes dominated by terrestrial C inputs? Aquat Microb Ecol 38: 103–111.
25. Gasol JM, Vázquez-Domínguez E, Vaqué D, Agustí S, Duarte CM (2009) Bacterial activity and diffusive nutrient supply in the oligotrophic Central Atlantic Ocean. Aquat Microb Ecol 56: 1–12. doi:10.3354/ame01310.
26. Carlsson P, Granéli E, Granéli W, Rodriguez EG, de Carvalho WF, et al. (2012) Bacterial and phytoplankton nutrient limitation in tropical marine waters, and a coastal lake in Brazil. J Exp Mar Bio Ecol 418–419: 37–45. doi:10.1016/j.jembe.2012.03.012.
27. Kritzberg ES, Cole JJ, Pace MM, Granéli W (2006) Bacterial Growth on Allochthonous Carbon in Humic and Nutrient-enriched Lakes: Results from Whole-Lake ^{13}C Addition Experiments. Ecosystems 9: 489–499. doi:10.1007/s10021-005-0115-5.
28. Cherrier J, Bauer JE, Druffel ER, Coffin RB, Chanton JP (1999) Radiocarbon in marine bacteria: Evidence for the ages of assimilated carbon. Limnol Oceanogr 44: 730–736.
29. Granéli W, Bertilsson S, Philibert A (2004) Phosphorus limitation of bacterial growth in high Arctic lakes and ponds. Aquat Sci 66: 430–439. doi:10.1007/s00027-004-0732-7.
30. Farjalla VF, Esteves FA, Bozelli RL, Roland F (2002) Nutrient limitation of bacterial production in clear water Amazonian ecosystems. Hydrobiologia 489: 197–205.
31. Farjalla VF, Azevedo DA, Esteves FA, Bozelli RL, Roland F, et al. (2006) Influence of Hydrological Pulse on Bacterial Growth and DOC Uptake in a Clear-Water Amazonian Lake. Microb Ecol 52: 334–344. doi:10.1007/s00248-006-9021-4.
32. Medina-Sánchez JM, Carrillo P, Delgado-Molina JA, Bullejos FJ, Villar-Argaiz M (2010) Patterns of resource limitation of bacteria along a trophic gradient in Mediterranean inland waters. FEMS Microbiol Ecol 74: 554–565. doi:10.1111/j.1574-6941.2010.00969.x.
33. Ortega-Retuerta E, Pulido-Villena E, Reche I (2007) Effects of Dissolved Organic Matter Photoproducts and Mineral Nutrient Supply on Bacterial Growth in Mediterranean Inland Waters. Microb Ecol 54: 161–169. doi:10.1007/s00248-006-9186-x.
34. Giorgi F, Lionello P (2008) Climate change projections for the Mediterranean region. Glob Planet Change 63: 90–104. doi:10.1016/j.gloplacha.2007.09.005.
35. IPCC (2013) Climate Change 2013. Summary for Policymakers. Stocker TF, Qin D, Plattner G-K, Tignor M, Allen SK, et al., editors Cambridge, United Kindom and New York, NY, USA: Cambridge University Press.
36. Xenopoulos MA, Lodge DM, Frentress J, Kreps TA, Bridgham SD, et al. (2003) Regional comparisons of watershed determinants of dissolved organic carbon in temperate lakes from the Upper Great Lakes region and selected regions globally. Limnol Oceanogr 48: 2321–2334.
37. Morales-Baquero R, Pulido-Villena E, Reche I (2006) Atmospheric inputs of phosphorus and nitrogen to the southwest Mediterranean region: Biogeochemical responses of high mountain lakes. Limnol Oceanogr 51: 830–837.
38. Reche I, Ortega-Retuerta E, Romera O, Pulido-Villena E, Morales-Baquero R, et al. (2009) Effect of Saharan dust inputs on bacterial activity and community composition in Mediterranean lakes and reservoirs. Limnol Oceanogr 54: 869–879.
39. Mladenov N, Sommaruga R, Morales-Baquero R, Laurion I, Camarero L, et al. (2011) Dust inputs and bacteria influence dissolved organic matter in clear alpine lakes. Nat Commun 2: 405–407. doi:10.1038/ncomms1411.

40. de Vicente I, Ortega-Retuerta E, Morales-Baquero R, Reche I (2012) Contribution of dust inputs to dissolved organic carbon and water transparency in Mediterranean reservoirs. Biogeosciences 9: 5049–5060. doi:10.5194/bg-9-5049-2012.
41. Medina-Sánchez JM, Delgado-Molina JA, Bratbak G, Bullejos FJ, Villar-Argaiz M, et al. (2013) Maximum in the Middle: Nonlinear Response of Microbial Plankton to Ultraviolet Radiation and Phosphorus. PLoS ONE 8: e60223. doi:10.1371/journal.pone.0060223.t001.
42. Smith VH (2003) Eutrophication of freshwater and coastal marine ecosystems a global problem. Environ Sci Pollut Res 10: 126–139. doi:10.1065/espr2002.12.142.
43. Vrede K (2005) Nutrient and Temperature Limitation of Bacterioplankton Growth in Temperate Lakes. Microb Ecol 49: 245–256. doi:10.1007/s00248-004-0259-4.
44. Saito MA, Goepfert TJ, Ritt JT (2008) Some thoughts on the concept of colimitation: Three definitions and the importance of bioavailability. Limnol Oceanogr 53: 276.
45. Elser JJ, Kyle M, Steger L, Nydick KR, Baron JS (2009) Nutrient availability and phytoplankton nutrient limitation across a gradient of atmospheric nitrogen deposition. Ecology 90: 3062–3073.
46. Allgeier JE, Rosemond AD, Layman CA (2011) The frequency and magnitude of non-additive responses to multiple nutrient enrichment. J Appl Ecol 48: 96–101. doi:10.1111/j.1365-2664.2010.01894.x.
47. Harpole WS, Ngai JT, Cleland EE, Seabloom EW, Borer ET, et al. (2011) Nutrient co-limitation of primary producer communities. Ecol Lett 14: 852–862.
48. Del Giorgio PA, Cole JJ (1998) Bacterial growth efficiency in natural aquatic systems. Annu Rev Ecol Syst: 503–541.
49. Elser JJ, Bracken MES, Cleland EE, Gruner DS, Harpole WS, et al. (2007) Global analysis of nitrogen and phosphorus limitation of primary producers in freshwater, marine and terrestrial ecosystems. Ecol Lett 10: 1135–1142. doi:10.1111/j.1461-0248.2007.01113.x.
50. Nürnberg GK (1996) Trophic State of Clear and Colored, Soft- and Hardwater Lakes with Special Consideration of Nutrients, Anoxia, Phytoplankton and Fish. Lake Reserv Manag 12: 432–447. doi:10.1080/07438149609354283.
51. Reche I, Pulido-Villena E, Conde-Porcuna JM, Carrillo P (2001) Photoreactivity of dissolved organic matter from high-mountain lakes of Sierra Nevada, Spain. Arct Antarctic Alp Res: 426–434.
52. Medina-Sánchez JM, Villar-Argaiz M, Carrillo P (2004) Neither with nor without you: A complex algal control on bacterioplankton in a high mountain lake. Limnol Oceanogr: 1722–1733.
53. Jessup CM, Kassen R, Forde SE, Kerr B, Buckling A, et al. (2004) Big questions, small worlds: microbial model systems in ecology. Trends in Ecology & Evolution 19: 189–197. doi:10.1016/j.tree.2004.01.008.
54. Benton TG, Solan M, Travis JMJ, Sait SM (2007) Microcosm experiments can inform global ecological problems. Trends in Ecology & Evolution 22: 516–521. doi:10.1016/j.tree.2007.08.003.
55. Coelho FJRC, Santos AL, Coimbra J, Almeida A, Cunha Â, et al. (2013) Interactive effects of global climate change and pollution on marine microbes: the way ahead. Ecol Evol 3: 1808–1818. doi:10.1002/ece3.565.
56. Martínez R, Canteras JC, Cruz L (1975) Limnologic study of the Cubillas reservoir chemical characteristics of the water. Cuadernos de Ciencias Biológicas Universidad de Granada 4: 223–238.
57. Villar-Argaiz M, Medina-Sánchez JM, Carrillo P (2002) Microbial plankton response to contrasting climatic conditions: insights from community structure, productivity and fraction stoichiometry. Aquat Microb Ecol 29: 253–266.
58. APHA (1992) Standard methods for the examination of water and wastewater, 18th ed. Washington: American Public Health Association.
59. Benner R, Strom M (1993) A critical evaluation of the analytical blank associated with DOC measurements by high-temperature catalytic oxidation. Mar Chem 41: 153–160.
60. Fuhrman JA, Azam F (1982) Thymidine incorporation as a measure of heterotrophic bacterioplankton production in marine surface waters: evaluation and field results. Mar Chem 66: 109–120.
61. Smith DC, Azam F (1992) A simple, economical method for measuring bacterial protein synthesis rates in seawater using 3H-leucine. Mar Microb Food Webs 6: 107–114.
62. Bell RT (1993) Estimating production of heterotrophic bacterioplankton via incorporation of tritiated thymidine. In: Kemp PF, Sherr BF, Sherr EB, Cole JJ, editors. Handbook of Methods in Aquatic Microbial Ecology. Lewis Publishers, Boca Raton, FL. 495–503.
63. Lee S, Fuhrman JA (1987) Relationships between Biovolume and Biomass of Naturally Derived Marine Bacterioplankton. Appl Environ Microbiol 53: 1298–1303.
64. Steemann-Nielsen E (1952) The use of Radio-active Carbon (C14) for measuring organic production in the sea. J Cons Int Explor Mer 18: 117–140. doi:10.1093/icesjms/18.2.117.
65. Carrillo P, Medina-Sánchez JM, Villar-Argaiz M (2002) The interaction of phytoplankton and bacteria in a high mountain lake: Importance of the spectral composition of solar radiation. Limnol Oceanogr: 1294–1306.
66. Anesio AM, Granéli W, Aiken GR, Kieber DJ, Mopper K (2005) Effect of Humic Substance Photodegradation on Bacterial Growth and Respiration in Lake Water. Appl Environ Microbiol 71: 6267–6275. doi:10.1128/AEM.71.10.6267-6275.2005.
67. Medina-Sánchez JM, Villar-Argaiz M, Carrillo P (2002) Modulation of the bacterial response to spectral solar radiation by algae and limiting nutrients. Freshwater Biol 47: 2191–2204. doi:10.1046/j.1365-2427.2002.00969.x.
68. Kamiya E, Izumiyama S, Nishimura M, Mitchell JG, Kogure K (2007) Effects of fixation and storage on flow cytometric analysis of marine bacteria. J Oceanogr 63: 101–112.
69. Zubkov MV, Burkill PH, Topping JN (2007) Flow cytometric enumeration of DNA-stained oceanic planktonic protists. J Plankton Res 29: 79–86. doi:10.1093/plankt/fbl059.
70. Gasol JM, Del Giorgio PA (2000) Using flow cytometry for counting natural planktonic bacteria and understanding the structure of planktonic bacterial communities. Sci Mar 64: 197–224.
71. Zubkov MV, Burkill PH (2006) Syringe pumped high speed flow cytometry of oceanic phytoplankton. Cytometry A 69A: 1010–1019. doi:10.1002/cyto.a.20332.
72. Rott E (1981) Some results from phytoplankton counting intercalibrations. Aquatic Sci 43: 34–62.
73. Lund J, Kipling C, Le Cren ED (1958) The inverted microscope method of estimating algal numbers and the statistical basis of estimations by counting. Hydrobiologia 11: 143–170.
74. Sun J, Liu D (2003) Geometric models for calculating cell biovolume and surface area for phytoplankton. J Plankton Res 25: 1331–1346. doi:10.1093/plankt/fbg096.
75. Olenina I, Hajdu S, Edler L, Andersson A, Wasmund N, et al. (2006) Biovolumes and size-classes of phytoplankton in the Baltic Sea. Baltic Sea environment proceedings 106: 1–144.
76. Rocha O, Duncan A (1985) The relationship between cell carbon and cell volume in freshwater algal species used in zooplanktonic studies. J Plankton Res 7: 279–294.
77. Putt M, Stoecker DK (1989) An experimentally determined carbon: volume ratio for marine "oligotrichous" ciliates from estuarine and coastal waters. Limnol Oceanogr: 1097–1103.
78. Tanaka JS (1987) How big is enough?: sample size and goodness fit in structural equation models with latent variables. Child Development: 134–146.
79. Bollen KA (1989) Structural Equations with Latent Variables. Barnett V, Bradley RA, Hunter JS, Kadane JB, Kendall DG, et al., editors Wiley Series in Probability and Mathematical Statistics.
80. Lohelin JC (1992) Latent variable models: An introduction to factor, path, and structural equation analysis. Mahwah NJ, editor Hillsdale, New Yersy: Lawrence Erlbaum.
81. Arrigo KR (2005) Marine microorganisms and global nutrient cycles. Nature 437: 349–355.
82. Danger M, Daufresne T, Lucas F, Pissard S, Lacroix G (2008) Does Liebig's law of the minimum scale up from species to communities? Oikos 117: 1741–1751. doi:10.1111/j.1600-0706.2008.16793.x.
83. Dodds WK, Cole JJ (2007) Expanding the concept of trophic state in aquatic ecosystems: It's not just the autotrophs. Aquat Sci 69: 427–439. doi:10.1007/s00027-007-0922-1.
84. Sherr EB, Sherr BF (2002) Significance of predation by protists in aquatic microbial food webs. Antonie Van Leeuwenhoek 81: 293–308.
85. Apple JK, Del Giorgio PA, Kemp W (2006) Temperature regulation of bacterial production, respiration, and growth efficiency in a temperate salt-marsh estuary. Aquat Microb Ecol 43: 243–254.
86. Felip M, Pace ML, Cole JJ (1996) Regulation of planktonic bacterial growth rates: the effects of temperature and resources. Microb Ecol 31: 15–28.
87. Hall EK, Cotner JB (2007) Interactive effect of temperature and resources on carbon cycling by freshwater bacterioplankton communities. Aquat Microb Ecol 49: 35–45. doi:10.3354/ame01124.
88. Cole JJ, Prairie YT, Caraco NF, McDowell WH, Tranvik LJ, et al. (2007) Plumbing the Global Carbon Cycle: Integrating Inland Waters into the Terrestrial Carbon Budget. Ecosystems 10: 172–185. doi:10.1007/s10021-006-9013-8.
89. Faithfull CL, Bergström AK, Vrede T (2011) Effects of nutrients and physical lake characteristics on bacterial and phytoplankton production: A meta-analysis. Limnol Oceanogr 56: 1703–1713. doi:10.4319/lo.2011.56.5.1703.
90. Walther GR (2010) Community and ecosystem responses to recent climate change. Philos Trans R Soc Lond B: Biol Sci 365: 2019–2024. doi:10.1016/j.tree.2009.06.008.
91. Woodward G, Perkins DM, Brown LE (2010) Climate change and freshwater ecosystems: impacts across multiple levels of organization. Philos Trans R Soc Lond B: Biol Sci 365: 2093–2106. doi:10.1016/j.tree.2005.04.005.

7

Microbial Functional Potential and Community Composition in Permafrost-Affected Soils of the NW Canadian Arctic

Béatrice A. Frank-Fahle[1¤], Étienne Yergeau[2], Charles W. Greer[2], Hugues Lantuit[1,3], Dirk Wagner[4*]

1 Alfred Wegener Institute, Helmholtz Centre for Polar and Marine Research, Research Unit Potsdam, Potsdam, Germany, **2** National Research Council, Montréal, Quebec, Canada, **3** University of Potsdam, Potsdam, Germany, **4** GFZ German Center for Geosciences, Section 4.5 Geomicrobiology, Potsdam, Germany

Abstract

Permafrost-affected soils are among the most obvious ecosystems in which current microbial controls on organic matter decomposition are changing as a result of global warming. Warmer conditions in polygonal tundra will lead to a deepening of the seasonal active layer, provoking changes in microbial processes and possibly resulting in exacerbated carbon degradation under increasing anoxic conditions. To identify current microbial assemblages in carbon rich, water saturated permafrost environments, four polygonal tundra sites were investigated on Herschel Island and the Yukon Coast, Western Canadian Arctic. Ion Torrent sequencing of bacterial and archaeal 16S rRNA amplicons revealed the presence of all major microbial soil groups and indicated a local, vertical heterogeneity of the polygonal tundra soil community with increasing depth. Microbial diversity was found to be highest in the surface layers, decreasing towards the permafrost table. Quantitative PCR analysis of functional genes involved in carbon and nitrogen-cycling revealed a high functional potential in the surface layers, decreasing with increasing active layer depth. We observed that soil properties driving microbial diversity and functional potential varied in each study site. These results highlight the small-scale heterogeneity of geomorphologically comparable sites, greatly restricting generalizations about the fate of permafrost-affected environments in a warming Arctic.

Editor: Jonathan H. Badger, J. Craig Venter Institute, United States of America

Funding: This study was funded by the "International Cooperation in Education and Research" program of the International Bureau of the German Federal Ministry of Education and Research (BMBF) and through a doctoral scholarship to Béatrice A. Frank-Fahle from the German Environmental Foundation (DBU). The funders had no role in study design, data collection and analysis, decision to publish, or preparation of the manuscript.

Competing Interests: The authors have declared that no competing interests exist.

* E-mail: dirk.wagner@gfz-potsdam.de

¤ Current address: Institute of Groundwater Ecology, Helmholtz Zentrum München – German Research Center for Environmental Health, Neuherberg, Germany

Introduction

Currently there is significant scientific attention being dedicated to the effect of climate warming on permafrost environments, given that they play a crucial role within the global carbon cycle. Current controls of decomposition in such ecosystems are already changing as a result of surface warming, potentially exposing large amounts of previously conserved carbon during the next few decades [1]. Permafrost contains up to 50% of the belowground organic carbon stocks [2], of which a significant part could be lost within the next 100 years due to globally increasing temperatures [3]. Peat wetlands with patterned ground features cover approximately 250,000 m^2 of permafrost environments [4]. Low-center ice-wedge polygons, which are created by cryogenic processes associated with strong seasonal freeze-thaw cycles [5] typically have a depressed center and an elevated rim, creating a microrelief which affects the hydrology and organic carbon content of the soils. A thickening of the seasonally thawed layer (active layer) of polygons and an ensuing release of previously stored organic matter can stimulate microbial decomposition of this organic carbon, resulting in a positive feedback-loop for global warming [6]. Indeed, peat wetlands are an ideal environment for increased methane production because of the waterlogged, anoxic conditions that prevail in seasonally increasing thawed layers [7]. In wet tundra soils, methanogenesis is the terminal step in the anaerobic decomposition of organic matter and is solely driven by members of *Euryarchaeota*, a subgroup of Archaea which has been identified in numerous permafrost environments [8], [9]. Active methanogenic communities make these environments a significant carbon source through the release of methane to the atmosphere. Methane-oxidizing Bacteria (MOB) present in the aerobic surface layers of the soil and in association with submerged mosses [10], [11] play an important role in the biological oxidation of the methane produced *in-situ*. MOB belong to the *Alpha-* and *Gamma-Proteobacteria* [12] and to the *Verrucomicrobia* [13]. They are able to oxidize up to 90% of the methane emitted in the deeper layers before it reaches the atmosphere [14], [15]. The balance between methane production and oxidation in Arctic environments is sensitive and non-linear as methanogens and methanotrophs have been observed to responds differently to temperature variations in Siberia [16], [17], [18] and Svalbard [19]. Furthermore, accurate predictions of the long-term rates of carbon and nitrogen cycling in Arctic soils, which in turn may determine total ecosystem carbon storage [20], plant productivity [21] and species composition [22], require a much greater understanding of microbial acclimation responses [23]. Yet despite their high relevance in the global climate equation [24], Arctic wetland ecosystems in the Canadian Western Arctic remain poorly characterized in terms of microbial diversity, abundance and the effect of global warming on microbial driven biogeochemical cycles that can result in the net release of greenhouse gases to the atmosphere.

Warmer temperatures in Northern latitudes and the ensuing deepening of the active layer of permafrost can create altered conditions in polygonal tundra, likely causing to major shifts in microbial community composition and their functional potential. This could lead to a strengthening of Arctic wetlands as a carbon source and further reinforce their positive feedback to climate warming.

Therefore, we investigated four polygonal tundra sites on Herschel Island and the Yukon Coast, Western Canadian Arctic. The objectives of our study were to: 1) describe and contrast the bacterial and archaeal diversity in a depth-resolved manner in the four geographically distant polygons; 2) define the essential soil properties shaping this microbial diversity and 3) thereby complement the Circum-Arctic knowledge on permafrost ecosystems in a region insufficiently studied until now. To attain these objectives, we compared and contrasted the microbial communities present in the active layer of four polygonal tundra sites using Ion Torrent sequencing and quantitative PCR. To understand abiotic factors driving this distribution, we described the main soil properties of the sampled profiles. Finally, we used multivariate statistics to elucidate the main biotic and abiotic factors driving the microbial community diversity and structure observed. The results presented give new insights into the distribution and function of microorganisms in a climate-sensitive environment.

Materials and Methods

Site description and sample collection

For sampling, the following permissions/permits were obtained: a Parks Canada Research Permit from Ivvavik National Park, Parks Canada; an Exemption from the Environmental Impact Screening Committee, Joint Secretariat of the Inuvialuit Settlement Region; a Research License from Herschel Island Qikiqtaruk Territorial Park, Yukon Parks and a Scientists and Explorers Licence from the Department of Tourism and Culture, Yukon Government.

Active layer samples were collected from four low-center polygons in the Western Canadian Arctic, on Herschel Island and the Yukon Coast: Drained Lake Polygon (DP, N 69°34′43", W 138°57′25"), Lake Polygon (LP, N 69°36′00.6", W 139°03′56.8"), King Point Polygon (KP, N 69°05′26", W 137°56′45") and Mainland Polygon (MP, N 69°28′23.7", W 139°11′06.3") (Figure 1). The soil at all four sites were characterized as *Hemic Glacistel* classified according to the U.S. Soil Taxonomy [25] with poor drainage and a loamy soil texture. The vegetation period on Herschel Island and the Yukon Coast spans yearly from mid-June to end of September. Average air temperatures vary annually between −26.3°C in February to 8.7°C in July with temperatures at the surface of the active layer ranging from −35°C in the winter to 25°C in the summer [26].

The sampling sites were characterized by an active layer (the layer of ground that is subject to annual thawing and freezing) consisting of a large peat horizon, with a depth varying between 25 cm and 36 cm as measured using a permafrost probe. Replicated sampling was conducted as follows: a large pit was dug in the center of each polygon down to the permafrost table,

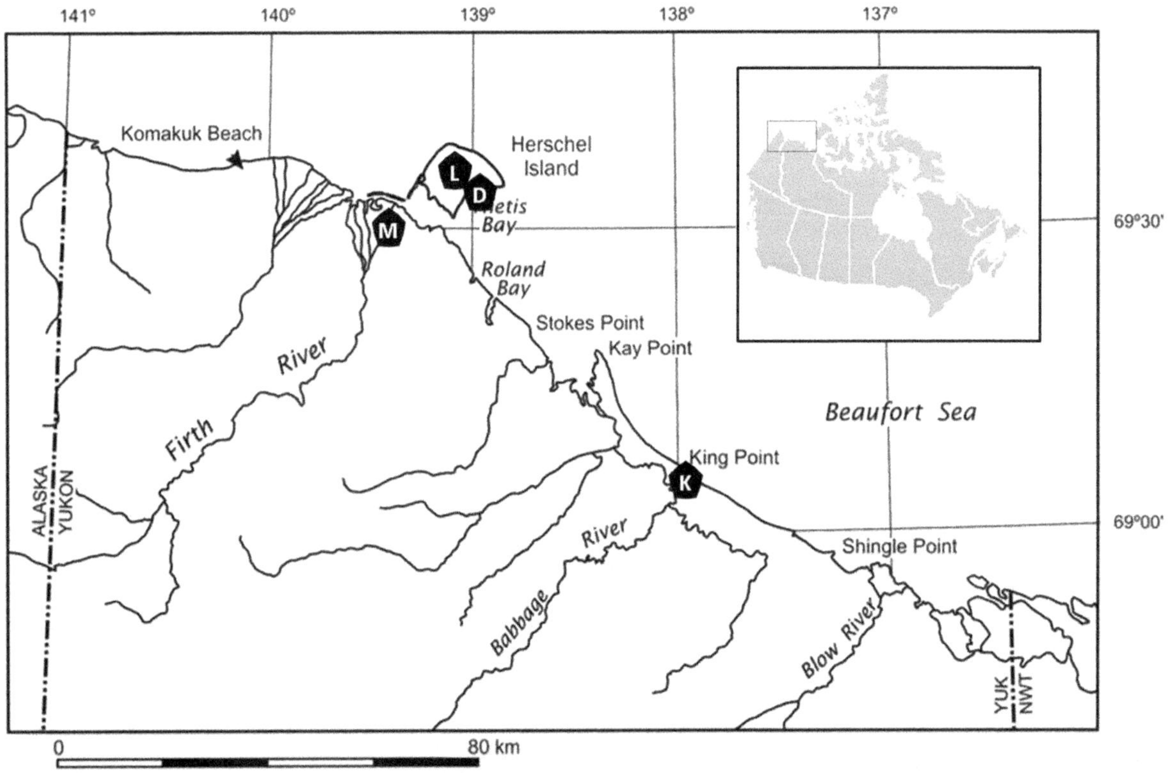

Figure 1. Geographical location of the polygons investigated on Herschel Island and along the Yukon Coast. Map modified from Burn and Zhang, 2009. DP: Drained Lake Polygon, LP: Lake Polygon, MP: Mainland Polygon and KP: King Point Polygon.

exposing a clean vertical profile of over a meter in width. At every 5 cm depth, soil samples were taken with a sharp sterile knife to cover small-scale horizontal heterogeneity. Samples obtained from each layer were then homogenized into sterile 125 ml Nalgene® screw-cap containers (Thermo Fischer Scientific Inc., Waltham, MA). The knife was wiped and sterilized with ethanol between different samples. Soil samples were frozen immediately after sampling and stored at −20°C upon arrival in the laboratory. All subsequent subsampling was performed under sterile and anaerobic conditions in an atmosphere-controlled glove box.

Soil properties

Gravimetric water content of soils was determined by weighing subsamples before and after freeze-drying until no moisture could be observed and the weight stayed constant.

Conductivity and pH were measured using a CyberScan PC 510 Bench Meter (Eutech Instruments Pte Ltd, Singapore) following the slurry technique which consists in mixing 1:2.5 mass ratio of samples and de-ionized water [27].

The percentage of total organic carbon (TOC) of dried, homogenized soils was measured in duplicate using a TOC analyzer (Elementar Vario max C, Germany) after HCl (10%) acid digestion to remove carbonates.

DNA extraction and PCR amplification

Total genomic DNA was extracted in duplicate from 0.6 g of soil using the PowerSoil™ DNA Isolation Kit (Mo Bio Laboratories, Carlsbad, CA). Duplicates were then pooled for downstream analyses. The obtained genomic DNA extracts were quantified using the PicoGreen dsDNA quantitation assay (Invitrogen, Carlsbad, CA). DNA was then stored at −20°C for further use. 16S rRNA gene amplification was carried out using multiplex identifiers as described by Bell *et al.* [28] for Bacteria and Yergeau *et al.* [29] for Archaea. Reactions were carried out in 25 µl volumes containing 2 µl of template DNA, 0.5 µM each primer and 12.5 µl of KAPA2G Robust HotStart ReadyMix (KAPA Biosystems, Woburn, MA). Cycling conditions involved an initial 5 min denaturing step at 95°C, followed by 35 cycles of 30 s at 95°C, 30 s at 57°C, and 45 s at 72°C, and a final elongation step of 10 min at 72°C. PCR products were purified on 2.5% w v^{-1} agarose gels using the QIAquick Gel Extraction Kit (Qiagen, Valencia, CA) and quantified using the PicoGreen dsDNA quantitation assay. All 20 amplification products from the different samples were pooled in an equimolar ratio (one pool for Bacteria and one pool for Archaea) and sequenced together. A total of 3.50×10^7 molecules were used in an emulsion PCR using the Ion OneTouch™ 200 Template Kit (Life Technologies, Carlsbad, CA) and the OneTouch™ instruments (Life Technologies) according to the manufacturer's protocol. Sequencing of the pooled library was done using the Personal Genome Machine™ (PGM) system and a 314 chip with the Ion Sequencing 200 kit according to manufacturer's protocol.

Quantitative PCR amplification

qPCR was performed in triplicate 20 µl volumes using the iQ SYBR green Supermix (Bio-Rad Laboratories, Hercules, CA, USA) on a Rotor-Gene 3000 apparatus (Corbett Life Science, Sydney, NSW, Australia) as previously described [30]. The primers and temperature conditions are shown in Table 1. Inhibition was checked by diluting the samples until three consecutive dilutions yielded the same results. Standards were made from 10-fold dilutions of linearized plasmids containing the gene fragment of interest that was cloned from amplified soil DNA.

Bioinformatic analyses

16S rRNA gene amplicons were pre-treated using the Ribosomal Database Project (RDP) Pyrosequencing pipeline [31]. For Bacteria, sequences having an average quality under 20, having Ns, not exactly matching the MID sequence or being shorter than 100 bp were discarded. For Archaea, sequences having an average quality under 17, having Ns, not exactly matching the MID sequence or being shorter than 75 bp were discarded. The remaining 289658 bacterial and 111596 archaeal sequences were submitted to the RDP classifier [32] using a 0.5 bootstrap cut-off as advised for short sequences [33]. The Shannon diversity index was calculated for each depth for all four study sites. For operational taxonomic unit (OTU) calculations, flowgrams from sff files were de-noised and clustered using AmpliconNoise [34]. Before performing AmpliconNoise calculations, datasets were normalized by randomly selecting 15,000 sequences for Bacteria and 2000 sequences for Archaea. A 97% similarity cutoff was used for OTU calculation, the most stringent OTU definition allowing for intragenomic 16S rRNA variation and PCR/sequencing errors [35].

Statistical analyses

Weighted-normalized UniFrac distances between each sample pair were calculated using the FastUniFrac website [36] based on the GreenGene core dataset. Most statistical analyses were performed in R (v2.13.2, The R foundation for statistical computing, Vienna, Austria). Spearman rank-order correlations (rs) were carried out using the "cor.test" function. Principal coordinate analyses (PCoA) were carried out using the "cmdcsale" function. Mantel tests based on Spearman correlations were performed using the "mantel" function while permANOVA was performed using the "adonis" function of the "vegan" package. Venny (http://bioinfogp.cnb.csic.es/tools/venny/index.html) was used to draw Venn diagrams from genus presence-absence data.

Table 1. Primers and temperature profiles used for quantitative PCR analysis.

Functional gene	Primer pair	Temperature (°C) (ann./read)	Reference
mcrA	MIF / MIR	55/83	[67]
pmoA	A189F / mb661R	55/82	[68]
archaeal *amoA*	amoA19F / crenamoA616r48x	50/72	[69]
bacterial *amoA*	amoA-1F / amoA2R-TC	52/81	[70]
nifH	PolF / PolR	55/83	[71]

Table 2. Main soils properties of the four polygons investigated.

Name of polygon GPS coordinates	GPS Coordinates	Depth (cm)	Water content (%)	N (%)	C (%)	C:N	TOC	pH	Conductivity TRef. 25°C (μS.cm^{-1})
Drained Lake	N 69°34′43″	5–10	79	1.79	26.65	14.89	24.48	5.3	68.1
Polygon (DP)	W 138°57′25″	10–15	77	1.89	30.63	16.21	28.24	5.46	54.9
		15–20	78	1.59	30.44	19.14	28.29	5.53	59.0
		20–25	76	1.18	24.12	20.44	25.20	5.44	62.8
		25–30	79	1.19	26.43	22.21	23.19	5.23	74.3
		30–35	83	1.25	27.62	22.1	26.04	5.29	85.9
King Point	N 69°05′26″	0–5	82	1.01	40.09	39.69	38.23	4.33	111.7
Polygon (KP)	W 137°56′45″	5–10	84	1.11	38.22	34.43	37.75	4.47	91.8
		10–15	86	1.11	35.91	32.35	34.31	4.96	101.7
		15–20	41	0.54	13.27	24.57	11.30	5.19	28.0
		20–25	42	0.25	6.04	24.16	9.77	5.08	39.2
Lake Herschel	N 69°36′00.6″	5–10	76	1.35	25.08	18.58	22.69	6.82	76.4
Polygon (LP)	W 139°03′56.8″	10–15	76	1.33	24.92	18.74	22.66	5.73	84.7
		15–20	69	1.14	20.62	18.09	16.56	5.83	72.5
		20–25	77	1.76	29.77	16.91	28.09	5.9	91.3
Mainland	N 69°28′23.7″	0–5	79	2.35	33.36	14.2	31.76	5.54	45.7
Polygon (MP)	W 139°11′06.3″	5–10	80	2.49	39.61	15.91	38.25	5.47	40.5
		10–15	80	2.68	42.69	15.91	39.20	5.46	38.6
		15–20	76	1.78	31.95	17.91	29.95	5.41	45.0
		20–25	68	1.45	26.32	18.15	24.08	5.44	53.4
		25–30	65	1.22	23.55	19.30	21.32	5.46	49.3
		30–36	78	2	35.96	17.98	34.57	5.39	57.2

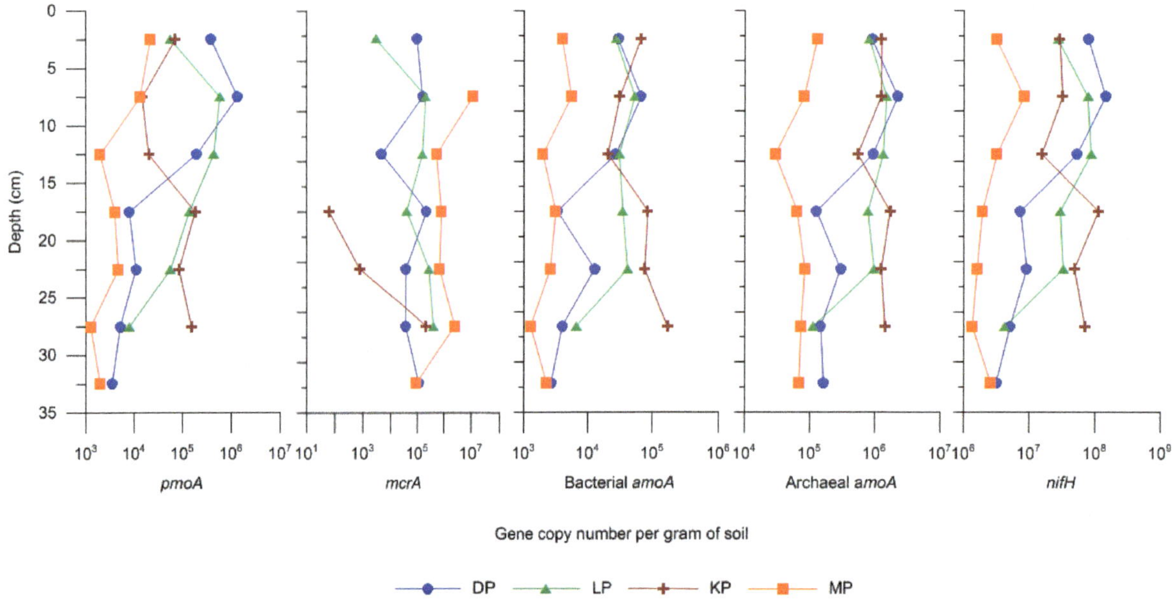

Figure 2. Abundance of selected functional genes in the active layer. Gene copy numbers per gram of soil in the active layer profile of each site investigated as determined by qPCR for *pmoA*, *mcrA*, bacterial *amoA*, archaeal *amoA* and *nifH* genes.

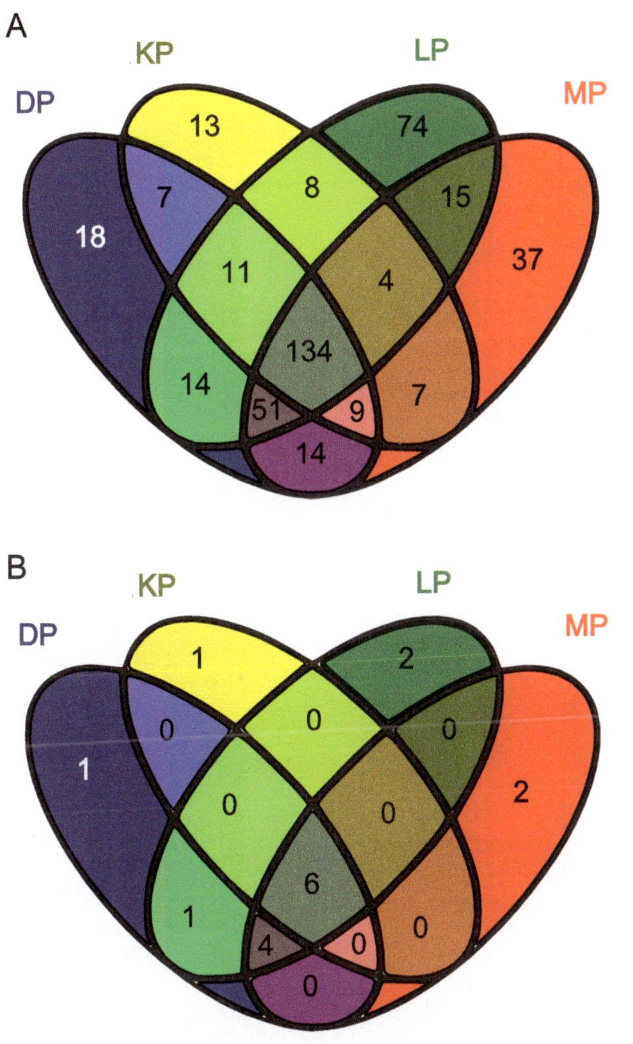

Figure 3. Venn diagrams illustrating the distribution of identified taxa between sampling locations. a) Distribution of classified bacterial genera between polygons. b) Distribution of classified archaeal genera between polygons.

Data deposition

All sequence data have been submitted to the NCBI sequence read archive (SRA) database under project SRP026122 (BioProject PRJNA208722).

Results

Characteristics of the soil

The permafrost table was measured at 36 cm depth for DP, 25 cm for KP, 26 cm for LP and 36 cm for MP in the period from July 27th to August 2nd, 2010. All polygon sites were visibly water saturated, with gravimetric water contents ranging from 65 to 86%, with the exception of the two lower depths of KP (41–42%). Within each polygon, water content did not significantly correlate with depth. The pH at polygons DP and MP was pH 5.4 on average, slightly more basic at LP (pH 5.7 to 6.8) and more acidic at KP (average pH 4.8), and was only significantly correlated with depth in the MP polygons ($r_s = -0.874$; $P = 0.0045$). Total organic carbon content (TOC) was high at all sites, ranging from 20 to 40%, with the exception again of the bottom layers of KP where TOC was lower (10–11%). TOC was only significantly correlated with depth in the KP polygons ($r_s = -1.00$; $P = 0.017$). Soil C:N ratio was significantly and positively correlated with depth in the DP ($r_s = 0.943$; $P = 0.017$) and MP ($r_s = 0.881$; $P = 0.0072$) polygons while it showed significant negative correlation with depth in the KP polygons ($r_s = -1.00$; $P = 0.017$). An overview of the soil properties of each polygon is given in Table 2.

Quantitative PCR

The functional genes investigated were *pmoA* (subunit A of particulate methane monooxygenase), *mcrA* (methyl coenzyme-M reductase), bacterial and archaeal *amoA* (ammonia monooxygenase) and *nifH* (nitrogenase reductase subunit). Overall, gene copy numbers per gram of wet soil were abundant for all of the functional genes investigated and in all four polygons. The only exception is the complete absence of an amplifiable *mcrA* signature in the first 15 cm depth in KP (Figure 2). *nifH* was consistently the most abundant of all of the genes analysed in all four polygons, reaching up to 10^8 copies per gram of wet soil (Figure 2). *nifH* abundance was negatively correlated with depth and C:N ratio in the DP polygons ($r_s = -0.952$; $P = 0.0011$ and $r_s = -0.886$; $P = 0.033$, respectively) and in the MP polygons ($r_s = -0.786$; $P = 0.028$ and $r_s = -0.881$; $P = 0.0072$, respectively). Furthermore, in the MP polygons, the abundance of *nifH* was significantly and positively correlated with the concentration of N and C ($r_s = 0.952.786$; $P = 0.0011$ and $r_s = 0.929$; $P = 0.0022$, respectively). Archaeal *amoA* was found to have between 10^5 to 10^6 copies.g^{-1}, while bacterial *amoA* was one or two orders of magnitude lower (10^3 to 10^4 copies.g^{-1}, Figure 2). In the DP polygons, both archaeal and bacterial *amoA* were negatively correlated with depth ($r_s = -0.762$; $P = 0.037$ and $r_s = -0.905$; $P = 0.0046$, respectively), in the KP polygons the abundance of bacterial *amoA* was negatively correlated with soil water content ($r_s = -1.00$; $P = 0.017$), while in the MP polygons, bacterial *amoA* was negatively correlated with depth and C:N ratio ($r_s = -0.786$; $P = 0.028$ and $r_s = -0.786$; $P = 0.028$, respectively). *mcrA* was more erratically distributed throughout the different depths and varied more between polygons, with copy numbers ranging anywhere from 10^1 to 10^7 copies g^{-1}. *mcrA* only showed significant correlations with soil properties in the KP polygons, where it was significantly correlated to depth, N, C and C:N ratio ($r_s = 0.941$; $P = 0.0051$, $r_s = -0.918$; $P = 0.028$, $r_s = -0.894$; $P = 0.041$ and $r_s = -0.894$; $P = 0.041$, respectively). *pmoA* only showed significant correlations in the DP polygons, where it was negatively correlated with depth and C:N ratio ($r_s = -0.952$; $P = 0.0011$ and $r_s = -0.886$; $P = 0.033$, respectively).

Biodiversity and richness index

The bacterial and archaeal diversity was high, with an overall Shannon index (H') of 5.8 and 4.7, respectively. In all polygons, the bacterial diversity was highest in the surface samples, then decreasing with depth (Table 3). Archaeal diversity was highest in the upper layers of the soils, the maximum diversity found between 0–15 cm depth (Table 3). LP showed the highest diversity overall with H'6.2 for Bacteria and 4.9 for Archaea. Based on OTU (0.03) calculations, the highest number of bacterial OTUs was found in DP, while the highest number of archaeal OTUs was found in LP (Table 3). Out of a total of 416 bacterial genera identified in all polygons combined, 134 were shared by all four sites. DP harboured 18 unique genera, KP 13, LP 74 and MP 37 (Figure 3a). However, due to the short length of the archaeal 16S rRNA sequences and the lower number of observed archaeal genera, only 12 genera could be identified, of which 6 were shared

Table 3. OTU (0.03) calculations and Shannon diversity index H′ (in parentheses) of bacterial and archaeal 16S rRNA gene sequences from the active layer of four low-center polygons on Herschel Island and the Yukon Coast.

	DP		LP		KP		MP	
Depth (cm)	**Bac.**	**Arch.**	**Bac.**	**Arch.**	**Bac.**	**Arch.**	**Bac.**	**Arch.**
0–5	8951	863	7889	1446	9084	N.a.	9207	1043
	(6.3)	(4.8)	(6.6)	(5.8)	(5.7)		(6.2)	(5.1)
5–10	9086	953	9215	1392	8586	N.a.	799	1023
	(6.2)	(5.0)	(6.4)	(5.7)	(5.7)		(6.1)	(5.1)
10–15	7453	1027	8317	1111	5379	N.a.	7892	1007
	(6.2)	(5.1)	(6.2)	(5.1)	(5.9)		(5.7)	(4.7)
15–20	9155	851	7856	971	6974	1040	7171	700
	(6.0)	(4.6)	(5.8)	(4.7)	(5.1)	(5.2)	(5.6)	(4.6)
20–25	8762	864	8829	767	5349	837	5923	933
	(5.8)	(4.3)	(6.3)	(4.3)	(5.1)	(4.7)	(5.7)	(4.5)
25–30	8258	937	N.d.	N.d.	N.d.	N.d.	7326	825
	(5.6)	(4.4)					(5.2)	(4.5)
30–35	8084	900	N.d.	N.d.	N.d.	N.d.	7802	821
	(5.5)	(4.2)					(5.5)	(4.6)
Average	**8512**	**887**	**7736**	**1064**	**7140**	**934**	**7706**	**869**
	(5.9)	**(4.5)**	**(6.2)**	**(4.9)**	**(5.4)**	**(4.8)**	**(5.7)**	**(4.7)**

N.d. refers to no sample being taken as the permafrost table has been reached. N.a. refers to no amplifiable sequence being recovered from those samples.

between all polygons (Figure 3.b., grey colored field). Interestingly, 4 genera were shared between all polygons except KP.

Community composition and vertical distribution of diversity

A bootstrap cutoff of 50% was shown to accurately classify sequences at the genus level for partial sequences of length longer than 50 bp and shorter than 250 bp, and to provide genus level assignments for higher percentage of sequences [33]. Bacterial sequences were classified to the phylum level, also detailing the proteobacterial classes. Archaea were classified to the genus level. The bacterial community composition at all four sites were found to be similar in terms of overall diversity, but the community evolved differently with depth (Figure 4). Sequences affiliated with *Proteobacteria* accounted for 40–50% of sequences in all sites, followed by *Bacteroidetes*-affiliated sequences at 20–40% and *Actinobacteria* at 10–15%.

Communities in all four polygons showed a similar evolution of community composition with depth: *Deltaproteobacteria*, *Bacteroidetes* and *Firmicutes* increasing in abundance and *Alpha-*, *Beta-*, *Gammaproteobacteria* and *Acidobacteria* decreasing with depth (Figure 4).

The class-level compositions of Archaea were similar in DP and MP (Figure 5). No archaeal sequences could be amplified in KP from the first 15 cm of the active layer. Sequences affiliated with methanogens (*Methanobacterium* and *Methanosarcina*) dominated in all sampling sites, representing up to 90% of all classified sequences. Within the classified sequences, *Methanobacterium* were found to clearly dominate in the surface layers of three out of the four polygons, representing up to 98% of classified sequences in DP. They decreased rapidly with depth, replaced by a majority of *Methanosarcina* and to a lesser extent *Methanosaeta*, representing more than 80% of the community in the deeper layers. The exception was LP, where the abundance of *Methanobacterium* remained relatively constant to a depth of 20 cm after which *Methanosarcina* became more abundant.

Overall, little variation was observed for the bacterial community between the polygons based on UniFrac values, however the variation within each polygon (i.e. with increasing depth) was higher. On the contrary, the archaeal community composition was more influenced by location, each polygon having an own, specific community. For Bacteria, the KP samples separated from all the other samples (Figure 6). Another interesting trend was the separation of the majority of deep samples (15–35 cm deep) versus the majority of the shallow samples (0–10 cm deep). This trend was also visible when looking at the average UniFrac distance between shallow and deep samples, which was relatively high. For Archaea, again the KP samples clustered separately from all other samples. The 15 to 35 cm deep samples tightly clustered together, as made visible by the relatively low UniFrac distance between samples from that depth. For both Archaea and Bacteria, the microbial communities were more similar within a polygon then between polygons (UniFrac distances), indicating some level of local adaptation.

To investigate potential relationships between microbial community structure and the underlying soil geochemistry, the sequence data were correlated to the soil properties data and illustrated by heat maps (Figure 7). Variables most consistently explanatory of bacterial sequence diversity patterns were depth, conductivity and C:N ratio. Archaeal community composition showed a stronger positive correlation to increasing pH, with depth and C:N also having a strong influence on community structure.

Discussion

Increasing temperatures in the Northern latitudes and the ensuing deepening of the active layer of permafrost will likely lead to an increase in the anoxic zone and thereby shift microbial

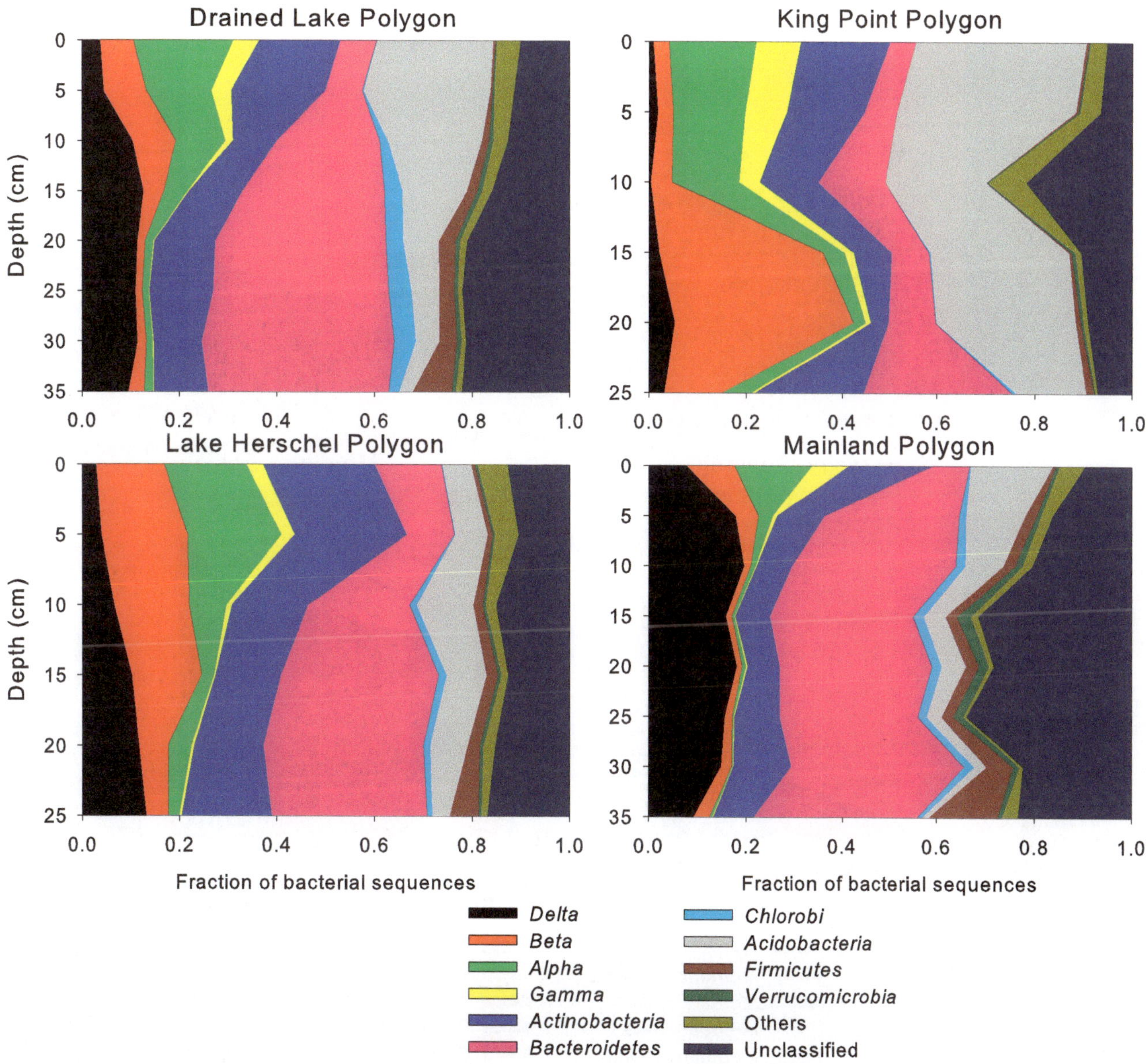

Figure 4. Distribution of bacterial phyla with depth based on 16S rRNA sequences in all four polygons. Alpha, Beta, Delta and Gamma stand for *Alphaproteobacteria*, *Betaproteobacteria*, *Deltaproteobacteria* and *Gammaproteobacteria*.

community composition and functional potential. Several studies have addressed aspects of this issue in Siberia (for example [6], [8], [37]), Svalbard [19], [38] and the Canadian High Arctic [9]; [39] but the communities of the Canadian Western Arctic remain underexplored [40]. This study represents the first local-scale comparative microbiological analysis of polygonal tundra in the North-American Arctic. We found that depth strongly influenced microbial community in terms of structure and function, this effect being related to an increase in C:N ratio (and thus nutrient limitation) in deeper layers. This depth-effect has previously been described by Wagner *et al.* [41] who observed a net decrease of bioavailable organic carbon and an increasing humification index with increasing soil depth in Siberian polygonal tundra. The authors measured the highest rates of methanogenesis at depths with the highest amounts of bioavailable carbon rather than total organic carbon. Their findings concur with ours, indicating that although organic carbon is highly accumulated in permafrost soils, there is a decrease in bioavailable carbon with increasing depth, creating substrate limitation close to the permafrost table. Looking at the microbial diversity in the polygons, a large number of microbial genera were shared between all sites despite the sites being up to 70 km apart as illustrated by the Venn diagrams (Figure 3). However, these diagrams also show that an important number of genera identified were unique to each polygon, suggesting that although the bulk of the community is similar across our study sites, each polygon contains a number of unique microorganisms whose presence is supported by specific local environmental characteristics. The shift in community composition with active layer depth at all sites suggests a certain level of local adaptation. Within each polygon, we saw highly parallel trends, with increasing *Bacteroidetes* and decreasing *Alpha-* and *Betaproteobacteria* in samples closer to the permafrost table, with the exception of increasing *Betaproteobacteria* with depth in KP. Although they represent 20 to 40% of sequences obtained in all

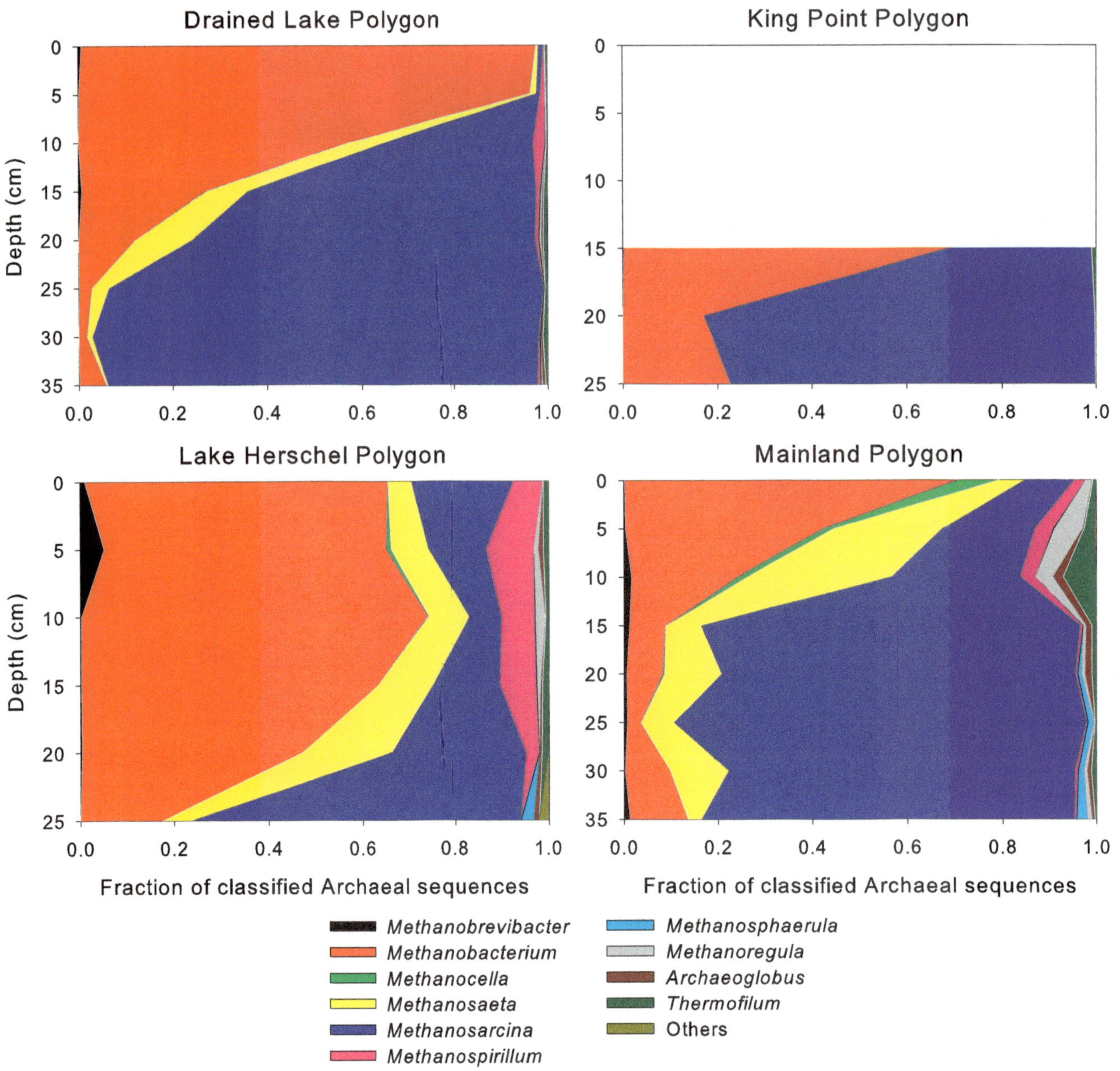

Figure 5. Distribution of archaeal genera with depth based on 16S rRNA sequences in all four polygons. No archaeal 16S rRNA sequence could be amplified between 0 and 15 cm depth in King Point Polygon (KP).

four polygons, the considerable variation within *Proteobacteria* with depth indicates consistent forcing mechanisms leading to the adaptation and specialization of the various groups to changes in temperature and soil properties. In deeper layers, the dominance of *Bacteroidetes* was not surprising as these microorganisms are known to include several psychrophilic members [42] which thrive under the stable cold conditions found in the deeper active layer close to the permafrost table. A study of polygonal tundra in the Lena Delta, Siberia also identified *Bacteroidetes* as dominant members of the soil microbial community [37]. Commonly found in permafrost-affected soil ecosystems [43], [44], *Actinobacteria* also made-up an important part of the bacterial community at our sites, their successful colonization of the entire active layer likely being supported by their ability to metabolize a wide range of substrates as sole carbon source and their adaptation to low temperatures [45], [46]. Overall, these results are consistent with other studies reporting that environmental conditions found in the active layer of permafrost-affected soils permit a diverse community of Bacteria, capable of utilizing a wide spectrum of substrates under varying redox conditions [44]. The redox potential in the active layer of permafrost environments has consistently been shown to decrease with depth, effectively shaping the structure and activity of the indigenous microbial community [37], [44], [47], [48].

In the active layer, the final step of organic carbon decomposition under anaerobic conditions is catalyzed by methanogenic Archaea. At our study sites the methanogens were distributed throughout the soil profile, this dominance being in line with the archaeal diversity described in peat wetlands in Siberia and Svalbard [8]; [49]. The archaeal community composition was

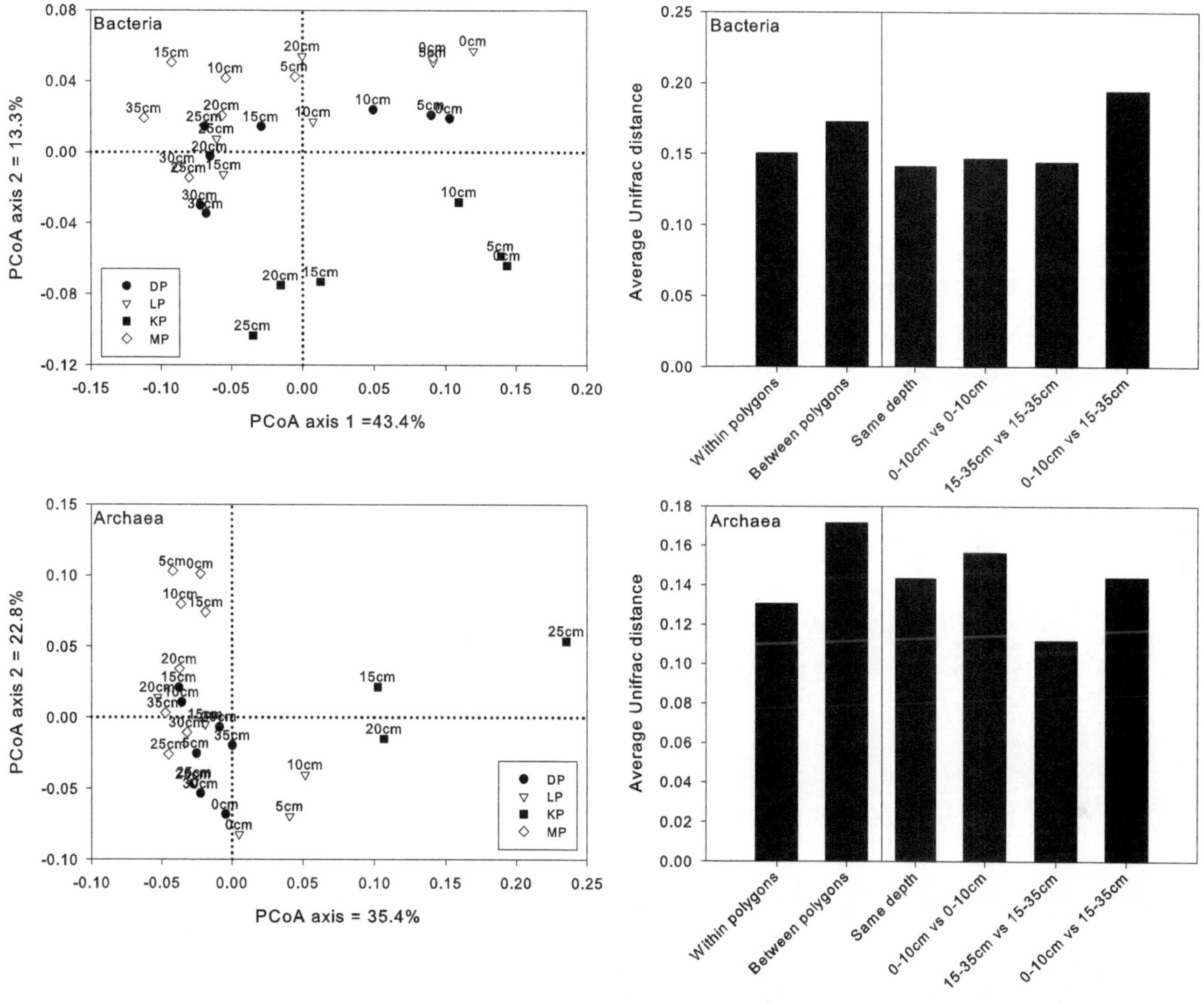

Figure 6. Unifrac analysis of 16S rRNA sequence variation within and between polygons.

similar at phylum level between DP, LP and MP but significantly different in KP. In all four polygons, it appears that the large shifts in archaeal community composition with depth observed had no apparent effect on the functional potential for methane emission as measured by quantitative PCR. This suggests a certain amount of functional redundancy, defined as species performing similar roles in communities and ecosystems and therefore being substitutable with little impact on ecosystem processes [50]. The strong difference in archaeal community structure observed at KP can be attributed to the lower pH, as suggested by the fact that only a few acidophilic methanogens have been cultured so far [51]; [52] and the positive correlation between increasing pH and archaeal community composition found in this study. The higher conductivity found in the top layers of KP due to increased sea spray because of the proximity of the sampling site to the Beaufort Sea could also have an inhibitory effect on Archaea in the surface of KP.

We found that both hydrogenotrophic and acetoclastic methanogenesis can occur in the active layer. The dominance of *Methanobacteria* in the top layers can be attributed to their use of hydrogen as a substrate, as hydrogenotrophy is especially dominant at higher temperatures [53]. The methanogenic community shifted to a *Methanosarcina* dominated community with depth in three of our study sites. *Methanosarcina*-related sequences are readily detected in permafrost-affected soils and sediments [17], [54], one reason for this ubiquity being the potential of *Methanosarcina*-like species to use all three metabolic pathways with a broad range of substrates, e.g. acetate, methanol and hydrogen [55]. This provides *Methanosarcina* with a high flexibility regarding changing substrate and environmental conditions. *Methanosarcina* species also show a high resistance against several abiotic and biotic stress factors, as recently illustrated by the new isolated strain *Methanosarcina soligelidi* [56], [57]. The sharp switch of the methanogenic community from a *Methanobacteria* to a *Methanosarcina* dominated community, i.e. a switch from an overall hydrogenotrophic to an acetoclastic methanogenic community, has been observed through cloning and T-RFLP analysis in the detailed study of a polygon in the same area [40] and in other Arctic tundra environments [8], [49], [58]. Our findings further support the hypothesis that the acetoclastic pathway is favored in cold environments [53] because of low temperatures inhibiting hydrogen-producing Bacteria. Furthermore in cold, anoxic conditions such as those encountered close to the permafrost table, hydrogenotrophic methanogenesis is disadvantaged because

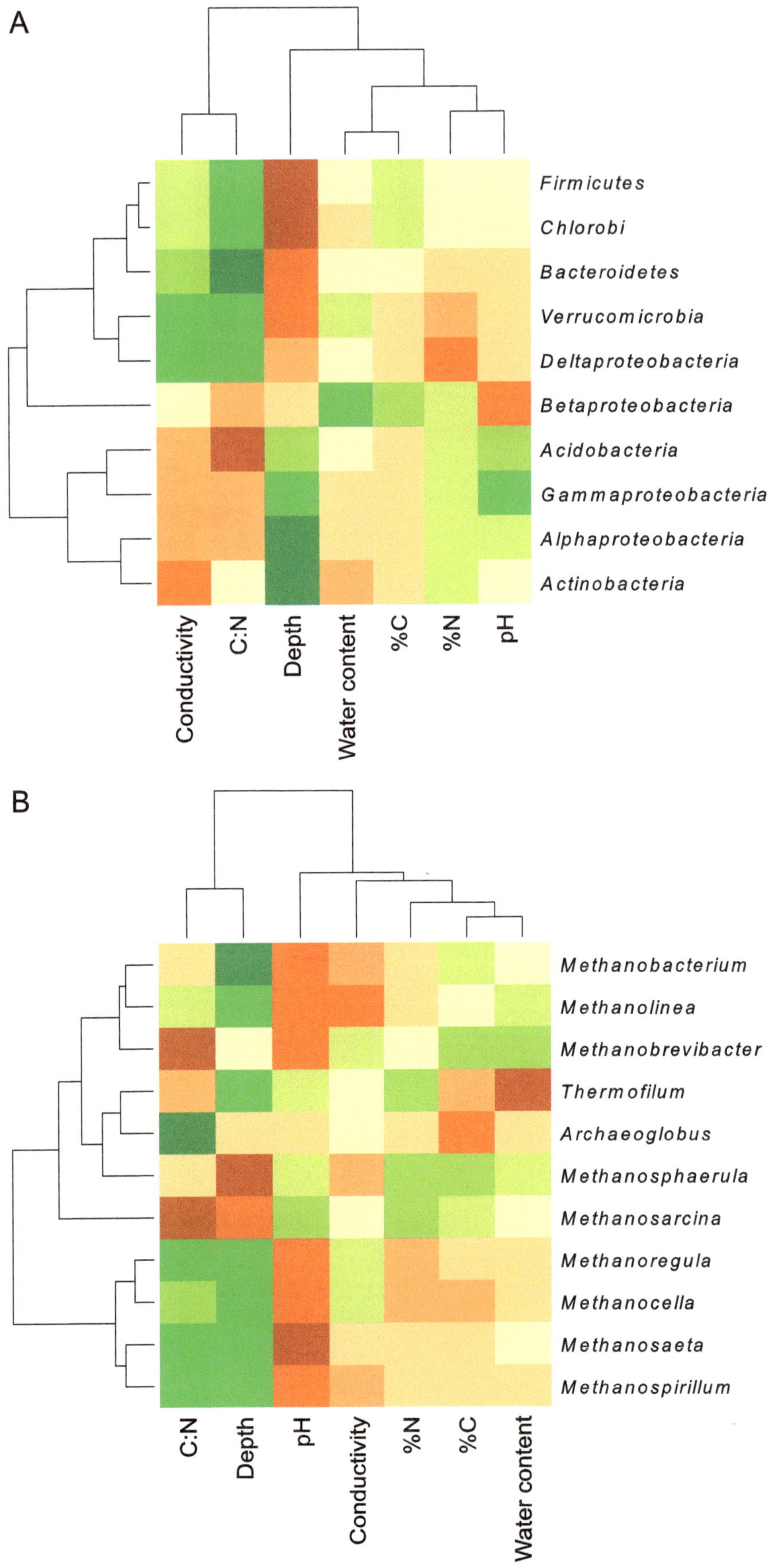
A
Firmicutes
Chlorobi
Bacteroidetes
Verrucomicrobia
Deltaproteobacteria
Betaproteobacteria
Acidobacteria
Gammaproteobacteria
Alphaproteobacteria
Actinobacteria
Conductivity
C:N
Depth
Water content
%C
%N
pH
B
Methanobacterium
Methanolinea
Methanobrevibacter
Thermofilum
Archaeoglobus
Methanosphaerula
Methanosarcina
Methanoregula
Methanocella
Methanosaeta
Methanospirillum
C:N
Depth
pH
Conductivity
%N
%C
Water content

Figure 7. Heat map correlation between the taxonomical distribution of microorganisms and the main properties of the soil. a) Correlations between soil properties and bacterial phyla. b) Correlations between soil properties and archaeal genera. A red colour indicates a positive correlation, while a green color indicates a negative correlation. The darker the colour, the stronger the correlation.

of competition for hydrogen and carbon dioxide with acetogenic bacteria (e.g. *Bacteroidetes*) which produce acetate as a precursor for acetoclastic methanogens [59]. The availability of low molecular substances (e.g. acetate) provided by the root system of the vegetation [60], [61] could also influence the composition of the methanogenic community.

The investigation of nitrogen and carbon cycling genes revealed the possible functional evolution of the microbial community with respect to increasing active layer temperatures. Several of the genes quantified decreased with increasing depth, suggesting a lower functional potential in the deeper layers of the polygons. The depth effect we observed was not caused by increasing water saturation in deeper layers as was expected. Rather, we observed that various soil factors were the most likely explanatory factors for microbial functional potential, and that this varied largely depending on the polygon. The decrease observed is proposed to be linked with the decreasing C:N ratio and the concomitant increased substrate limitation with active layer depth as discussed above for bacterial community composition. The abundance of functional genes involved in nitrogen fixation, ammonium oxidation, methane production and methane oxidation was found to be higher by one to two orders of magnitude compared what has been reported from Hess Creek, Alaska [62] and the Canadian High Arctic [9], [39]. The high abundance of nitrogen fixers and ammonia oxidizers in the active layer has already been described by previous studies [9], [63] especially the majority of archaeal compared to bacterial *amoA*, in our case by almost two orders of magnitude. The dynamics of nitrogen processes may be especially important, as most tundra ecosystems are considered to be nitrogen limited and changes in nitrogen supply can affect leaf development, carbon flux and biomass production [64], [65]. Genetic proxies for methane production and oxidation, *mcrA* and *pmoA* respectively, were both found in large amounts at virtually all depths of the active layer at all sites except KP. The presence of aerobic methane oxidizers also in deeper, water saturated layers has important implications for mitigating the release of methane to the atmosphere, their presence likely being supported through root exudates creating microaerobic conditions favorable for methane oxidation [66].

In summary, this study provides an important insight into the microbial functional role of Arctic permafrost-affected soils. We determined that soil water content does not have the expected influence on microbial communities, but that depth related changes in soil parameters have a strong influence in shaping microbial abundance and distribution. The future of carbon turnover in permafrost ecosystems largely depends on the structure and functional potential of the indigenous microbial community within the permafrost and the overlaying active layer. Other studies of permafrost-affected environments have determined that although these ecosystems are apparently similar in many aspects, their microbial community composition and constraints on organic matter turnover can vary greatly, emphasizing the need to conduct more circum-Arctic studies. Our results highlight the very large heterogeneity of polygons, as the microbial community composition and functional potential of each seemingly identical polygon responded to a different set of environmental controls. So, even if they appear to be highly similar, Artic polygon microbial communities and functions are under different environmental controls at each location. This heterogeneity makes the prediction of the fate of the carbon stocked in these ecosystems and generalizations across landscapes very difficult, if not impossible, and calls for more in depth studies like the current one.

Acknowledgments

We thank the German-Canadian field parties of the 2010 YUKON COAST expedition and the Herschel Island Rangers for help with field measurements and sampling. We particularly thank Sylvie Sanschagrin and Christine Maynard (NRC) for their expertise and help with Ion Torrent sequencing. We gratefully acknowledge three anonymous reviewers for constructive comments on the early versions of the manuscript.

Author Contributions

Conceived and designed the experiments: BFF EY CWG HL DW. Performed the experiments: BFF EY. Analyzed the data: BFF EY. Contributed reagents/materials/analysis tools: EY CWG HL DW. Wrote the paper: BFF EY DW.

References

1. Kuhry P, Dorrepaal E, Hugelius G, Schuur EAG, Tarnocai C (2010) Potential Remobilization of Belowground Permafrost Carbon under Future Global Warming. Permafrost Periglac 21: 208–214.
2. Tarnocai C, Canadell JG, Schuur EAG, Kuhry P, Mazhitova G, et al. (2009) Soil organic carbon pools in the northern circumpolar permafrost region. Global Biogeochem Cy 23: 1–11.
3. Knoblauch C, Beer C, Sosnin A, Wagner D, Pfeiffer EM (2013) Predicting long-term carbon mineralization and trace gas production from thawing permafrost of Northeast Siberia. Glob Change Biol 19: 985–1324.
4. Minke M, Donner N, Karpov NS, de Klerk P, Joosten H (2007) Distribution, diversity, development and dynamics of polygon mires: examples from NE Yakutia (NE Siberia). Peatlands Int 1: 36–40.
5. van Everdingen RO (ed) (2005) Multi-Language Glossary of Permafrost and Related Ground-Ice Terms, International Permafrost Association, University of Calgary, Calgary, Canada. Available: http://nsidc.org/fgdc/glossary. Accessed 2013 May 21.
6. Wagner D, Gattinger A, Embacher A, Pfeiffer EM, Schloter M, et al. (2007) Methanogenic activity and biomass in Holocene permafrost deposits of the Lena Delta, Siberian Arctic and its implication for the global methane budge. Glob Change Biol 13: 1089–1099.
7. Whalen SC, Reeburgh WS (1992) Interannual variations in tundra methane emission: a 4-year time series at fixed sites. Global Biogeochem Cy 6: 139–159.
8. Ganzert L, Jurgens G, Münster U, Wagner D (2007) Methanogenic communities in permafrost-affected soils of the Laptev Sea Coast, Siberian Arctic, characterized by 16S rRNA gene fingerprints. FEMS Microbiol Ecol 59: 476–488.
9. Yergeau E, Hogues H, Whyte LG, Greer CW (2010) The functional potential of high Arctic permafrost revealed by metagenomic sequencing, qPCR and microarray analyses. ISME J 4: 1206–1214.
10. Kip N, van Winden JF, Pan Y, Bodrossy L, Reichart G, et al. (2010) Global prevalence of symbiotic bacterial methane oxidation in peat moss ecosystems. Nat Geosci 3: 617–621.
11. Liebner S, Zeyer J, Wagner D, Schubert C, Pfeiffer E-M, et al. (2011) Methane oxidation associated with submerged brown mosses reduces methane emissions from Siberian polygonal tundra, J Ecol 99: 914–922.
12. Hanson RS, Hanson TE (1996). Methanotrophic bacteria. Microbiol. Rev. 60: 439.
13. Dunfield PF, Yuryev A, Senin P, Smirnova AV, Stott MB, et al. (2007) Methane oxidation by an extremely acidophilic bacterium of the phylum Verrucomicrobia. Nature 450: 879–882.
14. Le Mer J, Roger P (2001) Production, oxidation, emission and consumption of methane by soils: A review. Eur J Soil Biol 37: 25–50.
15. Wagner D, Liebner S (2009) Global warming and carbon dynamics in permafrost soils: methane production and oxidation. In: Margesin R, editor. Permafrost Soils, Springer, Berlin, Heidelberg. Vol. 16, pp. 219–236.

16. Knoblauch C, Zimmermann U, Blumenberg M, Michaelis W, Pfeiffer EM (2008) Methane turnover and temperature response of methane-oxidizing bacteria in permafrost-affected soils of northeast Siberia. Soil Biol Biochem 40: 3004–3013.
17. Koch K, Knoblauch C, Wagner D (2009) Methanogenic community composition and anaerobic carbon turnover in submarine permafrost sediments of the Siberian Laptev Sea. Environ Microbiol 11: 657–668.
18. Liebner S, Rublack K, Stuehrmann T, Wagner D (2009) Diversity of aerobic methanotrophic bacteria in a permafrost active layer soil of the Lena Delta, Siberia. Microb Ecol 57: 25–35.
19. Høj L, Olsen RA, Torsvik VL (2008) Effects of temperature on the diversity and community structure of known methanogenic groups and other archaea in high Arctic peat. ISME J 2: 37–48.
20. Hobbie SE, Schimel JP, Trumbore SE, Randerson JR (2000) Controls over carbon storage and turnover in high-latitude soils. Glob Change Biol 6: 196–210.
21. Van Wijk MT, Williams M, Shaver GR (2005) Tight coupling between leaf area index and foliage N content in arctic plant communities. Oecologia 142: 421–427.
22. Weintraub MN, Schimel JP (2005) Nitrogen cycling and the spread of shrubs control changes in the carbon balance of arctic tundra ecosystems. Bioscience 55: 408–415.
23. Hartley IP, Hopkins DW, Garnett MH, Sommerkorn M, Wookey PA (2008) Soil microbial respiration in arctic soil does not acclimate to temperature. Ecol Lett 11: 1092–1100.
24. Graham DE, Wallenstein MD, Vishnivetskaya TA, Waldrop MP, Phelps TJ, et al. (2012) Microbes in thawing permafrost: the unknown variable in the climate change equation. ISME J 6: 709–712.
25. Soil Survey Staff (1998) Keys to soil taxonomy, eighth edition, USDA-NRCS, Washington, DC.
26. Burn CR, Zhang Y (2009) Permafrost and climate change at Herschel Island (Qikiqtaruq), Yukon Territory, Can J Geophys Res 114: 1–16.
27. Edmeades DC, Pringle RM, Mansell GP, Shannon PW, Ritchie J, et al. (1985) Effects of lime on pasture production on soil in the North Island of New Zealand. 5. Description of a lime recommendation scheme. New Zeal J Exp Agr 13: 47–58.
28. Bell TH, Yergeau E, Maynard C, Juck D, Whyte LG, et al. (2013) Predictable bacterial composition and hydrocarbon degradation in Arctic soils following diesel and nutrient disturbance. ISME J 7: 1200–1210.
29. Yergeau E, Lawrence JR, Sanschagrin S, Waiser MJ, Korber DR, et al. (2012) Next-generation sequencing of microbial communities in the Athabasca River and its tributaries in relation to oil sands mining activities. Appl Environ Microbiol 78: 7626–7637.
30. Yergeau E, Arbour M, Brousseau R, Juck D, Lawrence JR, et al. (2009) Microarray and real-time PCR analyses of the responses of high Arctic soil bacteria to hydrocarbon pollution and bioremediation treatments. Appl Environ Microbiol 75: 6258–6267.
31. Cole JR, Wang Q, Cardenas E, Fish J, Chai B, et al. (2009) The Ribosomal Database Project: improved alignments and new tools for rRNA analysis. Nucleic Acids Res 37: D141–145.
32. Wang QG, Garrity M, Tiedje JM, Cole JR (2007) Naïve Bayesian Classifier for Rapid Assignment of rRNA Sequences into the New Bacterial Taxonomy. Appl Environ Microbiol 73: 5261–7.
33. Claesson MJ, O'Sullivan O, Wang Q, Nikkilä J, Marchesi JR, et al. (2009) Comparative analysis of pyrosequencing and a phylogenetic microarray for exploring microbial community structures in the human distal intestine. PLoS One 4: e6669.
34. Quince C, Lanzen A, Davenport RJ, Turnbaugh PJ (2011) Removing noise from pyrosequenced amplicons. BMC Bioinformatics 12: 38.
35. Lee CK, Barbier BA, Bottos EM, McDonald IR, Cary SC (2011) The Inter-Valley Soil Comparative Survey: the ecology of Dry Valley edaphic microbial communities. ISME J 6: 1046–1057.
36. Hamady M, Lozupone C, Knight R (2010) Fast UniFrac: facilitating high-throughput phylogenetic analyses of microbial communities including analysis of pyrosequencing and PhyloChip data. ISME J 4: 17–27.
37. Liebner S, Harder J, Wagner D (2008) Bacterial diversity and community structure in polygonal tundra soils from Samoylov Island, Lena Delta, Siberia. Int Microbiol 11: 195–202.
38. Graef C, Hestnes AG, Svenning MM, Frenzel P (2011) The active methanotrophic community in a wetland from the High Arctic. Environ Microbiol Rep 3: 466–472.
39. Martineau C, Whyte LG, Greer CW (2010) Stable isotope probing analysis of the diversity and activity of methanotrophic bacteria in soils from the Canadian high Arctic. Appl Environ Microbiol 76: 5773–5784.
40. Barbier BA, Dziduch I, Liebner S, Ganzert L, Lantuit H, et al. (2012) Methane-cycling communities in a permafrost-affected soil on Herschel Island, Western Canadian Arctic: active layer profiling of mcrA and pmoA genes. FEMS Microbiol Ecol 82: 287–302.
41. Wagner D, Lipski A, Embacher A, Gattinger A (2005) Methane fluxes in permafrost habitats of the Lena Delta: effects of microbial community structure and organic matter quality. Environ Microbiol 7: 1582–1592.
42. Shivja S, Ray MK, Shyamda Rao N, Saisree L, Jagannadham MV, et al. (1992) *Sphingobacterium antarcticus* sp. nov., a psychrotrophic bacterium from the soils of Schirmacher Oasis, Antarctica. Int J Syst Bacteriol 42: 102–106.
43. Steven B, Pollard WH, Greer CW, Whyte LG (2008) Microbial diversity and activity through a permafrost/ground ice core profile from the Canadian high Arctic. Environ Microbiol 10: 3388–3403.
44. Wagner D, Kobabe S, Liebner S (2009) Bacterial community structure and carbon turnover in permafrost-affected soils of the Lena Delta, northeastern Siberia. Can J Microbiol 55: 73–83.
45. Wynn-Williams DD (1996) Antarctic microbial diversity: the basis of polar ecosystem processes. Biodiver and Conserv 5: 1271–1293.
46. Aislabie JM, Chhour KL, Saul DJ, Miyauchi S, Ayton J, et al. (2006) Dominant bacteria in soils of Marble Point and Wright Valley, Victoria Land, Antarctica. Soil Biol Biochem 38: 3041–3056.
47. Fiedler S, Wagner D, Kutzbach L, Pfeiffer EM (2004) Element redistribution along hydraulic and redox gradients of low-center polygons, Lena Delta, northern Siberia. Soil Sci Soc Am J 68: 1002–1011.
48. Rivkina E, Shcherbakova V, Laurinavichius K, Petrovskaya L, Krivushin K, et al. (2007) Biogeochemistry of methane and methanogenic archaea in permafrost. FEMS Microbiol Ecol 61: 1–15.
49. Høj L, Olsen RA, Torsvik VL (2005) Archaeal communities in High Arctic wetlands at Spitsbergen, Norway (78°N) as characterized by 16S rRNA gene fingerprinting. FEMS Microbiol Ecol 53: 89–101.
50. Lawton JH, Brown VK (1993) Redundancy in ecosystems (pp. 255–270) Springer Berlin Heidelberg.
51. Bräuer SL, Cadillo-Quiroz H, Ward RJ, Yavitt JB, Zinder SH (2011) *Methanoregula boonei* gen. nov., sp. nov., an acidiphilic methanogen isolated from an acidic peat bog. Int J Syst Evol Micro 61: 45–52.
52. Cadillo-Quiroz H, Yavitt JB, Zinder SH (2009) *Methanosphaerula palustris* gen. nov., sp. nov., a hydrogenotrophic methanogen isolated from a minerotrophic fen peatland. Int J Sys Evol Micro 59: 928–935.
53. Conrad R, Schütz H, Babbel M (1987) Temperature limitation of hydrogen turnover and methanogenesis in anoxic paddy soil. FEMS Microbiol Lett 45: 281–289.
54. Bischoff J, Mangelsdorf K, Gattinger A, Schloter M, Kurchatova AN, et al. (2013) Response of methanogenic archaea to Late Pleistocene and Holocene climate changes in the Siberian Arctic. Global Biogeochem Cy 27: 305–317
55. Conrad R (2005) Quantification of methanogenic pathways using stable carbon isotopic signatures: a review and a proposal. Org Geochem 36: 739–752.
56. Morozova D, Wagner D (2007) Stress response of methanogenic archaea from Siberian permafrost compared with methanogens from nonpermafrost habitats. FEMS Microbiol Ecol 61: 16–25.
57. Wagner D, Schirmack J, Ganzert L, Morozova D, Mangelsdorf K (2013) *Methanosarcina soligelidi* sp. nov., a desiccation and freeze-thaw resistant methanogenic archaeon from a Siberian permafrost-affected soil. Int J Syst Evol Microbiol 63: 2986–2991.
58. Metje M, Frenzel P (2007) Methanogenesis and methanogenic pathways in a peat from subarctic permafrost. Environ Microbiol 9: 954–964
59. Kotsyurbenko OR (2005) Trophic interactions in the methanogenic microbial community of lowtemperature terrestrial ecosystems. FEMS Microbiol Ecol 53: 3–13.
60. Chanton JP, Bauer JE, Glaser PA, Siegel DI, Kelley CA, et al. (1995) Radiocarbon evidence for the substrates supporting methane formation within northern Minnesota peatlands. Geochim Cosmochim Act 59: 3663–3668.
61. Ström L, Ekberg A, Mastepanov M, Røjle Christensen T (2003) The effect of vascular plants on carbon turnover and methane emissions from a tundra wetland. Glob Change Biol 9: 1185–1192.
62. Mackelprang R, Waldrop M, DeAngelis K, David M, Chavarria K, et al. (2011) Metagenomic analysis of a permafrost microbial community reveals a rapid response to thaw. Nature 480: 368–371.
63. Leininger S, Urich T, Schloter M, Schwark L, Qi J, et al. (2006) Archaea predominate among ammonia-oxidizing prokaryotes in soils. Nature 442: 806–809.
64. Schimel JP, Kielland K, Chapin FS III (1996) Nutrient availability and uptake by tundra plants. J.F Reynolds, J.D Tenhunen (Eds.), Landscape Function and Disturbance in Arctic Tundra, Springer, Berlin (1996), pp. 203–221.
65. Shaver GR, Kummerow J (1992) Phenology, resource allocation, and growth of arctic vascular plants. Chapin FS III, Jefferies R, Reynolds J, Shaver G, Svoboda J (Eds.), Physiological Ecology of Arctic Plants: Implications for Climate Change, Academic Press, New York, pp. 193–212.
66. Liebner S, Wagner D (2007) Abundance, distribution and potential activity of methane oxidizing bacteria in permafrost soils from the Lena Delta, Siberia. Environ Microbiol 9: 107–117.
67. Luton PE, Wayne JM, Sharp RJ, Riley PW (2002) The *mcrA* gene as an alternative to 16S rRNA in the phylogenetic analysis of methanogen populations in landfill. Microbiology 148: 3521–3530.
68. Kolb S, Knief C, Stubner S, Conrad R (2003) Quantitative detection of methanotrophs in soil by novel pmoA-targeted real-time PCR assays. Appl Environ Microbiol 69: 2423–2429.
69. Le Roux X, Poly F, Currey P, Commeaux C, Hai B, et al. (2008) Effects of aboveground grazing on coupling among nitrifier activity, abundance and community structure. ISME J 2: 221–232.
70. Nicolaisen MH, Ramsing NB (2002) Denaturing gradient gel electrophoresis (DGGE) approaches to study the diversity of ammonia-oxidizing bacteria. J Microbiol Meth 50: 189–203.
71. Poly F, Monrozier LJ, Bally R (2001) Improvement in the RFLP procedure for studying the diversity of nifH genes in communities of nitrogen fixers in soil. Res Microbiol 152: 95–103.

8

Characterization of Epiphytic Bacterial Communities from Grapes, Leaves, Bark and Soil of Grapevine Plants Grown, and Their Relations

Guilherme Martins[1,2]*, Béatrice Lauga[3], Cécile Miot-Sertier[1], Anne Mercier[1], Aline Lonvaud[1], Marie-Louise Soulas[1], Guy Soulas[1], Isabelle Masneuf-Pomarède[1,2]

1 USC Oenologie-INRA, Université Bordeaux Segalen, ISVV, Villenave d'Ornon, France, **2** Bordeaux Science Agro, Gradignan cedex, Bordeaux, France, **3** Equipe Environnement et Microbiologie UMR IPREM 5254 IBEAS, Université de Pau et des Pays de l'Adour, Pau, France

Abstract

Despite its importance in plant health and crop quality, the diversity of epiphytic bacteria on grape berries and other plant parts, like leaves and bark, remains poorly described, as does the role of telluric bacteria in plant colonization. In this study, we compare the bacterial community size and structure in vineyard soils, as well as on grapevine bark, leaves and berries. Analyses of culturable bacteria revealed differences in the size and structure of the populations in each ecosystem. The highest bacteria population counts and the greatest diversity of genera were found in soil samples, followed by bark, grapes and leaves. The identification of isolates revealed that some genera – *Pseudomonas, Curtobacterium*, and *Bacillus* – were present in all ecosystems, but in different amounts, while others were ecosystem-specific. About 50% of the genera were common to soil and bark, but absent from leaves and grapes. The opposite was also observed: grape and leaf samples presented 50% of genera in common that were absent from trunk and soil. The bacterial community structure analyzed by T-RFLP indicated similarities between the profiles of leaves and grapes, on the one hand, and bark and soil, on the other, reflecting the number of shared T-RFs. The results suggest an interaction between telluric bacterial communities and the epiphytic bacteria present on the different grapevine parts.

Editor: Zaid Abdo, Agricultural Research Service, United States of America

Funding: This work was supported by the CIVB (Conseil Interprofessionnel du Vin de Bordeaux) and by Bordeaux Sciences Agro. The funders had no role in study design, data collection and analysis, decision to publish, or preparation of the manuscript.

Competing Interests: This work was supported by the CIVB (Conseil Interprofessionnel du Vin de Bordeaux) and by Bordeaux Sciences Agro.

* E-mail: oenopro.martins@gmail.com

Introduction

Plants sustain a complex microecosystem, which harbours a diverse array of bacteria, able to colonize different plant organs and tissues, including roots, leaves, flower clusters, seeds and fruits [1,2]. In so doing, plant-associated bacteria can affect crop health, due to their capacity to suppress or stimulate the colonization of tissues by plant pathogens [3]. In recent years, these bacteria have attracted considerable academic attention for their potential biotechnological applications [4,5].

So far studies of bacteria associated with grapevines have mainly focused on pathogens responsible for plant diseases, like Pierce's – caused by *Xylella fastidiosa* subsp. *fastidiosa* [6]; Crown gall – caused by *Agrobacterium tumefaciens* [7]; and bacterial blight, caused by *Xylophilus ampelinus* [8]. More recent investigations have studied the diversity of the grapevine endophytic bacteria, and have shown that *Pseudomonas* and *Bacillus* sp., can act as biological disease suppression agents, stimulating plant growth and health [9,10,11]. Few studies have investigated epiphytic bacteria on grapevines. Most research has concentrated on bacteria of oenological interest, like acetic and lactic acid bacteria [12,13,14,15,16], present in the microflora on grape berries. Little information is available about the effects of epiphytic bacteria on other plant parts, like leaves and bark. In a recent study, using high-throughput sequence analysis of 16s rRNA, Leveau and Tech [17], showed that the bacterial community on leaves differed, both in size and structure, from that on berries. They also reported a large diversity of bacteria associated with leaves and grapes, belonging to species known to be plant growth-promoters with significant activity against grapevine pathogens.

Therefore, grapevine bacteria play a key role not only in plant health, but also in crop quality and yields. The surface of grape berries represents a natural reservoir of bacterial microbiota that has various impacts on the sanitary quality of grapes and may influence the winemaking process, with major repercussions on wine quality [14,16,18]. Despite their importance, the diversity of epiphytic bacteria on grape berries remains poorly described, as is the role of other plant parts and vineyard soil in bacterial colonization. Soil microorganisms are able to colonize parts of the plant above the ground, including leaves and fruit [19,20,21]. Previous studies revealed the presence of pathogenic fungi and peronosporomycete, such as *Erysiphe necator* (formerly *Uncinula necator*) and *Plasmopara viticola* in vineyard soil [22,23], which also affect aerial parts of plants, like leaves, stems, flowers, and fruit.

Oenologically-important microorganisms, like *Saccharomyces cerevisiae*, the main yeast responsible for alcoholic fermentation, have been also isolated from vineyard soil [24,25,26].

Few studies have reported the presence in soil of bacterial species associated with grape berries. For instance, some lactic and acetic acid bacteria species found in wine environments have also been detected in vineyard soil [27].

These results suggest that vineyard soil may be a source of primary inoculum, able to participate in the structure of the microbial community on the aerial parts of the vine, including grapes. However, no comparative studies of the structure of bacterial communities on grapevine parts and in soil had previously been conducted. Consequently, it is difficult to evaluate the potential of plant- and soil-associated microflora to colonize grape berries.

In this study, culture-dependent and -independent methods were used to characterize and compare the bacterial community size and structure in vineyard soil, and on grapevine bark, leaves, and berries. Specific taxa from each ecosystem as well as taxa common to all ecosystems were identified. The results provide insights into the relations between epiphytic bacteria in different vineyard ecosystems.

Materials and Methods

Site description and sampling design

This study was performed in the Lussac St. Emilion wine-growing region, southwest France (44°57′15″N 0°06′12″ W, 77 m altitude), in 2010. Two vineyards were selected, spaced 400 meters apart. Both vineyards had very similar characteristics: grape variety (Merlot), age, pruning system, canopy management, and sun exposure.

Samples were collected at the beginning of berry ripening, corresponding to stage 34 of the modified E-L system for identifying major and intermediate grape vine growth stages [28]. Three sampling points were selected in each vineyard, each corresponding to five vines.

Ethics statement

All the samples in this study were collected on private proprieties, and the owners of the vineyards gave permission to conduct the study on these sites. No specific permissions were required for these locations, because there are no endangered or protected species in these areas, and this study did not involve endangered or protected species.

Soil sampling

Soil samples were collected in five randomly-chosen plots around each vine selected (15 to 25 cm from the trunk) at Ap horizon (0–5 cm depth) and mixed in sterile plastic bags. Samples were transported to the laboratory in refrigerated boxes and analyzed within 12 h after collection. Fresh soils were sieved (Ø<2 mm) to remove plant residues, soil macrofauna and stones. The soils had the following physico-chemical characteristics: 276 g/kg clay, 467 g/kg silt, 80 g/kg fine sand; 176 g/kg coarse sand; 39.8 g/kg OM, C/N 12, pH 6.59, CEC 13.1 cmol+/kg; in vineyard *I* and 354 g/kg clay, 373 g/kg silt, 80 g/kg fine sand; 189 g/kg coarse sand; 36 g/kg OM, C/N 11.5, pH 7.54, CEC 17.7 cmol+/kg in vineyard *II*.

Sampling grape berries, leaves and bark

At each sampling point, approximately 1 kg healthy, undamaged grapes, with their pedicels attached, was aseptically removed from several bunches and placed in sterile bags. Leaf and bark samples were collected from the same vines. Bark was collected avoiding damage to living tissue. To prevent cross contamination, sampling tools were sterilized with 75% ethanol before each sample. All samples were transported to the laboratory in refrigerated boxes and processed within 12 h.

Microbial biomass recovery

From each soil sample, aliquots of 1 g (dry weight equivalent) were taken, dispersed in a solution (10 ml) containing sodium hexametaphosphate (35 g/l) and sodium carbonate (7 g/l), and subjected to orbital shaking (Vibrax VXR Basic, 1200 rpm) for 1 h. These cell suspensions were used for downstream analysis.

Grape berry samples: 250 undamaged berries were randomly, aseptically removed from the bunches, placed in sterilized flasks with 500 ml isotonic solution containing 0.1% peptone and 0.01% Tween 80, and subjected to orbital shaking at 150 rpm for 1 h [29]. Leaves and bark: 50 g samples of each were placed in sterilized flasks and washed with 250 ml of the isotonic solution described above. These cell suspensions were used for downstream analysis. One part of the suspension was used to inoculate the culture medium and the rest was filtered through a 0.2 μm pore size, 47-mm diameter cellulose acetate filter (Sartorius AG, Göttingen, Germany) held in a stainless steel vacuum filtration unit (Millipore, Hertfordshire, UK).

DNA extraction

Soil DNA was extracted from 1 g (dry weight equivalent) soil using the UltraClean soil DNA isolation kit (MoBio Inc., Solana, CA, USA), as described by the manufacturer. DNA was extracted directly from the microbial biomass of grape berries, bark, and leaves retained on the membranes used to filter the cell suspensions.

The membranes were aseptically cut into small pieces and placed on the bead solution tubes provided by UltraClean soil DNA isolation kit. The extraction protocol was then continued, as described by the manufacturer.

Colony isolation and counting

Aerobic and aero-tolerant bacteria from the cell suspensions extracted from the soil, grape berries, leaves, and bark were cultured after spreading tenfold serial dilutions on 1/10 diluted LB culture medium (1 g/l bactotryptone, 1 g/l yeast extract, 0.5 g/l NaCl, and 20 g/l agar) with 150 mg/l biphenyl (Acros Organics, Belgium) to inhibit yeast and mould growth. Each dilution was prepared in triplicate.

Plates were incubated under aerobic conditions at 25°C. The number of colony-forming units was counted 5 days after inoculation. From each ecosystem in both vineyards, around 30 colonies were randomly picked from 1/10 diluted LB medium and purified by streaking onto fresh 1/10 diluted LB plates. The purity of each colony was verified and they were stored at −80°C on 33% glycerol stock for further genetic identification.

Identifying isolates by their 16 S rRNA gene sequence

The DNA was extracted from the isolates and stored using the FTA® CloneSaver™ card (Whatman® BioScience, USA), as described by Zott [30]. DNA was used as template for PCR amplification with 16S rRNA primers 8F (5′-AGAGTTTGATCCTGGCTCAG-3′) and 1063R (5′-ACGGGCGGTGTG TRC-3′) [31]. After Sanger sequencing of the amplicons (Plateforme Génomique Fonctionnelle, Université Victor Segalen, Bordeaux 2, France), the sequences were aligned and compared with references in the GenBank, using the NCBI Basic Local

Alignment Search Tools BLASTn program (http://www.ncbi.nlm.nih.gov/BLAST). The identification was considered valid when the identity of a contiguous sequence of 343 bp-989 bp was at least 98%. The 16 S rDNA sequences obtained were deposited in the EMBL Nucleotide Sequence Database under accession numbers HF566150 to F566373.

T-RFLP analysis of bacterial communities

DNA extracted from cell suspensions obtained from the soil, grape berries, leaves, and bark was amplified using nested PCR. A 1398 bp region of the 16S rRNA gene was first amplified using primers 8F and 1406R. In the second step, a 1055 bp region was amplified using primer 8F, fluorescently labelled at the 5′ end with 6-FAM (6-carboxyfluorescein), and 1063R (5′-CTCACGRCAC-GAGCTGACG-3′) [29]. The PCR was run in a final volume of 50 µl containing 10 mM Tris-HCl, 50 mM KCl, 1.5 mM $MgCl_2$, 0.2 mM dNTPs, 5% glycerol, 0.08% NP-40, 0.05% Tween-20, 25 units/ml Taq DNA polymerase, and 200 nM of each primer. PCR conditions for the first amplification step were: 95°C for 5 min, 20 cycles at 94°C and 58°C for 1 min each, 72°C for 1.5 min, and 72°C for 7 min. The same conditions were used for the second amplification step, reducing the number of cycles to 15. PCR products (100 ng) were purified using the Geneclean Turbo Kit (Qbiogene) and digested with 3 U *Hae* III or *Hinf* I enzymes (New England Biolabs). Fluorescently-labelled terminal-restriction fragments (T-RFs) were separated by capillary electrophoresis on an ABI prism 310 (Applied Biosystems). About 10 ng digested DNA was mixed with 9 µl de-ionised formamide and 0.5 µl 5-carboxytetramethylrhodamine size-standard TAMRA 500 (GenеScan^{TM}), denatured at 95°C for 5 min, and immediately chilled on ice prior to electrophoresis. After a 10 s injection step, electrophoresis was performed at a voltage of 15 kV for 30 min. T-RFLP peaks were analysed using GeneScan software (ABI), profiles were compiled to produce data matrices, background noise was reduced, and T-RF heights were normalized [32]. Only T-RFs between 50 bp and 500 bp were analysed. The profiles were aligned using T-align software [33], with a confidence interval of 1.0.

The 16 S rRNA gene of vine chloroplast was analyzed in order to eliminate operational taxonomic units (OTUs) of plant DNA. DNA was extracted from *in vitro*-grown merlot vine plantlets and used as a PCR template. PCR amplification and enzyme restriction with *Hinf*I and *Hae* III was carried out following the protocol described in 2.6. The profile generated revealed one peak with a molecular weight of 178 bp (*Hinf* I) and one with a molecular weight of 295 bp (*Hae* III). These two T-RFs were excluded from the sample profiles.

Statistical data analyses

Culture data. The statistical significance of the differences between bacterial counts in samples from the different ecosystems was tested by one way ANOVA followed by Tukey's honestly significant difference test (Tukey's HSD, $p<0.001$).

T-RFLP analysis. Statistical data were analyzed using Statistica V.7 software (Statsoft Inc., Tulsa, OK, USA). The phylotype richness (*S*) was calculated as the total number of distinct T-RF peaks in each normalized profile. Shannon–Weaver diversity indices *(H')* were calculated using peak heights as a metric of abundance for each T-RFLP profile, as described previously [34].

$$H' = \sum_{i=1}^{i=s} pi \ln(pi)$$

Shannon–Weaver evenness indices (*E*) were calculated as *H'*/*H'*max, where *H'*max = ln *(S)*. Principal component analysis of T-RFLP profiles was performed with samples as statistical observations and T-RFs as variables. Profiles were plotted as dots in the factor space and the original variables (each T-RF) were represented by arrows. The direction and length of the arrows indicate the contribution of the variables along the principal components.

To evaluate similarities between T-RFLP profiles, hierarchical clustering analysis was performed based on binary data, presence or absence of T-RFs, for all individuals. The Jaccard's index was used to quantify the similarity between datasets and UPGMA (Unweighted Pair Group *Method with* Arithmetic Mean) as a clustering method., The statistical significance of the differences between Shannon–Weaver diversity (H') and Evenness indices (E) from 16 S rRNA gene T-RFLP profiles was tested by one way ANOVA followed by Tukey's HSD test, ($p<0.001$).

Results

Size of culturable bacterial communities from the different plant parts and soil

Table 1 shows the culturable bacteria population densities. In both vineyards, the highest bacteria population counts were detected in the soil samples. The next highest was found in bark samples, followed by grapes and leaves. Tukey's honestly significant difference test (Tukey's HSD) was performed post-hoc. The results showed five homogeneous groups: soil from vineyards, bark from both vineyards, grapes from both vineyards, leaves from vineyards *I* and leaves from vineyard *II* respectively.

Diversity of culturable bacteria within different samples

From the randomly picked colonies, 5% and 7% did not survive subculturing, from vineyards *I* and *II*, respectively. From 16S

Table 1. Culturable bacteria populations in samples of soil, bark, leaves, and grape berries.

Samples	$\log_{10}$ CFU/g of fresh weight,	Tukey grouping				
		1	2	3	4	5
Leaves (vineyard I)	3,49 (±0.11)				****	
Leaves vineyard II	3,92 (±0.12)					****
Grapes (vineyard I)	4,52 (±0.04)	****				
Grapes vineyard II	4,65 (±0.09)	****				
Bark vineyard II	5,02 (±0.11)		****			
Bark (vineyard I)	5,15 (±0.12)		****			
Soil vineyard II	6,88 (±0.08)			****		
Soil (vineyard I)	6,97 (±0.08)			****		

Results expressed as $\log_{10}$ CFU/g of fresh weight, determined by plate counting on LB 1/10 medium. The values are the average of a triplicate experiment ± standard deviation. * denotes homogenous groups revealed by post-hoc tests (Tukey's HSD multiple-comparison test, $p<0.001$).

Table 2. Diversity of culturable bacteria within different samples.

Bacterial genera	Source of isolation				Source of isolation				
	Vineyard I				Vineyard II				
	Grapes	Leaves	Bark	Soil	Grapes	Leaves	Bark	Soil	Total
Acinetobacter sp.	0	0	0	1	0	0	0	0	**1**
Agrobacterium tumefaciens	0	0	1	1	0	0	1	1	**4**
Bacillus sp.	2	3	2	3	2	4	3	4	**23**
Brevibacterium sp.	0	0	1	0	1	0	1	0	**3**
Burkholderia sp.	1	0	2	1	1	0	0	1	**6**
Cellulomonas sp.	1	0	4	1	1	0	2	1	**10**
Clostridium sp.	0	0	0	5	0	0	0	4	**9**
Curtobacterium sp.	1	6	3	2	1	4	3	4	**24**
Enterobacter sp.	2	0	0	1	3	0	0	2	**8**
Massilia sp.	4	2	0	0	3	1	0	0	**10**
Micrococcus sp.	2	0	1	1	3	0	1	3	**11**
Paenibacillus sp.	0	0	0	0	0	0	0	2	**2**
Pantoea sp.	0	0	2	1	0	0	2	1	**6**
Pseudomonas sp.	9	10	5	3	7	10	4	3	**51**
Rhizobium sp.	0	0	1	2	0	0	0	2	**5**
Sphingomonas sp.	1	8	0	1	3	7	0	1	**21**
Staphylococcus sp.	0	0	2	1	0	0	2	1	**6**
Streptococcus sp.	0	0	0	1	0	0	0	0	**1**
Streptomyces sp.	0	0	1	1	0	0	1	1	**4**
Xanthobacter sp	0	0	2	1	0	0	3	1	**7**
Xanthomonas sp.	0	2	3	0	0	1	2	0	**8**
Xylanimonas sp.	0	0	1	0	0	0	0	0	**1**
Xylaphtlus sp.	0	0	1	0	0	1	0	0	**2**
Xylella sp.	0	0	1	0	0	0	0	0	**1**
Total	**23**	**31**	**33**	**27**	**25**	**28**	**25**	**32**	**224**

rDNA sequence analysis, the 224 remaining isolates were assigned to a specific genus, with a classification threshold above 98% (Table 2). The distribution of each genus varied according to the type of sample. However, there were similarities in the diversity and abundance of isolates from each ecosystem in both vineyards.

A total of 24 different genera were identified. The strains belonged to 6 different bacterial classes (*alpha-, beta-, and gamma-Proteobacteria, Actinobacteria, Clostridia,* and *Bacilli*).

Soil samples contained 16 and 17 genera, in vineyard *I* and *II*, respectively, followed by bark (14 and 12), grapes (9 and 10), and leaves (5 and 6). In grape samples, the genus *Pseudomonas* was the most abundant, followed by *Massilia, Micrococcus,* and *Bacillus*, but - unlike leaves, bark and soil – the occurrence of *Curtobacterium* sp. was low. In leaves, *Sphingomonas* and *Pseudomonas* were the most abundant genera, followed by *Bacillus* and *Curtobacterium*. Bark samples contained the highest percentage of *Xanthobacter*, *Xanthomonas*, and *Cellulomonas*. In soil samples, the predominant genus was *Clostridium*, followed by *Bacillus* and *Rhizobium*.

Some of the genera – *Pseudomonas*, *Curtobacterium* and *Bacillus* – were present in all ecosystems, although with varying abundance. Some other genera were ecosystem-specific, such as *Xylella* and *Xylanimonas* in bark and *Acinetobacter, Clostridium, Streptococcus,* and *Paenibacillus* in soil. Grape and leaf ecosystems did not contain any specific genera, but *Massilia* spp. was present in both grapes and leaves and absent from the other samples (soil and bark). Some genera (*Staphylococcus, Streptomyces, Rhizobium, Agrobacterium, Xanthobacter, Pantoea)* were always found together in soil and bark, but were absent from leaves and grapes. *Brevibacterium* were present in grapes and bark and *Enterobacter and Burkholderia sp.*in grapes and soil. Xylophilus was only present in leaves and bark. Between 42 and 50% (vineyard *I* and *II*, respectively) of bacterial genera identified in grape samples were also found on leaves. Around 55% of the genera identified in soil samples were also found in bark.

Taking into account some culture bias for the growth of specific microorganisms associated with culture-based detection method, we decided to use T-RFLP as community profiling tool to study the bacterial community structure from grapes, leaves, bark and soil.

Bacterial community structure analyzed by T-RFLP

Although the internal size marker ranged from 35 to 500 bp, no major T-RF peaks were found below 59 bp or above 492 bp. The samples showed considerable variability in the number of T-RFs and fluorescence intensity. In the profiles generated by *Hinf* I and *Hae* III digestion, a total of 69 and 72 different T-RFs, respectively, were conserved for analysis. Although there was only a slight difference in the number of unique T-RF peaks obtained using the two enzymes, the soil sample profiles obtained with *Hae* III generally exhibited more T-RF peaks than those obtained with

Table 3. Shannon–Weaver diversity and Evenness indices from 16S rRNA gene T-RFLP profiles.

A Vineyard *I*	**Evenness (E)**	**Tukey grouping**			**Diversity (H')**	**Tukey grouping**		
***(Hinf* I)**		**1**	**2**			**1**	**2**	
Leaves	0.57 (+/−0.08)	**			1.58 (+/−0.30)	**		
Grape berries	0.59 (+/−0.04)	**			1.56 (+/−0.26)	**		
Soil	0.68 (+/−0.04)		**		2.28 (+/−0.19)		**	
Bark	0.82 (+/−0.09)		**		2.68 (+/−0.1)		**	
(*Hae* III)		1	2	3		1	2	3
Leaves	0.42 (+/−0.09)	**			1.16 (+/−0.26)	**		
Grape berries	0.5 (+/−0.02)	**			1.33 (+/−0.36)	**		
Soil	0.89 (+/−0.03)		**		3.21 (+/−0.02)		**	
Bark	0.79 (+/−0.04)			**	2.68 (+/−0.11)			**
B Vineyard *II*	**Evenness (E)**	**Tukey grouping**			**Diversity (H')**	**Tukey grouping**		
***(Hinf* I)**		**1**	**2**			**1**	**2**	
Leaves	0.65 (+/−0.08)	**			1.89 (+/−0.15)	**		
Grape berries	0.43 (+/−0.09)	**			1.22 (+/−0.13)	**		
Soil	0.75 (+/−0.05)		**		2.44 (+/−0.1)		**	
Bark	0.7 (+/−0.02)		**		2.26 (+/−0.2)		**	
(*Hae* III)		1	2	3		1	2	3
Leaves	0.66 (+/−0.06)	**			1.94 (+/−0.28)	**		
Grape berries	0.6 (+/−0.03)	**			1.71 (+/−0.07)	**		
Soil	0.88 (+/−0.02)		**		3.17 (+/−0.02)		**	
Bark	0.83 (+/−0.03)			**	2.74 (+/−0.1)			**

Values in brackets represent standard deviation (n = 3). * denotes homogenous groups revealed by post-hoc tests (Tukey's HSD multiple-comparison test, $p < 0.001$) based on the comparisons between the (E) and (H') values of different samples (n = 6).

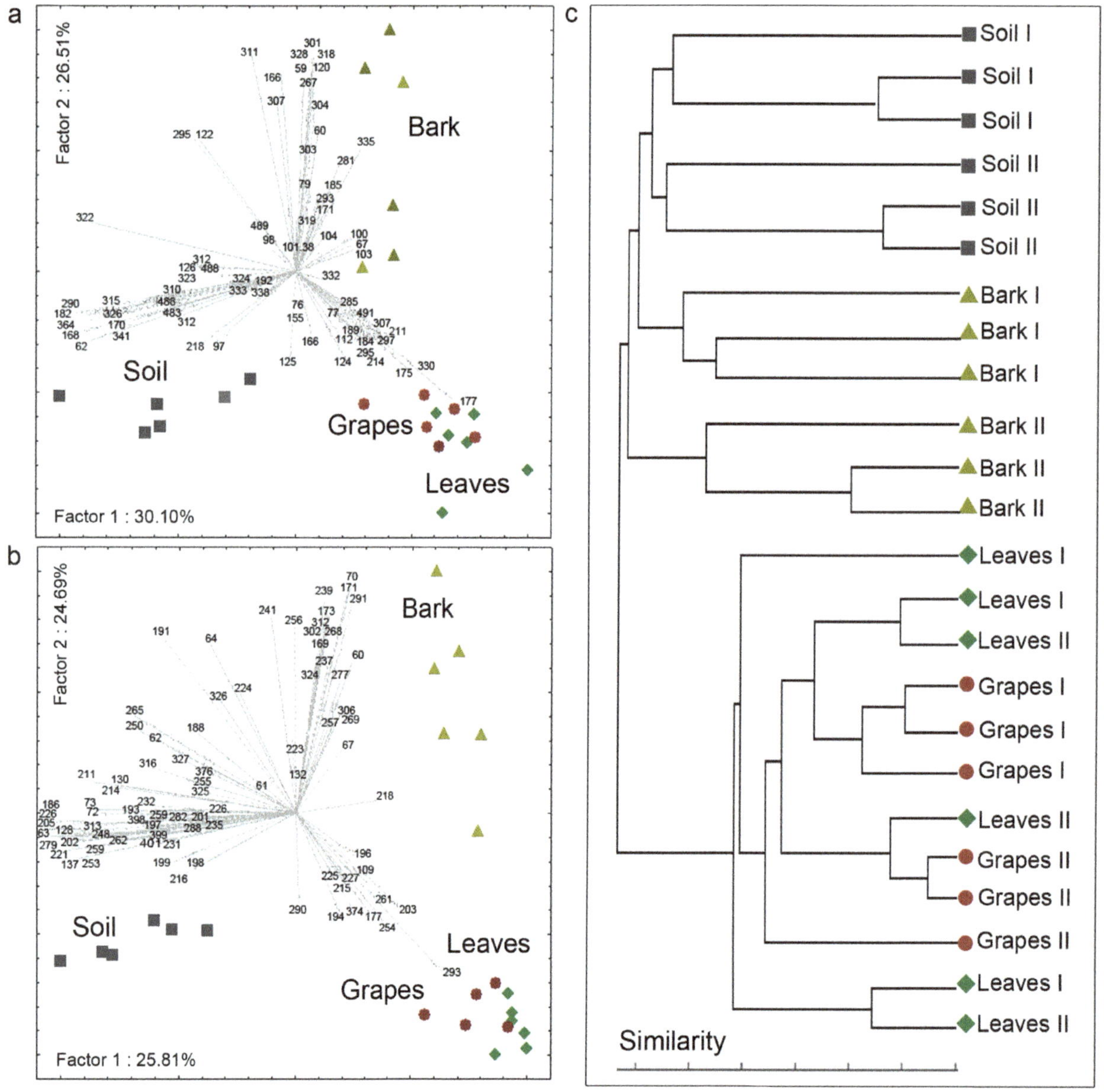

Figure 1. Comparison of bacterial colonization patterns in different ecosystems. Principal-component analysis based on bacterial community structure, assessed by 16S rRNA gene T-RFLP (including TRF size and relative abundance data), using Hinf I digest, from vineyard I (a) and vineyard *II* (b); the amount of variability accounted for by each factor is shown on the axes. (c) Clustering obtained by the hierarchical UPGMA method based on the arithmetical complement of the Jaccard similarity index of samples from both vineyards. Triangles (Δ) indicate bark samples; (◇) diamonds indicate leaf samples, squares (□) indicate soil samples; circles (○) indicate grape samples.

Hinf I. As to the evenness and diversity of the different profiles, there was a significant difference in terms of the (E) and (H') index (Table 3). Soil samples presented the highest (E) and (H') values, whereas leaves presented the lowest. Post-hoc tests revealed different homogeneous groups (Tukey's HSD, $P<0.001$). In both vineyards, the evenness and diversity values obtained with *Hinf* I revealed two groups: leaves and grapes on the one hand, and bark and soil on the other. The corresponding values obtained with *Hae* III fell into three different groups. Leaves and grapes exhibited the least diversity and evenness; bark had an intermediate degree; and soil had the highest degree of both. A Pearson correlation analysis revealed a positive correlation between (E) and (H') values ($r=0\ .79$, p-value <0.001).

Comparison of bacterial colonization patterns in different ecosystems

The results of PCA analysis of the T-RFLP profiles of the samples from each vineyard are shown in Figure 1 a and b. The samples were represented as four different types of dots, according to the Figures 2a and 2b represent the profiles obtained with *Hinf* I for samples from vineyards *I* and *II*. PCA demonstrated that the profiles obtained from leaves and grapes were very similar to each other. Samples from both vineyards were grouped together on the plot, indicating similarities between the bacterial communities present in these two ecosystems. Soil samples were clustered together on the opposite side of the plot from bark. In both vineyards *I* and *II* soil and bark samples were clearly distinguishable from those from leaves and grapes.

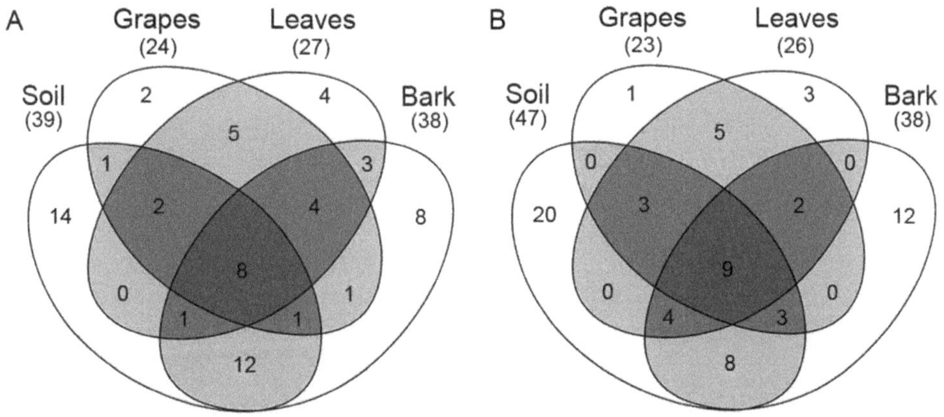

Figure 2. Four-way Venn diagram depicting the shared and unshared T-RFs between the four niches. Profiles obtained with *Hinf* I (A) and *Hae* III (B) endonucleases from samples of both vineyards *I* and *II*. Numbers in parenthesis indicate the total number of TRFs for each niche.

In order to assess the impact of the vineyard on the profiles, all the samples together were subjected to cluster analysis. Based on the profiles obtained using the *Hinf* I enzyme (Figure 1c), two major groups were defined – soil and bark versus leaves and grapes. In the soil and in bark sub-groups, there was a clear difference between vineyards *I* and *II*. Leaves and grapes were grouped together in several clusters, with no significant inter-vineyard differences. The only exception was grape samples from vineyard *I*, which formed a separate cluster. The results obtained using *Hae* III digestion were similar (data not shown).

The similarities between the leaf and grape profiles, on the one hand, and bark and soil, on the other, reflect the number of T-RFs shared by those samples. Soil and bark samples shared a large number of T-RFs, as did leaves and grapes. In contrast, neither soil nor bark shared many T-RFs with either leaves or grapes. A 4-way Venn diagram (Figure 2) of T-RFs revealed that soil and bark had more exclusive T-RFs than either leaves or grapes. Some T-RFs were present in all ecosystems; 8 and 9 T-RFs with *Hinf* I and *Hae* III digestion, respectively, were common to all profiles.

Phylogenetic assignment of T-RFs

PCA revealed correlations between T-RFs and each ecosystem. The T-RFs with larger factor loading in the direction of each cluster were selected. Their putative identities were predicted by *in silico* digestion with both *Hinf* I and *Hae* III enzymes, using the Phylogenetic Assignment Tool (PAT+) [35] provided by Microbial Community Analysis III (MiCA 3) (http://mica.ibest.uidaho.edu) [36], based on the RDP Release 10, 16S rRNA gene data base. One limitation of this kind of analysis is the inability to affiliate OTUs to phylogenetic groups with any degree of reliability, since T-RFs of the same size may yield phylogenetically disparate 16 S rRNA gene sequences [37]. In order to minimize this bias and obtain more reliable identification, profiles obtain with both enzymes were analyzed. Phylogenetic assignment of the T-RFs was grouped at the phylum level. Members of *Actinobacteria*, *Firmicutes*, and *Proteobacteria* corresponded to T-RFs present in all different clusters. The soil clusters were dominated by species classified as *Actinobacteria*, *Betaproteobacteria*, and *Clostridia*. Bark clusters presented T-RFs assigned to *Actinobacteria* and *Clostridia* and a considerable number of *Proteobacteria* (*Alphaproteobacteria*, *Epsilon-proteobacteria*, and *Gammaproteobacteria*). The majority of T-RFs from leaves and berries were classified as *Proteobacteria* (*Gammaproteobacteria* and *Alphaproteobacteria*).

Discussion

Recent studies indicate that bacterial populations on wine grapes are much more diverse than previously supposed [38,18]. Little is known about the diversity of the epiphytic bacteria associated with other grapevine parts, such as leaves or trunk bark, as well as bacteria living in vineyard soil. In this work, culture-dependent and independent methods were combined to compare the composition of epiphytic bacterial communities present on different plant parts and in soil. These approaches revealed differences in population density and diversity among different samples from the various ecosystems. Soil and bark hosted a greater diversity and species richness than grapes and leaves.

In addition, a comparison of all the profiles from both vineyards revealed similarities between leaves and grapes, on the one hand, and bark and soil, on the other.

Many factors are likely to be involved in determining the species composition of bacterial communities in plants and soil. They include the availability of immigrant inoculum [39,40], host plant phenology [41,38], physico-chemical environmental conditions [42,43], and nutritional characteristics of the phyllosphere or soil [4,44]. The variability in nutrient supply between these niches may partly explain the differences in bacterial community structure observed, as well as the diversity of culturable genera among the different plant parts and in soil.

Soil provides a large variety of carbon sources, including amino acids, organic acids, and carbohydrates, used by micro-organisms to obtain energy [45]. Bark contains starch, sugars [46,47], and other nutrients from xylem sap exudation [48]. These two nutrient-rich ecosystems are favourable to a greater number and more diverse range of bacteria.

The fact that bark contains high concentrations of cellulose, hemicellulose, lignin, and xylan [46] explains the abundance of *Xylanimonas*, *Xanthobacter*, *Xanthomonas*, and *Cellulomonas* found in these samples. These genera are associated with high-cellulose and xylan ecosystems, like decayed wood and bark [49,50,51].

In contrast to soil and bark, the micro-environments associated with leaves are generally considered nutrient-limited [52,4]. Grape skins, especially in the early stages of ripening, also provide a limited amount of nutrients to sustain bacterial growth [53,54,55]. Furthermore, the surface of leaves and grapes contains compounds such as stilbenes (resveratrol and derivatives), which are involved in plant and fruit defences against microbial activity [56,57]. In our study, the most abundant genera on leaves and grapes were

Pseudomonas and *Sphingomonas*. These ubiquitous genera are known for their ability to grow under low-nutrient conditions [58,59]. These results are in agreement with the findings of Leveau and Tech [17], that these two genera are among the most prevalent on vine leaves and grapes. Our results concerning the presence on grapes and leaves of members of genera commonly found in soil such as *Curtobacterium* and *Bacillus*, is also in agreement with that study. This observation and the fact that some T-RFs were shared by soil and all plant ecosystems, raised the hypothesis of a possible ecological link between telluric bacteria and epiphytic communities on the aerial parts of plants.

The colonisation of internal tissues by soil bacteria that thrive as endophytes and spread from the roots to different aerial plant parts through xylem vessels is well established [9]. While the endophytic continuum in grapevines is confirmed, the role of soil bacteria in epiphytic colonization of grapevine parts requires further elucidation. The physical proximity between soil and the various grapevine parts make this hypothesis very likely. In the most common grapevine training systems, the vine trunk and canopy are near the ground, facilitating the migration of micro-organisms from the soil to aerial parts of the plant through rain splash, high winds, insects, etc., [60,61,62,63]. Mechanical soil management, like tillage, a common practice used to control weeds in vineyards, may also contribute to the migration of telluric micro-organisms to the aerial part of the plant. By breaking up soil aggregates, tillage often generates dust that may deposit on leaves, berries, and trunk, inoculating them with bacteria.

The importance of trunk bark as a potential source of inoculum for leaves and grapes should be also considered. They are physically very close to each other and may even come into direct contact. Prior investigations revealed the presence in bark of grape berry pathogenic fungi, such as *Uncinula necator* [62,64,22], *Botrytis cinerea*, *Fusarium laterium*, *Penicillium* spp., *Phomopsis viticola*, [65,66], and the yeast-like fungus *Aureobasidium pullulans*, known for its biocontrol activity on grape berry pathogens. There is little information on the bacterial population in bark. However, Munkvold and Marois [65], found bacterial strains belonging to the *Pseudomonas* and *Bacillus* genera. Our study confirmed the presence of those genera on bark and grapes, suggesting that the bark community may influence leaf and grape-berry population structure.

The interaction between bacterial populations in vineyard soil and the epiphytic bacteria present on the various parts of grapevines, suggests that part of the plant epiphytic population may have a telluric origin. In addition, the bacterial population of the vegetative (leaf) and reproductive (fruit) structures of the vine may also be affected by trunk bark.

This first investigation is of particular importance, considering the role of bacteria in plant health and the fact that grape berries are the primary source of microbial communities that play a prominent role in the winemaking process and impact wine quality.

Further research is required to extend these observations and explore the ecological interaction between these different ecosystems. The use of other culture-independent approaches, like next-generation sequencing methods, will make it possible to present a complete survey of the bacterial communities on grape berries, leaves, and bark, and in vineyard soil.

Acknowledgments

The authors wish to thank the owners of Lussac and Pomerol vineyards for their kind cooperation in sampling.

Author Contributions

Conceived and designed the experiments: GM GS IM-P. Performed the experiments: GM BL CM-S AM M-LS. Analyzed the data: GM BL GS IM-P. Contributed reagents/materials/analysis tools: GM AL BL GS IM-P. Wrote the paper: GM AL GS IM-P.

References

1. Hallmann J, Quadt-Hallmann A, Mahaffee WF, Kloepper JW (1997) Bacterial endophytes in agricultural crops. Can J Microbiol 43: 895–914.
2. Gray EJ, Smith DL (2005) Intracellular and extracellular PGPR: commonalities and distinctions in the plant–bacterium signaling processes. Soil Biol Biochem 37: 395–412.
3. Lindow SE, Brandl MT (2003) Microbiology of the phyllosphere. Appl Environ Microbiol 69: 1875–1883.
4. Compant S, Duffy B, Nowak J, Clément C, Barka EA (2005) Use of Plant Growth-Promoting Bacteria for Biocontrol of Plant Diseases: Principles, Mechanisms of Action, and Future Prospects. Appl Environ Microbiol 71: 4951–4959.
5. Van Overbeek LS, Cassidy M, Kozdroj J, Trevors JT, van Elsas JD (2002) A polyphasic approach for studying the interaction between *Ralstonia solanacearum* and potential control agents in the tomato phytosphere. J. Microbiol. Methods 48: 69–86.
6. Newman KL, Almeida RPP, Purcell AH, Lindow SE (2003) Use of a Green Fluorescent Strain for Analysis of *Xylella fastidiosa* Colonization of *Vitis vinifera*. Appl Environ Microbiol 69: 7319–7327.
7. Burr TJ (1983) Isolation of *Agrobacterium tumefaciens* Biovar 3 from Grapevine Galls and Sap, and from Vineyard Soil. Phytopathology 73: 163.
8. Panagopoulos CG (1969) The disease "tsilik marasi" of grapevine: its description and identification of the causal agent (*Xanthomonas ampelina* sp. *nov.*). Ann Inst Phytopathol Benaki 9: 59–81.
9. Compant S, Mitter B, Colli-Mull JG, Gangl H, Sessitsch A (2011) Endophytes of grapevine flowers, berries, and seeds: identification of cultivable bacteria, comparison with other plant parts, and visualization of niches of colonization. Microb Ecol 62: 188–197.
10. West ER, Cother EJ, Steel CC, Ash GJ (2010) The characterization and diversity of endophytes of grapevine. Can J Microbiol 56: 209–216.
11. Bulgari D, Casati P, Brusetti L, Quaglino F, Brasca M, et al. (2009) Endophytic bacterial diversity in grapevine (Vitis vinifera L.) leaves described by 16S rRNA gene sequence analysis and length heterogeneity-PCR. J Microbiol 47: 393–401.
12. Joyeux A, Lafon-Lafourcade S, Ribéreau-Gayon P (1984) Evolution of acetic acid bacteria during fermentation and storage of wine. Appl Environ Microbiol 48: 153–156.
13. Lafon-Lafourcade S, Carre E, Ribéreau-Gayon P (1983) Occurrence of lactic acid bacteria during the different stages of vinification and conservation of wines. Appl Environ Microbiol 46: 874–880.
14. Barbe J-C, De Revel G, Joyeux A, Bertrand A, Lonvaud-Funel A (2001) Role of botrytized grape micro-organisms in SO_2 binding phenomena. J Appl Microbiol 90: 34–42.
15. Bae S, Fleet GH, Heard GM (2006) Lactic acid bacteria associated with wine grapes from several Australian vineyards. J Appl Microbiol 100: 712–727.
16. Nisiotou AA, Rantsiou K, Iliopoulos V, Cocolin L, Nychas G-JE (2011) Bacterial species associated with sound and *Botrytis*-infected grapes from a Greek vineyard. Int J Food Microbiol 145: 432–436.
17. Leveau JHJ, Tech JJ (2011) Grapevine microbiomics: Bacterial diversity on grape leaves and berries revealed by high-throughput sequence analysis of 16S rRNA amplicons. Acta Hort (ISHS) 905: 31–42.
18. Verginer M, Leitner E, Berg G (2010) Production of Volatile Metabolites by Grape-Associated Microorganisms. J Agric Food Chem 58: 8344–8350.
19. Compant S, Clément C, Sessitsch A (2010) Plant growth-promoting bacteria in the rhizo- and endosphere of plants: Their role, colonization, mechanisms involved and prospects for utilization. Soil Biol Biochem 42: 669–678.
20. Rekah Y, Shtienberg D, Katan J (2000) Disease Development Following Infection of Tomato and Basil Foliage by Airborne Conidia of the Soilborne Pathogens *Fusarium oxysporum f.* sp. *radicis-lycopersici* and *F. oxysporum f.* sp. *basilici*. Phytopathology 90: 1322–1329.
21. Horn BW (2005) Colonization of wounded peanut seeds by soil fungi: selectivity for species from *Aspergillus* section *Flavi*. Mycologia 97: 202–217.
22. Grove GG (2004) Perennation of *Uncinula necator* in Vineyards of Eastern Washington. Plant Dis 88: 242–247.
23. Kennelly MM, Gadoury DM, Wilcox WF, Magarey PA, Seem RC (2007) Primary Infection, Lesion Productivity, and Survival of Sporangia in the Grapevine Downy Mildew Pathogen *Plasmopara viticola*. Phytopathology 97: 512–522.
24. Davenport RR (1973) Vineyard yeasts – an environmental study. In: Lovelock DH (ed) Sampling – Microbiological Monitoring of Environments. Academic Press, London, pp.143–174.

25. Sabate J, Cano J, Esteve-Zarzoso B, Guillamón JM (2002) Isolation and identification of yeasts associated with vineyard and winery by RFLP analysis of ribosomal genes and mitochondrial DNA. Microbiol Res 157: 267–274.
26. Goddard MR, Anfang N, Tang R, Gardner RC, Jun C (2010) A distinct population of *Saccharomyces cerevisiae* in New Zealand: evidence for local dispersal by insects and human-aided global dispersal in oak barrels. Environ Microbiol 12: 63–73.
27. Yanagida F, Srionnual S, Chen Y-S (2008) Isolation and characteristics of lactic acid bacteria from koshu vineyards in Japan. Lett Appl Microbiol 47: 134–139.
28. Coombe BG (1995) Growth Stages of the Grapevine: Adoption of a system for identifying grapevine growth stages. Aust J Grape Wine Res 1: 104–110.
29. Prakitchaiwattana C, Fleet G, Heard G (2004) Application and evaluation of denaturing gradient gel electrophoresis to analyse the yeast ecology of wine grapes. FEMS Yeast Res 4: 865–877.
30. Zott K, Miot-Sertier C, Claisse O, Lonvaud-Funel A, Masneuf-Pomarede I (2008) Dynamics and diversity of non-*Saccharomyces* yeasts during the early stages in winemaking. Int J Food Microbiol 125: 197–203.
31. Wang Y, Qian P-Y (2009) Conservative Fragments in Bacterial 16S rRNA Genes and Primer Design for 16S Ribosomal DNA Amplicons in Metagenomic Studies. PLoS ONE 4: e7401.
32. Dunbar J, Ticknor LO, Kuske CR (2001) Phylogenetic Specificity and Reproducibility and New Method for Analysis of Terminal Restriction Fragment Profiles of 16S rRNA Genes from Bacterial Communities. Appl Environ Microbiol 67: 190–197.
33. Smith CJ, Danilowicz BS, Clear AK, Costello FJ, Wilson B, et al. (2005) T-Align a web-based tool for comparison of multiple terminal restriction fragment length polymorphism profiles. FEMS Microbiol Ecol 54: 375–380.
34. Hill TCJ, Walsh KA, Harris JA, Moffett BF (2003) Using ecological diversity measures with bacterial communities. FEMS Microbiol Ecol 43: 1–11.
35. Kent AD, Smith DJ, Benson BJ, Triplett EW (2003) Web-Based Phylogenetic Assignment Tool for Analysis of Terminal Restriction Fragment Length Polymorphism Profiles of Microbial Communities. Appl Environ Microbiol 69: 6768–6776.
36. Shyu C, Soule T, Bent S, Foster J, Forney L (2007) MiCA: A Web-Based Tool for the Analysis of Microbial Communities Based on Terminal-Restriction Fragment Length Polymorphisms of 16S and 18S rRNA Genes. Microb Ecol 53: 562–570.
37. Schmitt-Wagner D, Friedrich MW, Wagner B, Brune A (2003) Phylogenetic Diversity, Abundance, and Axial Distribution of Bacteria in the Intestinal Tract of Two Soil-Feeding Termites (*Cubitermes* spp.). Appl Environ Microbiol 69: 6007–6017.
38. Martins G, Miot-Sertier C, Lauga B, Claisse O, Lonvaud-Funel A, et al. (2012) Grape berry bacterial microbiota: Impact of the ripening process and the farming system. Int J Food Microbiol 158: 93–100.
39. Kinkel LL, Andrews JH, Berbee FM, Nordheim EV (1987) Leaves as islands for microbes. Oecologia 71: 405–408.
40. Lindemann J, Upper CD (1985) Aerial Dispersal of Epiphytic Bacteria over Bean Plants. Appl Environ Microbiol 50: 1229–1232.
41. Blakeman JB (1985) Ecological succession of leaf surface microorganisms in relation to biological control. In: Windels CE and Lindow SE (ed).Biological Control on the Phylloplane. Am. Phytopath. Soc. Press, Minnesota, pp.6–30.
42. O'Brien RD, Lindow SE (1989) Effect of plant species and environmental conditions on epiphytic population sizes of *Pseudomonas syringae* and other bacteria. Phytopathology 79: 619–627.
43. Berg G, Smalla K (2009) Plant species and soil type cooperatively shape the structure and function of microbial communities in the rhizosphere. FEMS Microbiol Ecol 68: 1–13.
44. Marschner P, Crowley D, Yang CH (2004) Development of specific rhizosphere bacterial communities in relation to plant species, nutrition and soil type. Plant Soil 261: 199–208.
45. Wawrik B, Kerkhof L, Kukor J, Zylstra G (2005) Effect of Different Carbon Sources on Community Composition of Bacterial Enrichments from Soil. Appl Environ Microbiol 71: 6776–6783.
46. Winkler AJ, Williams WO (1945) Starch and sugars of *Vitis vinifera*. Plant Physiol 20: 412–432.
47. Hunter JJ, Ruffner HP (2001) Assimilate transport in grapevines -effect of phloem disruption. Aust J Grape Wine Res 7: 118–126.
48. Glad C, Regnard JL, Querou Y, Brun O, Morot-Gaudry JF (1992) Flux and chemical composition of xylem exudates from chardonnay grapevines: temporal evolution and effect of recut. Am J Enol Vitic 43: 275–282.
49. Line MA (1997) A nitrogen-fixing consortia associated with the bacterial decay of a wooden pipeline. Lett Appl Microbiol 25: 220–224.
50. Rivas R, Sánchez M, Trujillo ME, Zurdo-Piñeiro JL, Mateos PF, et al. (2003) *Xylanimonas cellulosilytica* gen. nov., *sp.* nov., a xylanolytic bacterium isolated from a decayed tree (*Ulmus nigra*). Int J Syst Evol Microbiol 53: 99–103.
51. Goto M (1972) Survival of *Xanthomonas citri* in the bark tissues of citrus trees. Can J Bot 50: 2629–2635.
52. Andrews JH (1992) Biological control in the phyllosphere. Annu. Rev. Phytopathol 30: 603–635.
53. Padgett M, Morrison JC (1990) Changes in grape berry exudates during fruit development and their effect on mycelial growth of *Botrytis cinerea*. J Amer Soc Hort Sci 115: 269–273.
54. Coombe BG, McCarthy MG (2000) Dynamics of grape berry growth and physiology of ripening. Aust J Grape Wine Res 6: 131–135.
55. Martins G (2012) Communautés microbiennes de la baie de raisin: Incidence des facteurs biotiques et abiotiques.Ph.D. Thesis, Université Bordeaux 2.
56. Bavaresco L et al. (1997) Elicitation and accumulation of stilbene phytoalexins in grapevine berries infected by *Botrytis cinerea*. Vitis 36: 77–83
57. Harm A, Kassemeyer H-H, Seibicke T, Regner F (2011) Evaluation of Chemical and Natural Resistance Inducers Against Downy Mildew (*Plasmopara viticola*) in Grapevine. Am J Enol Vitic 62: 184–192.
58. Park JY et al. (2011) Draft genome sequence of the biocontrol bacterium *Pseudomonas putida* B001, an oligotrophic bacterium that induces systemic resistance to plant diseases. J Bacteriol 193: 6795–6796.
59. Hirano SS, Upper CD (2000) Bacteria in the Leaf Ecosystem with Emphasis on *Pseudomonas syringae*–a Pathogen, Ice Nucleus, and Epiphyte. Microbiol Mol Biol Rev 64: 624–653.
60. Venette JR, Kennedy BW (1975) Naturally Produced Aerosols of *Pseudomonas glycinea*. Phytopathology 65: 737–738.
61. Lacey J (1996) Spore dispersal – its role in ecology and disease: the British contribution to fungal aerobiology. Mycol Res 100: 641–660.
62. Behar A, Jurkevitch E, Yuval B (2008) Bringing back the fruit into fruit fly–bacteria interactions. Mol Ecol 17: 1375–1386.
63. Whipps JM, Hand P, Pink D, Bending GD (2008) Phyllosphere microbiology with special reference to diversity and plant genotype. J Appl Microbiol 105: 1744–1755.
64. Cortesi P, Bisiach M, Ricciolini M, Gadoury DM (1997) Cleistothecia of *Uncinula necator*–An Additional Source of Inoculum in Italian Vineyards. Plant Dis 81: 922–926.
65. Munkvold GP, Marois JJ (1993) Efficacy of Natural Epiphytes and Colonizers of Grapevine Pruning Wounds for Biological Control of Eutypa Dieback. Phytopathology 83: 624–629.
66. Barata A, Malfeito-Ferreira M, Loureiro V (2012) The microbial ecology of wine grape berries. Int J Food Microbiol 153: 243–259.

9

Bacteria and Fungi Respond Differently to Multifactorial Climate Change in a Temperate Heathland, Traced with ^{13}C-Glycine and FACE CO_2

Louise C. Andresen[1]*, Jennifer A. J. Dungait[2], Roland Bol[2,3], Merete B. Selsted[4], Per Ambus[4], Anders Michelsen[1]

1 Department of Biology, University of Copenhagen, Copenhagen, Denmark, 2 Sustainable Soils and Grassland Systems Department, Rothamsted Research-North Wyke, Okehampton, United Kingdom, 3 Institute of Bio- and Geosciences, Forschungszentrum Jülich, Jülich, Germany, 4 Chemical and Biochemical Engineering, Technical University of Denmark, Kongens Lyngby, Denmark

Abstract

It is vital to understand responses of soil microorganisms to predicted climate changes, as these directly control soil carbon (C) dynamics. The rate of turnover of soil organic carbon is mediated by soil microorganisms whose activity may be affected by climate change. After one year of multifactorial climate change treatments, at an undisturbed temperate heathland, soil microbial community dynamics were investigated by injection of a very small concentration (5.12 µg C g^{-1} soil) of ^{13}C-labeled glycine ($^{13}C_2$, 99 atom %) to soils *in situ*. Plots were treated with elevated temperature (+1°C, T), summer drought (D) and elevated atmospheric carbon dioxide (510 ppm [CO2]), as well as combined treatments (TD, TCO2, DCO2 and TDCO2). The ^{13}C enrichment of respired CO_2 and of phospholipid fatty acids (PLFAs) was determined after 24 h. ^{13}C-glycine incorporation into the biomarker PLFAs for specific microbial groups (Gram positive bacteria, Gram negative bacteria, actinobacteria and fungi) was quantified using gas chromatography-combustion-stable isotope ratio mass spectrometry (GC-C-IRMS). Gram positive bacteria opportunistically utilized the freshly added glycine substrate, *i.e.* incorporated ^{13}C in all treatments, whereas fungi had minor or no glycine derived ^{13}C-enrichment, hence slowly reacting to a new substrate. The effects of elevated CO_2 did suggest increased direct incorporation of glycine in microbial biomass, in particular in G^+ bacteria, in an ecosystem subjected to elevated CO_2. Warming decreased the concentration of PLFAs in general. The FACE CO_2 was ^{13}C-depleted ($\delta^{13}C = 12.2‰$) compared to ambient ($\delta^{13}C = \sim -8‰$), and this enabled observation of the integrated longer term responses of soil microorganisms to the FACE over one year. All together, the bacterial (and not fungal) utilization of glycine indicates substrate preference and resource partitioning in the microbial community, and therefore suggests a diversified response pattern to future changes in substrate availability and climatic factors.

Editor: Hauke Smidt, Wageningen University, Netherlands

Funding: The CLIMAITE experiment is sponsored by the Villum Kann Rasmussen Foundation with co-funding from Air Liquide, DONG Energy and SMC Pneutmatic A/S. The funders had no role in study design, data collection and analysis, decision to publish, or preparation of the manuscript.

Competing Interests: The authors declare that their commercial funders are Air Liquide, DONG Energy and SMC Pneutmatic A/S.

* E-mail: louise.andresen@gu.se

Introduction

Soils act as potential sinks or sources for atmospheric carbon dioxide (CO_2) depending on the balance between carbon (C) inputs from net primary production (NPP) and the C efflux by autotrophic and heterotrophic respiration [1]. The organomineral soils of heathlands in the Northern Hemisphere are essential for global C storage and it is vital to understand whether these ecosystems will be a net sink of C under predicted climate change scenarios. Their soils are characteristically of low pH (<5) and of low nutrient availability because of the nutrient poor litter produced by the plant ecotypes, and the positive feedback from symbiotic ericoid and ectomycorrhizal fungi which are highly developed to scavenge for labile nutrient sources [2]. A switch from fungal-to-bacterial dominance in these systems caused by climate change could trigger the release of vast amounts of C from the terrestrial pool. Microbial utilization of 'old' or 'new' carbon (C) from soil organic matter (SOM) may cause a release of C from previously sequestered C stocks (i.e. 'priming' [3,4]) or from recently assimilated CO_2 [5], which influences the stabilization of C in soils [6,7].

The quantity and quality of the organic matter (SOM) in soils may be strongly influenced by the combination of changes in atmospheric CO_2 concentration, air and soil temperature and moisture regime, which directly affects NPP [1,8–10], and is additionally regulated via other ecosystem processes, such as nutrient availability and herbivore abundance [11,12]. In C3 plants in temperate heathlands, CO_2 assimilation can be stimulated by elevated concentrations of CO_2 [13–15], but if plant metabolism is limited by other factors such as nutrient availability, stores of carbohydrates may accumulate leading to a total higher tissue C to nitrogen (N) ratio [16]. The relatively smaller N content partially persists through wilting, but along with alteration of lipids (e.g. carboxylic acids; [17]), and suberin and cutin concentrations [18], the litter input to SOM may be of slightly poorer quality for decomposing organisms. In addition, elevated CO_2 stimulates the efflux of labile organic compounds from roots into the soil, which diffuse passively due to the concentration gradient [19]. In plants under elevated CO_2, glycine may be routed into the rhizosphere [20] due to changes in the citric-acid cycle caused by low nutrient availability. Overall, elevated CO_2 may modify SOM quality through altered plant litter deposition and root exudation [21], while by contrast microbial residues in the SOM may not be affected [22]. Changes in the magnitude and chemistry of plant inputs to soils under climate change is, therefore, likely to affect the structure and function of the soil microbial community which drive the turnover of soil organic matter [23].

The net turnover of soil organic carbon in fungi dominated systems is slower than in bacterially dominated soils [24]. Only some fungi are able to complex organic polymers of the plant cell wall (i.e. lignocellulose), to access to access energy and nutrients, whilst bacteria generally rely on the availability of soluble substrates. Fungi also produce structures (i.e. chitin) that are relatively less available for decomposition [25] and have a higher C:N ratio (5 to 15) compared with bacteria (3 to 6; [26]). Moreover, the ramification of fungal hyphae through soils encourage soil aggregate formation and thereby physical protection of organic C from decomposition [27]. In *Calluna vulgaris* dominated ecosystems the ericoid mycorrhizal fungus *Hymenoscyphus ericae* is associated with heather roots and transfers nutrients (N and phosphorus) to the host plant in return for labile carbohydrates [2]. The mycorrhizal fungi additionally use litter as a substrate through the release of exo-enzymes [28]. In contrast, soil bacteria grow on carbohydrate substrates in the soil. Experimental addition of sugars to the soil generally increases soil respiration [29] leading to enhanced bacterial growth and substrate assimilation [30,31], while most fungal groups respond more slowly to addition of labile C [32]. This diversified utilization of substrates by different microbial groups suggests that changes in substrate input to the soil caused by changes in environmental drivers, may lead to altered composition of SOM [23]. This may have consequences for the wider soil food web because of the division of energy routed through fast bacterial or slow fungal channels [33].

Compound-specific stable isotope techniques have proven highly sensitive for detection of the selective incorporation of small concentrations of ^{13}C-labelled compounds into specific microbial phospholipid fatty acid (PLFA) biomarkers, and is an established method for assessment of soil microbial populations after applications of ^{13}C-labelled substrates [34,35], or of ^{13}C-depleted CO_2 [7,36,37]. Recent compound specific studies of substrate incorporation into fungal and bacterial PLFA biomarkers have revealed substrate specificity between microbial groups, i.e. with differences in the incorporation of organic compounds as glucose [30], amino acids [38,39], glycine [40], thymidine and leucine [31,41,42] or plant litter [43].

The effect of fluctuations in temperature and soil moisture content on soil microbial activity have been explored in many soils (e.g. [44,45]), and elevated CO_2 also causes shifts in soil microbial community compositions which is observed under FACE conditions (e.g. [23,46]) or in plant-free soil incubations [47]. However, the opportunity to explore the microbiology of soils under combinations of predicted climate change variables under field conditions is rare [48,49].

During three years of manipulation treatment with elevated CO_2, warming and summer drought at the experimental heathland site CLIMAITE [50], a consistent c. 30% increase in soil respiration was observed in response to elevated CO_2, and drought periodically reduced soil respiration [51]. Here, we attempt to explore the underlying mechanisms causing a possible shift in substrate utilization by bacterial and fungal groups, by identifying the group of soil microorganisms that respond to changes in availability of labile metabolites in root exudates. We explore the effect of controlled climate change manipulations at field scale on microbial community structure and diversity responses, and the effects of increased availability of glycine using changes in biomarker PLFA abundance and ^{13}C incorporation. Two ^{13}C tracing approaches were used to investigate the activity of soil microorganisms under different climate change conditions: (i) rapid response (24 h) of the microbial biomass to addition of the elicitor ^{13}C-labelled glycine to the soil and (ii) longer term microbial dynamics (one year) using indirect labeling via ^{13}C-depletion of the plant biomass and plant litter obtained by FACE fumigation. Our hypotheses were: I. Treatments with warming decrease the abundance of fungal PLFA biomarker; II. FACE treatment (elevated CO_2) primes SOM decomposition [52], and hereby increases bacterial glycine incorporation.

Materials and Methods

Site description

The field site was located in a public area owned by the Danish Government and under administration by the Danish Defence who granted permission to establish the experiment. The experiment took place at 'CLIMAITE' field site in Brandbjerg (55′53″N 11′58″E) c. 50 km NW of Copenhagen, Denmark. The site was a managed, dry, temperate heath on a hilly nutrient-poor sandy deposit, with an organic layer of c. 5 cm depth and a pH of 3.5 (in 0.01M $CaCl_2$). The vegetation was dominated by *Calluna vulgaris*, *Deschampsia flexuosa* and heathland mosses and herbs. The average annual precipitation was about 600 mm and the average temperature was 8°C.

Treatments

The climate manipulations at the 'CLIMAITE' field site started in 2005 [50] and consisted of eight treatments: elevated temperature (T), extended summer drought (D), elevated atmospheric CO_2 concentration (CO_2), all combinations of these treatments (TD, TCO_2, DCO_2 and $TDCO_2$) and untreated reference plots (A), each in six replicates. The field site covered an area of about 2 ha and the experimental plots were distributed in 12 seven meter diameter octagons arranged pair-wise in six blocks. The temperature was increased by passive night-time warming, by means of automatic curtains. The drought treatment was imposed by using automatic curtains to remove precipitation. The atmospheric CO_2 was increased by a regular FACE technique including feedback control on CO_2 concentrations from wind speed and wind direction. The $\delta^{13}C$ of CO_2 in the ambient atmosphere and in the atmosphere with elevated CO_2 concentration (FACE) were −8‰ and −12.2‰ respectively, averaged over the two years [53]. The temperature increase in 2 cm soil depth averaged 1°C, and the average CO_2 concentration in the FACE plots was 510 ppm. The drought treatment excluded 54 mm summer rain in July 2006, equivalent to 9% of annual rainfall. For further information about the experimental design see [50].

Isotope labeling

Dual-labeled glycine ($^{13}C_2{}^{15}N$-glycine, 99% labeled on both C atoms and 99% labeled with ^{15}N; Cambridge Isotope Laboratories) was added *in situ* in the end of September 2006 to 20 cm^2 subplots of undisturbed vegetation, providing 0.1 L of re-demineralized water with 0.027 g glycine, corresponding to 687 mg glycine·m^{-2} (223 mg C·m^{-2} or 0.016 mg glycine·g^{-1} DW soil). The label was injected into the soil in 16 points of a grid fixed in a transparent plastic plate, just below the soil surface, by using a syringe connected to a dispenser on a 1 L glass bottle [54,55]. The soil naturally contained free amino acids, the total total amino acid N of the free amino acids was 25.7±4.5 g N·m^{-2} in August 2005 [56].

Two types of ^{13}C label were mixed during this experiment in the undisturbed heath: **1.** the highly enriched ^{13}C label from glycine added to the soil in all plots and **2.** the slightly depleted ^{13}C label from the FACE CO_2 air in plots with elevated CO_2, which was sourced through the plant and made available for microbes in the soil.

Soil gas $\delta^{13}C$–CO_2

Soil gas was sampled at 1 cm depth in control plots at intervals before and after glycine label addition. A 20 ml sample was extracted directly from the soil by hand from four points with a 25 mL plastic syringe and immediately flushed through the septum of a crimp sealed 1.8 ml vial. The vials were left over-pressurized with c. 1 ml sample until analysis of $\delta^{13}CO_2$ at gas chromatograph (GC) (Hewlett-Packard 6890) in continuous flow mode with a Preparation Concentration unit (PreCon, Thermo Scientific, Germany) and stable isotope ratio mass spectrometer (IRMS) (Finnigan Delta PLUS, Thermo Scientific, Germany). These analyses were made at Technical University of Denmark.

Soil sampling

Soil cores were sampled at the CLIMAITE field site from the soil surface (including the organic top-layer) and cut at 5 cm depth one day (c. 24 hrs.) after labeling with glycine. Three soil cores from each plot were mixed to a composite sample from each depth and immediately sorted into soil and roots by hand (no sieving). The samples were kept on ice until further processing. A subsample of the fresh soil from each plot was freeze dried for PLFA extractions.

Soil organic matter content (SOM) was determined by burning soil at 550°C for 6 h (loss on ignition) at University of Copenhagen.

Just before the labeling was performed, additional soil samples were taken in adjacent subplots within the climate treated plots to obtain $\delta^{15}N$ and $\delta^{13}C$ natural abundance signatures.

Phospholipid fatty acid (PLFA) extraction

A modified Bligh Dyer extraction procedure was applied for the extraction of PLFA. Bligh Dyer solvent (potassium dihydrogen phosphate-buffered H_2O at pH 7.2, methanol and chloroform in 4:10:5 *v:v:v*) was added to freeze-dried soil in a 3:1 v:w ratio, and ultrasonicated prior to centrifugation. The supernatant was removed and saved and the extraction procedure repeated a further three times. The supernatants were combined and chloroform and buffered water added to break the aqueous and organic phases. The organic phase was removed and saved and the aqueous phase extracted (three times) further with chloroform. All organic fractions were combined as a total lipid extract (TLE) and solvent evaporated under N_2 atmosphere at 40°C. Silicic fractionation was used to isolate the phospholipid fraction from the TLE using the solvent system: chloroform (neutral lipids), acetone (glycolipids), methanol (polar lipids). Nonadecane in hexane was added to the polar fraction as the internal standard. After saponification (0.5 M methanolic NaOH; 1 h, 80°C) and acidification (1 M HCl), the liberated PLFAs were extracted with hexane, which was removed under N_2 (40°C). Acidified dry methanol was added to methylate the PLFAs (1 hour; 80°C), which were extracted as their methyl esters with diethyl ether after cooling and addition of DCM extracted double distilled water. The DCM was evaporated under N_2 (40°C), and the sample dissolved in hexane for GC analysis.

PLFA methyl esters (FAMEs) were quantified using a GC (Agilent 7890A, Agilent Technologies, Wokingham, UK) equipped with a DB-225 (30 m×25 mm (i.d.)×0.25 µm phase thickness). The temperature program was: 1 minute isothermal at 50°C, followed by a ramp to 150°C at 15°C minute^{-1}, followed by an increase to 240°C at 4°C minute^{-1} and then held isothermally at 240°C for 20 minutes.

FAMEs were identified using GC-MS analyses, with GC conditions the same as above, except He was used as the carrier gas in all cases. The interface was set to the maximum oven temperature, the ion source was set to 30°C below this and the quadropole MS scanned the range of m/z 50–650 at 1.7 scans per s. Data were acquired and processed using the Chemstation data system.

PLFA analysis GC and GC-C-IRMS

FAME samples were analyzed at The James Hutton Institute by GC-C-IRMS using a Trace GC Ultra gas chromatograph (Thermo Finnigan, Bremen, Germany) equipped with a GC PAL auto sampler (CTC Analytics AG, Zwingen, Switzerland) using the method described in [57]. The Trace GC Ultra contained an HP-5 column (50 m length, 0.2 mm i.d., film thickness of 0.33 µm; Agilent Technologies/J&W Scientific). The samples were injected in split-less mode via an inlet held at 250°C, and the He carrier gas was maintained at a constant flow rate of 1.5 mL min^{-1}. The GC oven was initially set at 100°C, held for 1 min, ramped at 20°C min^{-1} to 190°C, then at 1.5°C min^{-1} to 235°C, and finally at 20°C min^{-1} to 295°C; where the temperature was held for 15 min. A GC Combustion interface III was used to link the GC to a Delta V Advantage IRMS (both Thermo Finnigan). The oxidation reactor on the interface was maintained at 950°C and the reduction reactor at 650°C. Isodat 3.0 Gas Isotope Ratio MS Software (version 3.0; ThermoFisher Scientific) was used for data processing. For each peak the combined area of the peaks for all the ions (*m/z* 44, 45 and 46) after background subtraction was collected.

Calculations

Total bacteria were identified by the PLFAs: 15:0, 17:0, *i*15:0, *a*15:0, *i*17:0, *a*17:0, 10Me16:0, 10Me17:0, 10Me18:0, *cy*17:0, 18:1ω9c, *cy*19:0, and fungi were identified with the PLFA 18:2ω6,9c. Bacterial PLFAs were further grouped into gram positive (G^+) bacteria (*i*15:0, *a*15:0, *i*17:0, *a*17:0, 10Me16:0, 10Me17:0, 10Me18:0), actinobacteria (10Me16:0, 10Me17:0, 10Me18:0), and gram negative (G^-) bacteria (*cy*17:0, 18:1ω9c, *cy*19:0) [58,59].

The sum of the weighted mean frequency (PLFA diversity: Shannon-Weaver index for information density [24,61,62]) was calculated as the sum of each individual frequency weighted by the logarithm of the frequency (P):

$$H=\sum P_i \times \ln(P_i) \quad \text{(I)}$$

Bacterial stress index (BSI) [60] was calculated as:

$$BSI=\frac{cy17:0+cy19:0}{16:1\omega7c+18:1\omega7c} \quad \text{(II)}$$

The specific microbial activity (τ) calculated for each individual biomarker as:

$$\tau = \frac{PLFA\ ^{13}C\ enrichment}{PLFA\ concentration} \qquad \text{(III)}$$

Due to the ^{13}C-depleted labeling form the FACE air [1,53] each PLFA had a different baseline (^{13}C atom%) in ambient air plots compared to CO_2 treated plots [63]. This was used for calculation of ^{13}C enrichments for each biomarker individually. Hence the ^{13}C baseline from ambient was used for A, T, D and TD plots while the ^{13}C baseline from CO_2 treated plots was used for CO_2, TCO_2, DCO_2 and $TDCO_2$. The $\delta^{13}C$ of individual PLFA biomarkers ranged from 0.03 to 2.9‰ between A and CO_2 treatment and was significantly different for the bacterial biomarkers: 15:0 ($P = 0.0396$) and *a*15:0 ($P = 0.0472$).

Statistical methods

Linear mixed models (proc. mixed and GLM in SAS) were applied to analyze the effect of treatment on: PLFA concentrations and ^{13}C-enrichment of PLFAs. Random effect terms were block, octagon and plot, respecting the nested structure of the design. Main effect terms were the treatment factors: CO_2, temperature (T) and drought (D); in addition, all interaction terms between the factors CO_2, T and D were included. Root biomass was taken in as co-variate in proc. mixed. The models were gradually simplified, starting with third order interactions, taking out non-significant interactions until only significant terms were left. Statistics were carried out using SAS® 9.1 [64] and for all tests, a significance level of 0.05 was applied.

Results

Effect of climate change on soil community dynamics

The sum of the concentrations of the biomarker PLFA (Σ PLFA) was used to estimate differences in the size of the soil microbial community under each climate treatment, relative to the control plots (A):1.0(CO2), 1.0(D), 0.8(DCO2), 0.7(T), 0.5(TCO2), 0.4(TD), and 0.7(TDCO2). The warming treatments reduced Σ PLFA overall (Table 1 and Figure 1a) and the effect was similar for fungi (Figure 1b) and different bacterial groups (Figure 1 c and d). The sum of the weighted mean frequency (Equation I) of the bacterial and fungal PLFA biomarkers was significantly reduced by warming ($P = 0.0495$ for T) (*e.g.* index was $H = 1.03 \pm 0.19$ in A (ambient, no treatment) and $H = 0.787 \pm 0.15$ in T plots). There was no effect of treatment on BSI (Equation II; 7.0 in A (ambient, no treatment) and 8.5 in the CO_2 treatment) or on the PLFA biomarker concentrations before addition of glycine.

Effect of glycine addition on soil microbial community structure

The ratio of total bacterial to fungal biomarkers was 7.5 (±1.8) in A (ambient, no treatment) after glycine addition, with no significant effect of treatment or addition of glycine. The ratio of actinobacterial to G^- bacterial biomarkers was 3.8 (±0.1; in A) after glycine addition and that of all G^+ to G^- biomarkers was 5.4 (±0.1; in A) after glycine addition, and both increased as effect of glycine addition for A and CO_2 treatments ($P = 0.0002$ and 0.0056, respectively). In response to CO_2 treatments, actinobacteria to G− bacteria ratio (proc. mixed; CO_2: 0.0663) and all G^+ to G^- bacteria ratio (proc. mixed; CO_2: 0.0581 and T: 0.0998) tended to increase after glycine addition, but not in treatments including warming treatment.

Effect of climate change on incorporation of ^{13}C from glycine

The ^{13}C-CO_2 enrichment in soil air measured in a time-series beginning a few hours prior to addition of the ^{13}C-enriched glycine label till 168 hours after addition, fit a log-normal curve: $^{13}C\text{-}CO_2 = 1.108 + 1.359 \cdot \exp(-0.5 \cdot (\ln(T/3.269)/1.797)^2)/T$ ($R^2 = 0.8832$, Figure 2). This indicates microbial respiration during glycine acquisition with maximum microbial activity before 6 h, and this had declined by 70% after 24 hours. This indicated active up take of the ^{13}C-label within the timeframe of the experiment. Thereafter, the ^{13}C from glycine was respired at a stable rate through the next week (Figure 2). There was no significant effect of treatment on the ^{13}C-CO_2 enrichment.

The addition of $^{13}C_2{}^{15}N$ labeled glycine resulted in substantial microbial uptake during the first 24 hrs. The overall microbial enrichment with ^{15}N and ^{13}C correlated significantly in 0–5 cm depth, with $^{13}C = 1.74 \cdot {}^{15}N$ ($R^2 = 0.92$, $P < 0.0001$; [55], corresponding to N uptake with 87% in form of intact glycine.

^{13}C enrichment of PLFAs after the *in situ* incubation with ^{13}C-glycine suggested that the effects of climate treatment were significant. The largest ^{13}C enrichment of biomarker PLFAs from the ^{13}C-glycine label was found in the G^+ biomarker *i*16:0. The main effect of elevated CO_2 was significantly smaller ^{13}C enrichment (*i.e.* microbial uptake) in: the G^+ bacterial biomarker *i*17:0 (CO_2: $P = 0.0263$; CO_2*D: $P = 0.0499$, mixed model ANOVA with *Calluna* fine root biomass included as co-variate; Figure 3a) and larger ^{13}C enrichment in the non-specific PLFAs 16:0 (CO_2: $P = 0.0422$; Figure 3b), 16:1ω11c (CO_2: $P = 0.0298$) and 18:0 (CO_2: $P = 0.0333$ Figure 3c) and non-significant in the fungal biomarker 18:2ω6,9 (Figure 3d).

The specific microbial activity (τ; Equation III) was very different among individual biomarkers. The largest specific activity appeared in the non-specific biomarker 18:1ω9c with: $\tau = 151 \pm 94$ and the bacterial biomarkers 15:0 with: $\tau = 33 \pm 9$; and a15:0 with: $\tau = 15 \pm 4$. Small specific activity of $\tau = 5$ to 1 was observed in the bacterial biomarkers *i*15:0, 10Me16:0, *i*17:0, and 18:1ω9. Weak ($\tau < 1$) specific activity was observed in the fungal biomarker 18:2 ω6,9 and in 10Me17:0 and 16:0. There was no significant effect of treatment on τ.

Discussion

Experimental warming effects on soil microbes

The decrease in PLFA concentrations for bacteria and fungi in treatments with warming, reflects a change in the composition of the microbial community as found in parallel in a deciduous forest [65] and sub-arctic heath ecosystems [31]. Similar trends are frequent but not universal: In an old field ecosystem only the concentrations of fungal and G^- bacterial biomarkers decreased, while G^+ bacteria increased in response to the warming treatment [60]. In laboratory incubations of soil from a boreal *Picea abies* forest, temperature oscillations around 0°C decreased bacterial PLFA concentration but increased fungal biomarkers [38]. These observed ecosystem specific responses furthermore appeared at different treatment response times, i.e. after treatment through one year in the case of the *in situ* treated temperate heathland (this study); after 6 months in the laboratory incubations of forest soil [38]; and, after more than 10 years at the sub-arctic heath ecosystem and in the deciduous forest [41,65]. Storing of C in microbial membrane lipids, measured as soil PLFA concentrations, reflect the accumulated microbial response to the ecosystem manipulation treatments by: (i) quality: membrane lipid concentration can express physiological changes of the individual cell membranes which moderate the membrane fluidity (thickness and

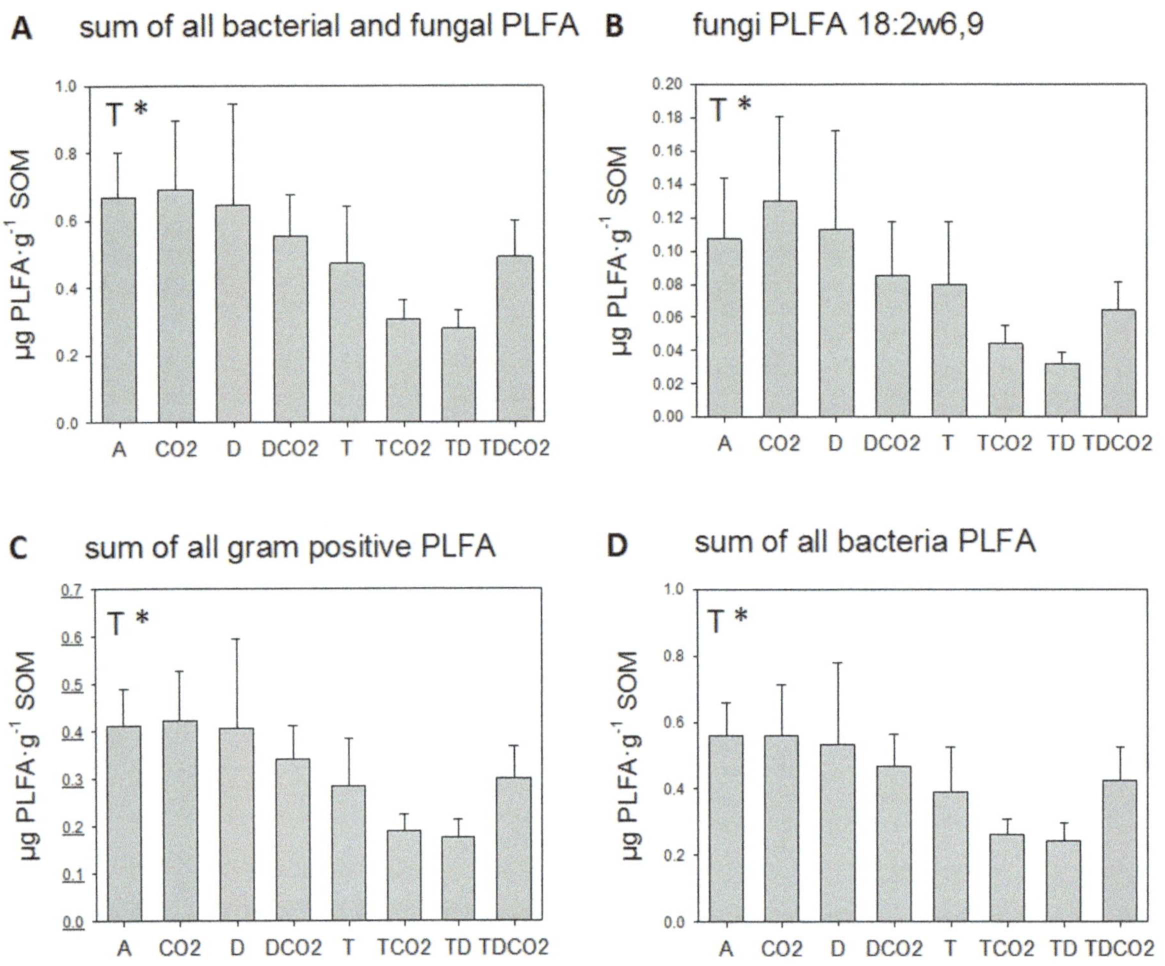

Figure 1. Concentration (µg PLFA·g^{-1} SOM) in top soil of PLFAs. Sum of all bacterial and fungal PLFA biomarkers (**A**); the fungal PLFA biomarker 18:2ω6,9 (**B**); all gram positive biomarkers (**C**) and all bacteria PLFA biomarkers (**D**); one day after addition of glycine to field plots subjected to *in situ* treatment through one year with the climate change factors: elevated atmospheric CO_2 concentration (CO2), summer drought (D) and warming (T), and those treatments in all combinations. A is ambient control = no treatment. Error bars represent standard error. Significant effects of treatments: * is $P<0.05$.

viscosity) to accommodate the environmental change [66,67]; and (ii) quantity: change of the number of cells and the overall biomass due to bacterial growth based on changes in the available substrate. Inferred from this, the reduced concentration of fungal PLFAs in warming treatment suggests that a future warmer climate will lead to a reduced storage of C in soil microbes, whereby C may be sourced out of the soil as efflux in the long term.

Bacteria and fungi use different C-substrates

Supplying an amino acid as an elicitor to activate the microbial community [68] and to trace the activity of microbes by ^{13}C-enrichment of the PLFA biomarkers, provided an accurate measure of the microbial response to the climate change treatment and the specific microbial activity on a short term scale [52]. The ^{13}C-enrichments and the high specific activity (τ) suggest that G^+ bacteria expressed opportunistic substrate use, representing organisms with the strategy to quickly colonize new substrates and on a short term time scale incorporate most of the amended glycine substrate, compared to fungi often utilizing more recalcitrant organic matter (a parallel to 'K' and 'r' strategists, see [69]). Similar results were obtained after glucose addition to grassland soil [70]. There was no general direction in the response to the climate change treatments for incorporation of ^{13}C from glycine across the analyzed biomarker PLFAs. However, the effects of elevated CO_2 did suggest increased direct incorporation of glycine in microbial biomass, in particular in G^+ bacteria, in an ecosystem subjected to elevated CO_2. Conclusively, the treatment with elevated CO_2 had mainly a stimulating effect on the bacterial usage of glycine on a short term scale. Hereby, elevated atmospheric CO_2 may in the future increase the routing of C through bacteria, which may source an elevated efflux of CO_2, as already observed in the experiment [51].

There is no universal effect of long-term treatment with elevated CO_2 on microbial community structure across ecosystems, as some systems respond with increase in fungal biomass, and others with increase in bacteria. At a desert ecosystem a shift in microbial community structure was observed in the form of as decreased bacterial and increased fungal abundance [63]. By contrast, in one grassland the fungal to bacterial ratio increased [71], while for a different grassland a higher PLFA concentration of both bacteria and fungi was found [46]. At our temperate heathland site, we here found a large (G^+) bacteria activity (τ) in response (glycine incorporation) to elevated CO_2, while there was no effect of

Table 1. Significant effect of climate change treatments (P-values) on PLFA ^{13}C enrichment from soils from the CLIMAITE experiment sampled in August 2007.

PLFA	Treatment effect	Biomarker
	P	Type
Σ(bacteria and fungi)	T: 0.0296	Σ(bacteria and fungi)
Σ(gram positive bacteria)	T: 0.0231	Σ(gram positive bacteria)
Σ(bacteria)	T: 0.028	Σ(bacteria)
Σ(actino bacteria)	T: 0.0267	Σ(actino bacteria)
18:2ω6,9	T: 0.0477	Fungi
*i*15:0	T: 0.0371	$gram^{+}$ bacteria
*a*15:0	T: 0.0059	$gram^{+}$ bacteria
*i*16:0	D^*CO_2: 0.0464	non-specific
10Me16:0	T: 0.0272	$gram^{+}$ bacteria (actino)
*i*17:0	T: 0.0232	$gram^{+}$ bacteria
17:0	T: 0.0479	Bacteria
10Me17:0	T: 0.0264	$gram^{+}$ bacteria (actino)
18:0	T: 0.0386	non-specific
18:1ω9	T: 0.0361	$gram^{-}$ bacteria

Soil was sampled one day after addition of $^{13}C_2$-glycine to field plots subjected to *in situ* treatment through one year with the climate change factors: elevated atmospheric CO_2 concentration (CO2) summer drought (D) and warming (T), and those treatments in all combinations. Linear mixed models (proc mixed and GLM in SAS) with main effects: CO_2, temperature (T) and drought (D) and interactions between the factors CO_2, T and D.

warming or drought. For fungi, the biomarker 18:2ω6,9 did not show large incorporation of the added ^{13}C-glycine and the specific activity was very small ($\tau<1$). Both suggest limited or no effective use of glycine by fungi in this soil. Interestingly, a contrasting conclusion was obtained at a ^{13}C-CO_2 pulse labeled grassland under elevated CO_2 treatment [72], however, arbuscular mycorrhiza was the principal fungal group in contrast to the dominance of ericoid mycorrhizal *Calluna vulgaris* in our experiment. Irresponsiveness by the fungal biomarker 18:2ω6,9 to labile carbohydrate utilization was also seen with ^{13}C incubation with labeled glucose, acetic acid, glycine, starch and vanillin in tundra soil [42,12], and with glucose with grassland soil [30]. By contrast, the fungal biomarker 18:2ω6,9 had the highest depletion in ^{13}C originating in the FACE (diff A to CO_2 treatments 2.9‰). From this we infer that out of the investigated biomarkers the fungi probably had the largest usage of plant derived substrate, i.e. rhizodeposits or litter components. Hence, we conclude that the main function of the dominant fungi in these soils was saprotrophic litter decomposition, and not the use of fresh, labile carbohydrate substrates, suggesting there was no direct priming effect of the fungi activity, but instead the fungi represented important microbial components for long-term circulation of C in soil organic matter [52].

Conclusions

The response to climate change treatments suggests that: i) bacterial usage of glycine will increase in ecosystems with elevated concentration of CO_2 on a short-term time-scale. Hereby, elevated atmospheric CO_2 may in the future increase the routing of C through bacteria, which may lead to an elevated efflux of soil CO_2; and ii) at longer-term time-scale fungi will through plant litter decomposition regulate the slow circulation of C in SOM; iii) the observed general decrease in concentrations and diversity of

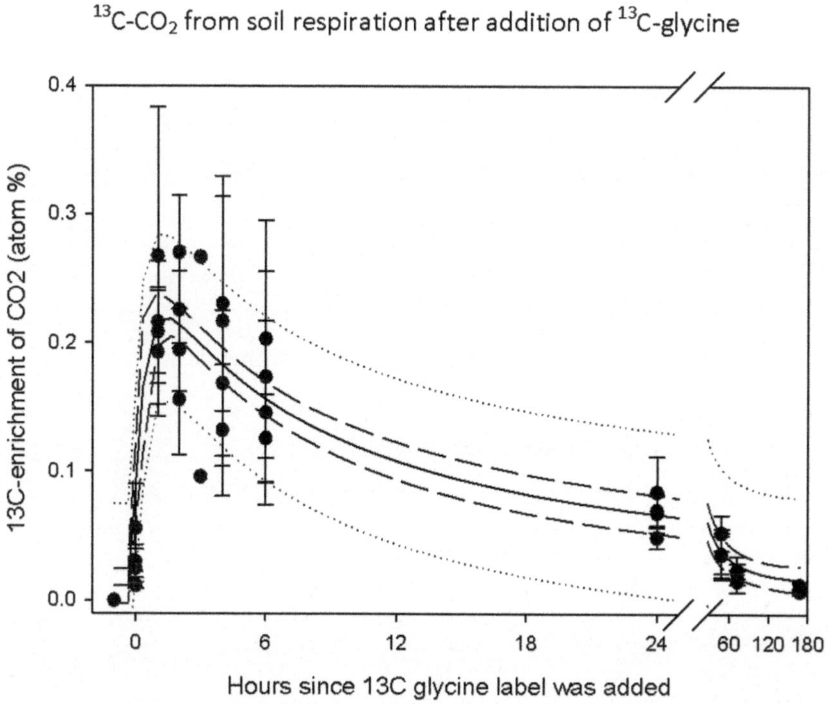

Figure 2. ^{13}C-CO_2 enrichment in soil air (atom %). A time-series beginning a few hours prior to addition of the ^{13}C-enriched glycine label till 168 hours after addition fit a log-normal curve: $^{13}C\text{-}CO_2 = 1.108+1.359^*\exp(-0.5^*(\ln(T/3.269)/1.797)^2)/T)$; ($R^2=0.8832$). Curve indicates average, 95% Confidence Band and 95% Prediction Band.

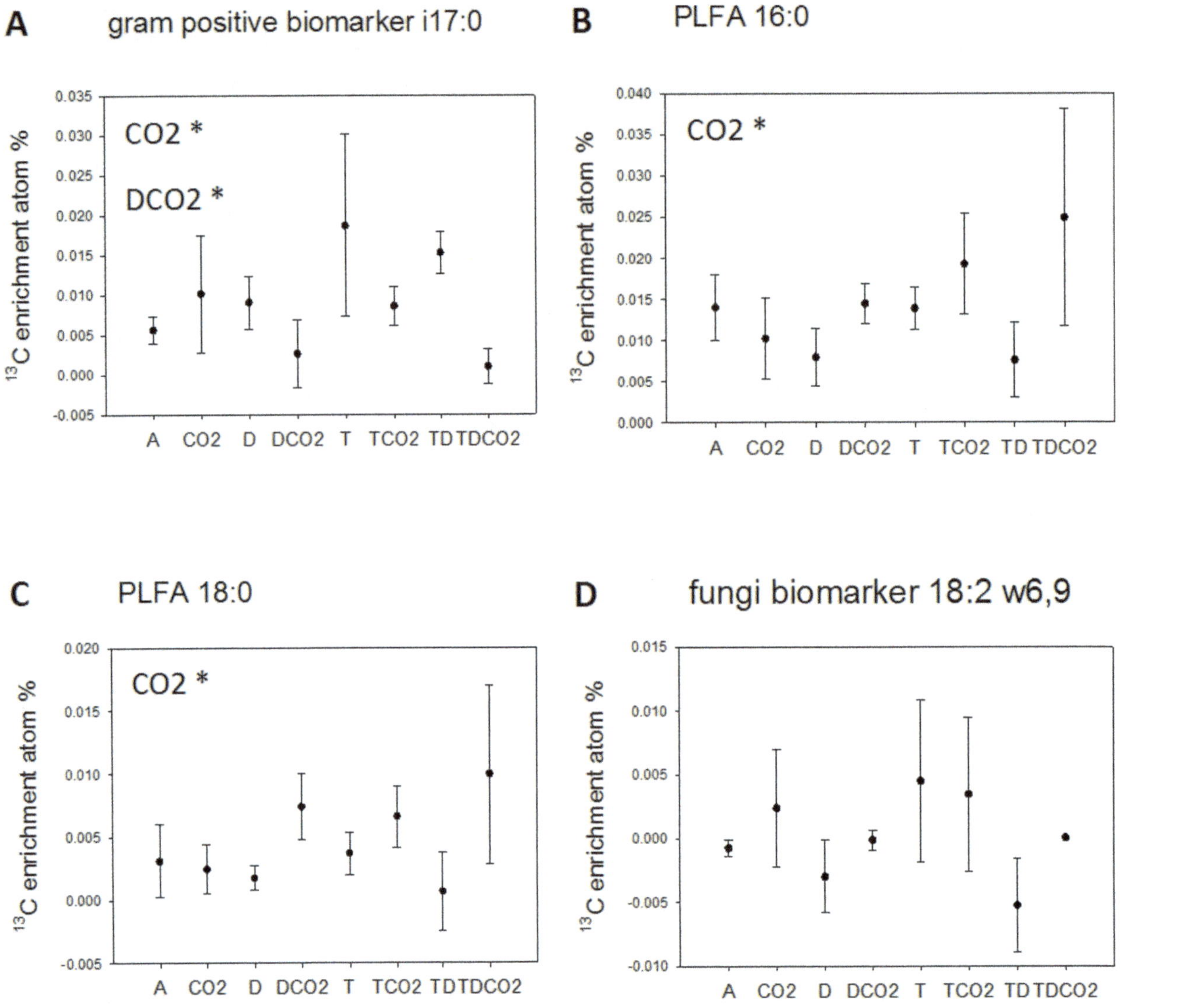

Figure 3. ^{13}C enrichment (atom%) in individual PLFAs. Gram positive biomarker *i*17:0 (**A**), the non-specific PLFA 16:0 (**B**), the non-specific 18:0 (**C**) and the fungi biomarker PLFA 18:2ω6,9 (**D**) one day after addition of $^{13}C_2$-glycine to field plots subjected to *in situ* treatment through one year with the climate change factors: elevated atmospheric CO_2 concentration (CO2), summer drought (D) and warming (T), and those treatments in all combinations. A is ambient control = no treatment. Error bars represent standard error. Significant effects of treatments: * is $P<0.05$.

PLFAs in response to warming (~1°C) might indicate stress-dieback of minority types or altered cell membrane properties, although the intensity of warming was modest in our experiment.

All together, the bacterial (and not fungal) utilization of glycine in our experiment indicated substrate preference and resource partitioning within the microbial community. This thereby suggested a potential diversified response pattern between microbial groups to future changes in substrate availability and climatic factors. While plant root exudation of glycine might increase under elevated CO_2, the resulting chemical composition of plant litter might decrease in quality for saprotrophic fungi. Herby, bacterial groups will advance their use of labile substrate (carbohydrates, amino acids), while saprotrophic fungi accessing recalcitrant substrate will rely on the poorer quality of SOM resource. Overall, we predict increased bacterial dominance in the soils of heathland ecosystems and a shift in functional decomposition of SOM at future climatic change.

Acknowledgments

WWFnovozymes research prize and the CLIMAITE centre.

Author Contributions

Conceived and designed the experiments: LCA MBS JD RB PA AM. Performed the experiments: LCA MBS JD RB PA AM. Analyzed the data: LCA MBS JD RB PA AM. Contributed reagents/materials/analysis tools: LCA MBS JD RB PA AM. Wrote the paper: LCA MBS JD RB PA AM.

References

1. Hobbie EA, Johnson MG, Rygiewicz PT, Tingey DT, Olszyk DM (2004) Isotopic estimates of new carbon inputs into litter and soils in a four-year climate change experiment with Douglas-fir. Plant and Soil 259: 331–343.
2. Read DJ, Perez-Moreno J (2003) Research review - Mycorrhizas and nutrient cycling in ecosystems – a journey towards relevance? New Phytologist 157: 475–492.

3. Bol R, Eriksen J, Smith P, Garnett MH, Coleman K et al. (2005) The natural abundance of ^{13}C, ^{15}N, ^{34}S and ^{14}C in archived (1923–2000) plant and soil samples from the Askov long-term experiments on animal manure and mineral fertilizer. Rapid Communications in Mass Spectrometry 19: 3216–3226.
4. Bardgett RD, Richter A, Bol R, Garnett MH, Bäumler R et al. (2007) Heterotrophic microbial communities use ancient carbon following glacial retreat. Biology Letters 3: 487–490.
5. Dorrepaal E, Toet S, van Logtestijn RSP, Swart E, van de Weg MJ et al. (2009) Carbon respiration from subsurface peat accelerated by climate warming in the subarctic. Nature 460: 616–620.
6. Luo YQ, Hui D, Zhang D (2006) Elevated CO_2 stimulates net accumulations of carbon and nitrogen in land ecosystems: A meta-analysis. Ecology 87: 53–63.
7. Amelung W, Brodowski S, Sandhage-Hofmann A, Bol R (2008) Combining biomarker with stable isotope analyses for assessing the transformations and turnover soil organic matter. Advances in Agronomy 100: 155–250.
8. Ineson P, Coward PA, Hartwig UA (1998) Soil gas fluxes of N_2O, CH_4 and CO_2 beneath *Lolium perenne* under elevated CO_2: The Swiss free air carbon dioxide enrichment experiment. Plant and Soil 198: 89–95.
9. Nowak RS, Ellsworth DS, Smith SD (2004) Functional responses of plants to elevated atmospheric CO_2 – do photosynthetic and productivity data from FACE experiments support early predictions? New Phytologist 162: 253–280.
10. Michelsen A, Rinnan R, Jonasson S (2012) Two Decades of Experimental Manipulations of Heaths and Forest Understory in the Subarctic. Ambio 41 (Suppl.3): 218–230.
11. Wookey PA, Aerts R, Bardgett RD, Baptist F, Bråthen KA et al. (2009) Ecosystem feedbacks and cascade processes: understanding their role in the responses of Arctic and alpine ecosystems to environmental change. Global Change Biology 15: 1153–1172.
12. Rinnan R, Michelsen A, Bååth E (2013) Fungi benefit from two decades of increased nutrient availability in tundra heath soil. PLoS ONE 8(2): e56532.
13. Pritchard E, Rogers HH, Prior SA, Peterson CM (1999) Elevated CO_2 and plant structure: a review Global Change Biology 5: 807–837.
14. Long SP, Ainsworth EA, Rogers A, Ort DR (2004) Rising atmospheric carbondioxide: Plants FACE the Future. Annual Reviews of Plant Biology 55: 591–628.
15. Albert KR, Ro-Poulsen H, Mikkelsen TN, Michelsen A, van der Linden L et al. (2011) Interactive effects of elevated CO_2, warming and drought on photosynthesis of *Deschampsia flexuosa* in a temperate heath ecosystem. Journal of Experimental Botany 62: 4253–4266.
16. Ainsworth EA, Long SP (2005) Transley review - What have we learned from 15 years of free-air CO_2 enrichment (FACE)? A meta-analytic review of the responses of photosynthesis, canopy properties and plant production to rising CO_2. New Phytologist 165: 351–372.
17. Wiesenberg GLB, Schmidt MWI, Schwark L (2008) Plant and soil lipid modifications under elevated atmospheric CO_2 conditions: I. Lipid distribution patterns. Organic Geochemistry 39: 91–102.
18. Feng X, Xu Y, Jaffé R, Schlesinger WH, Simpson M (2010) Turnover rates of hydrolysable aliphatic lipids in Duke Forest soils determined by compound specific ^{13}C isotopic analysis. Organic Geochemistry 41: 573–579.
19. Jones DL, Hodge A, Kuzyakov Y (2004) Plant and mycorrhizal regulation of rhizodeposition. New Phytologist 163: 459–480.
20. Lesuffleur F, Cliquet JB (2010) Characterization of root amino acid exudation in white clover (*Trifolium repens* L.). Plant and Soil 333: 191–201.
21. Wiesenberg GLB, Schwarzbauer J, Schmidt MWI, Schwark L (2008) Plant and soil lipid modification under elevated atmospheric CO_2 conditions: II. Stable carbon isotopic values (δ13C) and turnover. Organic Geochemistry 39: 103–117.
22. Glaser B, Millar N, Blum H (2006) Sequestration and turnover of bacterial- and fungal derived carbon in a temperate grassland soil under long-term elevated atmospheric pCO_2 Global Change Biology 12: 1521–1531.
23. Feng X, Simpson AJ, Schlesinger WH, Simpson M (2010) Altered microbial community structure and organic matter composition under elevated CO_2 and N fertilization in the duke forest. Global Change Biology 16: 2104–2116.
24. Bardgett RD, Leemans DK, Cook R, Hobbs PJ (1997) Seasonality of the soil biota of grazed and ungrazed hill grasslands. Soil Biology Biochemistry 29: 1285–1294.
25. Clemmensen KE, Bahr A, Ovaskainen O, Dahlberg A, Ekblad A et al. (2013) Roots and Associated Fungi Drive Long-Term Carbon Sequestration in Boreal Forest. Science 339: 1615–1618.
26. McGill WB, Shields JA, Paul EA (1975) Relation between carbon and nitrogen turnover in soil organic fractions of microbial origin. Soil Biol Biochem 7: 57–63.
27. Six J, Conant R, Paul E, Paustian K (2002) Stabilization mechanisms of soil organic matter: Implications for C-saturation of soils. Plant and Soil 241: 155–176.
28. Peretto R, Bettini V, Bonfante P (1993) Evidence of two polygalacturonases produced by a mycorrhizal ericoid fungus during its saprophytic growth. FEMS Microbiology Letters 114: 85–92.
29. Illeris L, JonassonS (1999) Soil and plant CO_2 emission in response to variations in soil moisture and temperature and to amendment with nitrogen, phosphorus, and carbon in Northern Scandinavia. Arctic, Antarctic and Alpine Research 31: 264–271.
30. Dungait JAJ, Kemmitt SJ, Michallon L, Guo S, Wen Q et al. (2011) Variable responses of the soil microbial biomass to trace concentrations of ^{13}C-labelled glucose, using ^{13}C-PLFA analysis. European Journal of Soil Science 62: 117–126.
31. Rinnan R, Michelsen A, Bååth E, Jonasson S (2007) Mineralization and carbon turnover in subarctic heath soil as affected by warming and additional litter. Soil Biology & Biochemistry 39: 3014–3023.
32. Schmidt IK, Ruess L, Bååth E, Michelsen A, Ekelund F et al. (2000) Long-term manipulation of the microbes and microfauna of two subarctic heaths by addition of fungicide, bactericide, carbon and fertilizer. Soil Biology & Biochemistry 32: 707–720.
33. Meidute S, Demoling F, Bååth E (2008) Antagonistic and synergistic effects of fungal and bacterial growth in soil after adding different carbon and nitrogen sources. Soil Biology and Biochemistry 40: 2334–2343.
34. Bull ID, Parekh NR, Hall GH, Ineson P, Evershed RP (2000) Detection and classification of atmospheric methane oxidizing bacteria in soil. Nature 405: 175–178.
35. Evershed RP, Crossman ZM, Bull ID, Mottram H, Dungait JAJ (2006) ^{13}C-labelling of lipids to investigate microbial communities in the environment. Current Opinion in Biotechnology 17: 72–82.
36. Paterson E, Midwood AJ, Millard P (2009) Through the eye of the needle: a review of isotope approaches to quantify microbial processes mediating soil carbon balance. New Phytologist 184: 19–33.
37. Paterson E, Thornton B, Midwood AJ, Osborne SM, Sim A et al. (2008) Atmospheric CO_2 enrichment and nutrient additions to planted soil increase mineralization of soil organic matter, but do not alter microbial utilization of plant- and soil C-sources. Soil Biology & Biochemistry 40: 2434–2440.
38. Haei M, Rousk J, Ilstedt U, Öquist M, Bååth E et al. (2011) Effects of soil frost on growth, composition and respiration of the soil microbial decomposer community. Soil Biology & Biochemistry 43: 2069–2077.
39. Dungait JAJ, Kemmitt SJ, Michallon L, Guo S, Wen Q et al. (2013) The variable response of soil microorganisms to trace concentrations of low molecular weight organic substrates of increasing complexity. Soil Biology & Biochemistry 64: 57–64.
40. Bardgett RD, Steeter TC, Bol R (2003) Soil microbes compete effectively with plants for organic-nitrogen inputs to temperate grasslands. Ecology 84: 1277–1287.
41. Rinnan R, Michelsen A, Bååth E, Jonasson (2007) Fifteen years of climate change manipulations alter soil microbial communities in a subarctic heath ecosystem. Global Change Biology 13: 28–39.
42. Rinnan R, Bååth E (2009) Differential utilization of carbon substrates by bacteria and fungi in tundra soil. Applied and Environmental Microbiology 75: 3611–3620.
43. Mehring M, Glaser B, de Camargo PB, Zech W (2011) Impact of forest organic farming change on soil microbial C turnover using ^{13}C of phospholipid fatty acids. Agronomy Sustainable Development 31: 719–731.
44. Frey SD, Elliott ET, Paustian K (1999) Bacterial and fungal abundance and biomass in conventional and no-tillage agroecosystems along two climatic gradients. Soil Biology & Biochemistry 31: 573–585.
45. Waldrop MP, Firestone MK (2004) Altered utilization patterns of young and old soil C by microorganisms caused by temperature shifts and N additions. Biogeochemistry 67: 235–248.
46. Drissner D, Blum H, Tscherkod D, Kandeler E (2007) Nine years of enriched CO_2 changes the function and structural diversity of soil microorganisms in a grassland. European Journal of Soil Science 58: 260–269.
47. Blagodatskaya E, Blagodatsky S, Dorodnikov M, Kuzyakov Y (2010) Elevated atmospheric CO_2 increases microbial growth rates in soil: results of three CO_2 enrichment experiments. Global Change Biology 16: 836–848.
48. Strickland MS, Rousk J (2010) Considering fungal:bacterial dominance in soils e Methods, controls, and ecosystem implications. Soil Biology & Biochemistry 42: 1385–1395.
49. Dieleman WJ, Vicca S, Dijkstra FA, Hagedorn F, Hovenden MJ et al. (2012) Simple additive effects are rare: a quantitative review of plant biomass and soil process responses to combined manipulations of CO_2 and temperature. Global Change Biology 18: 2681–2693.
50. Mikkelsen TN, Beier C, Jonasson S, Holmstrup M, Schmidt IK et al. (2008) Experimental design of multifactor climate change experiments with elevated CO_2, warming and drought: the CLIMAITE project. Functional Ecology 22: 185–195.
51. Selsted MB, van der Linden L, Ibrom A, Michelsen A, Larsen K et al. (2012) Soil respiration is stimulate by elevated CO_2 and reduced by summer drought: three years of measurements in a multifactor ecosystem manipulation experiment in a temperate heathland (Climaite). Global Climate Change 18: 1216–1230.
52. Kuzyakov Y (2010) Priming effects: Interactions between living and dead organic matter. Soil Biology & Biochemistry 42: 1363–1371.
53. Andresen LC, Konestabo HS, Maraldo K, Holmstrup M, Ambus P et al. (2011) Organic matter flow in the food web at a temperate heath under multifactorial climate change. Rapid Communications in Mass Spectrometry 25: 1485–1496.
54. Andresen L C, Michelsen A (2005) Off-season uptake of nitrogen in temperate heath vegetation. Oecologia 144: 585–597.
55. Andresen LC, Michelsen A, Jonasson S, Beier C, Ambus P (2009) Glycine uptake in heath plants and soil microbes responds to elevated temperature, CO_2 and drought. Acta Oecologica 313: 283–295.
56. Andresen LC, Michelsen A, Jonasson S, Ström L (2011) Seasonal changes in nitrogen availability, and root and microbial uptake of $^{15}N^{13}C_9$-phenylalanine

and ^{15}N-ammonium in situ at a temperate heath. Applied Soil Ecology 51: 94–101.
57. Thornton B, Zhang Z, Mayes RW, Högberg MN, Midwood AJ (2011) Can gas chromatography combustion isotope ratio mass spectrometry be used to quantify organic compound abundance? Rapid Communications in Mass Spectrometry 25: 2433–2438.
58. Frostegård A, Bååth E (1996) The use of phospholipid fatty acid analysis to estimate bacterial and fungal biomass in soil. Biology and Fertility of Soils 22: 59–65.
59. Zelles L (1999) Fatty acid patterns of phospholipids and lipopolysaccharides in the characterisation of microbial communities in soil: a review. Biology and Fertility of Soils 29: 111–129.
60. Gray SB, Classen AT, Kardol P, Yermakov Z, Miller M (2011) Multiple climate change factors interact to alter soil microbial community structure in an old-field ecosystem. Soil Science Society of America 75: 2217–2226.
61. Pielou EC (1975) Ecological diversity New York: Wiley InterScience. Pritchard E, Rogers HH, Prior SA, Peterson CM (1999) Elevated CO_2 and plant structure: a review Global Change Biology 5: 807–837.
62. Kaur A, Chaudhary A, Kaur A, Choudhary R, Kaushik R (2005) Phospholipid fatty acid – a bioindicator of environment monitoring and assessment in soil ecosystem. Current Science 89: 1103–1112.
63. Jin VL, Evans R D (2010) Microbial ^{13}C utilization patterns via stable isotope probing of phospholipid biomarkers in Mojave Desert soils exposed to ambient and elevated atmospheric CO_2 Global Change Biology 16: 2334–2344.
64. SAS Institute Inc (2004) The SAS system for windows.
65. Frey SD, Drijber R, Smith H, Mellilo J (2008) Microbial biomass, functional capacity, and community structure after 12 years of soil warming. Soil Biology & Biochemistry 40: 2904–2907.
66. Frostegård A, Tunlid A, Bååth E (2011) Use and misuse of PLFA measures in soils. Soil Biology and Biochemistry 43: 1621–1625.
67. Wixon DL, Balser TC (2013) Toward conceptual clarity: PLFA in warmed soils. Soil Biology and Biochemistry 57: 769–774.
68. Meli SM, Badalucco L, English LC, Hopkins DW (2003) Respiratory responses of soil micro-organisms to simple and complex organic substrates. Biology and Fertility of Soils 37: 96–101.
69. Begon M, Harper JL, Townsend CR (1996) Ecology. Milan: Blackwell Science.
70. Dungait JAJ, Bol R, Bull ID, Evershed RP (2009) Tracking the fate of dung-derived carbohydrates in a temperate grassland soil using compound-specific stable isotope analysis. Organic Geochemistry 40: 1210–1218.
71. Guenet B, Lenhart K, Leloup J, Giusti-Miller S, Pouteau V et al. (2012) The impact of long-term CO_2 enrichment and moisture levels on soil microbial community structure and enzyme activities. Geoderma 170: 331–336.
72. Denef K, Bubenheim H, Lenhart K, Vermeulen J, Van Cleemput O et al. (2007) Community shifts and carbon translocation within metabolically-active rhizosphere microorganisms in grasslands under elevated CO_2. Biogeosciences 4: 769–779.

Long-Term Nitrogen Amendment Alters the Diversity and Assemblage of Soil Bacterial Communities in Tallgrass Prairie

Joseph D. Coolon[1,2*⁹], Kenneth L. Jones[1,3⁹], Timothy C. Todd[1,4], John M. Blair[1], Michael A. Herman[1]

1 Kansas State University, Ecological Genomics Institute, Division of Biology, Manhattan, Kansas, United States of America, 2 University of Michigan, Department of Ecology and Evolutionary Biology, Ann Arbor, Michigan, United States of America, 3 University of Colorado School of Medicine, Department of Biochemistry and Molecular Genetics, Aurora, Colorado, United States of America, 4 Kansas State University, Department of Plant Pathology, Manhattan, Kansas, United States of America

Abstract

Anthropogenic changes are altering the environmental conditions and the biota of ecosystems worldwide. In many temperate grasslands, such as North American tallgrass prairie, these changes include alteration in historically important disturbance regimes (e.g., frequency of fires) and enhanced availability of potentially limiting nutrients, particularly nitrogen. Such anthropogenically-driven changes in the environment are known to elicit substantial changes in plant and consumer communities aboveground, but much less is known about their effects on soil microbial communities. Due to the high diversity of soil microbes and methodological challenges associated with assessing microbial community composition, relatively few studies have addressed specific taxonomic changes underlying microbial community-level responses to different fire regimes or nutrient amendments in tallgrass prairie. We used deep sequencing of the V3 region of the 16S rRNA gene to explore the effects of contrasting fire regimes and nutrient enrichment on soil bacterial communities in a long-term (20 yrs) experiment in native tallgrass prairie in the eastern Central Plains. We focused on responses to nutrient amendments coupled with two extreme fire regimes (annual prescribed spring burning and complete fire exclusion). The dominant bacterial phyla identified were Proteobacteria, Verrucomicrobia, Bacteriodetes, Acidobacteria, Firmicutes, and Actinobacteria and made up 80% of all taxa quantified. Chronic nitrogen enrichment significantly impacted bacterial community diversity and community structure varied according to nitrogen treatment, but not phosphorus enrichment or fire regime. We also found significant responses of individual bacterial groups including Nitrospira and Gammaproteobacteria to long-term nitrogen enrichment. Our results show that soil nitrogen enrichment can significantly alter bacterial community diversity, structure, and individual taxa abundance, which have important implications for both managed and natural grassland ecosystems.

Editor: Martha Trujillo, Universidad de Salamanca, Spain

Funding: Publication of this article was funded in part by the Kansas State University Open Access Publishing Fund. Joseph Coolon was supported by National Institutes of Health (NIH) National Research Service Award (NRSA) fellowship 1F32GM089009-02. This research was supported by National Science Foundation (NSF) EPSCoR grant EPS-0236913 to Michael Herman and Loretta Johnson, by NSF grant EF-0723862 to Michael Herman, and by NSF grant DEB-0823341 to the Konza Prairie LTER program. The funders had no role in study design, data collection and analysis, decision to publish, or preparation of the manuscript.

Competing Interests: The authors have declared that no competing interests exist.

* E-mail: jcoolon@umich.edu

⁹ These authors contributed equally to this work.

Introduction

Understanding biotic responses to anthropogenically-driven environmental changes is a central focus of many ecological studies, and is crucial for predicting the long-term consequences of contemporary and future human-induced changes in the Earth's ecosystems. Determining the responses of microbial communities is particularly important as they are responsible for the majority of energy and nutrient transformations in most ecosystems [1], are the most basal components of soil and aquatic food webs and exert widespread effects on ecosystem function. Many recent studies in a broad range of ecosystems indicate that the composition of soil microbial communities is sensitive to a variety of land use changes [2]. For example, wholesale land use conversions, such as deforestation [3,4] or tilling for agriculture [5,6], cause shifts in the bacterial community resulting in decreased overall diversity and/or disruption of biogeochemical processes, leading to alterations in ecosystem functioning. Even seemingly less intrusive soil perturbations such as periodic fires in forests and woodlands [3,4,7] or nitrogen enrichment in a variety of ecosystems [8,9,10,11,12] can alter microbial communities (but see [13]).

In the tallgrass prairie, where the predominant land-use change historically was conversion of native grasslands to agriculture [14], disturbances such as tilling [15,16,17], burning [15,18], fertilization and irrigation [17,19] elicit changes in soil biogeochemical processes and properties. Previous research in tallgrass prairie has documented a myriad of above- and below-ground responses to these disturbances. However, there have been fewer studies of soil microbial community responses to disturbance, and changes in the soil microbial community are often inferred based on biogeochemical changes or changes in frequency of microbial biomarkers in the soil [15,16,18,19]. Recently, researchers have turned to deep sequencing of microbial communities to assess changes in bacterial composition and diversity in response to perturbation in a variety of environmental contexts [10,20,21,22].

Our objective was to determine how bacterial communities respond to two important contemporary environmental changes common to tallgrass prairie ecosystems in the eastern Central Plains: changes in fire frequency and nitrogen enrichment. Periodic fires were historically common in these grasslands and prescribed burning is a common management practice for maintaining grass dominance and suppressing the growth and cover of woody vegetation. Fire suppression in these grasslands can lead to rapid transitions to shrubland and woodland. While fire has direct negative effects on some plant species (e.g., eastern red cedar), prescribed burning often takes place in the dormant season (late winter or early spring) and the indirect effects of fire on the soil environment and subsequent plant resource use and competition are generally considered to more important than direct effects. Surface fires are generally fast moving and, in contrast to forest, woodland or tundra fires, there is little heat penetration in soils and little direct effect on soils properties or soil communities. Instead, the major effects of fire in these grasslands are due to removal of accumulated standing dead and surface plant litter (detritus), the loss of nitrogen through volatilization of nitrogen in detritus, and resulting changes in the post-fire soil environment. Removal of accumulated detritus results in more rapid warming in the spring and a more favorable light environment for emerging plants. These post-fire environmental changes are important in structuring plant communities [23,24] and affecting soil biophysical properties and nutrient transformations [25,26]. In general, fire in tallgrass prairie enhances above- and belowground plant productivity [27] and increases the dominance of a few species of perennial C_4 (warm-season grasses) while reducing the cover of subdominant species and lowering overall plant species richness [23].

Nitrogen is generally considered to be a limiting nutrient in these grasslands, and N enrichment alters both plant productivity and community composition [28,29]. Although experimental nitrogen enrichment increases plant productivity, it also leads to declines in species richness and diversity [30]. Levels of nitrogen enrichment in these studies often exceed current atmospheric N deposition rates, but even relatively low levels of chronic nitrogen enrichment are a threat to grassland conservation [31]. Further, interactions between changes in fire frequency and N enrichment are expected, since one of the effects of frequent fire is to reduce soil nitrogen availability [25] and to increase conservative recycling of nitrogen in the plant-soil system [18]. Here we use long-term experimental manipulations of fire regimes and nutrient enrichment in native tallgrass prairie to address three hypotheses; 1) soil bacterial diversity is altered in response to nutrient enrichment and/or contrasting fire regimes (annually burned vs. unburned treatments), 2) soil microbial community structure is altered in response to nutrient enrichment and/or contrasting fire regimes, 3) individual taxa frequencies are altered in response to nutrient enrichment and/or contrasting fire regimes. To do this we used deep sequencing targeting the V3 region of the bacterial 16S rRNA gene to identify and quantify soil bacterial taxa from replicated factorial field experiment at the Konza Prairie Long-Term Ecological Research (LTER) site.

Materials and Methods

Study Design, Soil Sampling and DNA Extraction

This research utilized an ongoing experiment established and supported by the Konza Prairie Long-Term Ecological research (LTER) program, and was approved by the Konza Prairie Biological Station, a 3487-hectare area of native tallgrass prairie in the Flint Hills of northeastern Kansas. Konza Prairie Biological Station is jointly owned by Kansas State University and The Nature Conservancy, and managed by the KSU Division of Biology. No protected species were sampled in association with this study. We sampled soil bacterial communities from the Belowground Plot Experiment, which was initiated in 1986 to assess above- and below-ground responses of native tallgrass prairie communities to contrasting fire regimes and nutrient addition treatments. The treatments used here were: burning (annually burned vs. unburned); ammonium nitrate addition (10 g N/m^2 annually vs. no addition); and superphosphate addition (1 g P/m^2 annually vs. no addition). The experimental design consisted of eight large plots (50 m×25 m) subdivided into a total of 32, 12.5 m×12.5 m subplots (Figure S1). Treatments were arranged in a split-strip plot design with four replications per treatment combination. Burning treatment was assigned at the whole plot level and fertilization (nitrogen and phosphorus) additions were imposed as strip-plot treatments. Burning was performed in mid-March to mid-April of each year (burn applied March 28 in 2006) and fertilizer treatments were applied in late May to early June.

For each of the 32 plots sampled, two 2.5 cm diameter, 10 cm deep cores were collected post fire on April 11, 2006, pooled, and homogenized. We chose to sample to 10 cm as previous studies have indicated significant effects of contrasting fire regimes and nutrient additions on belowground plant biomass and soil properties at this depth [26,28]. From each composite sample, genomic DNA was extracted from a 10 g subsample using a standard soil DNA extraction kit (UltraClean Mega Soil DNA Kit, MoBio, Carlsbad, CA). After extraction, the DNA was diluted to ~5 ng/µl and stored at −80°C.

Soil Physico-chemical Characteristics

Selected soil characteristics were assessed by the Kansas State University Agronomy Soil Testing Lab from samples collected prior (samples collected in June 2004) to those used for microbial community analyses (Table 1). Briefly, soil for all measurements was prepared for all measurements by drying overnight at 50°C followed by passage through a 2-mm sieve. pH was measured directly using a 1:1 slurry of d soil and deionized water. Concentrations of available soil phosphorus were determined by Bray phosphorus test. Exchangeable cations (Ca_2^+, Mg_2^+, K^+, Na^+) were measured by flame atomic absorption or ISC spectroscopy following extraction with 1 M ammonium acetate (pH 7). Extractable ammonium (NH_4^+) and nitrate (NO_3^-) were determined by colorimetric assay using an automated analyzer. Total nitrogen and phosphorus were measured colorimetrically following a modified Kjeldahl digestion. Organic matter content (%) was measured by the Walkley-Black method (Table 1). For more details see the Kansas State University Agronomy Soil Testing Lab (http://www.agronomy.ksu.edu/soiltesting/).

16S rRNA Amplification for Deep Sequencing

DNA extracts of soil bacterial communities from each of the 32 plots sampled were amplified using universal bacterial primers (U341F and U533R, [32]) designed to amplify the third variable region (V3) of the 16S rRNA gene. The V3 region was chosen because preliminary analyses suggested that it had high primer site conservation and would maximize the information obtained using ~100 bp sequences from soil samples. This was subsequently confirmed by Vaseleiadis *et al.* [33] using an *in silico* approach that identified it as one of the most variable stretches of the 16S gene and results suggested that for sequence frequencies similar to those

Table 1. Average soil properties in the Belowground Plot Experiment.

Treatment	pH	Bray P	Ca	K	Mg	Na	NH_4−N	NO_3−N	N_{ext}	OM	Total N	Total P
Burn	6.26 (0.08)	4.25 (0.75)	3612.81 (153.96)	371.10 (18.58)	415.01 (36.90)	10.04 (0.88)	3.44 (0.30)	0.48 (0.28)	3.90 (0.45)	4.65 (0.40)	0.204 (0.012)	0.034 (0.0009)
Burn+Nitrogen	5.94 (0.11)	2.75 (0.48)	3203.75 (101.39)	283.74 (7.33)	512.66 (17.10)	10.68 (1.45)	3.58 (0.34)	0.95 (0.35)	4.53 (0.48)	4.23 (0.14)	0.192 (0.008)	0.032 (0.0007)
Control	6.21 (0.20)	4.25 (1.11)	3485.93 (277.16)	393.69 (79.93)	437.86 (69.08)	12.92 (2.39)	3.95 (0.53)	0.56 (0.21)	4.53 (0.57)	4.48 (0.20)	0.205 (0.008)	0.033 (0.0017)
Nitrogen	5.92 (0.23)	6.00 (3.03)	3690.17 (145.59)	335.69 (27.03)	416.81 (53.20)	12.45 (3.20)	4.80 (0.65)	3.87 (1.20)	8.68 (1.80)	4.35 (0.38)	0.210 (0.02)	0.032 (0.0009)
p-values from general linear model:												
Effect	**pH**	**Bray P**	**Ca**	**K**	**Mg**	**Na**	**NH_4−N**	**NO_3−N**	**N_{ext}**	**OM**	**Total N**	**Total P**
Burn	0.8504	0.3504	0.3419	0.4068	0.4626	0.305	0.0951	0.0382	0.0341	0.9352	0.4738	0.5641
Nitrogen	0.0933	0.9417	0.5833	0.1195	0.4414	0.9708	0.3175	0.0127	0.0341	0.3792	0.7939	0.1525
Burn X Nitrogen	0.9076	0.3504	0.1172	0.7407	0.2409	0.803	0.4706	0.0488	0.103	0.6275	0.5073	0.5996

Soil properties (and units) measured in the Belowground Plot Experiment included: pH, Bray Phosphorus (ppm), exchangeable Calcium (ppm), exchangeable Potassium (ppm), exchangeable Magnesium (ppm), exchangeable Sodium (ppm), NH_4−Nitrogen (ppm), NO_3−Nitrogen (ppm), extractable inorganic Nitrogen (ppm), Organic matter (%), total Nitrogen (% weight), and total Phosphorus (% weight). Treatment means and standard error is given for each parameter. P-values from general linear model testing effects listed are shown below.

found in soils, the V3 region had superior performance compared to other variable sites in the 16S gene [33]. These primer constructs (Table S1) also incorporated MPS sequencing primers [34] and a unique 5-bp sequence of DNA between the sequencing primer and the reverse 16S primer. These barcoded primer constructs were specifically designed to allow randomized sequencing from a mixture of multiple PCR products. PCR reactions (1x Amplitaq Gold polymerase reaction buffer (Applied Biosystems, Foster City, CA), 3.75 mM $MgCl_2$, 200 µM dNTP, 0.5 µM each forward and reverse $_b$MPS 16S primers, one unit Amplitaq Gold LD polymerase, and 5 ng soil extracted DNA) were run for 25 cycles at 95°C for 1 minute, 55°C for 1 minute, 74°C for 1 minute on an iCycler IQ real-time thermocycler (Bio-Rad Laboratories, Hercules, CA). PCR reactions were spiked with SYBR Green and run on a realtime thermocycler to assure that the PCR was stopped during log-phase. PCR amplification was done in triplicate, individual reactions pooled, and cleaned using an AMPure PCR cleanup kit (Agencourt Bioscience, Beverly, MA). We utilized a single MPS run with four blocks and randomly assigned 32 DNA samples to one of 32 different barcoded primers (Table S1) in one of the four sequencing blocks and amplified the V3 region of the 16S rRNA gene independently using the appropriate primers. For the sequencing block used for this study, 100 ng of each differentially barcoded PCR product were pooled and sequenced by 454 Life Sciences (Branford, CT; 2006) generating 71,769 sequences. Sequences generated for this study have been deposited in the NCBI Short Read Archive under accession SRP001373.

Bioinformatics and OTU Designation

Following previously described methodologies [35], raw sequences were searched for the occurrence of the primer barcode immediately preceding the 16S primer sequence. Where found, the barcode was removed and the corresponding plot designation was incorporated into the sequence name. Sequences were removed if they did not contain a valid primer or barcode sequence, contained more than one ambiguous base, or were shorter than 85 bp (low quality sequences) or longer than 125 bp (containing long homopolymers), to reduce the number of sequence errors [36]. An average of 1,584 sequences per plot that were on average 101 bases long were retained for analyses (Table S2). For the remaining sequences, CAP3 [37] was used to align sequences at each of 34 sequence identity levels (SILs, 66% to 99%) using a minimal overlap of 75 bp to generate operational taxonomic units (OTUs).

Rarefaction Curves

We calculated OTU accumulation curves (i.e., rarefaction curves) at all percent sequence identity levels (%SILs) in the sequence sampling range (66% to 99% sequence identity) for the main treatment effects and CAP3 results were pooled for each treatment. Within treatments, the pooled collection of observed sequences were sampled at increasing intervals of 1 (e.g., 1 sequence, 2 sequences, etc.) and the number of OTUs identified within each subsample of sequences was determined. Rarefaction curves were produced from ten replicate samplings at each of the sampling intervals and means reported for each interval for the pooled collection of sequences (Figure 1). In addition, in order to better evaluate the beta-diversity calculations we generated rarefaction curves for each plot at 93 and 97% SIL (Figure S2).

Diversity Indices

Overall taxonomic richness (S) was calculated by summing the number of OTUs at 93% SIL or 97% SIL, including singlets, which occurred within each plot. Simpson's Diversity ($\sum p_i^2$) and Shannon's Diversity ($e\sum p_i(\ln(p_i))$) were calculated for each plot, where p_i is the frequency of occurrence of each OTU. Evenness was calculated as the ratio of Shannon's Diversity and richness ($e\sum p_i(\ln(p_i)/S)$. A final index of diversity, Fisher's alpha log-series [38] was calculated iteratively using the equation $S/N = ((1-x)/x)(-\ln(1-x))$, where S is richness and N is the total number of sequences within the plot. Once x was solved, the diversity index alpha (α) was calculated as $N(1-x)/x$. Differences in diversity across treatments were analyzed using the MIXED procedure in SAS (SAS Institute Inc., Cary, NC).

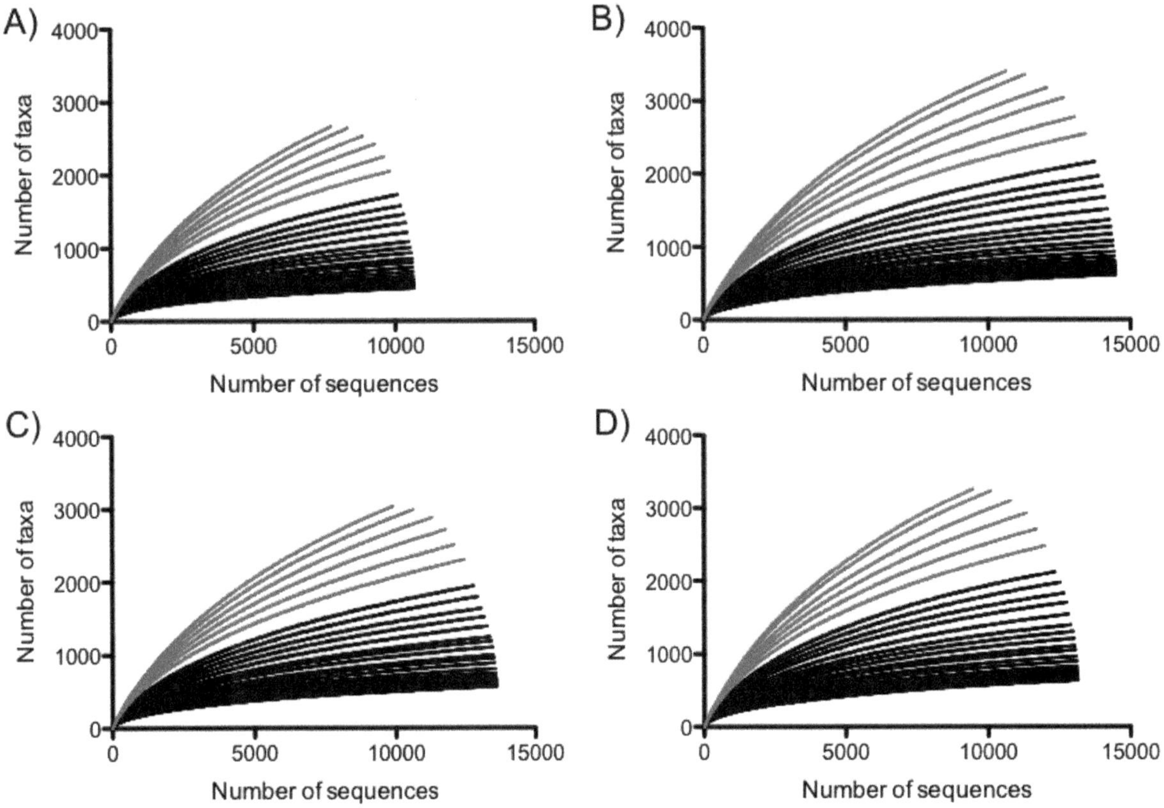

Figure 1. Rarefaction curves calculated independently for the cumulative treatments in the: A) burn+nitrogen, B) burn, C) nitrogen, and D) control. Black lines indicate curves for 66% SIL (bottom black line) to 93% SIL (top black line); gray lines indicate curves for 94% SIL (bottom gray line) to 99% SIL (top gray line).

OTU Frequency

Our rarefaction analysis indicated that 93% SIL was the most exclusive sufficiently sampled %SIL (Figure 1). However, since 97% SIL is commonly used in many bacterial community studies, we used both 93% and 97% SIL for OTU frequency analyses. For taxa at 93% or 97% SIL whose abundance was significantly different from zero (determined using a T-test in SAS; SAS Institute Inc., Cary, NC), a mixed model ANOVA was preformed to determine treatment effects, using the MIXED procedure in SAS. This procedure was included to reduce inclusion of erroneous rare taxa and increase the accuracy of taxa abundance measures. Phylum level taxa were also analyzed using the same model. Due to the multitude of simultaneously conducted tests, we used a false discovery rate (FDR) correction to account for the increase in type 2 error using the q-value package in R version 2.8.1 [39]. Those OTUs identified as having significant responses to treatments were identified with the naïve Bayesian classifier tool [40] in the Ribosomal Database Project (RDP 10.26, http://rdp.cme.msu.edu/, [41,42]) using a bootstrap cutoff of 50% as recommended for sequences in the length range generated for this study. Due to lengths of sequences generated for this study, we were not able to obtain genus-level taxonomic names for all taxa identified as significantly responsive to treatments. Therefore, we reported the taxonomic designation that was best supported by the naïve Bayesian classifier tool in the Ribosomal Database Project.

Fast UniFrac [43,44] was used to evaluate the relatedness of the bacterial communities within soil samples. We clustered sequences at 93% and 97% SIL using CAP3 and used one randomly selected sequence from each OTU as a representative. Sequences were aligned with MUSCLE [45] using the default parameters generating the input tree for UniFrac. Normalized Principle Coordinate Analyses (PCoA), weighted by OTU abundances, were performed to evaluate similarity between samples. The representative sequences from each OTU at 93% SIL used in the UniFrac analyses were also used to determine taxonomy with the naïve Bayesian classifier tool [40] in the Ribosomal Database Project and abundances of different groups were determined at the phylum level (Figure 2).

Results

Rarefaction Curves of Bar-coded Bacterial 16S rRNA Sequences

To quantify the bacterial community we used bar-coded 16S rRNA 454 sequencing. After removing aberrant, short and/or long sequences, a total of 47,508 sequences were retained for further analyses (Table S2). These quality-filtered sequences were assigned operational taxonomic units (OTUs) at each of 34% SIL (66% to 99%). At each level of sequence identity, sequences were parsed by plot and used to calculate the frequency of occurrence of all OTUs for each of the 32 plots sampled. To assess the quality of the data and determine where sampling of the microbial community was sufficient, we calculated rarefaction curves for the cumulative plots within each treatment. Figure 1 suggests that OTUs generated at 66–93% sequence identity (black lines in Figure 1) reached saturation indicating appropriate sampling intensity at lower sequence identity levels, while those generated at 94–99% sequence identity (gray lines in Figure 1) did not reach

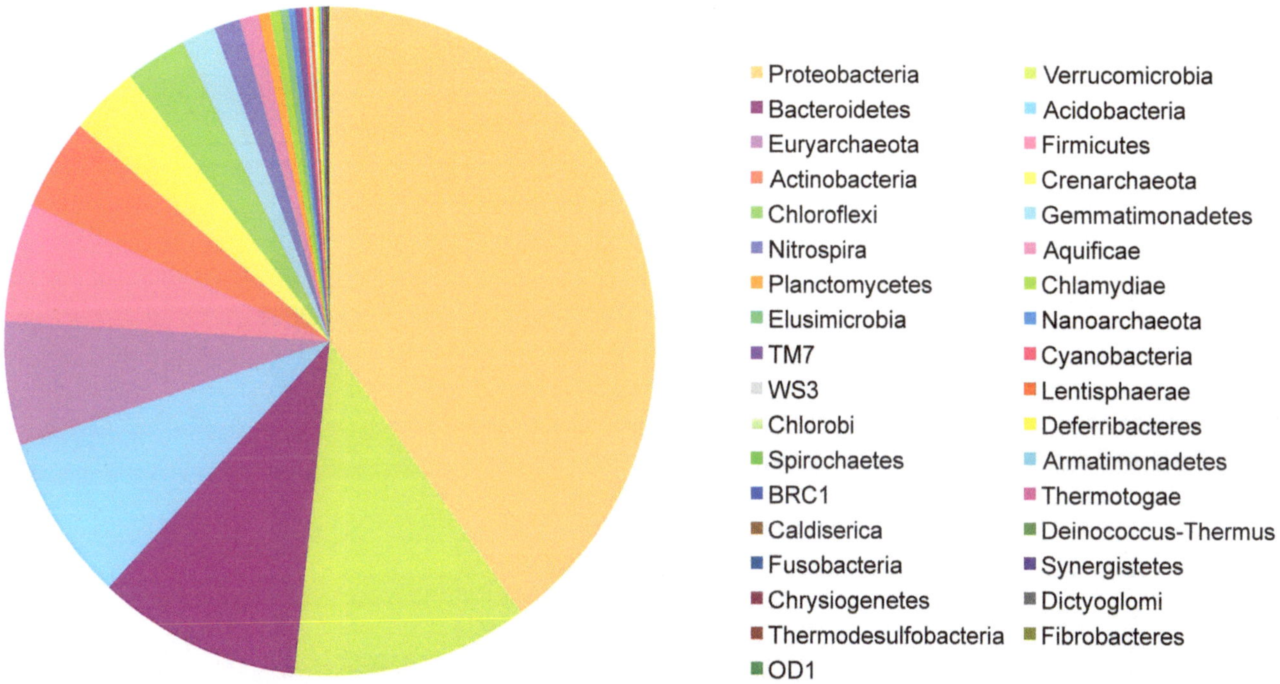

Figure 2. Bacterial phylum identified in soil samples. Bacterial frequency is shown for all plots sampled. Frequencies for each phylum in each treatment are show in Table S2 and plotted in Figure S3.

saturation. Thus our rarefaction analyses indicated that 93% SIL was the most taxonomically exclusive level for which a sufficient level of sampling was achieved (Figure 1). In addition, because 97% SIL is commonly used in the literature to determine and analyze OTU frequencies, we performed subsequent diversity analyses at both 93% and 97% SIL. In order to better evaluate the beta-diversity calculations we generated plot-level rarefaction curves for 93% and 97% SIL (Figure S2) and observed that the patterns were similar to treatment-level rarefaction curves (Figures 1, S2).

Bacterial Community Diversity in Different Treatments

To determine whether soil bacterial diversity was significantly altered in response to nutrient enrichment and contrasting fire regimes, we first determined the bacterial taxa present in the soils sampled. To do this we used OTU abundances and grouped the OTUs at the level of phylum (Table S3). Interestingly, we found that around 10% of the sequences generated were derived from Archaea, which is greater than previously reported values for grassland soils which are typically around 5% Archaea [46]. However this is consistent with reports of relatively high and variable Archaeal abundance in terrestrial ecosystems in general, and an increasing awareness of their potential roles in nutrient transformations in surface soils [46]. Proteobacteria were the most abundant (40%) of the bacterial phyla identified in the tallgrass prairie soils sampled, followed by Verrucomicrobia (11.7%), Bacteriodetes (10.1%), Acidobacteria (8.1%), Firmicutes (5.7%) and Actinobacteria (4.5%) with the remaining bacterial and Archaeal phylum summing to 20% (Figures 2, S3).

To determine treatment effects on the bacterial community, OTU frequencies generated at 93% and 97% SIL were used to calculate taxonomic richness, number of extremely rare or singlet taxa, dominance, diversity, and evenness for each plot and analyzed across treatments using ANOVA. Significant changes in alpha diversity were observed (Figures 3 and S4). The main effects of burning and phosphorus amendment (not shown) were not significant for any of the diversity indices at either %SIL ($p>0.05$ for all tests). The nitrogen treatment significantly reduced taxonomic richness of OTUs generated at 93% SIL ($p=0.008$) but not the OTUs generated at 97% SIL. However nitrogen treatment did significantly reduced Fisher's diversity for OTUs generated at 93% and 97% SIL ($p=0.001$ and 0.007, respectively) and significantly reduced the number of extremely rare or singlet taxa at 93% SIL ($p=0.05$), but not at 97% SIL. There was also a significant increase in Simpson's dominance in response to nitrogen addition at both SILs ($p=0.009$ and 0.02). The nitrogen addition treatment, however, did not significantly change Shannon diversity or taxonomic evenness (Figures 3, S4). In specific contrasts of the treatment combinations, we found significant differences between control and nitrogen treatments at both %SIL ($p=0.002$ and 0.013) and between control and burn+nitrogen treatments ($p=0.004$ and 0.02) for Fisher's diversity. We also found significant differences between control and burn+nitrogen treatments at 93% SIL ($p=0.031$); and between burn and burn+nitrogen treatments at both %SIL ($p=0.018$ and 0.04) for species richness, between control and nitrogen treatments at both %SIL ($p=0.007$ and 0.005) for Simpson's dominance, and between control and burn+nitrogen treatments at 93% SIL ($p=0.046$) in the number of singlet taxa (Figure 3). Thus, while there were some differences the analyses at 93% and 97% SIL were highly similar, indicating that our diversity results were robust to the %SIL level at which OTUs were determined.

Phylogenetically Informed Bacterial Community Analysis

To determine whether soil microbial community structure was altered by nutrient enrichment and/or contrasting fire regimes, we used the UniFrac metric [43,44] with Principle Coordinates

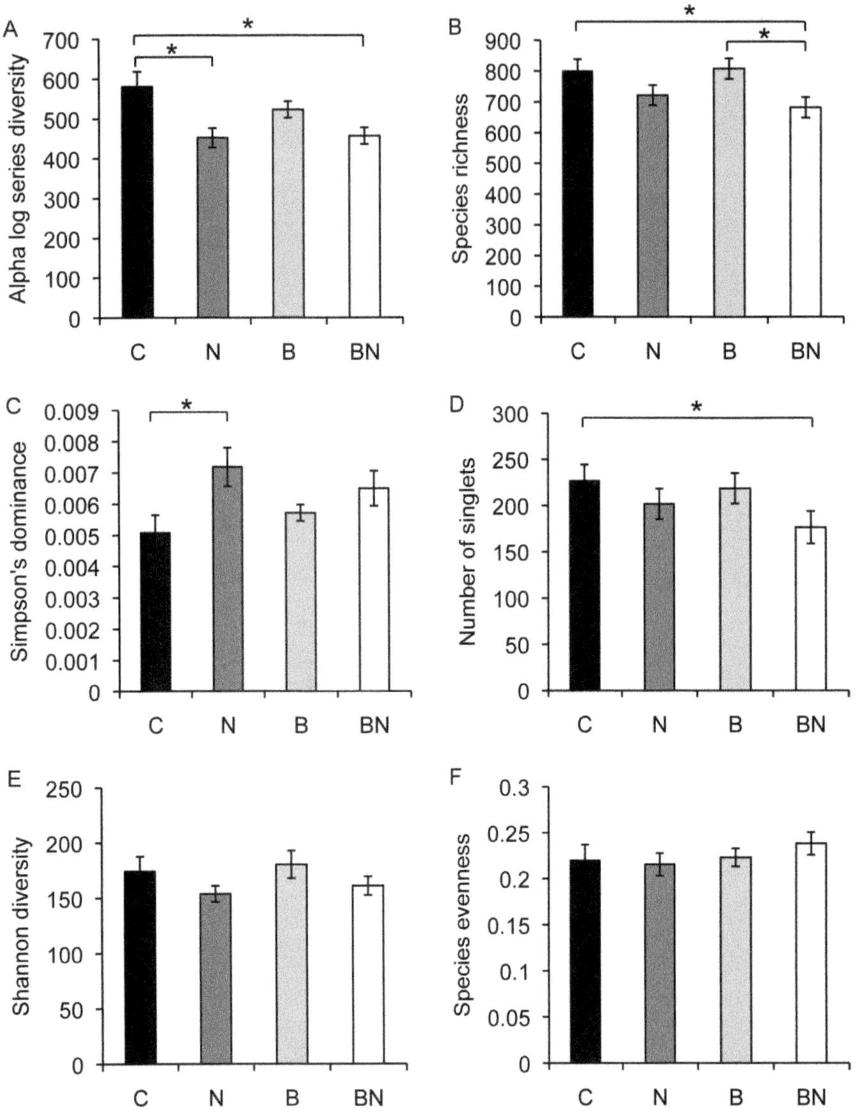

Figure 3. Mean alpha diversity for control (black), nitrogen amended (dark gray), burned (light gray), and burn+nitrogen treatment (white). A) Fisher's alpha, B) Taxonomic richness, C) Simpson's dominance, D) number of singlets (extremely rare taxa), E) Shannon diversity and F) taxonomic evenness are shown. A * indicates significant at $p<0.05$ in mixed model ANOVA. Error bars indicate standard errors.

Analysis (PCoA) to determine the relatedness of bacterial communities between pairs of samples. The UniFrac metric uses the overlap in branch lengths to estimate the phylogenetic distance, or relatedness, between pairs of bacterial communities [47] and the PCoA clusters related communities. Comparison of soil microbial community makeup using UniFrac PCoA for OTUs generated at 93% or 97% SIL showed distinct clustering by nitrogen treatment (Figures 4, S5), but there was no detectable clustering by burn treatment. Nitrogen amended plots were separated from non-nitrogen amended plots on the first two PCoA axes, which together represented 26% or 10% of the total variance for 93% or 97% SIL, respectively.

To better understand how the treatments could indirectly elicit changes in bacterial community structure we investigated how the different treatments altered soil chemical properties (Table 1). As expected, long-term nitrogen fertilization significantly increased the concentrations of extractable inorganic nitrogen in the soil ($p = 0.0341$), due primarily to a significant increase in nitrate concentration ($p = 0.0127$). We also found that burning significantly reduced extractable soil inorganic nitrogen concentrations ($p = 0.0382$) and we observed a significant interaction of burn and nitrogen treatments on nitrate concentrations ($p = 0.0488$). The increases in total extractable inorganic nitrogen and nitrate concentrations in response to nitrogen amendment alone were completely mitigated by the concurrent application of the annual burning treatment and associated N losses [25], making levels of extractable inorganic nitrogen in the control and burn+nitrogen treatments indistinguishable (Table 1). No other measured soil property differed by treatment.

Frequencies of Individual Bacterial Taxa in Different Treatments

Since our analyses indicated that nitrogen addition altered microbial diversity (Figure 3, Figure S4), we examined how individual taxa responded to these perturbations. To test the hypothesis that individual taxon frequencies were altered in

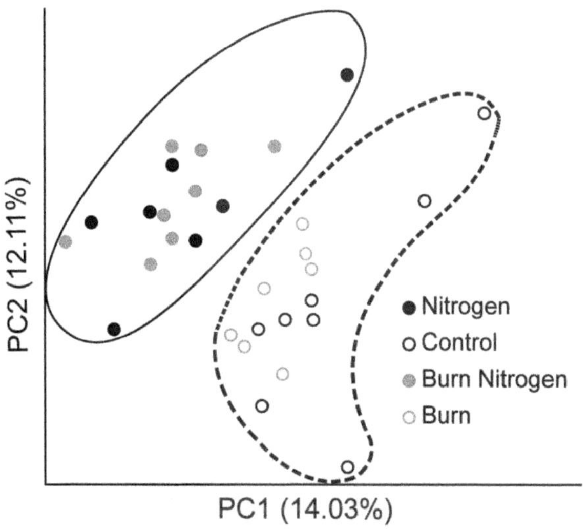

Figure 4. UniFrac Principle Coordinate Analysis plots for the different experiments in this study. Treatments indicated are control (open black), nitrogen addition (filled black), burned (open grey), and nitrogen+burn (filled grey) treatments were plotted. Clustering based on nitrogen treatment is outlined (nitrogen in solid line and no nitrogen addition in dashed line).

response to nutrient enrichment and/or contrasting fire regimes we performed a mixed model ANOVA on OTU frequency of occurrence at the phylum level, 93% SIL and 97% SIL. Interestingly, we found that phylum frequencies were quite similar across treatments and there were no significant treatment effects on phylum-level bacterial taxa frequencies (Figure S3). However, when we examined taxa at 93% SIL, we found that multiple taxa responded significantly to nitrogen amendment and to the interaction of nitrogen amendment and burning treatments, with most taxa showing significant response to treatment interactions (Table 2). We found that OTUs identified as Burkholderiales and Acidobacteria Gp1 responded significantly to nitrogen addition only, while all other taxa that exhibited a significant main effect response to nitrogen also exhibited a significant interactive response to nitrogen addition and burn treatments (Table 2). The Burkholderiales OTU was significantly reduced while the Acidobacteria Gp1 OTU significantly increased in response to nitrogen amendment (Table 2). OTUs corresponding to Nitrospira and Gammaproteobacteria were increased in response to nitrogen addition, with significant differences observed between burn and burn+nitrogen treatments. Similarly, an Acidobacteria Gp3 OTU increased in frequency in response to nitrogen addition, with significant differences observed between control and nitrogen amended plots in the fire exclusion treatment. An OTU classified as Acidobacteria Gp4 exhibited a significant interaction between nitrogen addition and burn treatments, with a decrease in response to nitrogen addition only in the presence of burning. An OTU classified as Ohtaekwangia and an OTU classified as Flavobacteriaceae showed more complex significant interactions between burn and nitrogen addition treatments and an Alphaproteobacteria OTU had multiple significant test results, including a significant main effect of nitrogen addition resulting in an increase in frequency and multiple significant interactions between burn and nitrogen addition treatments (Table 2). Analyses preformed at 97% SIL identified similar taxa with four of the nine significant responders at 93% SIL in the top ten most significant responders at 97% SIL (Table S4). However, because OTUs were broken up into more exclusive taxa, only two taxa were found to have significant responses after multiple testing corrections (FDR = 0.05). At 97% SIL, a Chloroflexi OTU had a significant interaction between burn and nitrogen with greater abundance in burn than burn+nitrogen treatments and a Gemmatimonas OTU had a significant main effect of nitrogen and with greater abundance in the presence on nitrogen enrichment (Table S4).

Discussion

An increasing number of studies are employing deep sequencing to quantify microbial communities, taking advantage of the large volume of sequences produced to determine community diversity (e.g. [21,48,49,50,51]), while others have extended this approach to allow examination of microbial community responses to a broad range of environmental perturbations [52,53,54,55,56]. The current study classified 16S rRNA sequences from soil bacteria as OTUs across long-term experimental treatments within a tallgrass prairie research site in the Central Plains (Konza Prairie LTER). We generated rarefaction curves at each %SIL to determine the best level for more detailed analyses of community diversity and taxonomic response to the treatments. We found that 93% SIL provided the greatest taxonomic resolution where we could be certain our sampling was sufficient (Figure 1). Furthermore, we tested the robustness of our analyses using the 93% SIL cutoff by also performing analyses using OTUs calculated at the more traditional 97% SIL and found that the results were highly similar. Because this study utilized the V3 region of the 16S rRNA gene, it is possible that our results provide a conservative estimate, perhaps an underestimate, of community diversity [57]. A recent study however, showed that the V3 region of the 16S performs well for sequence frequencies similar to those found in soils [33]. Although the sequences obtained here using the first generation of 454 sequencing technology were shorter and less numerous than those that can be obtained with the current technology, we were able to investigate the effect of replicated plot-level perturbation on both community diversilty and individual taxa responses. Within this system, we set out to address three hypotheses:

1) Soil bacterial community composition is altered in response to nutrient enrichment and/or contrasting fire regimes. We found that nitrogen amendment significantly reduced community richness and diversity, but the other treatments (phosphorus addition, contrasting fire regimes) had no effect (Figure 3). The lack of response to phosphorus additions is not surprising, and may be due to the relatively low levels of phosphorus added and to the high capacity of these calcium-rich soils to bind phosphorus. This is also consistent with the lack of response in plant productivity or species composition to added phosphorus in this experiment (*unpublished data*). The lack of bacterial community response to fire treatments was somewhat surprising, as previous studies have reported significant differences in plant community composition [23], belowground plant biomass [26], and soil carbon and nitrogen pools and fluxes [18,25,58] in annually burned and unburned tallgrass prairie. These results suggest that soil bacterial communities may not be affected as strongly as plant communities to changes in the soil environment associated with burning. In addition, the contrast in responses of plant and bacterial communities to fire treatments is consistent with other studies [10], and reinforces the suggestion that different factors may structure plant and microbial responses to the same treatments [59].

The decrease in taxonomic richness caused by nitrogen addition could result from consistent loss of particular taxa or an increase in

Table 2. Taxa with significant response to the addition of nitrogen at 93% SIL.

OTU #	number of seqs in OTU	Name	Effect	Contrast	Fold Change	stderr	*q*-value
213	818	Nitrospira	nitrogen	C vs N	7.3700	1.5596	0.0343
213	818	Nitrospira	burn*nitrogen	B vs BN	11.4001	2.2035	0.0204
690	304	Acidobacteria Gp1	nitrogen	C vs N	8.8649	1.6137	0.0148
982	177	Acidobacteria Gp4	burn*nitrogen	B vs BN	−10.1861	2.1063	0.0319
2499	50	Ohtaekwangia	burn*nitrogen	C vs BN	−9.4671	1.9053	0.0267
2770	653	Acidobacteria Gp3	nitrogen	C vs N	1.2752	0.238	0.0155
2770	653	Acidobacteria Gp3	burn*nitrogen	C vs N	1.9011	0.3366	0.0126
2784	93	Gammaproteobacteria	nitrogen	C vs N	7.5369	1.41	0.0155
2784	93	Gammaproteobacteria	burn*nitrogen	B vs BN	11.7232	1.9941	0.0094
4308	75	Burkholderiales	nitrogen	C vs N	−7.4906	1.5646	0.0319
4524	88	Flavobacterium	burn*nitrogen	N vs B	−11.1137	2.409	0.0418
4697	100	Alphaproteobacteria	nitrogen	C vs N	9.6399	1.2231	0.00058
4697	100	Alphaproteobacteria	burn*nitrogen	N vs B	−10.7196	1.671	0.00586
4697	100	Alphaproteobacteria	burn*nitrogen	B vs BN	10.4106	1.7297	0.00921
4697	100	Alphaproteobacteria	burn*nitrogen	C vs UBN	8.8693	1.7297	0.0204
4697	100	Alphaproteobacteria	burn*nitrogen	C vs BN	8.5602	1.7864	0.0319

Individual taxa with significant response to treatments are shown. The specific contrast in which a significant difference is indicated (C = control, N = nitrogen, B = Burn, BN = burn+nitrogen, UBN = unburned+nitrogen). Standard error (stderr) is shown as well as false discovery rate corrected p-values (q-values).

dominant taxa and concurrent loss of multiple rare taxa. Although we were able to identify individual taxa with significant increases and decreases in abundance in other analyses, the number of responsive taxa was insufficient to explain the apparent loss in taxonomic richness observed. Interestingly, we also found a significant increase in Simpson's dominance in response to nitrogen enrichment. Furthermore, we found that there was a significant reduction of extremely rare taxa in response to nitrogen addition. Taken together, our data suggest that an increase in species dominance, decrease in rare taxa and decrease in overall species richness result from nitrogen amendment of tallgrass prairie soils. Reduction in diversity in fertilized plots was also observed in arctic tundra [60] suggesting that nitrogen enrichment may reduce niche space and therefore community diversity. In contrast no reduction in diversity was observed in response to nitrogen enrichment in another grassland ecosystem (Cedar Creek), indicating that species richness responses to nitrogen enrichment may be site or soil specific [10].

2) Soil microbial community structure is altered in response to nutrient enrichment and/or contrasting fire regimes. We used a phylogenetic approach (UniFrac) to test this hypothesis. We found that the nitrogen enrichment treatment resulted in distinct bacterial communities as compared to control plots, whereas the burn treatments did not (Figure 3). This is somewhat surprising given that previous studies have shown significant biogeochemical changes in the soil in response to different fire frequencies [15,18,25,28]. One interpretation is that there are threshold responses in microbial community changes to enhanced soil nitrogen availability, and the changes associated with different fire treatments are not sufficient to elicit the responses observed to added nitrogen. The community structure responses we report in response to nitrogen enrichment are similar to that observed in a study of multiple terrestrial ecosystems where it was shown that nitrogen enrichment significantly shifted bacterial communities [12] suggesting that this may be a more general effect of nitrogen enrichment. When we tested for treatment elicited alterations in edaphic properties we found that nitrogen addition increased extractable inorganic nitrogen and nitrate concentrations by 0.92-fold and 6.9-fold, respectively, while the annual burn treatment reduced extractable soil inorganic nitrogen by 0.13-fold, and concurrent annual burning and nitrogen treatments resulted in no difference from control. These data are consistent with previous studies documenting lower inorganic nitrogen concentrations and lower net nitrogen mineralization rates in annually burned compared to unburned prairie, which results from the greater nitrogen limitation/immobilization that accompanies annual burning [18,25]. While these are long-term experiments with ongoing treatments conducted as part of the Konza Prairie LTER program, seasonal dynamics in both soil properties and in microbial communities have been documented [28,58] making it difficult to know how the measured soil properties directly relate to the measurements of the microbial community presented. In addition, we should point out the soil property data were not measured on the same samples from which DNA was extracted, but on alternate samples collected prior to those used for DNA extraction, thus correlative analyses were not appropriate. Subsequent studies with concurrent measurements of soil microbial communities and soil properties over time and on the same samples will be needed to more conclusively test for relationships between soil properties and microbial community structure.

3) Individual taxa frequencies are altered in response to nutrient enrichment and/or contrasting fire regimes. We found that individual bacterial taxa frequencies at 93% SIL were altered in response to nitrogen treatment and the interaction of nitrogen and burn treatments (Table 2) but bacterial phylum-level taxa frequencies were not significantly altered in response to either nitrogen or burn treatments. Interestingly, the taxa with frequencies significantly altered in response to treatments at 93% SIL were constituents of the phylum showing the largest, however non-significant changes in response to treatments (Figure S3). It seems

likely that at the resolution of bacterial phylum, the signal of individual taxa responses was masked because members of the phylum were responding differently, or that responsive members may have made up a minority of the constituents of the phylum. It is also possible that the significant shifts in individual taxa may be indicative of the active community of bacteria present, but that at the phylum level the dormant community of bacteria present in the soil masks these responses. Differentiating these two possibilities and others will be needed to understand why different taxonomic resolutions show different responsive taxa (Tables 2, S4) but do not show differences in community diversity and structure (Figures 3, 4, S3, S4, S5). Additionally, in all treatments the same five bacterial phylum were the most abundant: 1) Proteobacteria (40%) 2) Verrucomicrobia (11.7%), 3) Bacteriodetes (10.1%), 4) Acidobacteria (8.1%), 5) Firmicutes (5.7%) and 6) Actinobacteria (4.5%) with the remaining bacterial phylum combining to 20% (Figure 2) and these taxa have been previously found to be highly abundant in soils [61,62]. When we tested for individual taxon responses at 93% SIL we found that treatment interactions accounted for the majority of the significant effects, suggesting that the responses of bacterial taxa are context dependent. This is interesting because we found an OTU identified as Nitrospira increased in abundance in response to the addition of nitrogen fertilizer but only in the presence of the burn treatment (Table 2). In a recent study of nitrogen enrichment in two distinct study sites, Ramirez *et al.* [22] found reduced abundance of Nitrospira in response to increasing levels of nitrogen enrichment in an agricultural field at the Kellogg Biological Station. Together our data and that of Ramirez *et al.* [22] suggest that the responses of Nitrospira taxa to nitrogen enrichment may be highly context dependent, with a decrease in response to nitrogen alone in at least some ecosystems, but an increase in response to nitrogen with concurrent burn treatment. However, this raises questions about the generality of other bacterial species treatment responses and whether simplistic models of increases/decreases in response to a particular treatment are accurate. Furthermore, we found Gammaproteobacteria increased in response to nitrogen addition in the presence of the burn treatment (Table 2), which is consistent with a reported increase in Gammaproteobacteria in response to nitrogen enrichment in three other sites (grasslands at Cedar Creek, an agricultural field at the Kellogg Biological Station, arctic tundra [22,60]). This suggests that for at least Gammaproteobacteria taxa, the observed response may not be dependent on the presence of the burn treatment like that of Nitrospira taxa. More studies of bacterial taxa responses to treatment combinations are required before we know if the context dependant nature of Nitrospira responses are representative of bacterial taxa in general, or if the less context-dependent responses of bacteria like Gammaproteobacteria taxa are more typical.

Our results show that chronic soil nitrogen enrichment can significantly alter bacterial community diversity, structure, and individual taxa abundance, which has important implications for both managed and natural grassland ecosystems. Chronic nitrogen enrichment significantly impacted bacterial community diversity and community structure, with individual groups such as OTUs identified as Nitrospira and Gammaproteobacteria exhibiting changes in relative abundance under long-term nitrogen enrichment. Surprisingly, we found little response to burning by the soil microbial community, although contrasting fire regimes result in significant changes in the soil environment and in plant productivity and plant community composition in these grasslands. These results suggest that the factors controlling plant and microbial community responses to environmental change may differ, and that changes in plant communities may not correlate with changes in soil bacterial communities. Additional studies will be needed to more fully understand the factors underlying observed microbial community responses to nitrogen enrichment, and the consequences of changes in microbial community structure for ecosystem function.

Supporting Information

Figure S1 Experimental design for the Belowground Plot Experiment. Treatments are abbreviated: N = nitrogen addition, C = no nitrogen addition, P = phosphorous addition, NP = nitrogen and phosphorus addition; red outline = annually burned plots; shaded areas = treatments not sampled.

Figure S2 Rarefaction curves calculated independently for each plot. Rarefaction curves for A) 93% SIL and B) 97% SIL are shown.

Figure S3 Proportion of taxa represented in each phylum. The mean proportional abundance of each bacterial phylum identified is shown for control and treatments (nitrogen, burn and burn+nitrogen) as well as for the mean for all plots. In each case there was no significant effect of treatment on phylum proportional abundance.

Figure S4 Mean alpha diversity at 97% SIL. Treatments included were control (black), nitrogen amended (dark gray), burned (light gray), and burn+nitrogen treatment (white). A) Fisher's alpha, B) Taxonomic richness, C) Simpson's dominance, D) number of singlets (extremely rare taxa), E) Shannon diversity and F) taxonomic evenness are shown. An * indicates significant at $p<0.05$ in mixed model ANOVA. Error bars indicate standard errors.

Figure S5 UniFrac Principle Coordinate Analysis plot at 97% SIL. Treatments indicated are control (open black), nitrogen addition (filled black), burned (open grey), and nitrogen+burn (filled grey) treatments were plotted. Clustering based on nitrogen treatment is outlined (nitrogen in solid line and no nitrogen addition in dashed line).

Table S1 Primers for barcoded deep sequencing. The primers were produced by adding unique barcode sequences (underlined) between the "A" sequencing primer of Margulies *et al.* [34] and the reverse 16S primer U529R (**bold**) of Watanabe *et al.* [32]. As sequencing was done in only the reverse direction, no barcode was necessary within the "B" construct.

Table S2 Summary of read count per plot before and after quality filtering. Counts for the raw number of reads (Raw) generated per plot as well as the number that were retained after quality filtering (Final) are shown.

Table S3 Table of all phylum found in Belowground Plot Experiment samples. Mean frequency in percentages for each phylum for each treatment is indicated.

Table S4 Taxa with significant response to the addition of nitrogen at 97% SIL. The specific contrast for which a significant difference was observed is indicated (C = control,

N = nitrogen, B = Burn, BN = burn+nitrogen). Standard error (stderr) is shown as well as false discovery rate corrected p-values (q-values).

Acknowledgments

We wish to thank the many people of the Ecological Genomics Institute at Kansas State University for their insight and assistance during the planning and execution of this project. We also thank the Konza Prairie Biological Station and the Konza Prairie LTER program for the long-term maintenance of the research site and for allowing us to utilize their facilities. We specifically thank Ari Jumpponen and Kraig Stevenson for assistance with the manuscript.

Data deposition

Sequences are available from NCBI's Short Read Archive (accession number SRP001373).

Author Contributions

Conceived and designed the experiments: KLJ MAH. Performed the experiments: KLJ TCT. Analyzed the data: JDC KLJ TCT. Contributed reagents/materials/analysis tools: KLJ JDC TCT MAH. Wrote the paper: JDC KLJ TCT JMB MAH.

References

1. Falkowski PG, Fenchel T, Delong EF (2008) The microbial engines that drive Earth's biogeochemical cycles. Science 320: 1034–1039.
2. Bowen JL, Ward BB, Morrison HG, Hobbie JE, Valiela I, et al. (2011) Microbial community composition in sediments resists perturbation by nutrient enrichment. ISME 5: 1540–1548.
3. Baath E, Frostegard A, Pennanen T, Fritze H (1995) Microbial community structure and pH response in relation to soil organic-matter quality in wood-ash fertilized, clear-cut or burned coniferous forest soils. Soil Biology & Biochemistry 27: 229–240.
4. Pietikainen J, Fritze H (1995) Clear-cutting and prescribed burning in coniferous forest - Comparison of effects on soil fungal and total microbial biomass, respiration activity and nitrification. Soil Biology & Biochemistry 27: 101–109.
5. Allison VJ, Miller RM, Jastrow JD, Matamala R, Zak DR (2005) Changes in soil microbial community structure in a tallgrass prairie chronosequence. Journal of the Soil Science Society of America 69: 1412–1421.
6. Jangid K, Williams MA, Franzluebbers AJ, Sanderlin JS, Reeves JH, et al. (2008) Relative impacts of land-use, management intensity and fertilization upon soil microbial community structure in agricultural systems. Soil Biology & Biochemistry 40: 2843–2853.
7. Vazquez FJ, Acea MJ, Carballas T (1993) Soil microbial populations after wildfire. Fems Microbiology Ecology 13: 93–103.
8. Bittman S, Forge TA, Kowalenko CG (2005) Responses of the bacterial and fungal biomass in a grassland soil to multi-year applications of dairy manure slurry and fertilizer. Soil Biology & Biochemistry 37: 613–623.
9. Marschner P, Kandeler E, Marschner B (2003) Structure and function of the soil microbial community in a long-term fertilizer experiment. Soil Biology & Biochemistry 35: 453–461.
10. Fierer N, Lauber CL, Ramirez KS, Zaneveld J, Bradford MA, et al. (2012) Comparative metagenomic, phylogenetic, and physiological analyses of soil microbial communities across nitrogen gradients. ISME 6: 1007–1017.
11. Nemergut DR, Townsend AR, Sattin SR, Freeman K, Fierer N, et al. (2008) The effects of chronic nitrogen fertilization on alpine tundra soil microbial communities: implications for carbon and nitrogen cycling. Environmental Microbiology 10: 3093–3105.
12. Ramirez KS, Craine JM, Fierer N (2012) Consistent effects of nitrogen amendments on soil microbial communities and processes across biomes. Global Change Biology 18: 1918–1927.
13. Allison SD, Martiny JBH (2008) Resistance, resilience, and redundancy in microbial communities. Proc Nat Acad Sci 105: 11512–11519.
14. Samson F, Knopf F (1994) Prairie conservation in North America. BioScience 44: 418–421.
15. Groffman PM, Rice CW, Tiedje JM (1993) Denitrification in a tallgrass prairie landscape. Ecology 74: 855–862.
16. Sotomayor D, Rice CW (1996) Denitrification in soil profiles beneath grassland and cultivated soils. Journal of the Soil Science Society of America 60: 1822–1828.
17. Jangid K, Williams MA, Franzluebbers AJ, Blair JM, Coleman DC, et al. (2009) Development of soil microbial communities during tallgrass prairie restoration. Soil Biol Biochem 42: 302–312.
18. Dell CJ, Williams MA, Rice CW (2005) Partitioning of nitrogen over five growing seasons in tallgrass prairie. Ecology 86: 1280–1287.
19. Williams MA, Rice CW (2007) Seven years of enhanced water availability influences the physiological, structural, and functional attributes of a soil microbial community. Applied Soil Ecology 35: 535–545.
20. Acosta-Martinez V, Dowd S, Allen SV (2008) Tag-encoded pyrosequencing analysis of bacterial diversity in a single soil type as affected by management and land use. Soil Biology & Biochemistry 40: 2762–2770.
21. Roesch LFW, Fulthorpe RR, Riva A, Casella G, Hadwin AKM, et al. (2007) Pyrosequencing enumerates and contrasts soil microbial diversity. Isme Journal 1: 283–290.
22. Ramirez KS, Lauber CL, Knight R, Bradford MA, Fierer N (2010) Consistent effects of nitrogen fertilization on soil bacterial communities in contrasting systems. Ecology 91: 3463–3470.
23. Collins SL, Calabrese LB (2012) Effects of fire, grazing and topographic variation on vegetation structure in tallgrass prairie. Journal of Vegetation Science 23: 563–575.
24. Knapp AK, Seastedt TR (1986) Detritus accumulation limits productivity of tallgrass prairie. BioScience 36: 662–668.
25. Blair JM (1997) Fire, N availability, and plant response in grasslands: A test of the transient maxima hypothesis. Ecology 78: 2359–2368.
26. Kitchen DJ, Blair JM, Callaham Jr MA (2009) Annual fire and mowing alter biomass, depth distribution, and C and N content of roots and soil in tallgrass prairie. Palnt and soil 323: 235–247.
27. Briggs JM, Knapp AK (1995) Interannual Variability In Primary Production In Tallgrass Prairie - Climate, Soil-Moisture, Topographic Position, And Fire As Determinants Of Aboveground Biomass. American Journal of Botany 82: 1024–1030.
28. Turner CL, Blair JM, Schartz RJ, Neel JC (1997) Soil N and plant responses to fire, topography, and supplemental N in tallgrass prairie. Ecology 78: 1832–1843.
29. Clark CM, Cleland EE, Collins SL, Fargione JE, Gough L, et al. (2007) Environmental and plant community determinants of species loss following nitrogen enrichment. Ecology Letters 10: 596–607.
30. Suding KN, Collins SL, Gough L, Clark C, Cleland EE, et al. (2005) Functional and abundance based mechanisms explain diversity loss due to nitrogen fertilization. Proc Natl Acad Sci USA 102: 4387–4392.
31. Clark CM, Tilman D (2008) Loss of plant species after chronic low-level nitrogen deposition to prairie grasslands. Nature 451: 712–715.
32. Watanabe K, Kodama Y, Harayama S (2001) Design and evaluation of PCR primers to amplify bacterial 16S ribosomal DNA fragments used for community fingerprinting. Journal of Microbiological Methods 44: 253–262.
33. Vasileiadis S, Puglisi E, Arena M, Cappa F, Cocconcelli PS, et al. (2012) Soil Bacterial Diversity Screening Using Single 16S rRNA Gene V Regions Coupled with Multi-Million Read Generating Sequencing Technologies. PLoS One 7: e42671.
34. Margulies M, Egholm M, Altman WE, Attiya S, Bader JS, et al. (2005) Genome sequencing in microfabricated high-density picolitre reactors. Nature 437: 376–380.
35. Coolon JD, Jones KL, Narayanan S, Wisely S (2010) Microbial ecological response of the intestinal flora of *Peromyscus maniculatus* and *P. leucopus* to heavy metal contamination. Molecular Ecology 19: 67–80.
36. Huse SM, Huber JA, Morrison HG, Sogin ML, Welch DM (2007) Accuracy and quality of massively parallel DNA pyrosequencing. Genome Biology 8: R143.
37. Huang XQ, Madan A (1999) CAP3: A DNA sequence assembly program. Genome Research 9: 868–877.
38. Fisher RA, Corbet AS, Williams CB (1943) The relationship between the number of species and the number of individuals in a random sample of an animal population. Journal of Animal Ecology 12: 42–58.
39. Storey JD, Tibshirani R (2003) Statistical significance for genomewide studies. Proc Natl Acad Sci U S A 100: 9440–9445.
40. Wang Q, Garrity GM, Tiedje JM, Cole JR (2007) Naive Baysian Classifier for Rapid Assignment of rRNA Sequences into the New Bacterial Taxonomy. Appl Environ Microbiol 73: 5261–5267.
41. Cole JR, Wang Q, Cardenas E, Fish J, Chai B, et al. (2009) The Ribosomal Database Project: improved alignments and new tools for rRNA analysis. Nucleic Acids Res 37: D141–D145.
42. Cole JR, Chai B, Farris RJ, Wang Q, Kulam-Syed-Mohideen AS, et al. (2007) The ribosomal database project (RDP-II): introducing myRDP space and quality controlled public data. Nucleic Acids Res 35: D169–D172.
43. Hamady M, Lozupone C, Knight R (2010) Fast UniFrac: facilitating high-throughput phylogenetic analyses of microbial communities including analysis of pyrosequencing and PhyloChip data. ISME Journal 4: 17–27.
44. Lozupone C, Hamady M, Knight R (2006) UniFrac - An online tool for comparing microbial community diversity in a phylogenetic context. BMC Bioinfofmatics 7: 1–14.
45. Edgar RC (2004) MUSCLE: multiple sequence alignment with high accuracy and high throughput. Nucleic Acids Research 32: 1792–1797.
46. Bates ST, Berg-Lyons D, Caporaso JG, Walters WA, Knight R, et al. (2010) Examining the global distribution of dominant archaeal populations in soil. ISME 5: 908–917.

47. Hamady M, Knight R (2009) Microbial community profiling for human microbiome projects: Tools, techniques, and challenges. Genome Res 19: 1141–1152.
48. Konstantinidis KT, Tiedje JM (2007) Prokaryotic taxonomy and phylogeny in the genomic era: advancements and challenges ahead. Current Opinion in Microbiology 10: 504–509.
49. Leininger S, Urich T, Schloter M, Schwark L, Qi J, et al. (2006) Archaea predominate among ammonia-oxidizing prokaryotes in soils. Nature 442: 806–809.
50. Sogin ML, Morrison HG, Huber JA, Mark Welch D, Huse SM, et al. (2006) Microbial diversity in the deep sea and the underexplored "rare biosphere". Proceedings of the National Academy of Sciences of the United States of America 103: 12115–12120.
51. Trosvik P, Skanseng B, Jakobsen KS, Stenseth NC, Naes T, et al. (2007) Multivariate analysis of complex DNA sequence electropherograms for high-throughput quantitative analysis of mixed microbial populations. Applied and Environmental Microbiology 73: 4975–4983.
52. Miller SR, Strong AL, Jones KL, Ungerer MC (2009) Bar-Coded Pyrosequencing Reveals Shared Bacterial Community Properties along the Temperature Gradients of Two Alkaline Hot Springs in Yellowstone National Park. Applied And Environmental Microbiology 75: 4565–4572.
53. Jumpponen A, Jones KL (2009) Massively parallel 454-sequencing indicates hyperdiverse fungal communities in temperate Quercus macrocarpa phyllosphere. New Phytologist 184: 279–282.
54. Jumpponen A, Jones KL, Mattox J, Yaege C (2009) Massively parallel 454-sequencing of fungal communities in Quercus spp. ectomycorrhizas indicates seasonal dynamics in urban and rural sites. Molecular Ecology 19: 41–53.
55. Jumpponen A, Jones KL (2010) Seasonally dynamic fungal communities in the *Quercus macrocarpa* phyllosphere differ between urban and nonurban environments. New Phytologist 186: 496–513.
56. Jumpponen A, Jones KL, Blair JM (2010) Vertical distribution of fungal communities in tallgrass prairie soil. 102: 1027–1041.
57. Youssef N, Sheik CS, Krumholz LR, Najar FZ, Roe BA, et al. (2009) Comparison of Species Richness Estimates Obtained Using Nearly Complete Fragments and Simulated Pyrosequencing-Generated Fragments in 16S rRNA Gene-Based Environmental Surveys. Appl Environ Microbiol 75: 5227–5236.
58. Garcia FO, Rice CW (1994) Microbial Biomass Dynamics in Tallgrass Prairie. Soil Science Society of America Journal 58: 816–823.
59. Fierer N, Jackson RB (2006) The diversity and biogeography of soil bacterial communities. PNAS 103: 626–631.
60. Campbell BJ, Paulson SW, Hanson TE, Mack MC, Schuur EAG (2010) The effect of nutrient deposition on bacterial communities in Arctic tundra soil. Environmental Microbiology 12: 1872–1854.
61. Nacke H, Thurmer A, Wollherr A, C W, Hodac L, et al. (2011) Pyrosequencing-Based Assessment of Bacterial Community Structure Along Different Management Types in German Forest and Grassland Soils. PLoS One 6: e17000.
62. Evans SE, Wallenstein MD (2011) Soil microbial community response to drying and rewetting stress: does historical precipitation regime matter? Biogeochemistry 109: 101–116.

Microbial Community Profiling of Human Saliva Using Shotgun Metagenomic Sequencing

Nur A. Hasan[1,2], Brian A. Young[3], Angela T. Minard-Smith[3], Kelly Saeed[1], Huai Li[1], Esley M. Heizer[3], Nancy J. McMillan[3], Richard Isom[1], Abdul Shakur Abdullah[1], Daniel M. Bornman[3], Seth A. Faith[3], Seon Young Choi[1,2], Michael L. Dickens[3], Thomas A. Cebula[1,4], Rita R. Colwell[1,2,5,6*]

1 CosmosID, College Park, Maryland, United States of America, **2** Maryland Pathogen Research Institute, University of Maryland, College Park, Maryland, United States of America, **3** Battelle, Columbus, Ohio, United States of America, **4** Department of Biology, Johns Hopkins University, Baltimore, Maryland, United States of America, **5** University of Maryland Institute for Advanced Computer Studies, University of Maryland, College Park, Maryland, United States of America, **6** Bloomberg School of Public Health, Johns Hopkins University, Baltimore, Maryland, United States of America

Abstract

Human saliva is clinically informative of both oral and general health. Since next generation shotgun sequencing (NGS) is now widely used to identify and quantify bacteria, we investigated the bacterial flora of saliva microbiomes of two healthy volunteers and five datasets from the Human Microbiome Project, along with a control dataset containing short NGS reads from bacterial species representative of the bacterial flora of human saliva. GENIUS, a system designed to identify and quantify bacterial species using unassembled short NGS reads was used to identify the bacterial species comprising the microbiomes of the saliva samples and datasets. Results, achieved within minutes and at greater than 90% accuracy, showed more than 175 bacterial species comprised the bacterial flora of human saliva, including bacteria known to be commensal human flora but also *Haemophilus influenzae*, *Neisseria meningitidis*, *Streptococcus pneumoniae*, and Gamma proteobacteria. Basic Local Alignment Search Tool (BLASTn) analysis in parallel, reported *ca.* five times more species than those actually comprising the *in silico* sample. Both GENIUSand BLAST analyses of saliva samples identified major genera comprising the bacterial flora of saliva, but GENIUS provided a more precise description of species composition, identifying to strain in most cases and delivered results at least 10,000 times faster. Therefore, GENIUS offers a facile and accurate system for identification and quantification of bacterial species and/or strains in metagenomic samples.

Editor: Niyaz Ahmed, University of Hyderabad, , India

Funding: Funding for this study was provided, in part, by NIH Grants 1RO1A139129-01 and 2RO1A1039129-11A2. The funders had no role in study design, data collection and analysis, decision to publish, or preparation of the manuscript.

Competing Interests: Rita R. Colwell is Founder and Chairman of the Board, CosmosID®, and some of the other authors are employees of the company, a bioinformatics company. Rita R. Colwell is also Distinguished University Professor at the University of Maryland, College Park and at Johns Hopkins University Bloomberg School of Public Health.

* E-mail: rcolwell@umiacs.umd.edu

Introduction

The microbial flora of the human mouth has been extensively studied, providing an understanding of the role of bacterial species, not only in maintaining wellness, but also in dental caries and gingivitis[1,2,3]. The reductionist approach to understanding the microbiology of the human mouth that was followed in the early days of dental microbiology, in some ways inhibited achievement of full understanding of the oral flora as a microbial ecosystem. Recent research on the human microbiome has provided valuable information concerning a variety of processes, including interactions among microbial species in the human mouth. Next generation sequencing (NGS) permits an even more extensive characterization of the microbial ecology of the human body and has triggered an explosion in human microbiome discovery. As a result, the microbiome is now considered by some investigators to represent yet another organ of the human body, dictating health and well-being[4]. From the avalanche of data reported to date, spatio-temporal and host induced variations in microbiomes have been associated with a variety of human conditions, including colorectal carcinoma[5,6], cardiovascular disease[7], inflammatory bowel disease[8], obesity [9], white blood cell cancer [10], and even psychiatric conditions[11].

The oral cavity is a major gateway for bacterial entry to the human body and a natural route for passage to respiratory and digestive tracts and, ultimately, the blood stream. Historically, microorganisms in the oral cavity were found to comprise a diverse and complex community [12], comprised of hundreds of individual bacterial species[13,14]. Recent evidence shows that some bacterial species of the mouth microbiome are linked to oral disease, but are also important in the general health of an individual [12,15,16,17,18,19]. To date, the oral microbiome has been linked to many diseases, namely alveolar osteitis and tonsillitis [12,14,20,21,22,23], bacteremia [24], endocarditis [25], brain and liver abscesses [26,27], stroke [28], diabetes [29,30], pneumonia [31], and premature birth [32]. The mouth can also be considered an important site of genetic exchange among members of the bacterial flora, because of its high bacterial load and richness in species diversity, even where antibiotic resistant bacteria can become established through contact transmission

[33]. The oral microbiome is not homogenous but is made up of subpopulations inhabiting microenvironments within the mouth [15]. A primary example is saliva, which contains a specific bacterial community that helps maintain homeostasis of the mouth ecosystem. Thus, it is not surprising that the oral or salivary microbiome has attracted increased attention as a potential diagnostic tool [23,34,35,36]. Collection of saliva samples is simple and not invasive [15,16,37]. Saliva itself is clinically informative, containing many soluble biomarkers typically found in blood and urine [38,39] that are useful in prognosis of several systemic and oral conditions. In fact, the uniqueness of individual oral commensal flora provides a forensic tool, contributing to the development of the new discipline microbial forensics [40,41]. Because the salivary microbiome has diagnostic, epidemiological, and forensic value, we investigated the bacterial flora of saliva using direct whole genome shotgun (WGS) metagenomic sequencing, an unbiased metagenomic approach to determine the bacterial species and strain composition.

Historical microbiological studies employing conventional culture methods had shown that the human salivary microbiome is comprised of a complex assemblage of bacteria, viruses, fungi, and parasites, with less than half of the bacterial species cultured [12,13,14,42]. 16S rRNA-based identification revealed extensive bacterial diversity, providing results more quickly than traditional culture [23]. However, 16S rRNA is a single gene-centric method, providing less resolution in differentiating closely related species. It also suffers from limitations imposed by non-uniform distribution of sequence dissimilarity among taxa, presence of multiple copies of the 16S rRNA gene [43], failure of target amplification of polymerase chain reaction (PCR) primers [44], and generation of chimeric sequences [45,46].

The accuracy and robustness of any identification methods is dependent on the quality and breadth of the reference database [12]. The popular sequence alignment tool, BLAST [47] relies on the NCBI public database from which even NCBI removes sequences and genomes due to errors [48]. The known limitations of 16S rRNA sequencing for microbial identification has prompted investigators to use WGS sequencing for characterization and resolution of metagenomic communities [12,36], The large amounts of data produced by WGS sequencing, however, present significant challenges in data analysis and interpretation [49]. There are many approaches that have been devised for analysis of WGS data, including alignment, assembly, binning, and gene prediction based methods [50,51]. Read reference alignment or mapping performs reasonably well, but these are computationally very expensive [52]. Compositional binning tools, i.e. MEGAN [53] and MG-RAST [54], are also computationally expensive and do not resolve closely related taxa with the short reads as generated by Illumina and Life Technologies NGS platforms [50].

NGS has progressed today to being able to deliver highly accurate sequences economically and with fast turn-around. As a result, whole genome shotgun (WGS) metagenomic sequencing emerged as a powerful tool for studying the human microbiome [55]. At present, WGS metagenomic data comprise millions to billions of short reads, aiding necessary sequencing depth as needed as well as offering an unprecedented opportunity to identify individual species at or near strain level and determine their relative abundance. In this study, WGS metagenomics was employed, in combination with GENIUS algorithms, to identify and quantitate bacterial species comprising the salivary microbiome.

Materials and Methods

Sample Collection and DNA Isolation

Total DNA was collected from saliva samples provided by two anonymous healthy adult donors following the approved protocol of the Battelle Memorial Institute Internal Review Board. DNA was purified from saliva employing the Oragene-DNA isolation kit (DNA Genoteck, Kanata, ON, Canada), following the manufacturer's recommended protocol.

Next Generation Sequencing and Filtering

DNA samples for metagenomics were prepared for 150 bp and 100 bp single-end sequencing using the Illumina GAIIx and HiSeq 2000 instrument (Illumina, San Diego, CA), respectively. Numerically coded aliquots of approximately 0.5–1 μg DNA per sample were used to create sequencing libraries. First, genomic DNA was fragmented using a Covaris^{TM} S220 Sonicator (Covaris, Inc., Woburn, MA) to approximately 300 base pairs (bp). Fragmented DNA was used to synthesize indexed sequencing libraries using the TruSeq DNA Sample Prep Kit V2 (Illumina, Inc., San Diego, CA), according to manufacturer's recommended protocol. Cluster generation was performed on the cBOT using the TruSeq PE Cluster Kit v3 – cBot – HS (Illumina). Libraries were sequenced with an Illumina HiSeq 2000 at Nationwide Children's Hospital (NCH) Biomedical Genomics Core (Columbus, Ohio) using the TruSeq SBS Kit v3 reagents (Illumina) for paired end sequencing with read lengths of 100 base pairs (bps) (200 cycles) and at CosmosID with an Illumina GAIIx for 150 base pairs (bps) single read using the TruSeq SBS Kit v5 reagents (Illumina). Primary analysis (image analysis and basecalling) were performed using HiSeq Control Software (HCS) version 1.5.15.1 and Real Time Analysis (RTA) version 1.13.48. Secondary Analysis (demultiplexing) was performed using Illumina CASAVA Software v1.6 Post processing of GAIIx reads was performed with RTA/SCS v1.9.35.0 and CASAVA 1.8.0 software. High throughput sequencing reads were quality filtered using the fastq_quality_filter program provided with the FASTX-Toolkit (http://hannonlab.cshl.edu/fastx_toolkit/index.html) (v. 0.0.13). Only those reads with a quality score ≥17 for at least 80% of the read length (i.e., probability of correct base call ~98%) were retained. Ion Torrent (Life Technologies, NY) sequencing was also performed using amplicons specific to the V4 region of the 16S rRNA gene. Sequence reads are available under NCBI BioProject ID PRJNA231652.

BLASTn Analysis

Sequence data were compared to the NCBI RefSeq database (v. May 19, 2012), but restricted to microbial gis, the NCBI 16S database (v. October 30, 2012), using BLASTn [47] (top hit only) (v. 2.2.25, National Library of Medicine, Bethesda, MD). Table 1 provides details of analyses carried out for each sample with data bases used. Resulting BLASTn hits were filtered to retain only those hits with percent identity ≥97%. An additional filter was applied to the BLASTn hit report to reduce false positives (i.e., reads whose corresponding taxonomic identifier (taxid) appeared ≤0.01% (1:1000)). This was accomplished using a custom script. The Krona (v. 2.2) [56] program, ClassifyBLAST.pl, was also used within a custom script, to obtain a list of organisms identified with read counts associated with each taxon. Krona ImportBLAST.pl program was used to provide interactive visualization of identified bacterial species.

Table 1. Multi-platform BLASTn analysis of two salivary samples against various databases.

Sample	Sequencing Platform		
	Illumina HiSeq 2000	Illumina GAIIx	Ion Torrent (16S, V4)
	Reference Database used with BLASTn		
VFD10-018	Greengenes 16s[1]	Greengenes 16s[4]	Greengenes 16s[3]
	NCBI 16s	NCBI 16s[4]	NCBI 16s[3]
	NCBI RefSeq (microbial subset)[1]	NCBI RefSeq (microbial subset)[4]	NCBI RefSeq (microbial subset)[3]
VFD12-006	Greengenes 16s[1]	Greengenes 16s[2]	Greengenes 16s[3]
	NCBI 16s[1]	NCBI 16s[2]	NCBI 16s[3]
	NCBI RefSeq (microbial subset)[1]	NCBI RefSeq (microbial subset)[2]	NCBI RefSeq (microbial subset)[3]

[1]Illumina HiSeq2000 sequencing carried out at Nationwide Children's Hospital, Columbus, OH.
[2]Illumina GAIIx sequencing carried out at CosmosID.
[3]Ion Torrent sequencing performed by SeqWright, Inc., Houston, TX.
[4]Illumina GAIIx sequencing performed at The Ohio State University, Columbus, OH.

GENIUS Analysis

Raw unassembled WGS short reads generated by the Illumina GAIIx and HiSeq 2000 platforms were analyzed using GENIUS software package for rapid identification of bacterial species and relative abundance. GENIUS creates sample libraries from unassembled short WGS reads using two algorithms, 5VCE and NmerCE, and utilizes GeneBook® reference libraries derived from curated genomic databases to assign taxonomic membership of sample libraries, employing probabilistic matching. Identification is achieved at species, sub-species, and/or strain level, depending on adequate representation of relevant reference genomes in the GeneBook® libraries

In silico Metagenome Construction

A synthetic metagenome was created comprising a total of 5.5 M reads from ten bacterial species and the human genome (Table 2). The reads, each 100 nucleotides in length, were created using a custom R script from each of the ten bacterial genome and the human chromosome 21 using Illumina sequencing error model.

Results and Discussion

Description of the human microbiome has been made possible by NGS with its significant reduction in cost and improvement in throughput. Metagenomics, as a result, is moving from a16S rRNA gene-centric approach to WGS metagenomic approach. To date, 16S rRNA gene sequencing has been used to identify major taxa and explore the microbial diversity of the human salivary microbiome [23,35,57] linking composition of the microbiome with oral health and/or systemic disease. In this study, WGS metagenomics was used, along with several bioinformatics analysis methods e.g., BLASTn, mapping, and GENIUS, to determine relative performances of taxonomic assignment, and identification of community composition and structure, thereby achieving improved understanding of the human salivary microbiome.

GENIUS 5VCE algorithm was employed to determine the bacterial species composition of a human saliva sample, VFD10-018, sequenced by the Illumina GAIIx (150 bp, ~22 M reads) platform. A total of 26 bacterial genera and 58 species were identified with majority of genera and species previously identified

Table 2. Species composition and simulation statistics of the synthetic metagenomic dataset.

Simulations Statistics				BLAST (RefSeq_genomic)
Species	Genome Coverage	Number of Reads	Relative Abundance	Number of Reads
Homo sapiens	0.08	4034394	72.81	3349
Rothia dentocariosa ATCC 17931	47.21	537919	9.71	554
Prevotella melaninogenica ATCC 25845	52.69	430335	7.77	440
Fusobacterium nucleatum subsp. nucleatum ATCC 25586	5.44	53791	0.97	58
Streptococcus oralis Uo5	12.08	107583	1.94	106
Streptococcus mitis B6	11.02	107583	1.94	88
Veillonella parvula DSM 2008	11.1	107583	1.94	83
Peptostreptococcus stomatis DSM 17678	0.99	46261	0.83	NA
Peptostreptococcus anaerobius 653-L	0.94	46261	0.83	NA
Porphyromonas gingivalis W83	0.84	46261	0.83	46
Mycoplasma pneumoniae FH	1.21	23130	0.42	23

The right most column represents number of reads from the sample that BLAST was able to assign to species. NA: Not Assigned due to lack of RefSeq entries.

as members of the human salivary and/or oral microbiome (http://www.homd.org/index.php, Fig. S1). BLASTn (microbial subset) and short read mapping (CLC genomic workbench, using same genome database as GENIUS 5VCE) identified 45 and 102, and 67 and 108, bacterial genera and species in this data set respectively, indicating a much larger microbial community compared to that identified by GENIUS 5VCE (Fig. S1). A global 16S metagenic survey of saliva samples collected from 120 healthy individuals in 12 geographically different locations reported that an individual salivary microbiome typically contains six to 30 bacterial genera [41], an observation in agreement with GENIUS 5VCE identification. Dominant genera identified by GENIUS 5VCE (*Streptococcus,Prevotella, Veillonella, Mycoplasma, Rothia, Haemophilus, Fusobacterium etc.)* were also in agreement with genera identified in the global survey [41]. It is concluded that bacterial taxa identified by BLASTn and short read mapping produces an overestimation of diversity. GENIUS 5VCE performed favorably in both speed, at least 10,000X faster than BLAST, and accuracy in identifying the most likely bacterial community of this dataset with significantly reduced false prediction.

As the actual microbial composition of the saliva samples from the volunteers was unknown, only comparative analysis between orthogonal methods was possible. Therefore, an *in silico* sample (Table 2) containing a composite of 5.5 M reads from ten bacterial species and the human genome was prepared to measure accuracy of the metagenomic analysis. Results show that GENIUS accurately identified bacterial species composition with a negligible computation time (2 minutes for NmerCE and 29 minutes for 5VCE) (Figs. 1 and S2). Despite the fact that the number of sequencing reads for each of the genomes comprising the test set was small, the GENIUS algorithms identified the bacterial species with appropriate strain designation. Identification obtained using GENIUS 5VCE and NmerCE algorithms were in agreement, with one possible false positive identification (*Streptococcus pneumoniae*) by 5VCE. When statistical analysis was carried out, which is a built in function in GENIUS 5VCE algorithm, to provide point estimates for genome coverage and detection confidence limits using random k-mers for each of the identified species, *S. pneumoniae* had the lowest confidence interval (Fig. S2). Furthermore, considering the marginal coverage of strain specific attributes (only 1.6% of unique identifiers), error rate of the sequencing platform, and abundant presence of other *Streptococcus* species in the dataset, this call, at best, would have been tallied as dubious. In contrast, results of BLASTn analysis reported 48 species, even though the sample contained only ten bacterial species, confirming what has been suspected by other investigators as over-estimation of diversity by BLASTn analysis. In contrast, GENIUS was efficient in filtering out false positive signals caused by high genomic similarity among closely related genera and species. Briefly, GENIUS was successful in identifying species of the saliva microbial community, even when only limited sequencing data were available, accomplishing identification that required only minimum computational time. Precise identification was achieved in most cases even with only a small number of reads, i.e., fractional coverage (<1%) of the genome (i.e.,*Peptostreptococcus stomatis, Peptostreptococcus anaerobius*, and *Porphyromonas gingivalis*) were available and also when the target pathogen (i.e., *Mycoplasma pneumoniae)* was present at very low concentration (<1%).

For comparative purposes, 16S rRNA (V4 region)sequencing was carried out using Ion Torrent PGM for salivary samples VFD10 and VFD12 and the sequence data were analyzed by BLAST, using the NCBI 16S ribosomal database (v. 10/30/2012) to enable direct comparison of bacterial communities inferred by both 16S metagenic and WGS metagenomics analysis. Comparison of the accuracy of identification between the two methods was assessed by the number of overlapping genera, since extrapolation of 16S data beyond genus is very limited [43,58,59]. Saliva microbiomes VFD10 and VFD12 showed high concordance (~80%), with respect to genera identified by GENIUS algorithms 5VCE and NmerCE (Fig. 2). Concordance was shared, to a large extent, with 88 genera identified by BLAST-16S (Fig. 2a). GENIUS algorithms 5VCE and NmerCE identified 27 and 26 genera, respectively, in VFD10. Of the 21 genera identified by both algorithms, 17 were also identified by BLAST-16S. The six additional genera identified by5VCE (n = 2) and NmerCE (n = 4) were also identified by BLAST-16S. However, five genera were identified by GENIUS, either by5VCE (n = 4) or NmerCE (n = 1) that were not detected by BLAST-16S. Relative abundance (≥ 1%)of genera determined by GENIUS algorithms and byBLAST-16S (Table S1) were in agreement, with respect to dominant genera, with26 genera comprising 97–99% of the bacterial community.

Results for saliva sample VFD12 are shown in Fig. 2b, with comparison of relative abundance showing 22 genera accounting for 98–99% of the bacterial community (Table S2), providing very little evidence for the large number of genera (n = 82) identified by BLAST-16S. In fact, the global survey of saliva samples [41]reported individual salivary microbiomes contained six to 30 bacterial genera, reinforcing overestimation of diversity by 16S analysis. Based on the literature,16S rRNA sequencing can be biased by unequal amplification of 16S rRNA genes [46] and by taxon-specific biases arising from the primer set used [60]. Generation of chimeric sequences also can skew and inflate diversity estimates significantly [45,46]. Chimera formation is most pronounced when 16S rDNA amplicons are present in low amount, making identification of minor species suspect without supporting data [61].

WGS metagenomics sequencing reads (GAIIx) were compared with the NCBI RefSeq database (microbial subset, v. 05/19/2012), using BLASTn for comparison of species identified by GENIUS algorithms and BLASTn. Fig. 3 shows estimated relative abundance of species at ≥2% for saliva sample VFD10, determined by both GENIUS and BLASTn analyses. A phylogenetic tree of 23 species (depicted as shaded squares) in Fig. 4 shows relative abundance estimated by each method. Most prevalent were *Streptococcus* and *Prevotella*. Similar results were obtained for saliva sample VFD12 (Fig. 4).

A vexing problem in metagenomics is sample diversity, that is, whether sequencing captures the total diversity of a given sample. Libraries generated for saliva sample VFD10 were sequenced using all eight lanes of a GAIIx flow cell, with 18 to 24 million reads (average ~22 million) generated per lane (Fig. S3). GENIUS analysis, with respect to number of species identified, showed good concordance in both identification and relative abundance of bacterial species, for all eight lanes (Fig. S3). Thus, use of an entire flow cell for a single sample library did not influence the diversity estimate.

Illumina HiSeq 2000 allows *ca.*180 million reads per lane, with TruSeq v3 chemistry, compared to *ca.* 40 million reads per lane, with GA v5 chemistry and the GAIIx instrument. When the larger number of reads generated by HiSeq (66–75 million) was analyzed, the number of species identified and percent of each species increased for both saliva samples VFD10 and VFD12, indicating that a larger number of HiSeq reads will contribute breadth and depth in genome coverage (Fig. S4). These results suggest that identification of bacterial species, particularly those present in low number, can be improved with a larger number of

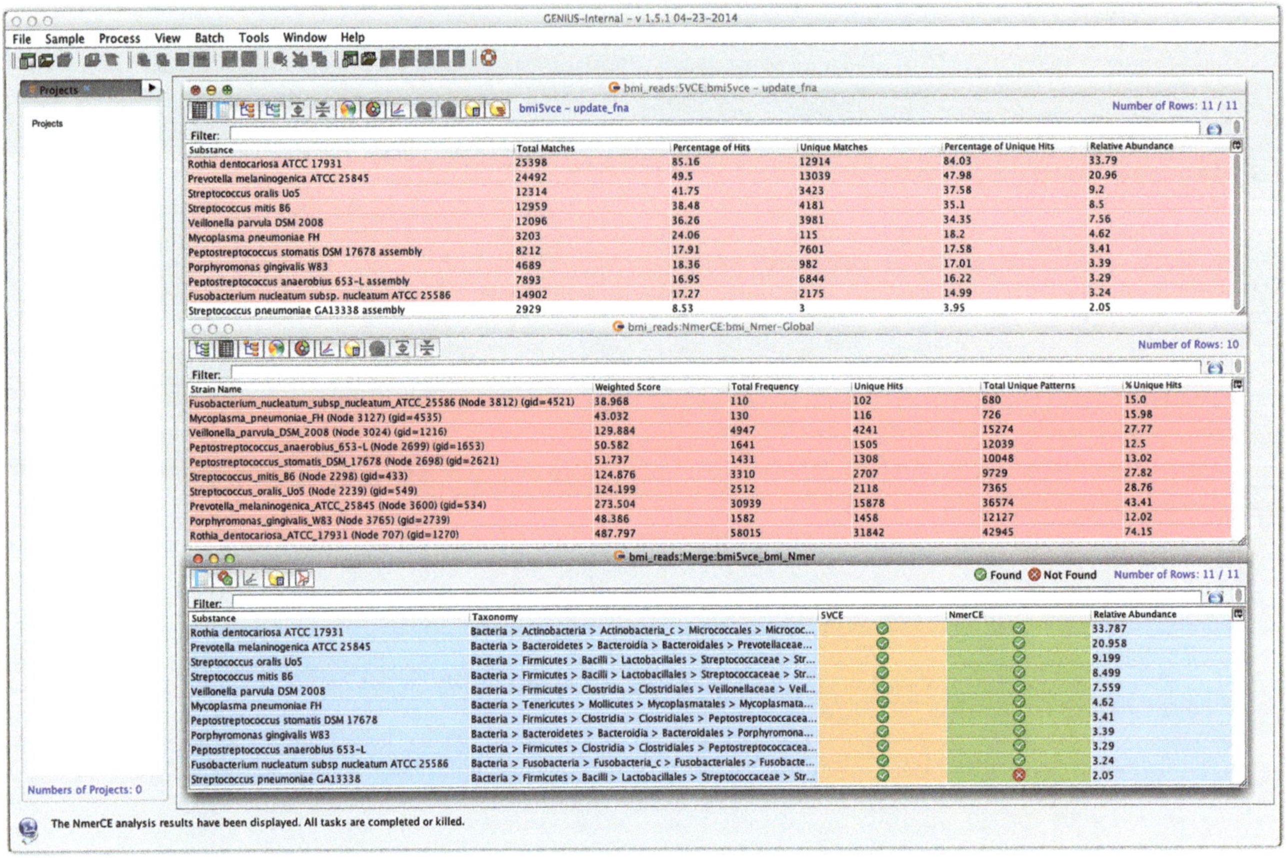

Figure 1. Screenshot of GENIUS client software displaying tabular output of *in silico* metagenomic data. Left panel indicates projects loaded to this graphical user interface. Three table views to the rightmost panel represent output of 5VCE (a)NmerCE (b) and merged output from both 5VCE and NmerCE (c) algorithms, respectively.

reads, especially in the case of samples containing large a mounts of human DNA. Analysis of saliva sample VFD10, using GAIIx sequence data, showed approximately 97% of the sequenced reads was from human DNA. Since background human DNA can range from less than 1% in stool samples to greater than 99% in nasal and vaginal samples [62], the effect of having an increased number of reads to capture species diversity will vary according to host DNA content, as well as complexity of the bacterial population. Improved sampling, extraction, and library construction methods, therefore, should be considered for maximum coverage of species diversity.

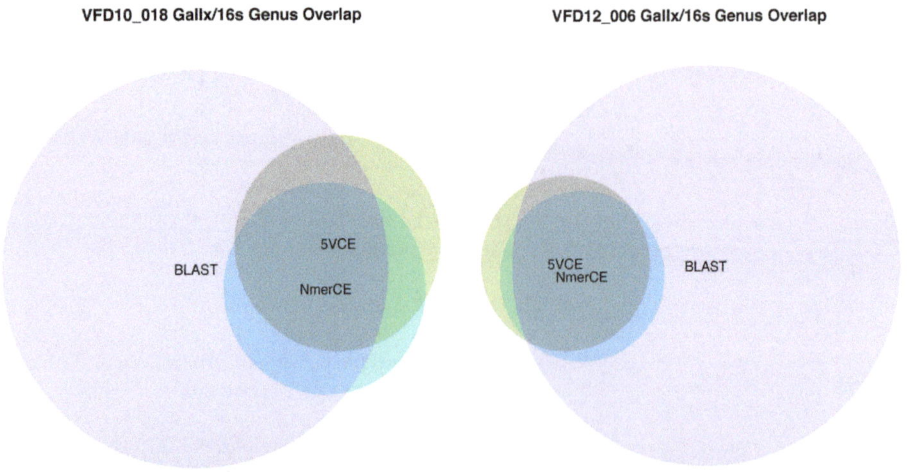

Figure 2. Genus overlap for sample VFD10-018 (a) and VFD12-006 (b) estimated by 16S sequencing/NCBI 16S BLAST and GAIIx sequencing/5VCE-NmerCE.

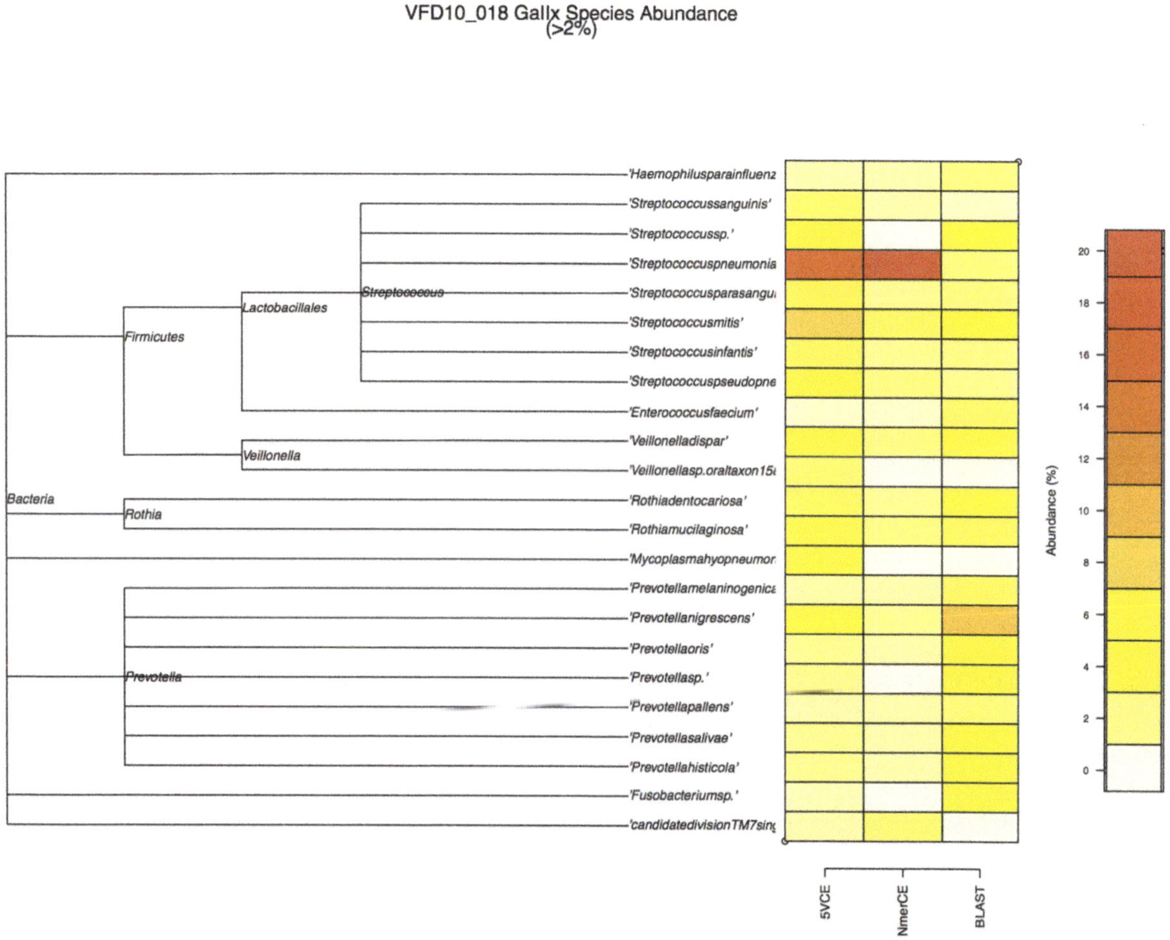

Figure 3. Relative abundance of species in VFD10-018 estimated by GAIIx sequencing and BLAST (microbial reference database), 5VCE, and NmerCE algorithms.

GENIUS was used to analyze salivary datasets from the Human Microbiome Project (HMP) (http://hmpdacc.org/HMASM/). Five human salivary microbiomes were analyzed and nine major phyla were identified by GENIUS 5VCE with Firmicutes, Bacteroidete, Actinobacteria, and Proteobacteria most abundant, Fusobacteria and TM7moderately abundant, and Spirochaetes, Synergistetes, and Tenericutes least abundant. Sixty seven bacterial genera belonging to nine phyla were identified, with eleven genera, *Streptococcus, Prevotella, Veillonella, Neisseria, Haemophilus, Campylobacter, Fusobacterium, Rothia, Mycoplasma,Actinomyces,and Aggregatibacter* comprising ~90% of the bacterial community. Relative abundance estimates of phyla and genera varied (Fig. S5). Overall abundance and distribution of phyla and genera were in agreement with results of studies of human saliva reported by other investigators [34,63,64]. Interestingly, ***Streptococcus*** was observed to be prevalent in most of the HMP datasets, whereas *Prevotella* predominated in the saliva samples, VFD12 and SRS014692, analyzed in this study (Fig. S5). That is, a readily distinguishable abundance profile of ***Prevotella*** species and strains was observed in saliva samples VFD10, VFD12, and SRS014692 (Fig.S6). Greater abundance of ***Prevotella*** in caries-active, **compared to** healthy, individuals has been reported with caries-active individuals often carrying a mixture of *Prevotella* species different from normal healthy individuals [65].

GENIUS identified more than 175 bacterial species, including bacteria commensal to the human salivary microbiome ([65],HOMD, www.homd.org) but also others not usually found in the saliva flora of healthy individuals, including *Haemophilus influenzae*, *Neisseria meningitidis*, *Streptococcus pneumoniae* and Gammaproteobacteria. Several bacterial species, including ***Aggregatibacter actinomycetemcomitans, Porphyromonas gingivalis,*** *Treponema denticola, Fusobacterium nucleatum, Campylobacter rectus, Parvimonas micra, Eikenella corrodens, Prevotella melaninogenica, Prevotella nigrescens, Eubacterium saburreum*, and *Eubacterium yurii*, associated with periodontitis [12,66,67,68], were also identified in varying abundance and distribution (Fig. S7). Although presence of these bacterial species may indicate disease, i.e., periodontitis, it has been shown that periodontitis can be attributed to genetic factors [69]. Therefore, diagnosis of periodontitis cannot yet be made by bacteria present in saliva.

Principal component analysis (PCA) of the five HMP saliva samples and the two saliva samples sequenced in this study showed the bacterial species composition comprised two major clusters (Fig. 5). Four saliva samples, VFD10, VFD12, SRS015055, and SRS014692, clustered separately from SRS09210, SRS013942, and SRS014468. Distinction between clusters, as well as variation within each population, was apparent from a double hierarchical dendrogram showing abundance and distribution of the bacterial species (Fig. 6). Even though two major clusters were observed,

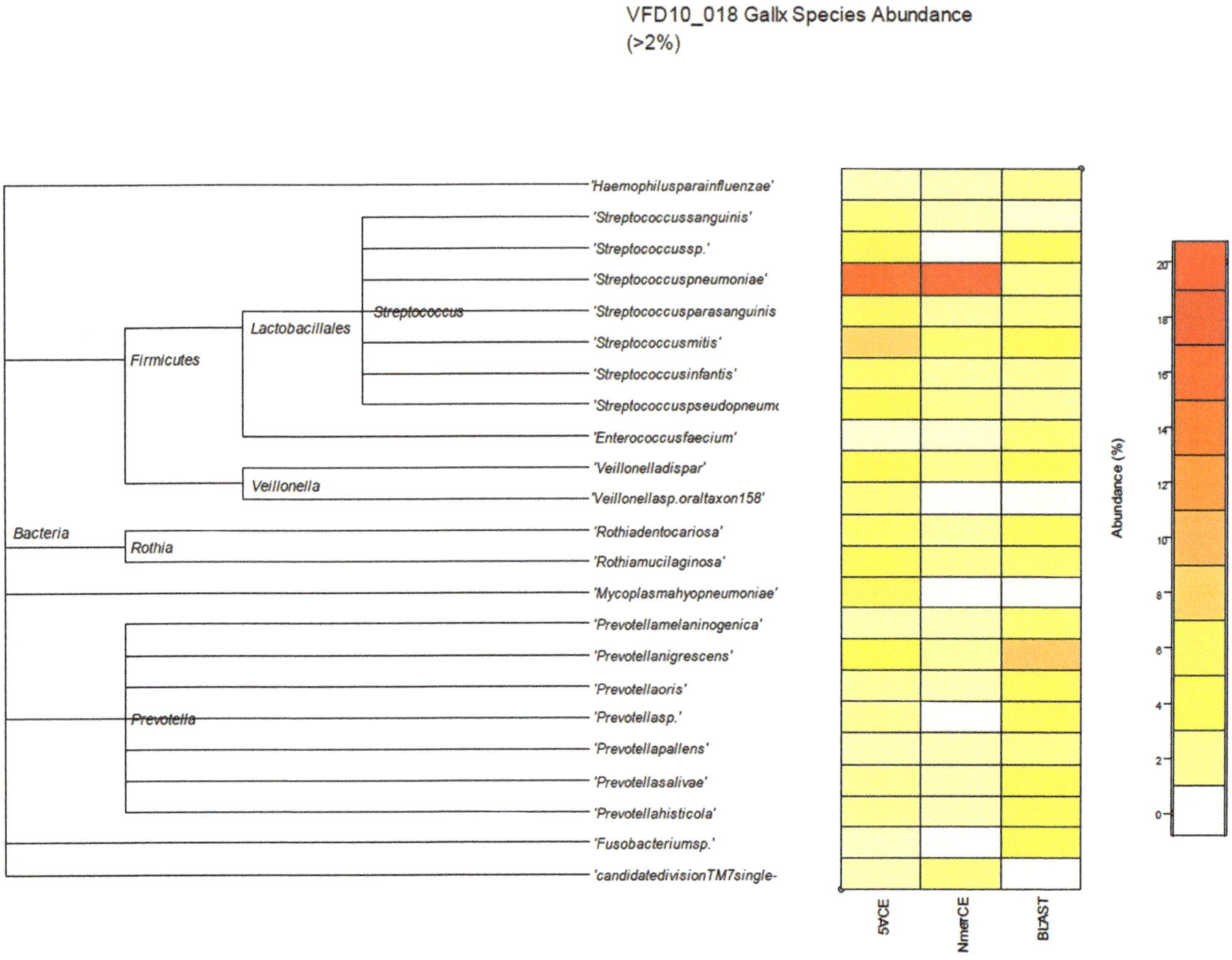

Figure 4. Relative abundance of species in VFD12-006 estimated by GAIIx sequencing and BLAST (microbial reference database), 5VCE, and NmerCE algorithms.

clearly diversity and species assemblage of the saliva samples were not identical. Such differences reflect diet, hygiene, and/or family and culture, all of which influence the oral microbiome. Centroid classification [70] within the two major groupings showed four genera, *Haemophilus*, *Streptococcus*, *Neisseria*, and *Aggregatibacter*, were over-represented in cluster B (Fig. S8). Over representation of *Haemophilus* species, i.e., *H. influenzae*, *H. parainfluenzae*, and *Aggregatibacter aphrophilus*(née, *Haemophilus aphrophilus*)in cluster B is particularly interesting, since these bacterial species have been shown to be associated with *Haemophilus* endocarditis [71,72]. Although *Haemophilus* species can cause adult endocarditis (0.8–1.3%) [73], the presence of a significant number of each of these species represent a skewing from healthy human saliva.

Conclusion

WGS metagenomics applied to the human microbiome has provided useful information applicable to public health and personalized medicine, especially as high-throughput ultra-deep sequencing approaches real time and becomes cost effective. However, post sequencing processing and analysis of data generated by WGS metagenomics are extremely challenging. While traditional BLAST analysis is hindered by factors like time for analysis, low resolution, and large computational requirements, a marker gene approach will speed detection, but sacrifices resolution. Genome mapping and reconstruction ensure precise identification but takes a long 'time to identification' and require powerful computational infrastructures and skilled manpower. In this study, GENIUS algorithms and WGS metagenomic data were used to identify the bacterial community composition of human saliva. Compared to16S metagenic sequencing and analysis, WGS metagenomics provided greater accuracy, both in identification and quantitation of bacterial species and less biased estimate of diversity, when GENIUS algorithms were used. Superior speed, accuracy, and precision in identification were achieved compared to 16S which significantly overestimated diversity. GENIUS algorithms provide high specificity and accurate identification of species, even those present in low abundance and with fractional genome coverage. WGS metagenomics employing GENIUS algorithms is proposed as method of choice for rapid, accurate, and user friendly bacterial identification and metagenomics.

In this study, it has been demonstrated that WGS metagenomics provides a practical approach in answering questions about the human salivary microbiome. Therefore, metagenomic analysis of clinical samples, not only of the salivary microbiome, but also other microbial flora, in general, offers greater power of decision making, precision, and speed compared to traditional methods.

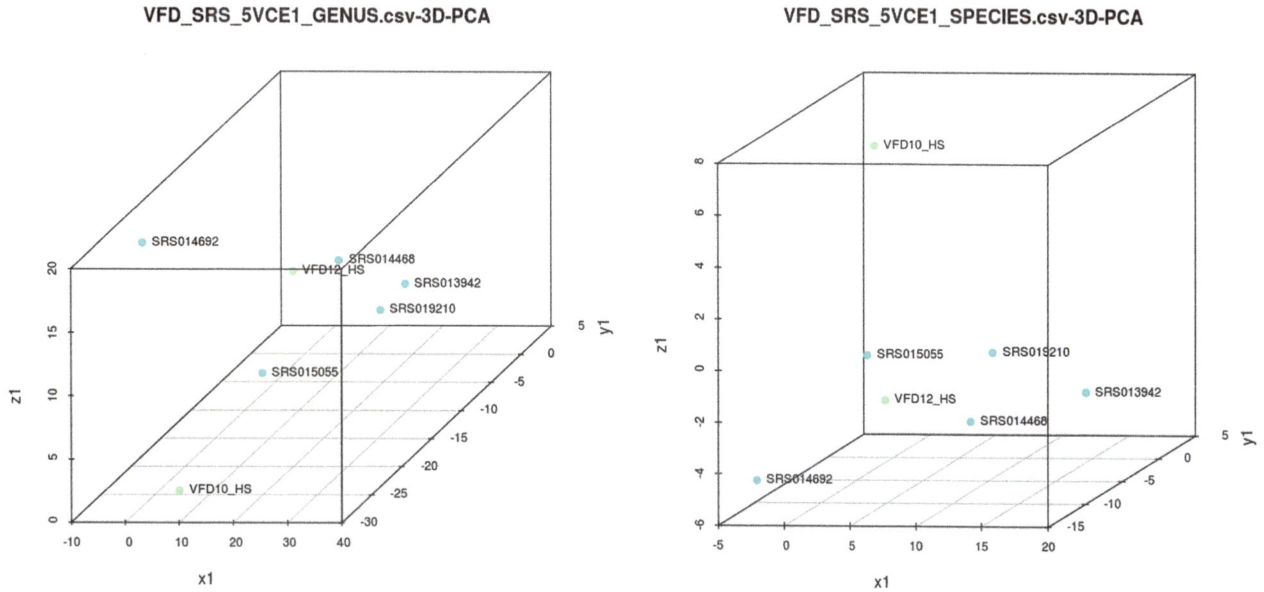

Figure 5. Principal component analysis of data for seven saliva samples analyzed by GENIUS.

The typical "time to answer" for culture-based methods requires weeks for completion, whereas sequencing approaches can reduce the timeline to a few days. For WGS metagenomics for microbial detection and identification, laboratory protocols require a fraction of time compared to culture-based methods, especially since culturing is not required prior to library construction. The time required for a sample being processed and sequenced is approximately two days if MiSeq, 454 GS junior, Ion Torrent PGM, NextSeq or HiSeq X platforms are used for sequencing throughput. However, this timeline could be reduced through automation and will soon be less than a day. The main difference is actually the data analysis, when GENIUS is used, it takes only half an hour or less for analysis of metagenomic data derived from any routine clinical samples and does not require time-consuming alignment or mapping.

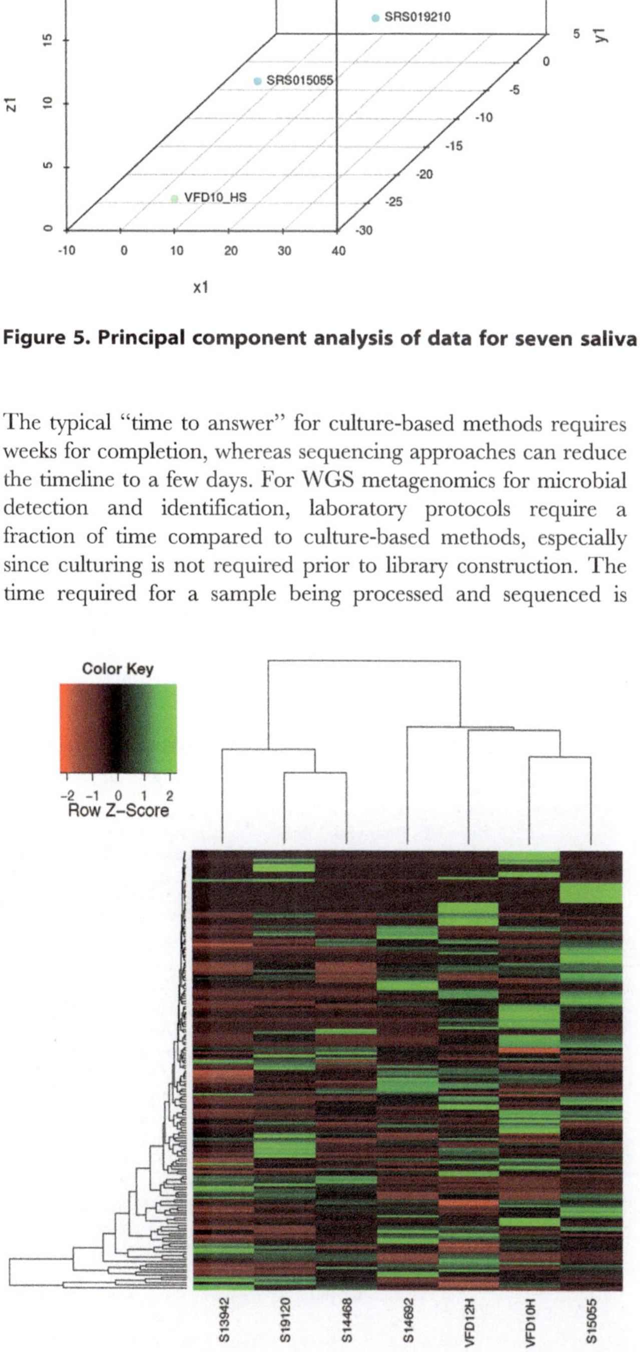

Figure 6. Double hierarchical dendrogram showing bacterial distribution at the species level for seven saliva samples. The relative values for bacterial species are depicted by color intensity, with legend indicated at the top of the figure.

The cost of NGS has reduced dramatically and continues to decrease. It is clear that sequencing is not yet the least expensive method, but considering the amount of information obtained from NGS and the depth of resolution of the analyses, it is proving to be more cost-effective because of the greater breadth and depth of information provided compared to traditional methods involving culture and other bioassays, with battery of tests and reagents required. Metagenomics also provides opportunity to interrogate the same dataset against multiple databases (i.e, GeneBook libraries) for detection of bacteria, viruses and their virulence factors and/or antibiotic resistance in a single assay. It is concluded that application of GENIUS and GeneBook libraries can be utilized effectively for wider application in the clinical laboratory.

Supporting Information

Figure S1 Comparative analysis of human saliva sample VFD10 sequenced by Illumina GAIIx using GENIUS 5VCE, BLAST (NCBI, microbial subset) and short read mapping.

Figure S2 Statistical analysis of confidence interval by GENIUS and visualization of the metagenomic community using the Krona visualization tool.

Figure S3 GENIUS 5VCE prediction of species relative abundance in eight lanes of an Illumina GAIIx flowcell. The smaller chart to the upper right corner represents the number of

Figure S4 GENIUS 5VCE prediction of percent of total hits for identified bacterial species in VFD10 and VFD12, sequenced by both GAIIx and HiSeq 2000. The smaller chart to the upper right corner shows number of reads generated for two samples by GAIIx and HiSeq 2000.

Figure S5 Abundance of bacterial phyla and genera identified in the salivary microbiome by GENIUS.

Figure S6 Distribution and abundance of different *Prevotella* spp. and strains in salivary microbiomes.

Figure S7 Occurrence and relative abundance of bacterial species associated with periodontal disease.

Figure S8 The centroid classification analysis of cluster A and B salivary samples. The top ranking markers can differentiate cluster B (green) and cluster A (red) by the centroid scores. Species and genera with nonzero components in each class are almost mutually exclusive.

Table S1 Comparison of relative abundance of genera in VFD10-018 using different methods.

Table S2 Comparison of relative abundance of genera in VFD12-006using different methods.

Acknowledgments

The authors wish to extend thanks to the Nationwide Children's Hospital Biomedical Genomics Core, Columbus, Ohio for sequencing services.

Author Contributions

Conceived and designed the experiments: TAC RRC NAH BAY DMB SAF. Analyzed the data: NAH HL KS RI ASA SYC DMB ATM-S EMH NJM SAF BAY. Contributed reagents/materials/analysis tools: RRC HL BAY MLD. Wrote the paper: NAH RRC TAC HL RI KS SYC ATM-S NJM EMH BAY MLD. Critical review of the paper for intellectual content: RRC NAH TAC BAY MLD.

References

1. Socransky SS, Haffajee AD, Smith GL, Dzink JL (1987) Difficulties encountered in the search for the etiologic agents of destructive periodontal diseases. J Clin Periodontol 14: 588–593.
2. Haffajee AD, Socransky SS (1994) Microbial etiological agents of destructive periodontal diseases. Periodontol 2000 5: 78–111.
3. Marcotte H, Lavoie MC (1998) Oral microbial ecology and the role of salivary immunoglobulin A. Microbiol Mol Biol Rev 62: 71–109.
4. Weinstock GM (2012) Genomic approaches to studying the human microbiota. Nature 489: 250–256.
5. Castellarin M, Warren RL, Freeman JD, Dreolini L, Krzywinski M, et al. (2012) Fusobacterium nucleatum infection is prevalent in human colorectal carcinoma. Genome Res 22: 299–306.
6. Kostic AD, Gevers D, Pedamallu CS, Michaud M, Duke F, et al. (2012) Genomic analysis identifies association of Fusobacterium with colorectal carcinoma. Genome Res 22: 292–298.
7. Wang Z, Klipfell E, Bennett BJ, Koeth R, Levison BS, et al. (2011) Gut flora metabolism of phosphatidylcholine promotes cardiovascular disease. Nature 472: 57–63.
8. Garrett WS, Gordon JI, Glimcher LH (2010) Homeostasis and inflammation in the intestine. Cell 140: 859–870.
9. Turnbaugh PJ, Ley RE, Mahowald MA, Magrini V, Mardis ER, et al. (2006) An obesity-associated gut microbiome with increased capacity for energy harvest. Nature 444: 1027–1031.
10. Yamamoto ML, Maier I, Dang AT, Berry D, Liu J, et al. (2013) Intestinal bacteria modify lymphoma incidence and latency by affecting systemic inflammatory state, oxidative stress, and leukocyte genotoxicity. Cancer Res 73: 4222–4232.
11. Gonzalez A, Stombaugh J, Lozupone C, Turnbaugh PJ, Gordon JI, et al. (2011) The mind-body-microbial continuum. Dialogues Clin Neurosci 13: 55–62.
12. Wade WG (2013) The oral microbiome in health and disease. Pharmacol Res 69: 137–143.
13. Paster BJ, Boches SK, Galvin JL, Ericson RE, Lau CN, et al. (2001) Bacterial diversity in human subgingival plaque. J Bacteriol 183: 3770–3783.
14. Dewhirst FE, Chen T, Izard J, Paster BJ, Tanner AC, et al. (2010) The human oral microbiome. J Bacteriol 192: 5002–5017.
15. Mager DL, Haffajee AD, Devlin PM, Norris CM, Posner MR, et al. (2005) The salivary microbiota as a diagnostic indicator of oral cancer: a descriptive, non-randomized study of cancer-free and oral squamous cell carcinoma subjects. J Transl Med 3: 27.
16. Slots J, Slots H (2011) Bacterial and viral pathogens in saliva: disease relationship and infectious risk. Periodontol 2000 55: 48–69.
17. Belda-Ferre P, Alcaraz LD, Cabrera-Rubio R, Romero H, Simon-Soro A, et al. (2012) The oral metagenome in health and disease. ISME J 6: 46–56.
18. Watanabe T, Shibata K, Yoshikawa T, Dong L, Hasebe A, et al. (1998) Detection of Mycoplasma salivarium and Mycoplasma fermentans in synovial fluids of temporomandibular joints of patients with disorders in the joints. FEMS Immunol Med Microbiol 22: 241–246.
19. Wu T, Trevisan M, Genco RJ, Dorn JP, Falkner KL, et al. (2000) Periodontal disease and risk of cerebrovascular disease: the first national health and nutrition examination survey and its follow-up study. Arch Intern Med 160: 2749–2755.
20. Socransky SS, Haffajee AD, Cugini MA, Smith C, Kent RL Jr (1998) Microbial complexes in subgingival plaque. J Clin Periodontol 25: 134–144.
21. Faveri M, Mayer MP, Feres M, de Figueiredo LC, Dewhirst FE, et al. (2008) Microbiological diversity of generalized aggressive periodontitis by 16S rRNA clonal analysis. Oral Microbiol Immunol 23: 112–118.
22. Colombo AP, Boches SK, Cotton SL, Goodson JM, Kent R, et al. (2009) Comparisons of subgingival microbial profiles of refractory periodontitis, severe periodontitis, and periodontal health using the human oral microbe identification microarray. J Periodontol 80: 1421–1432.
23. Crielaard W, Zaura E, Schuller AA, Huse SM, Montijn RC, et al. (2011) Exploring the oral microbiota of children at various developmental stages of their dentition in the relation to their oral health. BMC Med Genomics 4: 22.
24. Poveda-Roda R, Jimenez Y, Carbonell E, Gavalda C, Margaix-Munoz MM, et al. (2008) Bacteremia originating in the oral cavity. A review. Med Oral Patol Oral Cir Bucal 13: E355–362.
25. Parahitiyawa NB, Jin LJ, Leung WK, Yam WC, Samaranayake LP (2009) Microbiology of odontogenic bacteremia: beyond endocarditis. Clin Microbiol Rev 22: 46–64, Table of Contents.
26. Schiff E, Pick N, Oliven A, Odeh M (2003) Multiple liver abscesses after dental treatment. J Clin Gastroenterol 36: 369–371.
27. Franca AV, Martinelli A, Silva OC Jr (2004) Brain metastasis of hepatocellular carcinoma detected after liver transplantation. Arq Gastroenterol 41: 199–201.
28. Joshipura KJ, Hung HC, Rimm EB, Willett WC, Ascherio A (2003) Periodontal disease, tooth loss, and incidence of ischemic stroke. Stroke 34: 47–52.
29. Genco RJ, Grossi SG, Ho A, Nishimura F, Murayama Y (2005) A proposed model linking inflammation to obesity, diabetes, and periodontal infections. J Periodontol 76: 2075–2084.
30. Saremi A, Nelson RG, Tulloch-Reid M, Hanson RL, Sievers ML, et al. (2005) Periodontal disease and mortality in type 2 diabetes. Diabetes Care 28: 27–32.
31. Awano S, Ansai T, Takata Y, Soh I, Akifusa S, et al. (2008) Oral health and mortality risk from pneumonia in the elderly. J Dent Res 87: 334–339.
32. Buduneli N, Baylas H, Buduneli E, Turkoglu O, Kose T, et al. (2005) Periodontal infections and pre-term low birth weight: a case-control study. J Clin Periodontol 32: 174–181.
33. Palacios G, Druce J, Du L, Tran T, Birch C, et al. (2008) A new arenavirus in a cluster of fatal transplant-associated diseases. N Engl J Med 358: 991–998.
34. Lazarevic V, Whiteson K, François P, Schrenzel J (2010) The salivary microbiome, assessed by a high-throughput and culture-independent approach. Journal of Integrated OMICS 1: 28–35.
35. Cephas KD, Kim J, Mathai RA, Barry KA, Dowd SE, et al. (2011) Comparative analysis of salivary bacterial microbiome diversity in edentulous infants and their mothers or primary care givers using pyrosequencing. PLoS One 6: e23503.
36. Lazarevic V, Whiteson K, Gaia N, Gizard Y, Hernandez D, et al. (2012) Analysis of the salivary microbiome using culture-independent techniques. J Clin Bioinforma 2: 4.
37. Hu S, Loo JA, Wong DT (2007) Human saliva proteome analysis and disease biomarker discovery. Expert Rev Proteomics 4: 531–538.
38. Malamud D (2011) Saliva as a diagnostic fluid. Dent Clin North Am 55: 159–178.
39. Refulio Z, Rocafuerte M, de la Rosa M, Mendoza G, Chambrone L (2013) Association among stress, salivary cortisol levels, and chronic periodontitis. J Periodontal Implant Sci 43: 96–100.

40. Young EA, Aggen SH, Prescott CA, Kendler KS (2000) Similarity in saliva cortisol measures in monozygotic twins and the influence of past major depression. Biol Psychiatry 48: 70–74.
41. Nasidze I, Quinque D, Li J, Li M, Tang K, et al. (2009) Comparative analysis of human saliva microbiome diversity by barcoded pyrosequencing and cloning approaches. Anal Biochem 391: 64–68.
42. Chen T, Yu WH, Izard J, Baranova OV, Lakshmanan A, et al. (2010) The Human Oral Microbiome Database: a web accessible resource for investigating oral microbe taxonomic and genomic information. Database (Oxford) 2010: baq013.
43. Garrity GM, Thompson LM, Ussery DW, Paskin N, Baker D, et al. (2009) Studies on monitoring and tracking genetic resources: an executive summary. Stand Genomic Sci 1: 78–86.
44. Venter JC, Remington K, Heidelberg JF, Halpern AL, Rusch D, et al. (2004) Environmental genome shotgun sequencing of the Sargasso Sea. Science 304: 66–74.
45. Quince C, Lanzen A, Curtis TP, Davenport RJ, Hall N, et al. (2009) Accurate determination of microbial diversity from 454 pyrosequencing data. Nat Methods 6: 639–641.
46. Shah N, Tang H, Doak TG, Ye Y (2011) Comparing bacterial communities inferred from 16S rRNA gene sequencing and shotgun metagenomics. Pac Symp Biocomput: 165–176.
47. Altschul SF, Gish W, Miller W, Myers EW, Lipman DJ (1990) Basic local alignment search tool. J Mol Biol 215: 403–410.
48. Tatusova T, Ciufo S, Fedorov B, O'Neill K, Tolstoy I (2014) RefSeq microbial genomes database: new representation and annotation strategy. Nucleic Acids Res 42: D553–559.
49. Kunin V, Copeland A, Lapidus A, Mavromatis K, Hugenholtz P (2008) A bioinformatician's guide to metagenomics. Microbiol Mol Biol Rev 72: 557–578, Table of Contents.
50. Thomas T, Gilbert J, Meyer F (2012) Metagenomics - a guide from sampling to data analysis. Microb Inform Exp 2: 3.
51. Mavromatis K, Ivanova N, Barry K, Shapiro H, Goltsman E, et al. (2007) Use of simulated data sets to evaluate the fidelity of metagenomic processing methods. Nat Methods 4: 495–500.
52. Martin J, Sykes S, Young S, Kota K, Sanka R, et al. (2012) Optimizing read mapping to reference genomes to determine composition and species prevalence in microbial communities. PLoS One 7: e36427.
53. Huson DH, Auch AF, Qi J, Schuster SC (2007) MEGAN analysis of metagenomic data. Genome Res 17: 377–386.
54. Meyer F, Paarmann D, D'Souza M, Olson R, Glass EM, et al. (2008) The metagenomics RAST server - a public resource for the automatic phylogenetic and functional analysis of metagenomes. BMC Bioinformatics 9: 386.
55. Didelot X, Bowden R, Wilson DJ, Peto TE, Crook DW (2012) Transforming clinical microbiology with bacterial genome sequencing. Nat Rev Genet 13: 601–612.
56. Ondov BD, Bergman NH, Phillippy AM (2011) Interactive metagenomic visualization in a Web browser. BMC Bioinformatics 12: 385.
57. Lazarevic V, Whiteson K, Hernandez D, Francois P, Schrenzel J (2010) Study of inter- and intra-individual variations in the salivary microbiota. BMC Genomics 11: 523.
58. Kuczynski J, Costello EK, Nemergut DR, Zaneveld J, Lauber CL, et al. (2010) Direct sequencing of the human microbiome readily reveals community differences. Genome Biol 11: 210.
59. Chakravorty S, Helb D, Burday M, Connell N, Alland D (2007) A detailed analysis of 16S ribosomal RNA gene segments for the diagnosis of pathogenic bacteria. J Microbiol Methods 69: 330–339.
60. Shakya M, Quince C, Campbell JH, Yang ZK, Schadt CW, et al. (2013) Comparative metagenomic and rRNA microbial diversity characterization using archaeal and bacterial synthetic communities. Environ Microbiol 15: 1882–1899.
61. Wang GC, Wang Y (1997) Frequency of formation of chimeric molecules as a consequence of PCR coamplification of 16S rRNA genes from mixed bacterial genomes. Appl Environ Microbiol 63: 4645–4650.
62. Gevers D, Pop M, Schloss PD, Huttenhower C (2012) Bioinformatics for the Human Microbiome Project. PLoS Comput Biol 8: e1002779.
63. Keijser BJ, Zaura E, Huse SM, van der Vossen JM, Schuren FH, et al. (2008) Pyrosequencing analysis of the oral microflora of healthy adults. J Dent Res 87: 1016–1020.
64. Ling Z, Kong J, Jia P, Wei C, Wang Y, et al. (2010) Analysis of oral microbiota in children with dental caries by PCR-DGGE and barcoded pyrosequencing. Microb Ecol 60: 677–690.
65. Yang F, Zeng X, Ning K, Liu KL, Lo CC, et al. (2012) Saliva microbiomes distinguish caries-active from healthy human populations. ISME J 6: 1–10.
66. Zhou M, Rong R, Munro D, Zhu C, Gao X, et al. (2013) Investigation of the effect of type 2 diabetes mellitus on subgingival plaque microbiota by high-throughput 16S rDNA pyrosequencing. PLoS One 8: e61516.
67. Inagaki S, Onishi S, Kuramitsu HK, Sharma A (2006) Porphyromonas gingivalis vesicles enhance attachment, and the leucine-rich repeat BspA protein is required for invasion of epithelial cells by "Tannerella forsythia". Infect Immun 74: 5023–5028.
68. Kinane DF (2000) Aetiology and pathogenesis of periodontal disease. Ann R Australas Coll Dent Surg 15: 42–50.
69. Michalowicz BS, Diehl SR, Gunsolley JC, Sparks BS, Brooks CN, et al. (2000) Evidence of a substantial genetic basis for risk of adult periodontitis. J Periodontol 71: 1699–1707.
70. Tibshirani R, Hastie T, Narasimhan B, Chu G (2002) Diagnosis of multiple cancer types by shrunken centroids of gene expression. Proc Natl Acad Sci U S A 99: 6567–6572.
71. Ratnayake L, Olver WJ, Fardon T (2011) Aggregatibacter aphrophilus in a patient with recurrent empyema: a case report. J Med Case Rep 5: 448.
72. Lynn DJ, Kane JG, Parker RH (1977) Haemophilus parainfluenzae and influenzae endocarditis: a review of forty cases. Medicine (Baltimore) 56: 115–128.
73. Pai RK, Pergam SA, Kedia A, Cadman CS, Osborn LA (2004) Pacemaker lead infection secondary to Haemophilus parainfluenzae. Pacing Clin Electrophysiol 27: 1008–1010.

12

The Soil Bacterial Communities of South African Fynbos Riparian Ecosystems Invaded by Australian *Acacia* Species

Etienne Slabbert[1], Shayne Martin Jacobs[2], Karin Jacobs[1]*

1 Department of Microbiology, Stellenbosch University, Stellenbosch, Western Cape, South Africa, 2 Department of Conservation Ecology and Entomology, Stellenbosch University, Stellenbosch, Western Cape, South Africa

Abstract

Riparian ecosystem along rivers and streams are characterised by lateral and longitudinal ecological gradients and, as a result, harbour unique biodiversity. Riparian ecosystems in the fynbos of the Western Cape, South Africa, are characterised by seasonal dynamics, with summer droughts followed by high flows during winter. The unique hydrology and geomorphology of riparian ecosystems play an important role in shaping these ecosystems. The riparian vegetation in the Western Cape has, however, largely been degraded due to the invasion of non-indigenous plants, in particular *Acacia mearnsii, A. saligna* and *A. dealbata*. This study investigated the effect of hydrology and invasion on the bacterial communities associated with fynbos riparian ecosystems. Bacterial communities were characterised with automated ribosomal intergenic spacer analysis (ARISA) and 454 16S rDNA pyrosequencing. Chemical and physical properties of soil within sites were also determined and correlated with community data. Sectioning across the lateral zones revealed significant differences in community composition, and the specific bacterial taxa influenced. Results also showed that the bacterial community structure could be linked to *Acacia* invasion. The presence of invasive *Acacia* was correlated with specific bacterial phyla. However, high similarity between cleared and pristine sites suggests that the effect of *Acacia* on the soil bacterial community structure may not be permanent. This study demonstrates how soil bacterial communities are influenced by hydrological gradients associated with riparian ecosystems and the impact of *Acacia* invasion on these communities.

Editor: Zaid Abdo, Institution and Department: Agricultural Research Service, United States of America

Funding: This research was supported by the South African Water Research Commission (WRC) and the National Research Foundation (NRF). The funders had no role in study design, data collection and analysis, decision to publish, or preparation of the manuscript.

Competing Interests: The authors have declared that no competing interests exist.

* E-mail: kj@sun.ac.za

Introduction

Riparian ecosystems are broadly classified as the interface between terrestrial and freshwater aquatic ecosystems [1,2,3]. Riparian ecosystems of the South African fynbos play a crucial role in the health and functioning of the diverse fynbos biome as a whole [4]. Fynbos riparian ecosystems have received very little scientific attention compared to the rest of the South African fynbos biome, although they have been shown to contribute disproportionately to ecological processes considering their relatively small land area [5,6,7]. These ecosystems are also the site of several important physical and biochemical processes [3,8]. Riparian ecosystems are unique in the landscape and often exhibit different rates of microbial mediated soil processes compared to upland areas [9]. The most important ecosystem service provided by riparian ecosystems is the supply of clean water, which is greatly, influenced by soil microbial processes [10,11].

Riparian ecosystems associated with the fynbos biome can be easily distinguished from the terrestrial fynbos based on hydrology, geomorphology and the structure of the vegetation [12,13]. The lateral zones commonly occurring in the fynbos riparian ecosystems are classified as the dry bank and wet bank zones [12] (Figure 1). The dry bank is infrequently inundated, typically only during periods of high flooding, which happen every few years [14]. The water from the river influences the dry bank in the form of ground water during low flow. In the upper catchments (the mountain fynbos riparian ecosystems), high flow and flooding is likely to occur during the rainy winter season [14]. On the other hand, the wet bank zones are classified as the area at the river's edge, which is under constant influence of the river throughout the year, and is always likely to be moist to wet [12].

The importance of riparian ecosystems and associated plant and microbial biota are now becoming clear due to the increased degradation of riparian ecosystems worldwide [15], [16]. The riparian ecosystems of the Western Cape fynbos biome are largely degraded due to the invasion of non-native *A. mearnsii and A. longifolia* originating from Australia [17] which have, to a large extent, displaced the native riparian vegetation [18]. *Acacia* species such as *A. saligna*, have been shown to be associated with nitrogen-fixing bacteria, which may have a dramatic effect on soil nitrogen stocks and soil microbial community structure [15,19]. The recovery of the native fynbos vegetation occurs very slowly where Acacias are cleared [20,21]. This poor recovery sparked interest in the soil dynamics of the fynbos riparian ecosystems, and it has been suggested that soil microbial dynamics may have been

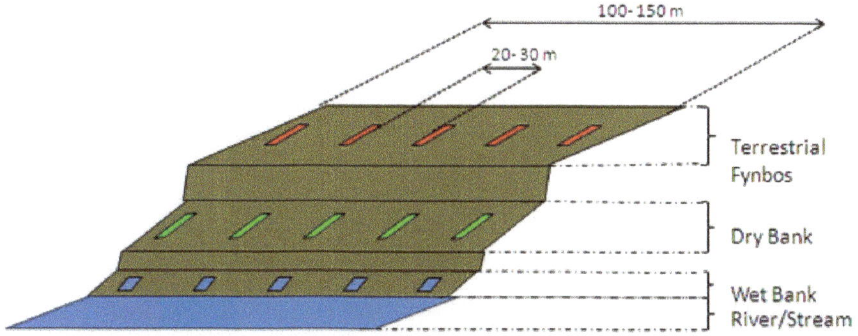

Figure 1. Schematic representation of the lateral zones of the riparian sites and the positions of the sample plots. The extent of each lateral zone is variable depending on the specific dimensions of the riparian zone but ranged from 2m to 10m wide.

modified to the detriment of the restoration of fynbos ecosystem function and structure [22].

Studies have shown that certain bacterial species belonging to the order Rhizobiales may play an important role in the invasiveness of *Acacia* species. [19]. The presence of invasive *Acacias* in South Africa can further be linked to the prevalence of known root associated bacteria, namely *Bradyrhizobium japonicum*, *B. elkanii*, *Rhizobium leguminosarum*, and *R. tropici* that were found to occur in symbiosis with *A. mearnsii* [23]. The effect of invasive *Acacias* on soil microbes has also been observed in other Mediterranean ecosystems [24]. However, most studies, although referring to the importance of soil microbes in the dynamics of invasion by non-native *Acacia*, they have not specifically studied the soil microbial community structure [15].

The main goal of the study was to investigate the effect of non-indigenous invasive *Acacia* spp. on the microbial community structure in riparian zones in the fynbos biome of South Africa. Sharp ecological gradients, laterally across riparian sites, are also included in the study to determine their influence on the soil bacterial community structure.

Materials and Methods

Experimental Sites

The ten experimental sites are located in the Western Cape, South Africa, and include natural, cleared and invaded riparian ecosystems. A permit was obtained from the conservation authority CapeNature. Permit number: (AAA005-00137-0028). Sampling sites were selected within the upper foothill and adjacent mountain stream areas. Sites with similar vegetation structure and geomorphology according to [18], were selected in order to standardise sampling conditions. The sites included three pristine reference sites flanking the upper Eerste River (UE), lower Eerste River (LE) and the upper Dwars River (UD) (Table 1, Figure 2b). The riparian vegetation of the Molenaars River (UM), Sir Lowry's River (S) and Jakkals River (UJ) included sites that were clear of non-indigenous vegetation for at least five years (Table 1, Figure 2c). The cleared Jakkals River site experienced a fire event during the winter season and as a consequence those samples were removed from the analysis. The riparian ecosystem adjacent to the Wit River (W) and Jakkals River (LJ) were invaded by *A. mearnsii* and that of the Dwars River (LD) by *A. longifolia* and *A. mearnsii* (Table 1, Figure 2d). The invaded sites were mostly overgrown with woody *Acacias* and the natural fynbos riparian vegetation was almost entirely displaced. Experimental sites adjacent to the rivers were divided into three lateral zones: the riparian wet bank zone, the riparian dry bank zone and the upland terrestrial fynbos zone (Figure 1). The riparian wet bank zone was the area at the edge of the river that was under direct influence of the river water and was characterized by riparian vegetation [13]. This zone is characteristically inundated during all seasons, but dries out to a certain extent during summer. The riparian dry bank zone was also under the influence of the water from the river, although usually by means of groundwater. Typically, the river water only sporadically inundates the dry bank zone during flooding. The terrestrial zone occurs outside the direct influence of the river and excludes plants typical of riparian vegetation [13]. All upland terrestrial sites in the experimental zones consisted of mountain fynbos due to their resistance to invasion and were characterized by Ericoids and Proteas [25].

Soil Sampling

The first soil samples were collected during March 2010 during the end of the dry autumn season. Sampling plots were positioned to form transects of five, 5 m×5 m plots, located in each lateral zone (Figure 1). Three core samples were taken in the vicinity of the vegetation from each of the five plots using a 25 mm diameter steel cylinder at a depth of 10 cm. This process was repeated for all the plots in the lateral zones and all the experimental sites, resulting in 45 samples per site. The samples were homogenised and DNA extracted within 24h of sampling. The sampling protocol was repeated in August 2010 at the end of the rainy season and again during January 2011 during the dry summer season.

Abiotic Soil Properties

The abiotic soil properties measured included particle size, available phosphate, pH,, nitrate content, ammonium content, total available nitrogen, total soil nitrogen and soil carbon content. Soil samples were sieved (2 mm) to remove roots and organic debris. The nitrate content, ammonium content and moisture content were determined on fresh soil. The total nitrogen and carbon, and pH were determined on air-dried soil. Bray-2 extractable inorganic P (Pi) was determined on fresh soil with the method described by Bray and Kurtz (1945) [26]. Concentrations of Bray-2 Pi, NO_3-N, and NH_4-N were determined colorimetrically with a Genesys 20 spectrophotometer (Thermo Scientific, Waltham, USA). NO_3-N and NH_4-N were extracted with 0.5M K_2SO_4. For concentrations of NO_3-N and NH_4-N, 10 g of soil (<2mm) were placed in 50 ml plastic vials and 25 ml extractant added. The vials were shaken at medium speed for one hour and filtered. NH_4-N was analysed based on the Berthelot reaction involving phenol [27] and NO_3-N by nitration of salicylic acid [28]. Total nitrogen and total carbon were analysed by the

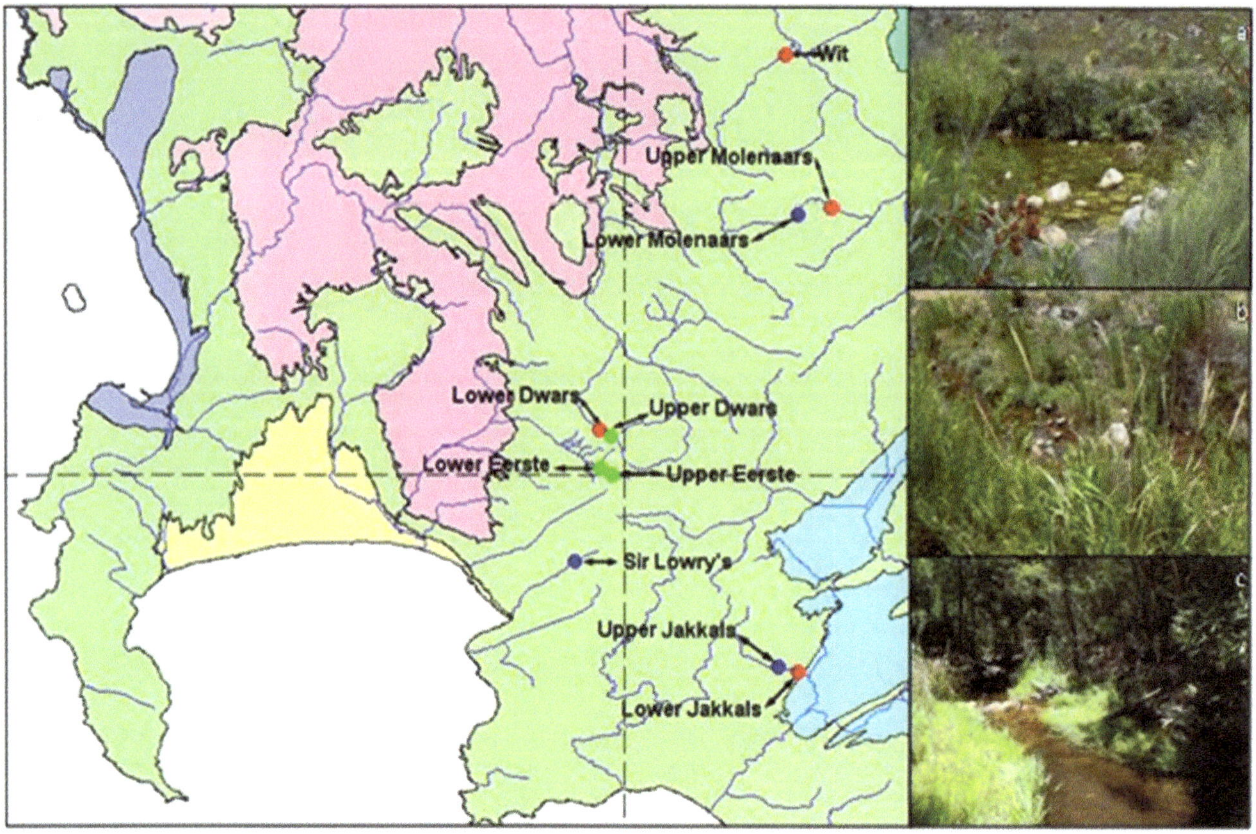

Figure 2. Location of sampling sites with invaded sites indicated in red, cleared sites in blue and pristine sites in green. 2a: Natural riparian vegetation at the lower Eerste River site. 2b: Cleared riparian zone from the Sir Lowry's and 2c: Riparian zones invaded by Acacia mearnsii at the lower Jakkals River site.

dry combustion elemental analyser method. Soil pH was measured electrometrically with the Hanna 211 Microprocessor (Hanna Instruments, Woonsocket, USA) in a 1:2 (w/v, soil: deionised water) slurry [29].

All the multivariate data analysis methods were done using Statistica software v.10 (Statsoft, Tulsa USA). The environmental variables used to compare the sites included available P (ug/g), pH (H_20), nitrate concentration (ug/g), total available N (ug/g), total soil nitrogen (%), total soil carbon (%) and the C:N ratio.

Automated Ribosomal Intergenic Spacer Analysis (ARISA)

DNA was extracted from 0.35 g of fresh soil using the ZR Soil Microbe DNA kit (Zymo Research, California, USA) and the presence of genomic DNA was checked on a 1% agarose gel

Table 1. Summary of the sample sites indicating the location, position within the catchment, invasive status of the site and the time when the sites were sampled, adapted form Reinecke et al. 2007.

Sample site	GPS		Invaded status	Season 1	Season 2	Season 3
Upper Eerste River (UE)	S 33° 59.803	E 18° 59.056	Natural	√	√	√
Lower Eerste River (LE)	S 33° 59.511	E 18° 58.326	Natural	√	√	√
Lower Dwars River (LD)	S 33° 57′ 14.1″	E 18° 58′ 43,5″	Natural	x	√	√
Upper Jakkals River (UJ)	S 33° 59′ 81.4″	E 18° 59′ 05.3″	Cleared	√	√	√
Upper Molenaars River (UM)	S 33° 44.366′	E 19° 06.805′	Cleared	√	√	√
Sir Lowry's River (S)	S 34° 05′ 41.5″	E 18° 56′ 39.7″	Cleared	√	√	√
Upper Dwars River (UD)	S 33° 57′ 15.0″	E 18° 58′ 45.1″	Invaded	√	√	√
Wit River (W)	S 33 32′ 18.5″	E 19 10 55.6″	Invaded	√	√	√
Lower Jakkals River (LJ)	S 34° 12.870′	E 19° 11.880′	Invaded	x	√	√
Lower Molenaars River (LM)	S 33° 42.188	E 19° 13.844	Invaded	x	√	x

stained with ethidium bromide PCR reactions were performed using bacterial specific primers for application in ARISA analysis. Bacterial specific primers, ITSReub and FAM (carboxy-fluorescein) labelled ITSF, were used [30]. PCR reactions were done using a GeneAmp PCR System 9700 (AppliedBiosystems, USA). The reaction mixture contained 0.5 μl (±50 ng/ul) of the purified genomic DNA, 500 nM of each primer and 5 μl of 2× KapaTaq Readymix (KapaBiosystems, South Africa) in a total volume of 10 μl. The PCR conditions consisted of an initial denaturing step of 4 min at 95°C followed by 36 cycles of 94°C, for 30 s, 56°C for 45 s and 72°C for 70 s. The reaction was completed with a final extension at 72°C for 5 min and then cooled and held at 4°C. PCR samples were separated on a 1% agarose gel, stained with ethidium bromide and visualized using ultraviolet light. The amplicons from the bacterial specific PCR were run on an ABI 3010×l Genetic analyser to obtain an electropherogram of the different fragment lengths and fluorescent intensities. ARISA samples were run with the ROX 1.1 size standard which varied from 60–1120 bp [31]. ARISA data was analysed using Genemapper 4.1 software (Applied Biosystems), which converted fluorescence data to an electropherogram that represents fragments of different sizes. Only peak height larger than 0.5% of the total fluorescence, ranging from 120 to 1000 base pairs in length, were considered for analysis. A bin size of 3 bp for fragments below 700 bp and 5 bp for fragments above 700 bp was employed to minimise the inaccuracies in the ARISA profiles [32,31].

ARISA Data Analysis

The number and relative abundance of OTUs observed with ARISA was used to calculate the Shannon diversity index of all the samples [33]. The Shannon diversity indices were compared using 3 way ANOVA to test for any significant interaction between seasons and the lateral zones and the lateral zones and the invasive status. In cases where significant differences were observed, post-hoc analysis was performed using Tukey's HDS test for unequal number of samples. ARISA data were used to calculate a Bray-Curtis dissimilarity matrix [34]. The distance matrices were analysed using non-metrical multidimensional scaling (NMDS). The Scree-test was performed to determine the number of dimensions used for every NMDS analysis. The ordination of the NMDS was accepted as informative only when the stress value was below 0.15. The significance and degree of dissimilarity between the Bray-Curtis distances from the samples was analysed based on *a priori* grouping of samples based on the lateral zone and invasive status using ANOSIM [35]. Homogeneity of multivariate dispersion was tested by PERMADISP [36].

The environmental and community data were standardized and log transformed respectively before applying multivariate analysis [37]. Canonical correspondence analysis (CCA) were used to analyze the relationship between abiotic soil variables on the bacterial community structure [38]. The significance of the relationship between soil variables and the community structure obtained from the CCA ordination was tested using 200 permutations [38].

Pyrosequencing

Due to recognized limitation associated with community fingerprinting methods, rare phylotypes are generally not detected [39]. Consequently, in addition to ARISA, 454 pyrosequencing was used. Based on results from the ARISA analysis, representative samples from invaded, pristine and cleared sites and including all three lateral zones were selected. The samples which were sequenced fell within the 75% confidence limit for the groupings and were randomly selected. Sequencing was done using samples from all three seasons. The hyper-variable regions of V1 to V3 in the bacterial 16S rRNA gene were amplified from extracted community DNA from all samples using the universal bacterial primers 27F and 340R (Table S1) [40]. PCR mixtures of 50 μl volume were prepared using 25 μl KAPA HiFi™ HotStart ReadyMix (Kapa Biosystems, South Africa), 1 μl of each primer (50 *p*mol), 1 μl of the template DNA and 22 μl sterile MilliQ water. PCR reactions were performed using GeneAmp PCR system 9700 (Applied Biosystems, South Africa). The initial PCR was performed under the following conditions: initial denaturation (94°C; 5 min), 25 cycles of denaturation (98°C; 20 sec), annealing (60°C; 15 sec), extension (72°C; 1 min), and a final extension (72°C; 7 min). The PCR products were purified using the DNA clean and concentrator kit (Zymo Research, USA) which served as template for the second round of PCR. The PCR mixture in the second round of PCR was the same as that used for the first reaction. However, during this reaction, the reverse primer 340RA was used, and contained the sequencing adaptor A, an identification key, a unique multiplex identifier (MID) and the universal bacterial primer 340F (Table S1). The forward 27FB primer contained the sequencing adaptor B, an identification key, and the universal bacterial primer 27FB (Table S1). PCR conditions were the same as those used for the first reaction. The PCR amplicons were gel purified using the BioSpin Gel Extraction Kit (BioFlux, Japan). Equimolar concentrations of the PCR amplicons with the different sample-specific barcode sequences (MIDs) were multiplexed and submitted for pyrosequencing (Inqaba Biotec, South Africa) using the 454 GS FLX Titanium Sequencing System (454 Life Sciences, a Roche Company, Branford, USA). The sequence data was submitted to the European Nucleotide Archive (ENA) as PRJEB4470.

Pyrosequencing Data Analysis

Sequences were filtered, trimmed and assigned to the samples according to their MID sequences. Reads with no undefined base pairs (N's) and length >50 nt were retained. The MID sequence, key and the reverse and forward primers were trimmed prior to further analysis by using Mothur [41]. Sequences less than 300 nt after trimming were discarded in the analysis. The sequences was screened for chimeras using UCHIME [42]. The sequences were binned into OTUs at 97% sequence similarity, which served as OTUs using cd-hit (Li and Godzik, 2006 [43]). Shannon diversity index and bootstrap and Chao1 richness estimators were calculated with EstimateS [44]. The Shannon diversity indices of the different samples were compared using Kruskal-Wallis ANOVA. A representative sequence for each OTU was selected and aligned using NAST with a minimum alignment length of 150nt and sequence identity of 70% using lanemask PH to screen out hypervariable regions (http://greengenes.lbl.gov/). The clustered OTUs were assigned to taxonomic groups using RDPII taxonomy from the ribosomal database project (RDPII) [45]. RDP training set nine was used, based on nomenclatural taxonomy of Bergey's Manual, using a minimum confidence threshold of 60%. The alignment was used to construct a distance matrix using the DNAdist function from the PHYLIP 3.6 package [46]. The distance matrix was used to construct a phylogenetic tree using the Jukes–Cantor model. UniFrac analysis was used to overcome the constraints when analysing communities at singular levels of taxonomic classification [47]. The phylogenetic composition of the sample was compared. The Unifrac distances were used to calculate a phylogenetic metric of community similarity. UniFrac distances and measured environmental parameters were correlated using Mantel test. The relative abundances and frequencies of the taxa were used to determine which taxa deliniated the different hydrolohical zones within the riparian ecosytems by performing indicator species analysis with R software using the labdsv package [48]. In addition, indicator analysis was performed using invasive status as the categorical variable.

Results

Bacterial OTU Diversity

When comparing the mean Shannon diversity indices within the different lateral zones from autumn, winter and summer samples, no significant differences were observed ($p>0.05$). The results from the two factorial ANOVA, however, showed that the bacterial diversity of the wet bank zones were significantly lower ($p<0.05$) when *Acacia* invasion occurred (Figure 3). This decreased diversity was also observed with the pyrosequencing data, which showed a significantly lower Shannon diversity index in the invaded wet bank zones according to Kruskal-Wallis ANOVA ($F=2.54$, $p=0.048$) (Figure S1).

Bacterial Community Structure (ARISA)

The NMDS and ANOSIM, comparing all samples, showed significant groupings of the terrestrial, dry bank and wet bank samples over all three seasons (Figure 4, Figure S2). Multivariate analysis, showed significant similarity between samples from the same sites, thus indicating a significant location effect. The effect of lateral zoning was however, the most important factor determining the bacterial community structure (Table 2). PERMADISP dispersions showed that no significant differences in dispersion occurred between sites, however a significant difference in dispersal occurred between the different lateral zones. The ANOSIM post-hoc test showed that the dissimilarity between the terrestrial and wet bank samples were the largest and remained this way over all three seasons (Table 2). Although the dissimilarity between wet bank and dry bank zones was smaller it was still significant ($R=0.28$, $p<0.05$). The dissimilarity between samples from the dry bank and the wet bank increased during the winter season with a increase in the R value from 0.277 to 0.32 ($p<0.05$) (Table 2). In season three the dissimilarity between the dry bank and wet bank zone again increased to the values seen in season one resulting in a R value of 0.29 between the wet bank and the dry bank zones. The largest dissimilarity still occurred between the terrestrial and the wet bank samples ($R=0.48$, $p<0.05$).

ANOSIM analysis revealed significant groupings based on the presence of *Acacia* invasion within the dry bank and wet bank zones (Figure 5). The ANOSIM post-hoc test showed that significant differences occurred between the natural and invaded sites and between the cleared and invaded sites (Table 2). The largest difference was seen between the natural wet bank zones and invaded wet bank zones ($R=0.4$, $p<0.05$), followed by the cleared wet bank zone and the invaded wet bank zones ($R=0.39$, $p<0.05$) (Table 2). The differences between the cleared and natural dry bank and wet bank zones were small and only significant for some comparisons (Table 2).

Pyrosequencing Data

The quality filtering of the pyrosequencing reads resulted in 3260 to 12918 quality trimmed sequences per sample, with an average of 5627 reads per sample (Table S2). The average length of the quality filtered pyrosequencing reads was 350 nt. The general composition of the bacterial community on a phylum level was similar, with 18 phyla detected in all the soil samples, six of which were abundant in all the samples. Only eight of the 18 phyla occurred at levels higher than 1% of the total number of reads (Figure S3). The five most abundant phyla accounted for an average of 88.1% of sequence reads. The average representation of phyla in the soil samples included Actinobacteria (32.83%), Alphaproteobacteria (32.2%), Acidobacteria (12.64%), Betaproteobacteria (6.03%) and Planctomycetes (4.39%). When comparing the distribution of phyla between sites (Figure S3), a significantly higher proportion of *Alphaproteobacteria* was observed in the samples from the invaded sites. The proportion of *Actinobacteria* represented in the samples from the invaded sites was, however, lower. The NMDS plot and ANOSIM analysis of the pairwise UniFrac distances showed strong clustering of samples based on the samples' hydrological location, and not according to invasive status (Table 2, Figure 6).

The indicator analysis revealed the significant comparative indicator phyla, values for the different lateral zones (Figure 7a). The genera that were significantly overrepresented in the terrestrial zones were *Gp4*, *Gemmatimonas*, *Isosphaera* and *Oceaniba-*

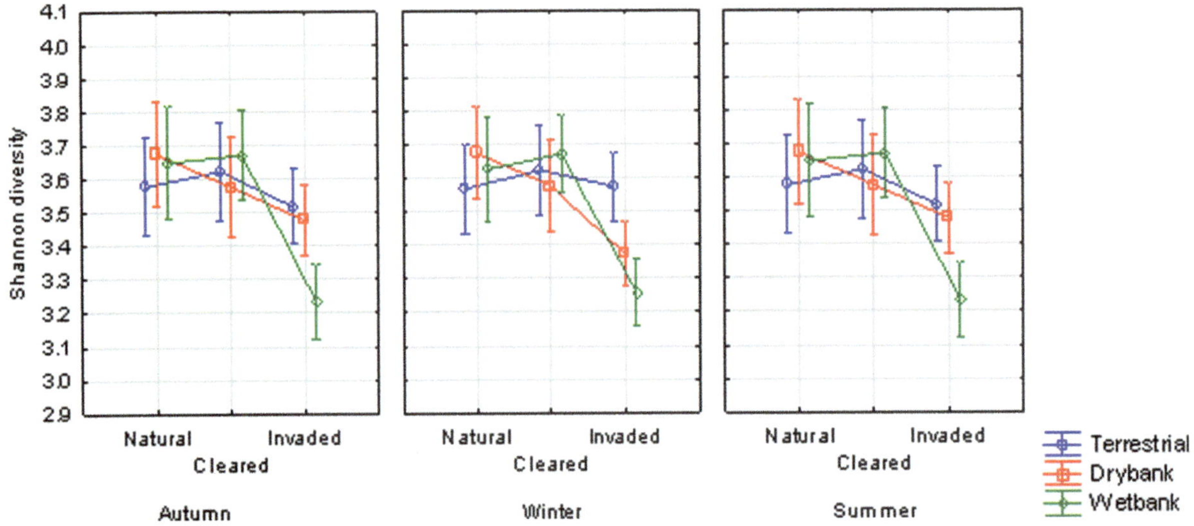

Figure 3. The Shannon's diversity index based on the bacterial ARISA profiles comparing natural, cleared and invaded lateral zones for all three seasons. The interaction effect between hydrological zones and the invasive status was significant ($F=2.7191$, $p=.00580$).

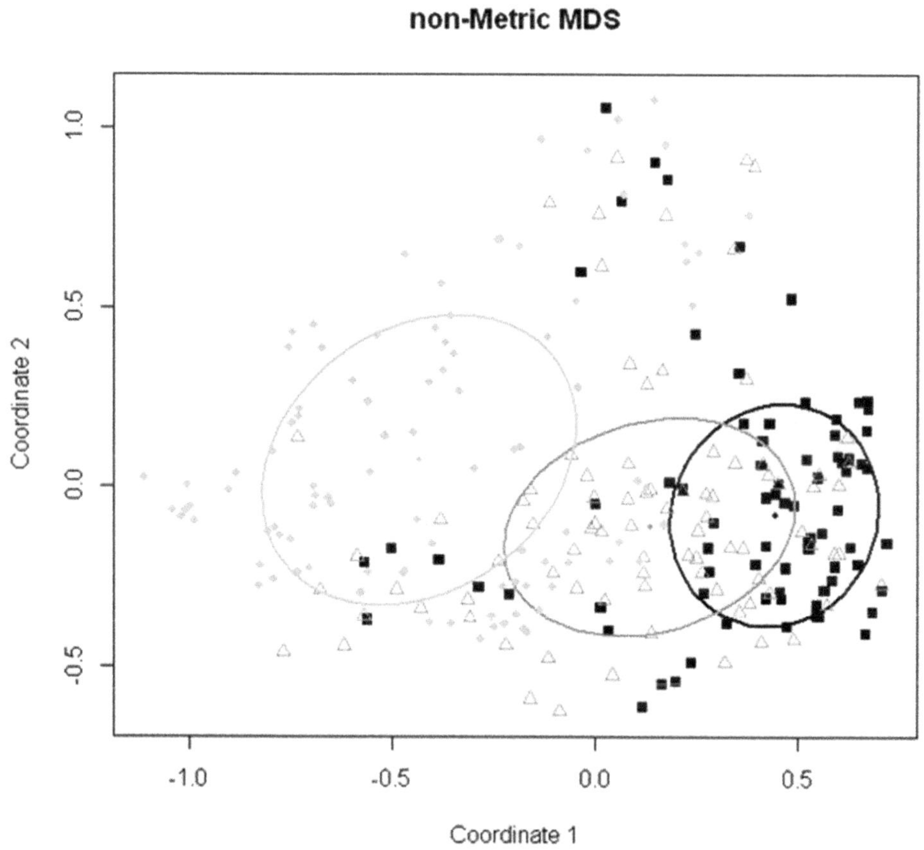

Figure 4. Non-metric multi-dimensional scaling ordination plot of the distance between bacterial communities based on the Bray-Curtis distance. The ellipses represent the samples which were within 75% confidence limit of the centroids and included the wet bank (circles), dry bank (triangles) and the terrestrial samples (squares) (Stress = 0.12).

culum. The genus that was overrepresented in both terrestrial and dry bank samples was *Pseudonocardia*. The genera that distinguished the dry bank from all other samples were *Streptomyces* and *Paracraurococcus*. The wet bank samples were characterised by the overrepresentation of the genera *Methylocystis*,, *Acidocella*, *Anaeromyxobacter* and *Geothrix*. The phylum *Actinobacteria* was represented to a lesser extent in all the wet bank samples. The genera in this phylum, which were underrepresented in comparison to other lateral zones, include *Conexibacter*, *Mycobacterium* and *Blastococcus*.

Table 2. R values of the ANOSIM comparisons made between hydrological zones and invasion status within dry bank and wet bank zones.

R value	Season 1	Season 2	Season 3
Terrestrial×Wet Bank	0.47	0.48	0.47
Terrestrial×Dry Bank	0.28	0.33	0.27
Dry Bank×Wet Bank	0.32	0.27	0.29
Dry Bank (Natural×Cleared)	0.18*	0.21	0.2
Dry Bank (Natural×Invaded)	0.29	0.22*	0.26
Dry Bank (Cleared×Invaded)	0.26	028	0.2*
Wet Bank (Natural×Cleared)	0.21	0.19*	0.2*
Wet Bank (Natural×Invaded)	0.38	0.4	0.39
Wet Bank (Cleared×Invaded)	0.36	0.37	0.39

Non-significant p values >0.05 are indicated by an asterisk.

When indicator analysis was performed using the invasion status of the sites, the genus *Bradyrhizobium* occurred at higher frequencies in invaded sites compared to the pristine and cleared sites. The genus *Methylocapsa*, an obligatory methanotrophic gram-negative bacterium [49], also occurred at lower frequencies in the invaded sites (Figure 7b). The phototrophic genus *Rhodopila* [50] showed the same trend. Some groups were present at higher levels in invaded soils compared to pristine and cleared soils. These include the genera *Microvirga* and *Rhizobium*, which are root-nodule forming bacteria [51], *Methylosinus*, and *Methylobacterium*, which are obligatory methanotrophic gram-negative bacteria [52] and *Acidicaldus*, a moderately acidophilic thermophile [53].

Relationship between Bacterial Community Structure and Abiotic Variables

The PCA analysis showed that total phosphate, nitrate concentration, denitrification rate and total soil carbon explained most of the variation seen in environmental variables between sites (Figure S4, Table S3). The most significant correlations between spesific bacterial genera and environmental variables were with available phosphate, pH and the C: N ratio. These correlations were, however, relatively small (Table S3). The CCA analysis indicated that no significant relationship could be observed between the structure of the bacterial community structure and the soil properties (F = 0.953, p = 0.74) (Figure S5).

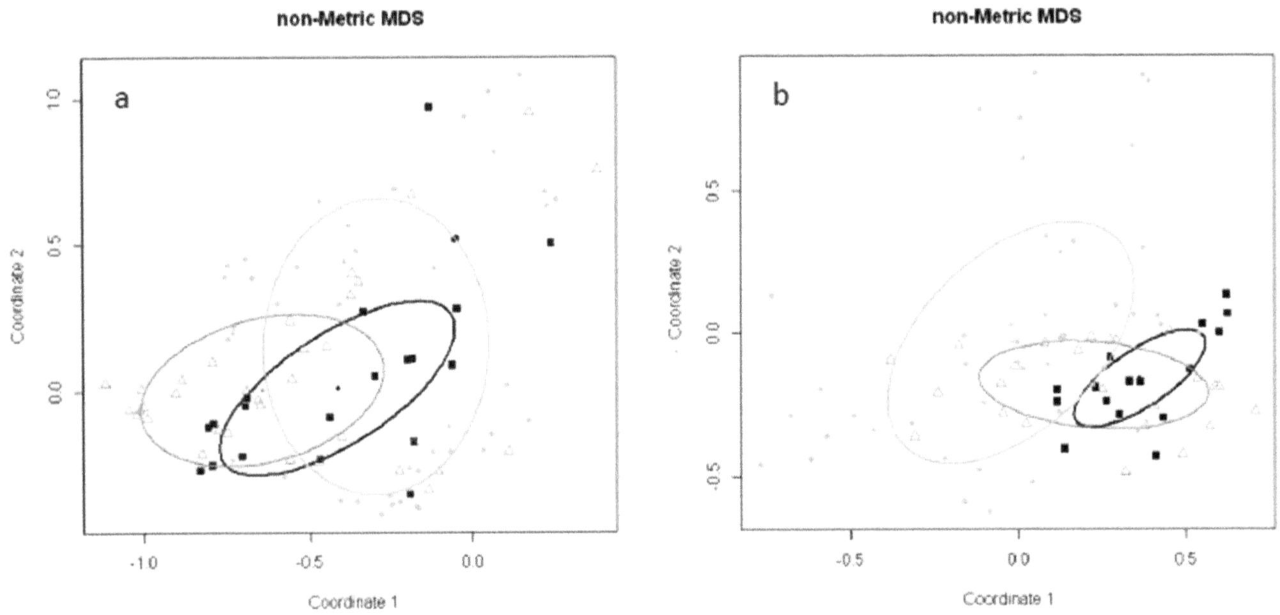

Figure 5. NMDS plot representing the Curtis-Bray distances of the ARISA profiles from the dry bank (a) and wet bank (b) zones. The ellipses represent the samples which were within 75% confidence limit of the centroids and included Pristine (squares), Cleared (triangles) and Invaded (circles) sites (Stress (a) = 0.08, (b) = 0.07).

Discussion

Bacterial OTU Diversity and Richness

The diversity of the bacterial communities (measured as Shannon diversity index) of terrestrial and dry bank lateral zones during all seasons did not differ significantly. The most prominent difference in bacterial diversity was observed in the invaded wet bank zones. In addition to reduced diversity, there was also a reduction in the number of OTUs that could be observed in the invaded sites with the dominance of certain OTUs. This indicates that *A. mearnsii* invasion reduced bacterial diversity, but only under the conditions that occurred in the wet bank zones. The presence of *Acacia* is associated with a decrease in plant diversity which would be expected to cause a reduction in bacterial diversity [54]. The wet bank zones are regularly subjected to flooding and sediment erosion. Studies have shown that such frequent

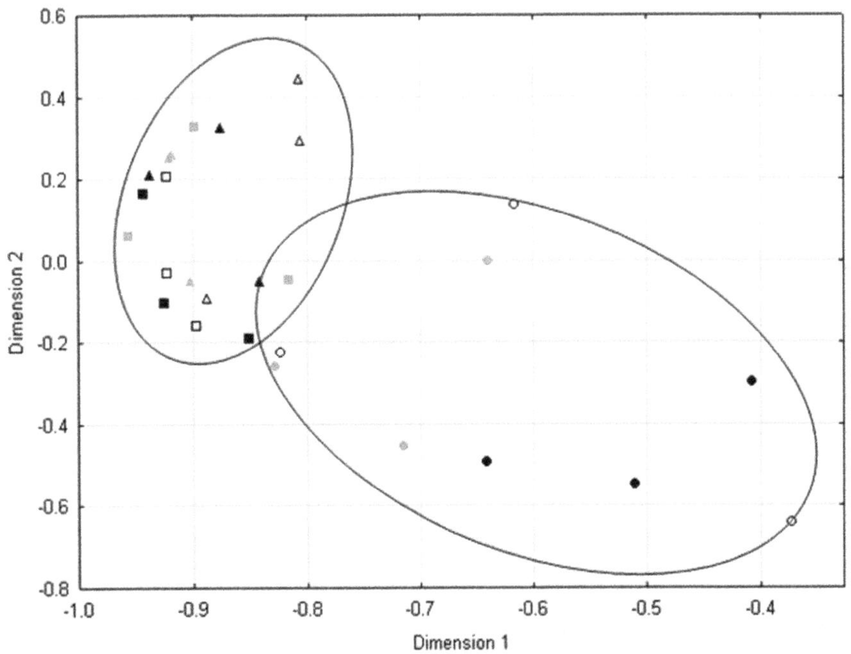

Figure 6. The terrestrial (squares) and dry bank (triangles) form a cluster, which samples showed under dispersion (measured with BETADISP).

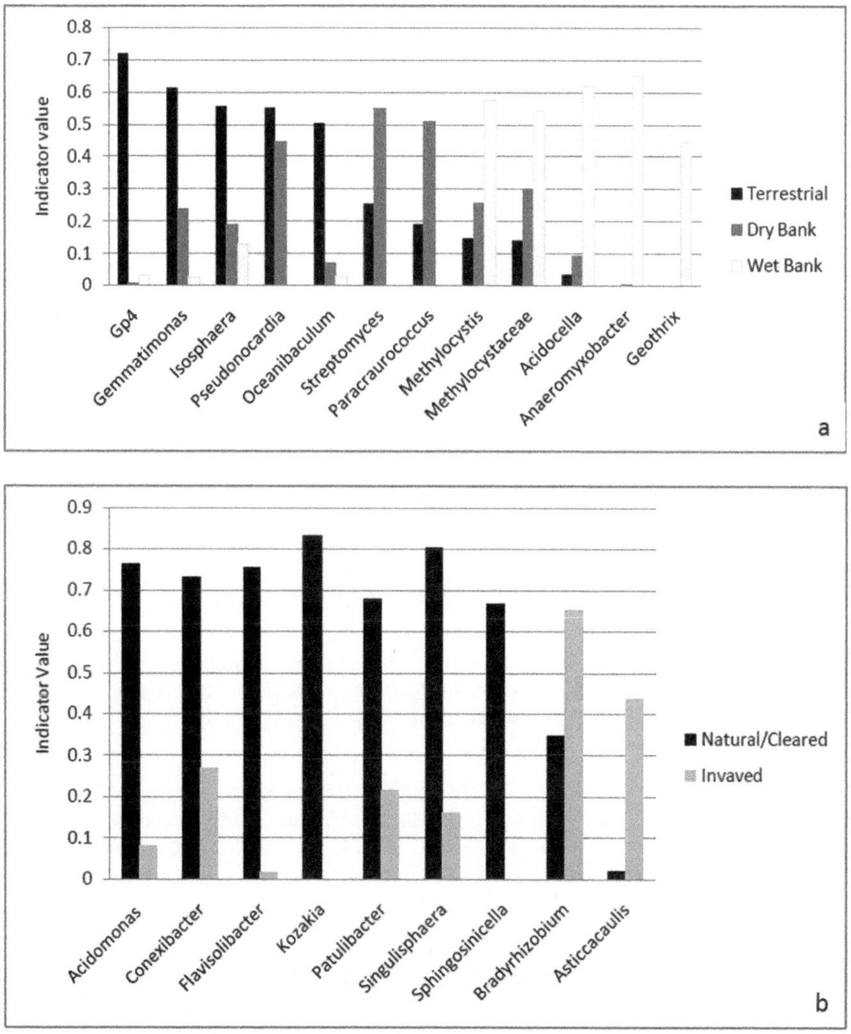

Figure 7. a: The indicator values of the indicator phyla in the community between lateral zones (p<0.05). b: The indicator values and the relative abundance of the indicator phyla in the community when comparing invasive status (p<0.05).

ecological disturbances results in a reduction of bacterial diversity [55].

Bacterial Community Structure

The environmental gradients across riparian lateral zones, had significant structuring effects on the bacterial communities These included the vegetation structure and hydrology, which involves different frequencies of inundation between the dry bank and wet bank. The wet bank harbour more variable bacterial communities, as showed by the dispersion analysis. The wet bank zone is frequently disturbed by high river flows and previous studies have shown that higher levels of disturbance increases the variability of a community in contras to the more stable communities of the terrestrial samples [56,57,58]. The structure of the bacterial community of the dry bank zone during the winter season was more similar to that of the wet bank zone, implying that seasonal changes, is this case higher levels of disturbance by flooding, have an effect on the bacterial community structure. The impact of the ecological gradients associated with riparian lateral zones proved to have a larger effect on the structure of the bacterial communities compared to *Acacia* invasion. Landscape-scale ecological influences have previously been shown to be a determining factor of the microbial community structure [59]. The hydrology of the rivers in this study constitutes a large-scale influence that had a similar structuring effect on the microbial community despite spatial separation and some ecological differences between sampling sites. The ecological influence of the river hydrology is a relatively long term and consistent effect although seasonal. These effects, which shape the different riparian lateral zones, are reflected in the differences in the microbial community structure. The bacterial communities within the three lateral zones remain relatively similar over seasons (Figure S2). This suggests that the long term natural dissimilarity between riparian lateral zones is more important in determining the bacterial community structure than the seasonal differences in this system.

The invasive Acacias affected the structure of the bacterial community of the dry bank and wet bank zones during all the seasons. In invaded sites, the structuring effect of lateral zoning was stronger than the structuring effect of the invasive status (Table 2). This was, however, not observed in the cleared and natural sites, indicating that the invasive *Acacias* are linked with this community shift. During wet conditions, the effect of invasion is visible in the bacterial community structure of the invaded sites.

The similarity of cleared and natural sites may be an indication that the community structure of invaded sites shifts back.

Bacterial Classification

The lower frequency of representatives from the phylum *Actinobacteria* is expected when considering the hydrological properties of the wet bank. The wet banks are generally less well aerated and the phyla *Actinobacteria* and *Acidobacteria* consist of a substantial number of obligatory aerobic bacteria, which was evident in the indicator analysis. Sequence analysis showed that members of the phylum *Alphaproteobacteria* were the dominant group in all the soil samples. *Alphaproteobacteria* was, however, significantly overrepresented in the invaded sites. Although function cannot be inferred on bacterial OTUs when observing bacterial genera, higher frequencies of nitrogen-fixing root associated *Alphaproteobacteria* genera were observed in the invaded wet bank samples [60,61]. The bacteria most commonly associated with *Acacia* invasion belong to the genus *Bradyrhizobium* [19]. This is consistent with the *Acacia's* root associated bacteria's ability to fix nitrogen, which has been suggested to play a role in its invasive ability [15,62]. The bacterial genera shown to be abundant in the invaded sites also occurred in the natural and cleared sites, although at lower levels.

Conclusion

Natural fynbos riparian ecosystems are characterised by bacterial communities, which are conspicuously structured by the large-scale hydrology of the area. The structuring effect of the lateral zones is only weakly linked with soil environmental parameters. The experimental design however allowed for the observation the effect of large-scale ecological gradients on the microbial communities structure. Indeed, the general macro-ecological characteristics of the lateral zones across the fynbos riparian ecosystems best explain the bacterial community structure that was observed. Invasion by *Acacia* affected both the diversity and the community structure within invaded wet banks and dry banks. The removal of invasive *Acacia* individuals resulted in the shift of the bacterial communities to their reference or natural state. This is evidence that the bacterial communities within the soil may return to a natural structure if the site remains clear of invasive *Acacias*.

Supporting Information

Figure S1 Shannon diversity index of the lateral zones with different invasive status based on pyrosequencing data. The Shannon diversity of the invaded wet bank zones was significantly lower according to Kruskal-Wallis ANOVA ($F = 2.54$, $p = 0.048$).

Figure S2 NMDS plot representing the bacterial community structure terrestrial (red), dry bank (green) and wet bank (blue) samples in autumn(triangle), winter (cross) and summer (squares) (Stress = 0.1).

Figure S3 Summary of the distribution frequency of bacterial phyla between natural, cleared and invaded hydrological zones. Only phyla occurring at levels higher than 1% of total reads are shown.

Figure S4 PCA of the soil property data with the bacterial genera plotted as supplementary values (red).

Figure S5 The CCA analysis indicated that no significant relationship could be observed between the structure of the bacterial community and the soil properties ($F = 0.953$, $p = 0.74$).

Table S1 Pyrosequencing primers used for 454 FLX titanium sequencing. Primers 340R A1 to A9 contained the universal bacterial 16S rRNA primer 340R, sequencing primer A, an identification key and 9 different multiplex identifiers. Primer 27FB consisted of the universal bacterial 16S rRNA 27F, an identification key and sequencing primer B.

Table S2 Table indicating the number of sequences, nonsingleton OTUs, the Chao1 richness estimator, the bootstrap richness estimator and the Shannon diversity index of pyrosequenced samples.

Table S3 Significant Pearson correlations ($p<0.05$) between enviroenmetal variables and bacterial genera.

Acknowledgments

We would like to acknowledge Minette Naude for the soil environmental and chemical analyses.

Author Contributions

Conceived and designed the experiments: ES SMJ KJ. Performed the experiments: ES SMJ. Analyzed the data: ES KJ. Contributed reagents/materials/analysis tools: SMJ KJ. Wrote the paper: ES.

References

1. Gregory SV, Swanson FJ, Mckee WA, Cummins KW (1991) An Ecosystem Perspective of Riparian Zones. Bioscience 41: 540–552.
2. Naiman RJ, Décamps H (1990) Aquatic-terrestrial ecotones: Summary and recommendations. In: Naiman RJ, DéCamps H (eds.). Ecology and Management of Aquatic-Terrestrial Ecotones. UNESCO, Paris, and Parthenon Publishing Group: Carnforth, 295–301.
3. Naiman RJ, Decamps H (1997) The ecology of interfaces: Riparian Zones. Annu Rev Ecol Syst 28: 621–658.
4. Richardson DM, Holmes PM, Esler KJ, Galatowitsch SM, Stromberg JC, et al. (2007) Riparian vegetation: degradation, alien plant invasions, and restoration prospects. Diversity Distrib 13: 126–139.
5. Brinson MM, Verhoeven J (1999) Riparian forests. In: Hunter ML, Jr, (ed). Maintaining biodiversity in forest ecosystems. Cambridge University Press: Cambridge, pp265–299.
6. Ferreira LV, Stohlgren TJ (2012) Effects of river level fluctuation on plant species richness, in a floodplain and distribution forest in Central Amazonia diversity, and distribution in a floodplain forest in Central Amazonia. Oecologia 120: 582–587.
7. Crow TR, Baker ME, Barnes BV (2000) Diversity in riparian landscapes. In: Verry ES, Hornbeck JW, Dolloff CA (eds). Riparian Management in Forests of the Continental Eastern United States. Lewis Publishers: Boca Raton, 43–65.
8. Ranalli AJ, Macalady DL (2010) The importance of the riparian zone and in-stream processes in nitrate attenuation in undisturbed and agricultural watersheds – A review of the scientific literature. J Hydrol 389: 406–415.
9. Décamps H, Pinay G, Naiman RJ, Petts GE, Mcclain ME, et al. (2004) Riparian zones: where biogeochemistry meets biodiversity in management practice. Pol J Ecol 52: 3–18.
10. Fischer RA, Fischenich JC (2000) Design recommendations for riparian corridors and vegetated buffer strips. Technical Note ERDC-TN-EMRRP-SR-24, Army Engineer Waterways Experiment Station, Vicksburg, MS.

11. Brauman KA, Daily GC, Duarte TK, Mooney HA (2007) The nature and value of ecosystem services: An overview highlighting hydrological services. Annu Rev Environ Resour 32: 67–98.
12. Sieben EJJ, Mucina L, Boucher C (2009) Scaling hierarchy of factors controlling riparian vegetation patterns of the Fynbos Biome at the Western Cape, South Africa. J Veg Sci 20: 1–10.
13. Rowntree KM (2001) Geomorphological classification and study sites. Chapter 3 in JM King, DM Schael, Assessing the ecological relevance of a spatially nested geomorphological hierarchy for river management. Water Research Commission: Pretoria.
14. Boucher C (2002) Flows as determinants of riparian vegetation zonation patterns in selected Southern African Rivers. In: Environmental Flows 2002, Proceedings of the International Conference on Environmental Flows for River Systems, incorporating the 4th International Ecohydraulics Symposium, Cape Town.
15. Yelenik SG, Stock WD, Richardson DM (2004) Ecosystem Level Impacts of invasive *Acacia saligna* in the South African Fynbos. Restor Ecol 12: 44–51.
16. Esler KJ, Holmes PM, Richardson DM, Witkowski ETF (2008) Riparian vegetation management in landscapes invaded by alien plants: Insights from South Africa. S Afr J Bot 74: 397–400.
17. Holmes PM, Richardson DM, Esler KJ, Witkowski ETF, Fourie S (2005) A decision-making framework for restoring riparian zones degraded by invasive alien plants in South Africa. S Afr J Sci 101: 553–564.
18. Reinecke MK, King JM, Holmes P, Blanchard PR, Malan H (2007) The Nature and Invasion of Riparian Vegetation Zones in the South-western Cape, Water Research Commission: Pretoria.
19. Rodríguez-Echeverría S, Le Roux JJ, Crisóstomo JA, Ndlovu J (2011) Jack-of-all-trades and master of many? How does associated rhizobial diversity influence the colonization success of Australian Acacia species? Diversity Distrib 17 (5): 946–957.
20. Daehler CC (2003) Performance comparisons of co-occuring native and alien invasive plants: implications for conservation and restoration. Ann Rev Ecol Evol Systs 34: 183–211.
21. Holmes PM, Cowling RM (1997) The effects of invasion by *Acacia saligna* on the guild structure and regeneration capabilities of South African fynbos shrublands. J Appl Ecol 34: 317–332.
22. Holmes PM, Richardson DM (1999) Protocols for restoration based on recruitment dynamics, community structure, and ecosystem function: Perspectives from South African fynbos. Restor Ecol 7: 215–230.
23. Joubert C (2003) Rhizobia Associated with Australian Acacia species (Acacia mearnsii, Acacia dealbata and Acacia decurrens) in South Africa as Determined by Sodium Dodecyl-Sulphate Polyacrylamide Gel Electrophoresis. Msc Thesis. University of Pretoria, Pretoria.
24. Lorenzo P, Gonzalez L, Reigosa M (2010) The genus Acacia as invader: the characteristic case of Acacia dealbata Link in Europe. Ann For Sci 67: 101–112.
25. Mucina L, Rutherford MC, Powrie LW (2006) Inland azonal vegetation. In: Mucina L, Rutherford MC (Eds). The vegetation of South Africa, Lesotho and Swaziland. SANBI: Pretoria, 617–657.
26. Bray RH, Kurtz LT (1945) Determination of total, organic, and available forms of phosphorus in soils. Soil Sci. 59: 39–45.
27. Keeney DR, Nelson DW (1982) Nitrogen inorganic forms. In: Page AL (ed). Methods of soil analysis Part 2. American Society of Agronomy: Madison, 643–698.
28. Cataldo DA, Haroon M, Schrader LE, Youngs VL (1975) Rapid colorimetric determination of nitrate in plant tissue by nitration of salicylic acid. Commun. Soil Sci. Plant. An. 6: 71–80.
29. Robertson GP, Bledsoe CS, Coleman DC, Sollins P (1999) Standard soil methods for long-term ecological research. Oxford University Press: New York.
30. Cardinale M, Brusetti L, Quatrini P, Borin S, Puglia AM, et al. (2004) Comparison of different primer sets for use in automated ribosomal intergenic spacer analysis of complex bacterial communities. Appl Environ Microbiol 70: 6147–6156.
31. Slabbert E, Kongor RY, Esler KJ, Jacobs K (2010) Microbial diversity and community structure in Fynbos soil. Mol Ecol 19: 1031–1041.
32. Ranjard L, Poly F, Nazaret S (2000) Monitoring complex bacterial communities using culture-independent molecular techniques: application to soil environment. Res Microbiol 151: 167–177.
33. Shannon C (1948) A mathematical theory of communication. AT&T Tech J 27: 379–423.
34. Bray JR, Curtis JT (1957) An ordination of the upland forest communities of Southern Wisconsin. Ecol Monogr 27: 325–349.
35. Clarke KR (1993) Non-parametric multivariate analyses of changes in community structure. Australian Journal of Ecology 117–143.
36. Anderson MJ (2006) Distance-based tests for homogeneity of multivariate dispersions. Biometrics 62: 245–253.
37. Li T (2002) Robust and consistent estimation of nonlinear errors-in-variables models. J Econometrics 110: 1–26.
38. Legendre P, Legendre L (1998) Numerical Ecology. 2nd English ed. Elsevier.
39. Bent SJ, Forney LJ (2008) The tragedy of the uncommon: understanding limitations in the analysis of microbial diversity. ISME J 2: 689–695.
40. Lane DJ (1991) 16S/23S rRNA sequencing. In: Nucleic acid techniques in bacterial systematics Stackebrandt, E, and Goodfellow, M., eds., John Wiley and Sons, New York, NY, 115–175.
41. Schloss PD, Westcott SL, Ryabin T, Hall JR, Hartmann M, et al. (2009) Introducing mothur: open-source, platform-independent, community-supported software for describing and comparing microbial communities. Applied and environmental microbiology 75: 7537–7541.
42. Edgar RC, Haas BJ, Clemente JC, Quince C, Knight R (2011) UCHIME improves sensitivity and speed of chimera detection. 27: 2194–2200.
43. Li W, Godzik A (2006) Cd-hit: a fast program for clustering and comparing large sets of protein or nucleotide sequences. Bioinformatics, 22, 1658–1659.
44. Colwell R (2005) EstimateS: statistical estimation of species richness and shared species from samples. Version 7.5. Available: http://purl.oclc.org/estimates.
45. Cole JR, Chai B, Farris RJ, Wang Q, Kulam-Syed-Mohideen AS, et al. (2007) The ribosomal database project (RDP-II): introducing myRDP space and quality controlled public data. Nucleic Acids Res 35: 169–172.
46. Felsenstein J (2005) PHYLIP (Phylogeny Inference Package) version 3.6. Department of Genome Sciences, University of Washington, Seattle.
47. Lozupone C, Hamady M, Knight R (2006) UniFrac–an online tool for comparing microbial community diversity in a phylogenetic context. BMC Bioinformatics 7: 371.
48. Roberts DW (2012) Labdsv: Ordination and Multivariate Analysis for Ecology. Version 1.4–1. Available: http://CRAN.R-project.org/package = labdsv.
49. Dedysh SN, Panikov NS, Tiedje JM (1998) Acidophilic methanotrophic communities from Sphagnum peat bogs. Appl Environ Microbiol 64: 922–929.
50. Imhoff JF, Truper HG, Pfennig N (1984) Rearrangement of the species and genera of the phototrophic "purple nonsulfur bacteria". Int J Syst Bacteriol 34: 340–343.
51. Ardley JK, Parker MA, De Meyer S, O'Hara GW, Reeve WG, et al. (2010) Species of *Microvirga* are novel alpha-proteobacterial root nodule bacteria that specifically nodulate *Lotononis angolensis* and *Lupinus texensis*. In: 9th European Nitrogen Fixation Conference, 6–10 September, Geneva, Switzerland.
52. McDonald IRMC, Kenna EM, Murrell JC (1995) Detection of Methanotrophic Bacteria in Environmental Samples with the PCR. Appl Environ Microbiol 61: 116–121.
53. Johnson BD, Hallberg KB (2009) Carbon, iron and sulfur metabolism in acidophilic micro-organisms. In: Poole RK (ed). Advances in microbial physiology. Academic Press: Amsterdam, 201–255.
54. Zhou J, Xia B, Treves DS, Wu LY, Marsh TL, et al. (2002) Spatial and resources factors influencing high microbial diversity in soil. Appl Environ Microbiol 68: 326–334.
55. Muller AK, Westergaard K, Christensen S, Sorensen SJ (2002) The diversity and function of soil microbial communities exposed to different disturbances. Microbial Ecology 44: 49–58.
56. Jiang L, Patel SN (2008) Community assembly in the presence of disturbance: a microcosm experiment. Ecology 89: 1931–1940.
57. Ferrenberg S, O'Neill SP, Knelman JE, Todd B, Duggan S, et al. (2013). Changes in assembly processes in soil bacterial communities following a wildfire disturbance. ISME J 7: 1102–1111.
58. Wang J, Shen J, Wu Y, Tu C, Soininen J, et al. (2013) Phylogenetic beta diversity in bacterial assemblages across ecosystems: deterministic versus stochastic processes. ISME J. 7: 1310–1321.
59. Krause S, Meima-Franke M, Hefting MM, Bodelier PLE (2013) Spatial patterns of methanotrophic communities along a hydrological gradient in a riparian wetland. FEMS microbiology ecology 1: 1–12.
60. Bontemps C, Elliott GN, Simon MF, Dos Reis Júnior FB, Gross E, et al. (2010) Burkholderia species are ancient symbionts of legumes. Molecular ecology 19: 44–52.
61. Weir BS, Turner SJ, Silvester WB, Park DC, Young JM (2004) Unexpectedly diverse *Mesorhizobium* strains and *Rhizobium leguminosarum* nodulate native legume genera of New Zealand, while introduced legume weeds are nodulated by *Bradyrhizobium* species. Appl Environ Microbiol 70: 5980–5987.
62. Morris TL, Esler KJ, Barger NN, Jacobs SM, Cramer MD (2011) Ecophysiological traits associated with the competitive ability of invasive Australian Acacias. Diversity Distrib 17: 898–910.

13

NaOH-Debittering Induces Changes in Bacterial Ecology during Table Olives Fermentation

Luca Cocolin[1]*, Valentina Alessandria[1], Cristian Botta[1], Roberta Gorra[1], Francesca De Filippis[2], Danilo Ercolini[2], Kalliopi Rantsiou[1]

1 Dipartimento di Scienze Agrarie, Forestali e Alimentari, Università degli Studi di Torino, Grugliasco, Italy, **2** Dipartimento di Agraria, Università degli Studi di Napoli Federico II, Portici, Italy

Abstract

Limited information is available on the impact of the NaOH treatment on table olive fermentations, and for this reason a polyphasic approach has been adopted here to investigate its effect on the fermentation dynamics and bacterial biodiversity. The microbial counts of the main groups involved in the transformation have not shown any differences, apart from a more prompt start of the fermentation when the olives were subjected to the NaOH treatment. The data produced by culture-independent analyses highlighted that the fermentation of table olives not treated with NaOH is the result of the coexistence of two different ecosystems: the surface of the olives and the brines. A sodium hydroxide treatment not only eliminates this difference, but also affects the bacterial ecology of the olives to a great extent. As proved by high-throughput sequencing, the fermentation of the olives not treated with NaOH was characterized by the presence of halophilic bacteria, which were substituted by *Lactobacillus* at the later stages of the fermentation, while enterobacteria were dominant when the olives were treated with sodium hydroxide. Higher biodiversity was found for *Lactobacillus plantarum* isolated during untreated fermentation. Different biotypes were found on the olive surface and in the brines. When the debittering process was carried out, a decrease in the number of *L. plantarum* biotypes were observed and those originating from the surface of the olive did not differentiate from the ones present in the brines.

Editor: Paul D. Cotter, Teagasc Food Research Centre, Ireland

Funding: The research conducted to obtain these results received funding from the EU (FP7/2007–2013), under grant agreement no. 243471- PROBIOLIVES (www.probiolives.eu). The funders had no role in study design, data collection and analysis, decision to publish, or preparation of the manuscript.

Competing Interests: The authors have declared that no competing interests exist.

* E-mail: lucasimone.cocolin@unito.it

Introduction

Table olives are an important fermented food in Mediterranean countries and they constitute a fundamental food production sector. Olives are the fruit that is produced by the olive tree (*Olea europea*), a species that is successfully cultivated in all Mediterranean countries. Estimations made by the International Olive Council for the 2011/2012 campaign indicate a world production of 3.1 million tons, with a 3% increase compared to the previous year. About 3 quarters of this production is from the European Union (EU), with a production of 2.18 million tons. Spain is the first producing country in the EU, producing 1.35 million tons (62% of the production), and this is followed by Italy (20%), Greece (14%), Portugal (3%) and other EU countries (1%) [1].

The olive fruit, which has a strong bitter taste that comes from glucoside oleuropein, is subjected to a number of transformations, through fermentation, which make them edible, stable and safe. Most table olive fermentation processes start spontaneously and are influenced to a great extent by the olive cultivar itself, its indigenous microbiota [2] and methodological factors, such as fermentation temperature and the salt concentration in the brines [3]. It is widely accepted that the main microbiota responsible for table olive fermentations are lactic acid bacteria (LAB) members, namely *Lactobacillus* spp., and yeasts [4,5].

There are several ways of treating table olives. However, the green Spanish type and the Greek type are those that account for most of the world's production, with the Spanish type covering about 60% [6]. The first method consists of treating the fruit with a diluted NaOH solution (2–3%) to reduce the bitterness, through the degradation of oleuropein and polyphenols, but also to increase the permeability of the olive pericarp. The debittering treatment is followed by a water wash to remove the excess alkali. The olives are then placed under brine (initial concentration of 8–12%), where they undergo lactic acid fermentation [7]. Instead, in the Greek production, the natural or untreated olives (green or naturally black) are directly brined after picking. In brine, the olives undergo a mixed-acid fermentation until they at least partially lose their bitterness. The fermentation period therefore depends on the physico-chemical conditions, such as the type of cultivar, the salt content and the temperature [3,4,6].

The scientific community recognizes that the use of methods that rely on the cultivation of microorganisms (culture-dependent techniques) do not offer a complete profile of the microbial diversity that is present in a specific ecosystem [8]. Culture-independent techniques have attracted the attention of many scientists from different investigation domains, ranging from environmental microbiology to food fermentation. At present, high-throughput sequencing has emerged as a new culture-independent tool to quantitatively investigate the structure of

Table 1. Microbial counts on the olive surfaces and in brines during the fermentation processes examined in this study.

	Days of fermentation						
	3	**8**	**10**	**15**	**30**	**60**	**90**
Olive surfaces							
Lactic acid bacteria							
Not treated	2.54 ± 0.08	4.79 ± 1.96	5.51 ± 0.50	4.21 ± 0.73	6.07 ± 0.42	7.06 ± 0.54	5.65 ± 1.44
Treated	6.01 ± 0.04	6.13 ± 0.40	5.69 ± 0.87	7.10 ± 0.44	7.24 ± 0.12	7.22 ± 0.24	6.08 ± 0.13
Sig.[a]	**	NS	NS	NS	*	NS	NS
Enterobacteria							
Not treated	2.91 ± 2.03	5.29 ± 1.53	5.38 ± 0.47	3.92 ± 1.06	5.98 ± 1.53	<5[b]	<5
Treated	5.39 ± 0.10	5.54 ± 1.33	6.20 ± 0.23	7.15 ± 0.80	4.71 ± 1.27	<5	<5
Sig.	*	NS	NS	NS	NS	NS	NS
Yeasts							
Not treated	4.01 ± 0.57	4.33 ± 0.04	4.34 ± 0.23	3.72 ± 0.60	7.25 ± 1.25	7.07 ± 0.69	6.29 ± 1.70
Treated	5.65 ± 0.32	6.27 ± 0.01	4.13 ± 0.80	6.40 ± 0.74	7.66 ± 0.09	7.44 ± 0.33	5.47 ± 0.60
Sig.	**	**	NS	NS	NS	NS	NS
Molds							
Not treated	<50	<50	<50	<50	<50	<50	5.16 ± 1.55
Treated	<50	<50	<50	<50	<50	<50	4.70 ± 0.00
Sig.	NS	NS	NS	NS	NS	NS	NS
Brines							
Lactic acid bacteria							
Not treated	4.23 ± 0.11	5.74 ± 0.30	5.81 ± 0.24	6.06 ± 0.25	5.92 ± 0.18	5.44 ± 0.66	6.03 ± 0.35
Treated	5.14 ± 0.92	5.54 ± 0.24	5.67 ± 0.53	5.72 ± 0.37	6.48 ± 0.12	6.57 ± 0.24	6.93 ± 0.49
Sig.	NS	NS	NS	NS	*	*	NS
Enterobacteria							
Not treated	3.00 ± 0.40	3.85 ± 1.29	4.73 ± 0.88	5.27 ± 0.56	4.39 ± 0.34	<1	<1
Treated	5.33 ± 0.49	5.71 ± 0.21	6.36 ± 0.16	6.67 ± 0.04	5.83 ± 0.10	2.04 ± 0.06	<1
Sig.	**	NS	**	**	**	NS	NS
Yeasts							
Not treated	4.22 ± 0.07	5.67 ± 1.04	5.64 ± 0.21	5.89 ± 0.02	6.17 ± 0.29	6.19 ± 0.59	5.37 ± 0.39
Treated	5.09 ± 0.33	6.12 ± 0.18	4.33 ± 0.04	5.88 ± 0.59	6.59 ± 0.19	6.91 ±0.44	6.42 ± 0.26
Sig.	*	NS	NS	NS	NS	NS	*

Table 1. Cont.

	Days of fermentation						
	3	8	10	15	30	60	90
Molds							
Not treated	<10	<10	<10	<10	4.25 ± 0.49	5.25 ± 0.92	5.35 ± 0.57
Treated	<10	<10	<10	3.62 ± 0.54	5.30 ± 0.43	5.30 ± 1.41	5.14 ± 0.22
Sig.	NS	NS	NS	NS	NS	NS	NS

Values are expressed as average Log_{10} colony forming units per g or ml ± standard deviation.
[a] *, ** and NS indicate significance at $P<0.05$, $P<0.01$ and no significant differences, respectively.
[b], values presented as "<" are expressed in colony forming units per g or ml.

microbial communities, several applications can be already found in food microbiology [9–14] and the related critical aspects have been recently reviewed [15]. This and other molecular techniques were employed here for the evaluation of the microbial diversity during olives fermentation.

The aim of this study was to evaluate the changes in bacterial ecology in table olives during fermentation and to assess the effect of NaOH treatment on the bacterial ecology and dynamics.

Methods

Table olive fermentations

In November 2009, two table olive fermentations were conducted in duplicate and studied in a small enterprise (SME) in Sicily, Italy. Vats containing 140 kg of the Nocellare etnea variety were prepared. Two of them contained olives treated with NaOH (1%, w/v), while the other two were loaded with olives rinsed with tap water only. In this study, a short 30 min lye treatment was employed, as it is the normal procedure the SME employs in the preparation of the olives prior to fermentation. I should be mentioned that normally this treatment lasts several hours, until the NaOH solution reaches 2/3 of the flesh [16]. Vats were filled with 60 L of brine (8% NaCl, w/v) and the fermentation was carried out at room temperature (without temperature control) for a period of 3 months. The activities described here did not require any specific permission and they were the expression of interest of the SME, which commissioned the study after receiving the funding from the European Commission (FP7/2007–2013), under grant agreement no. 243471- PROBIOLIVES (www.probiolives.eu). The study did not involve endangered or protected species.

Microbiological analysis and pH measurement

Samples, constituted of 25 g of olives and 50 ml of brines, were collected at 3, 8, 10, 15, 30, 60 and 90 days from the center of the tank, using a sterile dipper. The olives (after a vigorous rinse with Ringer's solution to remove unattached microorganisms and homogenization in 25 ml of Ringer) and brines of both fermentations were sampled separately. Appropriate decimal dilutions were prepared for microbial enumeration of the predominant populations and plated, in triplicate, on different media: lactic acid bacteria (LAB) on de Man Rogosa Sharp agar (MRS, Oxoid, Milan, Italy) using the double layer technique and incubated at 30°C for 48 h; yeasts and molds on malt extract agar (Oxoid) supplemented with a tetracycline solution (1µg/mL, Sigma-Aldrich, Milan, Italy) incubated at 30°C for 48 h; and enterobacteria on Violet Red Bile Glucose Agar (Oxoid) incubated at 37°C for 24 h. After the incubation time, the colonies were counted and the means and standard deviations were calculated. A total of 10 colonies were randomly isolated from the MRS plates at each sampling point and, after purification and growth in MRS broth for 24 h at 30°C, they were stored at −80°C with 20% (w/v) glycerol (Sigma).

The pH of the brine was measured, for both types of fermentation, using a Basic 20 pH meter (Crison, Modena, Italy).

Identification of the LAB isolates

After DNA extraction, as described by Cocolin et al. [17], and normalization at 100 ng/ µL, the LAB isolates were identified by multiplex PCR analysis of the *rec*A gene with species-specific primers for *Lactobacillus pentosus*, *L. plantarum* and *Lactobacillus paraplantarum*, according to the protocol described by Torriani et al. [18].

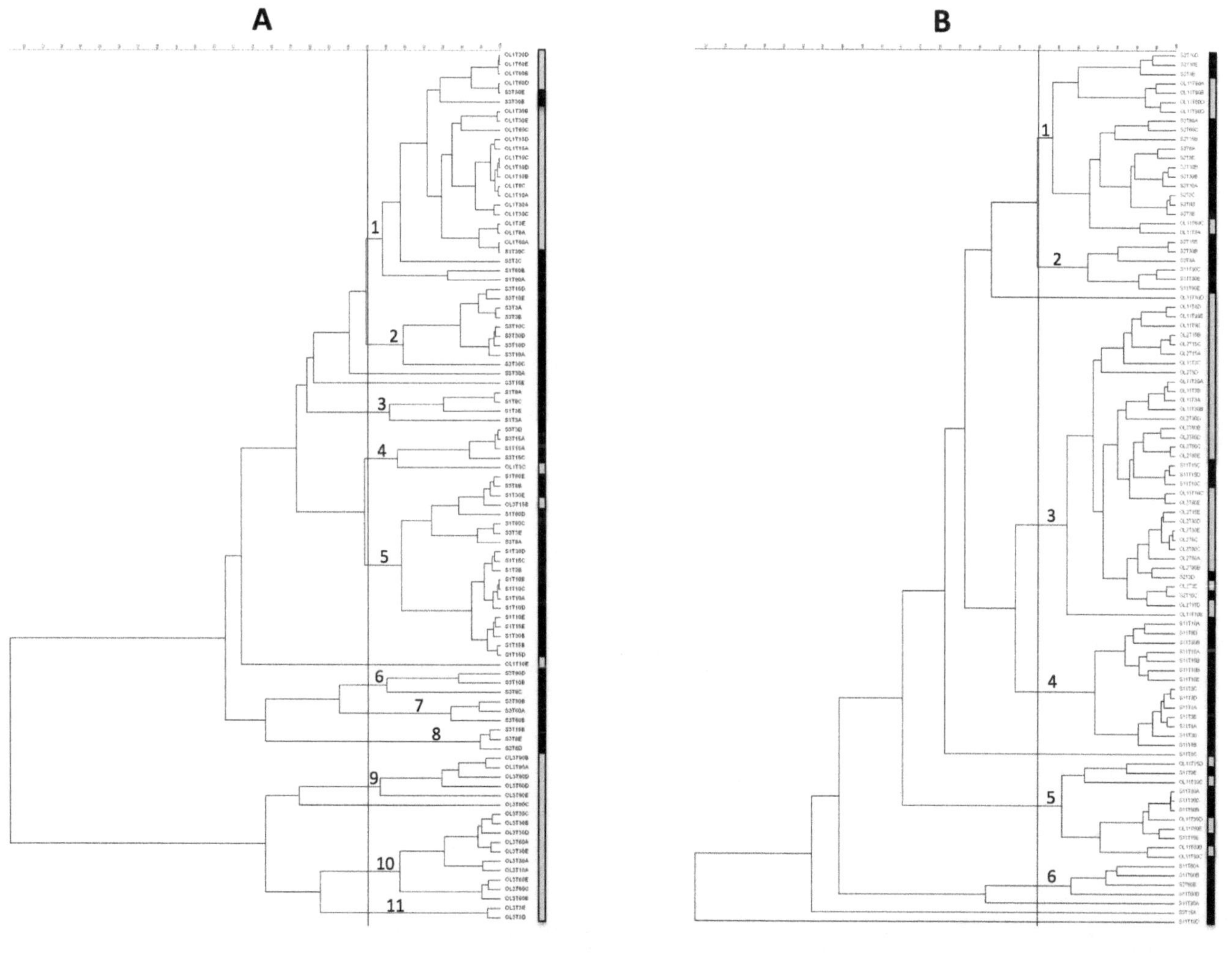

Figure 1. Dendrograms obtained comparing the Rep-PCR profiles of *Lactobacillus plantarum* isolated from the untreated (A) and NaOH treated (B) table olive fermentations. An arbitrary coefficient of similarity of 85% was selected to produce the clusters indicated with numbers. The lines on the right of the dendrograms indicate the origin of the isolates: gray, from the olives; black, from the brines.

Rep-PCR fingerprinting of the *Lactobacillus plantarum* isolates

Rep-PCR was performed on DNA extracted from *L. plantarum* isolates with the single oligonucleotide primer $(GTG)_5$ (5′-GTGGTGGTGGTGGTG-3′) [19] using the conditions described by Dal Bello et al. [20]. Rep-PCR products were electrophoresed in a 2% agarose gel for 4 h at a constant voltage of 120 V in a 1X TBE buffer (10 mmol/L Tris-borate, 1 mmol/L EDTA, pH 8.0) and externally stained using ethidium bromide (0.5 mg/mL, Sigma). A 1 Kb DNA ladder (Sigma) was used as a molecular size marker. Rep-PCR profiles were visualized under ultraviolet light, and this was followed by digital image capturing using a CCD UVI pro Platinum 1.1 (Eppendorf, Milan, Italy). The resulting fingerprints were analyzed using the BioNumerics 4.6 software package (Applied Maths, Sint-Martens-Latem, Belgium). Similarity among the digitized profiles was calculated

Table 2. Molecular identification of the LAB isolates obtained from treated and untreated table olive fermentations.

	Type of fermentation												Totals	%
	without NaOH treatment						with NaOH treatment							
	olives	%	brines	%	Totals	%	olives	%	brines	%	Totals	%		
L. pentosus	2	4.76	10	15.87	12	11.43	1	2.33	5	8.77	6	6.00	**18**	**8.78**
L. plantarum	40	95.24	53	84.13	93	88.57	42	97.67	52	91.23	94	94.00	**187**	**91.22**
Totals	**42**	**100.00**	**63**	**100.00**	**105**	**100.00**	**43**	**100.00**	**57**	**100.00**	**100**	**100.00**	**205**	**100.00**

Number of isolates and percentages obtained from the surface of the olives and the brines for each treatment are reported in the table.

using the Pearson correlation, and an average linkage (UPGMA) dendrogram was derived from the profiles.

DGGE analysis

One ml of the olives homogenate and 1 ml of brines were subjected to centrifugation at 13,400×*g* at 4°C and direct nucleic acid extraction was performed from the resulting pellets as described by Cocolin et al. [17]. In order to investigate the dominant bacterial species, the variable V3 region of the 16S rRNA gene was amplified with 338f (5′-ACT CCT ACG GGA GGC AGC AGC AG-3′) and 518r (5′- ATT ACC GCG GCT GCT GG-3′) primers. A GC clamp (5′- CGC CCG CCG CGC GCG GCG GGC GGG GCG GGG GCA CGG GGG G-3′) was attached to the 5′ end of primer 338f for DGGE analysis [21]. The PCR and reverse transcription (RT)-PCR reactions were performed in a final volume of 25 µL, as previously described [22]. The Dcode universal mutation detection system (BioRad, Milan, Italy) was used for DGGE analysis. DGGE was carried out as previously described [23], with a gradient from 30 to 60%. Electrophoresis was conducted at 200 V for 5 h (with an initial 10 mins at 80 V) at 60°C in a 1X TAE buffer. Gels were stained for 20 min in 1X TAE containing 1X SYBR Green I (Sigma) and then analyzed under UV using UVI pro platinum 1.1 Gel Software (Eppendorf). The DGGE profiles were subjected to image analysis using the Bionumerics 4.6 software, as described above. DGGE bands of interest were excised, re-amplified by PCR, sequenced [24] and sequence similarities were searched for in the National Center for Biotechnology Information (NCBI) database using nucleotide Basic Local Alignment Search Tool (BLAST) analysis [25].

Pyrosequencing

DNA (standardized at 20 ng/µl) and cDNA (prepared by reverse transcription of 100 ng of RNA in RT reactions containing random hexamers [Promega, Milan, Italy]) were amplified with Gray28f (5′-TTTGATCNTGGCCTCAG-3′) and Gray519r (5′-GTNTTACNGCGGCKGCTG-3′) primers [26] and a 520 bp fragment of the 16S rRNA gene was generated. Tag-encoded FLX amplicon pyrosequencing analysis was carried out in the premises of the Research and Testing laboratories (RTL, Lubbock, TX, USA) using a Roche 454 FLX instrument with Titanium reagents and procedures. Raw reads were analyzed and filtered by using QIIME 1.6.0 software [27]. In order to guarantee a higher level of accuracy in terms of Operational Taxonomic Units (OTUs) detection, after the split library script performed by QIIME, the reads were excluded from the analysis if they had an average quality score lower than 25, if they were shorter than 250 bp and if there were ambiguous base calls. Sequences that passed the quality filters were subjected to denoising and chimera checking, as previously described [28,29]. Filtered sequences were then clustered in operational taxonomic units (OTU) with 96.5% identity using USEARCH [30] and queried against a database of high quality sequences derived from NCBI with a distributed BLASTn algorithm [31] as previously described [9,26]. The abundance (%) of each OTU was calculated on the basis of the number of sequence reads obtained in each sample. The OTU taxonomy table was used to produce a heat map by using the clustering software TMeV v 4.8 [32]. Alpha diversity was evaluated through QIIME to generate rarefaction curves, Good's coverage, Chao1 richness [33] and Shannon diversity indices [34]. The OTU taxonomy table and the sequence phylogenetic tree were used to generate the weighted UniFrac distance matrix [35].

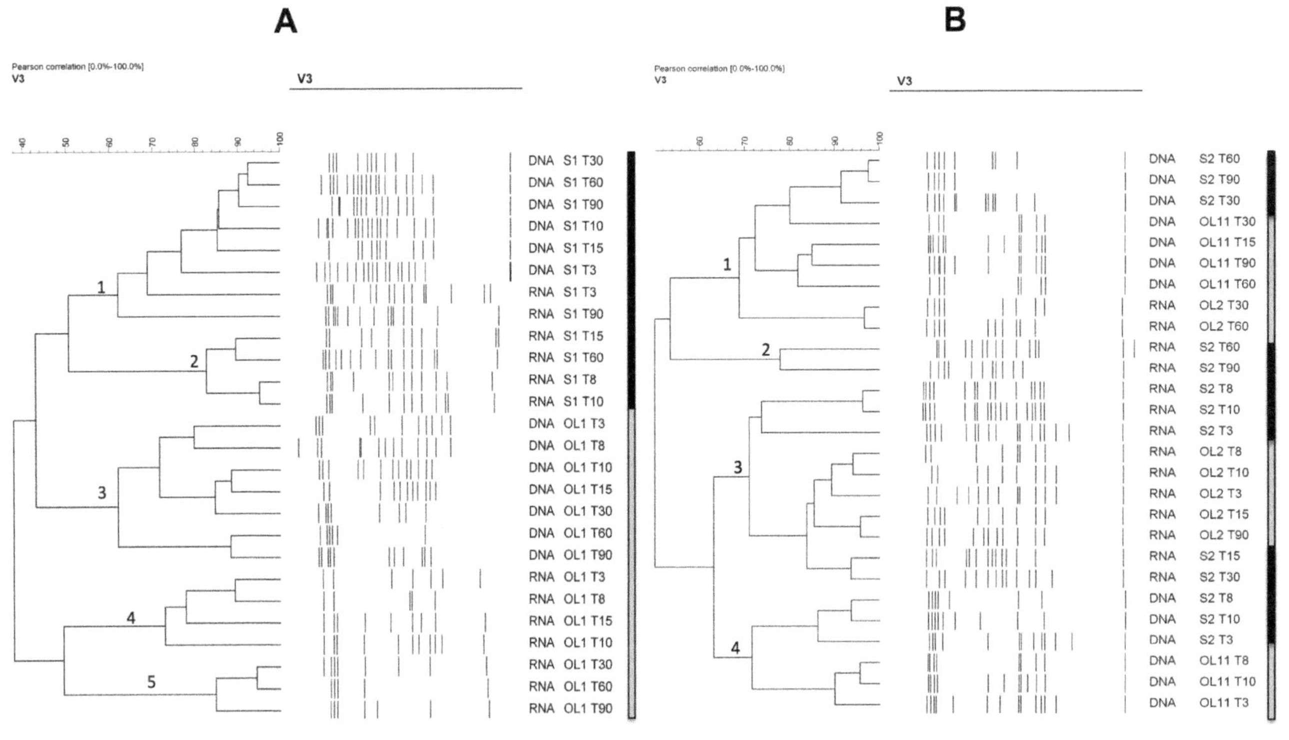

Figure 2. Digitalized DGGE profiles of the DNA and RNA extracted directly from olive surfaces and brines during the fermentation processes. Panel A, untreated table olive fermentation; panel B, NaOH treated table olive fermentation. The identified clusters are indicated with numbers. The lines on the right of the dendrograms indicate the origin of the samples: gray, from the olives; black, from the brines. OL stands for olive surface and S for brine. The day of fermentation and the nucleic acid analyzed is also indicated in the code.

Table 3. Identification, at a genus level, of selected bands from DGGE gels used to analyze the PCR and RT-PCR products obtained from the amplification of the olive and brine samples.

						Untreated table olive fermentation		NaOH treated table olive fermentation	
						DNA			
Closest described relative	**Identity %**	**Phylum**	**Class**	**Order**	**Family**	**Olive**	**Brine**	**Olive**	**Brine**
Marinilactibacillus	99–100	*Firmicutes*	*Bacilli*	*Lactobacillales*	*Carnobacteriaceae*		*		*
Halomonas	98–100	*Proteobacteria*	*Gamma*	*Oceanospirillales*	*Halomonadaceae*	*	*		
Salinicola	99–100	*Proteobacteria*	*Gamma*	*Oceanospirillales*	*Halomonadaceae*		*		
Lactobacillus	99–100	*Firmicutes*	*Bacilli*	*Lactobacillales*	*Lactobacillaceae*	*		*	
Acinetobacter	99–100	*Proteobacteria*	*Gamma*	*Pseudomonadales*	*Moraxellaceae*				*
Enterobacter	99–100	*Proteobacteria*	*Gamma*	*Enterobacteriales*	*Enterobacteriaceae*			*	*
Citrobacter	99–100	*Proteobacteria*	*Gamma*	*Enterobacteriales*	*Enterobacteriaceae*			*	
						RNA			
Marinilactibacillus	99–100	*Firmicutes*	*Bacilli*	*Lactobacillales*	*Carnobacteriaceae*		*		*
Halomonas	98–100	*Proteobacteria*	*Gamma*	*Oceanospirillales*	*Halomonadaceae*	*	*		*
Salinicola	99–100	*Proteobacteria*	*Gamma*	*Oceanospirillales*	*Halomonadaceae*	*	*		
Lactobacillus	99–100	*Firmicutes*	*Bacilli*	*Lactobacillales*	*Lactobacillaceae*	*		*	*
Pediococcus	100	*Firmicutes*	*Bacilli*	*Lactobacillales*	*Lactobacillaceae*	*			*
Acinetobacter	99–100	*Proteobacteria*	*Gamma*	*Pseudomonadales*	*Moraxellaceae*				*
Enterobacter	100	*Proteobacteria*	*Gamma*	*Enterobacteriales*	*Enterobacteriaceae*	*	*	*	*
Pseudomonas	99	*Proteobacteria*	*Gamma*	*Pseudomonadales*	*Pseudomonadaceae*				*

The presence in the specified sample is indicated by an asterisk.

Samples were clustered using UPGMA (Unweighted Pair Group Method with Arithmetic mean, also known as average linkage). Sequences are available at the Sequence Read Archive (SRP019475).

Statistical analysis

Statistical analysis of the microbial counts during fermentation was performed using the Statistica software package (version 7.1, StatSoft Inc., Tulsa, OK). The Tukey test was used in order to establish any statistical differences by one-way analysis of variance (ANOVA) between NaOH treated and not treated table olive fermentations.

Results

Microbial trends determined by plate counts and pH measurement

The counts performed during the fermentations followed in this study are presented in Table 1. The main populations involved in the transformation process were once again confirmed to be LAB and yeasts, which dominated the ecosystem from the very beginning. Significant differences were observed between the NaOH treated and untreated table olive fermentations, especially at the beginning of the transformations (for LAB and yeasts $P<0.01$ and for enterobacteria $P<0.05$ on the olive surface and for enterobacteria $P<0.01$ and yeasts $P<0.05$ in brines, Table 1). The populations present on the surface of the untreated olives were delayed at the beginning of the fermentation and only at day 30 did the LAB and yeasts exceed 10^6 colony forming units (cfu)/g. In the NaOH treated olives, these counts were already reached at day 15. Such a difference was not so evident in the brines, and in both cases, after 10 days of fermentation, the LAB and yeasts counts were above 10^5 cfu/mL. The molds only showed an increase in the counts on the olives between day 60 and 90, while they reached final values of about 10^5–10^6 cfu/mL in the brines, with the NaOH treated fermentation already showing an increase in molds after day 10, compared to day 30, for the untreated process. No significant differences in mold counts were observed throughout the fermentation period. Like the molds, the enterobacteria also presented a similar trend in all the fermentations. They were not counted from day 60 on the olives, while the transformation without NaOH allowed the elimination of enterobacteria after 60 days in the brines, compared to the 90 days required in the treated fermentations. Significant differences in enterobacteria counts ($P<0.05$) were detected in brines from day 10 to day 30.

Both fermentations started with a pH of the brines of 5.7–5.8 and, after a slow decrease in the first 10 days, a drop to values of around 4 – 4.5 was observed. A difference of about 0.5 units of pH was recorded at the end of the fermentation between the two processes. The fermentation of olives treated with NaOH reached pH values of about 4, while the other never went below 4.5 (data not shown).

Molecular identification and characterization of the LAB isolates

A total of 205 LAB were isolated and successfully identified by applying the method suggested by Torriani et al. [18]. The results of the identification are presented in Table 2. The vast majority was identified as *L. plantarum* (187 isolates), while the rest was *L. pentosus* (18 isolates). *Lactobacillus plantarum* was more frequently isolated (94.00%) compared to untreated fermentations (88.57%). The results of the Rep-PCR of *L. plantarum* are presented in Figure 1. The isolates from the surface of the olive and the brines are indicated in the dendrograms with a gray and black box, respectively. Using an arbitrary coefficient of similarity of 85%, isolates from the untreated table olive fermentations (Fig. 1A) clustered in 11 different groups, while in the case of the NaOH treatment, only 6 clusters could be observed (Fig. 1B). The isolates from the untreated fermentations clustered above all according to the source of isolation (as described by the gray and black boxes in Fig. 1). As shown in Figure 1A, *L. plantarum* isolated from the surface of the olives formed 4 clusters, namely 1, 9, 10 and 11, among which only in cluster 1 were six isolates from brines included. A homogeneous distribution of the isolates from the olive surfaces and brines was observed in the NaOH treated table olive fermentations. Only clusters 2 and 6, grouping 6 and 4 isolates, respectively, were composed exclusively of *L. plantarum* from the brines (Fig. 1B). Considering the distribution of the *L. plantarum* isolated from the surface of the olives, it can be observed that, in the case of the untreated table olive fermentations, 4 biotypes were detected, compared to the 3 observed in the fermentation with the NaOH treatment.

DGGE analysis

The digitalized profiles of the PCR-DGGE and RT-PCR-DGGE gels, as well as the dendrograms obtained from the cluster analysis performed with the Bionumerics software, are shown in Figure 2. Samples from olive surfaces and brines are indicated with gray and black boxes, respectively. The figure reports one fermentation per type (not treated with NaOH, panel A, and treated with NaOH, panel B). No differences in the profiles were observed for the duplicate fermentations or within each sampling point (data not shown). Generally, the DGGE profiles were characterized by high complexity with some samples containing up to 15 bands. With regards to the fermentations in which the olives were not treated with NaOH, the profiles clustered in 5 groups. Clusters 1 and 2 were related to samples extracted from brines, while clusters 3, 4 and 5 were formed by DGGE profiles produced by nucleic acids on the surface of the olive. Within the clusters of the brines and olives, it was also possible to differentiate samples from DNA and RNA, respectively, which grouped separately. As can be observed, considering the olive surface, clusters 4 and 5 were formed by RNA samples, while cluster 3 grouped the DNA samples. This evidence was not confirmed in the case of the NaOH treated table olive fermentation, in which the dendrogram showed 4 clusters grouping samples regardless of the source and the analyzed nucleic acid. None of the clusters contained samples from only the olive surface or brines, or samples from only the DNA and RNA. Moreover, samples from the beginning and the end of the fermentations generally clustered differently, underlining a change in the bacterial population.

The results of the analysis of the excised and sequenced bands are reported in Table 3. *Marinilactobacillus* and *Lactobacillus* genera were found in both fermentations at DNA and RNA levels, and they represented the most common populations. *Halomonas* and *Salinicola* were mainly detected in fermentations in which the olives were not treated with sodium hydroxide, whereas *Enterobacter*, *Citrobacter*, *Acinetobacter* and *Pseudomonas* were mostly observed in the NaOH treated table olive fermentations. The species most frequently detected in DGGE gels were *Marinolactobacillus piezotolerans*, *L. plantarum*, *Acinetobacter johnsonii*, *Citrobacter freundii* and *Pseudomonas mendocina* (data not shown).

Pyrosequencing

The DNA and RNA extracted from the olive and brine samples of both fermentations examined in this study were selected from the DGGE profiles presented in Figure 2 and were subjected to

Table 4. Number of sequences analyzed, observed diversity and estimated sample coverage for 16S rRNA amplicons from olives fermentations.

		Reads	Observed OTUs	Shannon	Chao1	ESC (%)
Not treated olives fermentation						
Olive surfaces						
Day of fermentation	Target					
3	RNA	1697	116	4.68	128.5	98.53
3	DNA	4258	74	2.43	78.33	99.67
8	RNA	5957	70	2.09	80.91	99.73
8	DNA	5199	78	2.92	101.1	99.58
90	RNA	1831	49	2.7	73	99.13
90	DNA	6325	73	2.79	88.55	99.70
Brines						
Day of fermentation	Target					
8	RNA	1894	39	2.09	54.6	98.81
8	DNA	3980	56	2.47	77.38	99.52
90	RNA	2894	68	3.03	83	99.27
90	DNA	3763	77	3.02	89.05	99.39
NaOH treated olives fermentation						
Olive surfaces						
Day of fermentation	Target					
3	RNA	1131	40	2.22	57.14	98.59
3	DNA	3306	81	2.6	110.08	99.15
8	RNA	2813	51	2	82.63	99.18
8	DNA	3171	53	2.06	70	99.43
90	RNA	2945	91	3.07	128.71	98.88
90	DNA	1465	80	3.59	107.19	97.95
Brines						
Day of fermentation	Target					
8	RNA	7834	127	3.38	156.75	99.55
8	DNA	2772	101	3.56	125	98.81
90	RNA	5851	82	2.75	111.25	99.54
90	DNA	6428	112	3.27	137.83	99.52

pyrosequencing. The nucleic acids extracted from the samples at the beginning (day 8) and at the end (day 90) of the fermentations were selected because of their different clustering profiles. Moreover, the DNA and RNA from olives treated or not treated with NaOH were also subjected to pyrosequencing at day 3. A total of approximately 74,000 reads were analyzed with an average number of about 3,700 reads per sample. The rarefaction analysis and the diversity indexes indicated that there was a satisfactory coverage of the diversity for all the samples analyzed with Good's coverage values above 97% (Table 4); this result was also confirmed by the analysis of rarefaction curves (Figure S1). Overall, despite the diversity of sequencing depth between samples, the rarefaction analysis indicated that a number of reads above 1,500 per sample was satisfactory to obtain a good coverage (Table 4).

The results obtained for the DNA correlated well with those from RNA. The composition of the bacterial consortium in most of the analyzed samples was the same for both nucleic acids, with a difference of less than 15%. Only in the case of the fermentation in which the olives were not treated with NaOH, was a population of *Pseudomonas* found on the olive surface, but only at an RNA level, with a prevalence of about 50% (Table S1).

The results obtained after sequence identification of the DNA samples are shown in Figure 3, where only OTUs (genus level) that represented at least 5% of the total sequence reads in each sample are shown. Panel A reports the population profiles on the olive surface, while panel B shows the changes in the bacterial ecology of the brines. The entire set of identifications is reported in Table S2. The surface of the olives not treated with NaOH showed a different bacterial colonization in the first 8 days of fermentation compared to the last day of sampling. The initial fermentation stage was characterized by a high level of halophilic bacteria, namely *Chromohalobacter* and *Halomonas*, which represented about 60 and 50% of the total bacterial population at days 3 and 8, respectively. After 90 days of fermentation, the structure of the microbiota changed dramatically, and *Lactobacillus* represented the main bacterial population present on the olive surface (Fig. 3A). This switch in bacterial ecology was not observed on the surface of the olives treated with sodium hydroxide. In fact, enterobacteria already represented the main components of the bacterial

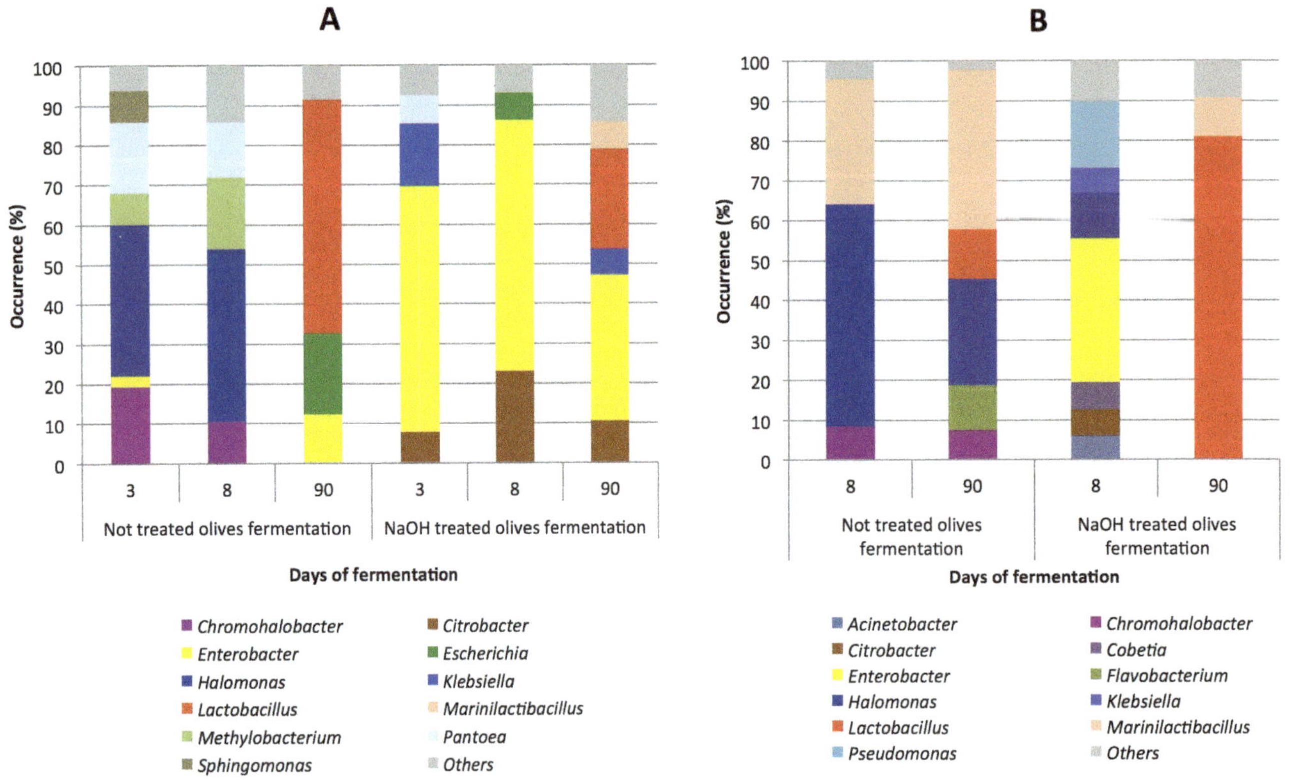

Figure 3. Occurrence (%) of genera obtained by pyrosequencing. Only genera above 5% occurrence are reported. Reference can be made to Table 1S for the entire data set. Panel A, bacterial ecology on the olive surfaces; panel B, bacterial ecology in the brines.

consortium at day 3, and basically remained stable until the end of the fermentation. It is worth noticing that the *Lactobacillus* genus was once again only detected at the last sampling point, although with low prevalence (25%) compared to the olives not treated with NaOH (Fig. 3A). Considering the results obtained from the brines (Fig. 3B), a different bacterial ecology could once again be described as being influenced by the NaOH treatment. When the untreated olives were fermented, the brines were characterized by an abundant presence of halophiles throughout the process. *Chromohalobacter*, *Halomonas* and *Marinilactibacillus* accounted for more than 90% of the total population, and only at day 90 were the *Flavobacterium* and *Lactobacillus* genera detected at percentages of about 10%. A complex bacterial ecosystem was described in the brines, when the olives subjected to the NaOH treatment were fermented. Samples taken at day 8 contained 7 genera, among which *Enterobacter* and *Pseudomonas* were the most numerous. At the end of the fermentation (day 90), an overturned bacterial ecology was observed. This was characterized by a relevant presence (80%) of the *Lactobacillus* genus, which surpassed all the other components of the consortium detected at day 8. *Chromohalobacter salarius and M. piezotolerans* were identified among the halophilic genera, while *L. plantarum* and *L. pentosus* were the main species belonging to the *Lactobacillus* genus, with the first one being the most abundant in all the fermentation processes, confirming the results obtained by traditional isolation and molecular identification. Selecting the minimum incidence of the 0.1% in at least 1 sample we considered 19 OTUs and used their abundance in each sample to generate the hierarchical clustering reported in Figure 4. NaOH treated and untreated manufactures could be clearly distinguished. The untreated olives and brines were characterized by the occurrence of halophilic bacteria regardless of the fermentation time, whereas the NaOH treated samples were separated from the others on the basis of the abundance of *Enterobacteriaceae* such as *Enterobacter*, *Citrobacter*, *Escherichia* and *Klebsiella*. Finally, untreated olives and treated brines after 90 days of fermentation clustered separately because of the high incidence (above 50%) of *Lactobacillus* (Fig. 4).

Discussion

The available literature on table olive fermentations is rather extensive and there is international scientific consensus on the fact that LAB and yeasts are the main microbial populations, which are active throughout the transformation process. The significance and the role of these two groups have recently been reviewed in two comprehensive papers by Hurtado et al. [5] for LAB and by Arroyo-Lopez et al. [36] for yeasts.

As pointed out by Botta and Cocolin [37], in the field of table olive fermentations, scientists have only taken advantage of molecular approaches in the last few years and most studies currently available are above all related to the use of molecular methods for the identification of isolated strains [38–44]. To the authors' knowledge, culture independent methods have rarely been applied to study the dynamic changes that take place during table olive fermentation. In 2006, Ercolini et al. [45] employed fluorescence in situ hybridization to detect the *L. plantarum* group on olives used in natural fermentations, while Abriouel et al. [46] and Randazzo et al. [47] exploited DGGE to study the diversity of microbial populations in brines during fermentation of Aloreña olives and to assess the influence of inoculated starter cultures on bacterial dynamics during table olive fermentations.

For the first time in this study, olives and brines were treated as different samples in order to investigate the differences in

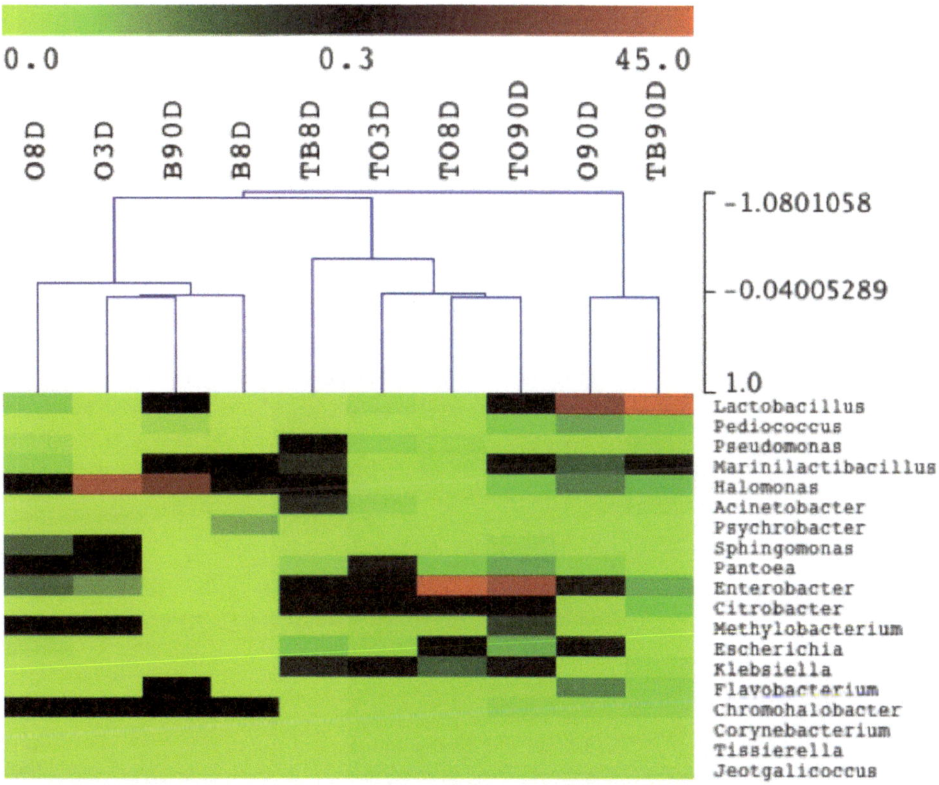

Figure 4. Heat map depicting bacterial diversity and relative abundance in treated (T) and untreated olives (O) and brines (B) during fermentation. Numbers in the samples identity indicate the days of fermentation. Hierarchical dendrogram shows distribution of samples based on average linkage clustering calculated with Pearson correlation. Legend and color scale shown in the upper part of the figure represent colors in the heat map associated with the relative percentage of each OTU within the samples.

microbial ecology on the fruit and in the liquid brine during fermentation. Appropriate succession of the microbiota and fermentation is reported as necessary to obtain safe and good quality table olives [7]. However, most of the studies focus on the microbiota of the brine, useful for pH drop and food protection, while the microbiota developing on the surface of the olives can be still responsible for changes in the olive texture and sensory properties [48]. So far, the microbiological analysis of table olive fermentations has only been performed by spreading the brines over culture media, and only in a few cases has the microbiological investigation specifically been performed on the olives in order to study the formation of biofilms on the surface of the fruit [2,16] and to assess adherence of the probiotic strain *L. paracasei* IPMC2.1 inoculated as a starter culture [49]. Therefore in this study olives and brines were handled as two different bacterial ecosystems.

The results obtained from the plate counts have once again highlighted the technological importance of the LAB and yeast populations. When comparing the counts of the fermentations of olives treated or not treated with NaOH, it was interesting to observe a more prompt start of the growth after the sodium hydroxide wash. This evidence was particularly obvious in the case of the olive surface. As can be seen in Table 1, without the NaOH treatment, the counts only reached values of just over 10^6 cfu/g after 30 days, while those counts were already surpassed at day 8 for the treated olives. More than 90% of the isolated colonies from MRS were identified as *L. plantarum*, regardless of the NaOH treatment. These results correlate well with previous studies of LAB ecology in fermented table olives, in which *L. plantarum* and *L. pentosus* were among the main lactobacilli responsible for the fermentation process [5].

Only a few studies focused on the effect of the NaOH treatment on the olive microbial populations. Arroyo-Lopez et al. [50] reported a total destruction of the initial micobiota in Spanish style processed olives when heavy NaOH treatments were applied, whereas Bevilacqua et al. [51] emphasized that yeasts can survive NaOH processing and colonize olives throughout the fermentation. As far as the brines and the olives are concerned, Hernández et al. [52] found a greater presence of yeast cells in the brine after treatment than on the olive surfaces and also a different predominant species.

The culture independent methods used in this study, i.e. DGGE and pyrosequencing, underlined important differences in the bacterial ecology in NaOH treated or not table olives fermentations. It should be pointed out that while DGGE is only able to pick up major populations, having a limit of detection of about 10^3 cfu/g or ml [37], pyrosequencing can potentially detect large minor populations and define the relative abundance of the OTUs [15]. Independently from this difference, both methods were able to highlight the changes in the succession of the bacterial populations when the olives are treated with NaOH.

Overall, NaOH modified the composition of the table olive ecosystem to a great extent and promoted a fermentation process that was different, in terms of bacterial species and strain. This evidence could allow us to speculate that the debittering process could influence the number of species present on the surface of olives and affect the biodiversity of the fermentation system. This hypothesis is supported by both the results of the DGGE performed at DNA and RNA levels, in which samples from the olive surfaces and brines clustered in two different groups in untreated table olive fermentations, while they were mixed in the case of the NaOH wash, but also by considering the diversity of *L. plantarum* at a strain level. As observed for the DGGE analysis, in the case of the Rep-PCR characterization, isolates from the untreated olive surfaces again always clustered separately from those in the brines, underlining the presence of different biotypes that colonize the two considered ecosystems (Fig. 1A). This was not found for the *L. plantarum* isolates from the treated table olive fermentation. As reported in Figure 1B, no specific separation was obtained on the basis of the isolation source. Finally, the molecular characterization of the isolates also indicates a reduction in *L. plantarum* biodiversity. This aspect can be seen by simply analyzing the obtained number of clusters, that is, 11 for the untreated and 6 for the treated table olive fermentations, respectively.

The results obtained from pyrosequencing confirmed the different structure of the microbiota of olives and brines during fermentation and definitively strengthened the need to treat them as separate bacterial ecosystems. This was further supported by the UniFrac analysis and clustering of the treated and untreated samples (Figure S2). The effect of the NaOH treatment on the bacterial ecology of the olive surface was remarkable. The halophilic populations found in the untreated olives, were replaced by enterobacteria, which remained stable until the end of the fermentation in the treated olives, and this evidence was also confirmed in the brines at the beginning of the fermentation (Fig. 3 and 4). This is of particular relevance for table olive fermentation because it is well established that enterobacteria are involved in the olive spoilage process described as gas pockets [48,53]. The presence of halophiles, such as *Marinilactibacillus*, *Halomonas* and *Chromohalobacter*, could be explained by the use of marine salt in the preparation of the brines. Previously, *Halomonas* was detected by means of high-throughput sequencing in the rind of artisanal cheeses produced in Ireland and subjected to a salting process [12], and *Marinilactobacillus* has been used to control *Listeria* spp in the cheese rind ecosystem [54]. Pyrosequencing highlighted relevant differences in the localization of *Lactobacillus*, the main genus of technological importance. *Lactobacillus* was revealed by sequencing only after 90 days of fermentation (Fig. 3). However, *Lactobacillus* colonized the surface of the olives more in untreated than treated fermentations, while it was the dominant population in the brines only when the olives had been treated by NaOH.

This paper has highlighted the effect of NaOH treatment on the bacterial ecology of olives fermentation and the necessity of investigating table olive fermentations as two different ecosystems: the surface of the olives and the brine. Future studies on these transformation processes should take into consideration the evidence reported here, and also try to match the evolution of the microbiota with the changes in sensory quality of the table olives in order to appropriately define the most suitable fermentation conditions to enhance the product quality while assuring its safety.

Supporting Information

Figure S1 Rarefaction curves obtained by QIIME for untreated (A) and treated (B) olives and brine samples. OL stands for olive surface and S for brine. The day of fermentation and the nucleic acid analyzed is also indicated in the code.

Figure S2 UPGMA clusters based on weighted UniFrac distance matrix obtained by QIIME for the untreated (A) and treated (B) olives and brine samples. OL stands for olive surface and S for brine. The day of fermentation and the nucleic acid analyzed is also indicated in the code.

Table S1 Differences in the occurrence of the genera obtained by pyrosequencing between DNA and RNA samples. Positive values mean a higher prevalence at DNA level, negative values indicate a higher prevalence at RNA level.

Table S2 Complete results of the identification, at genus level, of the sequences obtained by pyrosequencing and expressed as percentage.

Author Contributions

Conceived and designed the experiments: LC KR. Performed the experiments: VA CB RG. Analyzed the data: DE FDF. Contributed reagents/materials/analysis tools: LC. Wrote the paper: LC DE.

References

1. UNAPROL (2012) Monitoraggio delle aziende olivicole e dei frantoi. Available: http://www.unaprol.it. Accessed 2012 Oct 1.
2. Nychas GJE, Panagou EZ, Parker ML, Waldron KW, Tassou CC (2002) Microbial colonization of naturally black olives during fermentation and associated biochemical activities in the cover brine. Lett Appl Microbiol 34: 173–177.
3. Tassou CC, Panagou EZ, Katsaboxakis KZ (2002) Microbiological and physicochemical changes of naturally black olives fermented at different temperatures and NaCl levels in the brines. Food Microbiol 19: 605–615.
4. Arroyo-López FN, Querol A, Bautista-Gallego J, Garrido-Fernández A (2008) Role of yeasts in table olive production. Int J Food Microbiol 128: 189–196.
5. Hurtado A, Reguant C, Bordons A, Rozès N (2012) Lactic acid bacteria from fermented table olives. Food Microbiol 31: 1–8.
6. Panagou EZ, Schillinger U, Franz CMAP, Nychas GJE (2008) Microbiological and biochemical profile of cv. *Conservolea* naturally black olives during controlled fermentation with selected strains of lactic acid bacteria. Food Microbiol 25: 348–358.
7. Garrido-Fernández A, Fernandez- Diez MJ, Adams MR (1997) Table olives: production and processing. Chapman & Hall, London, UK.
8. Hugenholtz P, Goebel BM, Pace NR (1998) Impact of culture-independent studies on the emerging phylogenetic view of bacterial diversity. J Bacteriol 180: 4765–4774.
9. Ercolini D, Ferrocino I, Nasi A, Ndagijimana M, Vernocchi P, et al. (2011) Monitoring of microbial metabolites and bacterial diversity in beef stored under different packaging conditions. Appl Environ Microbiol 77: 7372–7381.
10. Humblot C, Guyot JP (2009) Pyrosequencing of tagged 16S rRNA gene amplicons for rapid deciphering of the microbiomes of fermented foods such as pearl millet slurries. Appl Environ Microbiol 75: 4354–4361.
11. Quigley L, O'Sullivan O, Beresford TP, Ross RP, Fitzgerald GF, et al. (2012) High-throughput sequencing for detection of subpopulations of bacteria not previously associated with artisanal cheeses. Appl Environ Microbiol 78: 5717–5723.
12. Ercolini D, De Filippis F, La Storia A, Iacono M (2012) A "remake" of the microbiota involved in the production of water buffalo mozzarella cheese by high throughput sequencing. Appl Environ Microbiol 78: 8142–8145.
13. Masoud W, Takamiya M, Vogensen FK, Lillevang S, Al-Soud WA, et al. (2011) Characterization of bacterial populations in Danish raw milk cheeses made with different starter cultures by denaturing gradient gel electrophoresis and pyrosequencing. Int Dairy J 21: 142–148.
14. Bokulich NA, Joseph CML, Allen G, Benson AK, Mills DA (2012) Next-generation sequencing reveals significant bacterial diversity of botrytized wine. PLoS ONE 7: e36357.
15. Ercolini D (2013) High-throughput sequencing and metagenomics: moving forward in the culture-independent analysis of food microbial ecology. Appl Environ Microbiol 79: 3148–3155.

16. Arroyo-López FN, Bautista-Gallego J, Domínguez-Manzano J, Romero-Gil V, Rodriguez-Gómez F, et al. (2012) Formation of lactic acid bacteria-yeasts communities on the olive surface during Spanish-style Manzanilla fermentations. Food Microbiol 32: 295–301.
17. Cocolin L, Rantsiou K, Iacumin L, Urso R, Cantoni C, et al. (2004) Study of the ecology of fresh sausages and characterization of populations of lactic acid bacteria by molecular methods. Appl Environ Microbiol 70: 1883–1894.
18. Torriani S, Felis GE, Dellaglio F (2001) Differentiation of *Lactobacillus plantarum*, *L. pentosus*, and *L. paraplantarum* by *recA* gene sequence analysis and multiplex PCR assay with *recA* gene-derived primers. Appl Environ Microbiol 67: 3450–3454.
19. Versalovic J, Schneider M, de Bruijn FJ, Lupski JR (1994) Genomic fingerprinting of bacteria using repetitive sequence based PCR (rep-PCR). Method Mol Cell Biol 5: 25–40.
20. Dal Bello B, Rantsiou K, Zeppa G, Ambrosoli R, Cocolin L (2010) Microbial ecology of artisanal products from the Piedmont region (North West Italy) and antimicrobial activity of the autochthonous populations. LWT- Food Sci Technol 43: 1151–1159.
21. Muyzer G, De Waal ED, Uitterlinden AG (1993) Profiling of complex microbial populations by denaturing gradient gel electrophoresis analysis of polymerase chain reaction-amplified genes coding for 16S rRNA. Appl Environ Microbiol 59: 695–700.
22. Rantsiou K, Urso R, Dolci P, Comi G, Cocolin L (2008) Microflora of Feta cheese from four Greek manufacturers. Int J Food Microbiol 126: 36–42.
23. Webster G, Parkes RJ, Cragg BA, Newberry CJ, Weightman AJ, et al. (2006) Prokaryotic community composition and biogeochemical processes in deep subseafloor sediments from the Peru Margin. FEMS Microbiol Ecol 58: 65–85.
24. O'Sullivan LA, Webster G, Fry JC, Parkes RJ, Weightman AJ (2008) Modified linker-PCR primers facilitate complete sequencing of DGGE DNA fragments. J Microbiol Meth 75: 579–581.
25. Altschul SF, Madden TL, Shaffer AA, Zhang J, Zhang Z, et al. (1997) Gapped BLAST and PSI-BLAST: a new generation of protein database search programs. Nucleic Acids Res 25: 3389–3402.
26. Andreotti R, Pérez de León AA, Dowd SE, Guerrero FD, Bendele KG, et al. (2011) Assessment of bacterial diversity in the cattle tick *Rhipicephalus (Boophilus) microplus* through tag-encoded pyrosequencing. BMC Microbiol 11: 6.
27. Caporaso JG, Kuczynski J, Stombaugh J, Bittinger K, Bushman FD, et al. (2010) QIIME allows analysis of high-throughput community sequencing data. Nat Methods. 7: 335–336.
28. Quince C, Lanzen A, Davenport RJ, Turnbaugh PJ (2011) Removing noise from pyrosequenced amplicons. BMC Bioinformatics 12: 38.
29. Edgar RC, Haas BJ, Clemente JC, Quince C, Knight R (2011) UCHIME improves sensitivity and speed of chimera detection. Bioinformatics 27: 2194–2200.
30. Edgar RC (2010) Search and clustering orders of magnitude faster than BLAST. Bioinformatics 26: 2460–2461.
31. Dowd SE, Zaragoza J, Rodriguez JR, Oliver MJ, Payton PR (2005) Windows.NET Network Distributed Basic Local Alignment Search Toolkit (W.ND-BLAST). BMC Bioinformatics 6: 93.
32. Saeed AI, Sharov V, White J, Li J, Liang W, et al. (2003) TM4: a free, open-source system for microarray data management and analysis. Biotechniques 34: 374–378.
33. Chao A, Bunge J (2002) Estimating the number of species in a stochastic abundance model. Biometrics. 58: 531–539.
34. Shannon CE, Weaver W (1949) The mathematical theory of information. AT&T Technical J 27: 359–423.
35. Lozupone C, Knight R (2005) UniFrac: a new phylogenetic method for comparing microbial communities. Appl Environ Microbiol 71: 8228–8235.
36. Arroyo-López FN, Romero-Gil V, Bautista-Gallego J, Rodríguez-Gómez F, Jiménez-Díaz R, et al. (2012) Yeasts in table olive processing: desirable or spoilage microorganisms? Int J Food Microbiol 160: 42–49.
37. Botta C, Cocolin L (2012) Microbial dynamics and biodiversity in table olive fermentation: culture-dependent and -independent approaches. Front Microbiol 3: 245.
38. Hurtado A, Ben Othman N, Chammem N, Hamdi M, Ferrer S, et al. (2011) Characterization of *Lactobacillus* isolates from fermented olives and their bacteriocin gene profiles. Food Microbiol 28: 1514–1518.
39. Hurtado A, Reguant C, Bordons A, Rozès N (2010) Evaluation of a single and combined inoculation of a *Lactobacillus pentosus* starter for processing cv. Arbequina natural green olives. Food Microbiol 27: 731–740.
40. Chamkha M, Sayadi S, Bru V, Godon JJ (2008) Microbial diversity in Tunisian olive fermentation brine as evaluated by small subunit rRNA – single strand conformation polymorphism analysis. Int J Food Microbiol 122: 211–215.
41. Silva T, Reto M, Sol M, Peito A, Peres CM, et al. (2011) Characterization of yeasts from Portuguese brined olives, with a focus on their potentially probiotic behavior. LWT – Food Sci Technol 44: 1349–1354.
42. Bautista-Gallego J, Rodríguez-Gómez F, Barrio E, Querol A, Garrido-Fernández A, et al. (2011) Exploring the yeast biodiversity of green table olive industrial fermentations for technological applications. Int J Food Microbiol 147: 89–96.
43. Aponte M, Ventorino V, Blaiotta G, Volpe G, Farina V, et al. (2010) Study of green Sicilian table olive fermentations through microbiological, chemical and sensory analyses. Food Microbiol 27: 162–170.
44. Doulgeraki AI, Pramateftaki P, Argyri AA, Nychas GJE, Tassou CC, et al. (2013) Molecular characterization of lactic acid bacteria isolated from industrially fermented Greek table olives. LWT – Food Sci Technol 50: 353–356.
45. Ercolini D, Villani F, Aponte M, Mauriello G (2006) Fluorescence in situ hybridisation detection of *Lactobacillus plantarum* group on olives to be used in natural fermentations. Int J Food Microbiol 112: 291–296.
46. Abriouel H, Ben Omar N, Lucas R, Gálvez A (2011) Culture-independent study of the diversity of microbial populations in brines during fermentation of naturally-fermented Aloreña green table olives. Int J Food Microbiol 144: 487–496.
47. Randazzo CL, Ribbera A, Pitino I, Romeo FV, Caggia C (2012) Diversity of bacterial population of table olives assessed by PCR-DGGE analysis. Food Microbiol 32: 87–96.
48. Sanchez-Gomez AH, Garcia Garcia P, Navarro LR (2006) Trends in table olive production. Grasas y Aceites 57: 86–94.
49. De Bellis P, Valerio F, Sisto A, Lonigro SL, Lavermicocca P (2010) Probiotic table olives: microbial populations adhering on olive surface in fermentations sets inoculated with the probiotic strains lactobacillus paracasei IMPC2.1 in an industrial plant. Int J Food Microbiol 140: 6–13.
50. Arroyo-Lopez FN, Duran-Quintana MC, Ruiz-Barba JL, Querol A, Garrido-Fernandez A (2006) Use of molecular methods for the identification of yeast associated with table olives. Food Microbiol 23: 791–796.
51. Bevilacqua A, Perricone M, Cannarsi M, Corbo MR, Sinigaglia M (2009) Technological and spoiling characteristics of the yeast microflora isolated from Bella di Cerignola table olives. Int J Food Sci Technol 44: 2198–2207.
52. Hernández A, Martín A, Aranda E, Pérez-Nevado F, Córdoba MG (2007) Identification and characterization of yeast isolated from the elaboration of seasoned green table olives. Food Microbiol 24: 346–351.
53. Arroyo-López FN, Bautista-Gallego J, Rodríguez-Gómez F, Garrido-Fernández A (2010) Predictive microbiology and table olives. In: Méndez-Vilas A, ed. Current Research, Technology and Education Topics in Applied Microbiology and Microbial Biotechnology, Formatex Research Centre, Spain. 1452–1461.
54. Roth E, Schwenninger SM, Eugster-Meier E, Lacroix C (2011) Facultative anaerobic halophilic and alkaliphilic bacteria isolated from a natural smear ecosystem inhibit Listeria growth in early ripening stages. Int J Food Microbiol 147: 26–32.

14

A Metagenomic Approach to Characterization of the Vaginal Microbiome Signature in Pregnancy

Kjersti Aagaard[1,3]*, Kevin Riehle[3], Jun Ma[1,3], Nicola Segata[4], Toni-Ann Mistretta[2], Cristian Coarfa[3], Sabeen Raza[2], Sean Rosenbaum[1], Ignatia Van den Veyver[1], Aleksandar Milosavljevic[3], Dirk Gevers[5], Curtis Huttenhower[4], Joseph Petrosino[6], James Versalovic[2]

1 Department of Obstetrics & Gynecology, Division of Maternal-Fetal Medicine, Baylor College of Medicine and Texas Children's Hospital, Houston, Texas, United States of America, **2** Department of Pathology & Immunology, Baylor College of Medicine and Texas Children's Hospital, Houston, Texas, United States of America, **3** Bioinformatics Research Laboratory, Baylor College of Medicine and Texas Children's Hospital, Houston, Texas, United States of America, **4** Harvard School of Public Health, Harvard University, Cambridge, Massachusetts, United States of America, **5** Broad Institute, Massachusetts Institute of Technology, Cambridge, Massachusetts, United States of America, **6** Department of Molecular Virology and Microbiology, Baylor College of Medicine and Texas Children's Hospital, Houston, Texas, United States of America

Abstract

While current major national research efforts (*i.e.,* the NIH Human Microbiome Project) will enable comprehensive metagenomic characterization of the adult human microbiota, how and when these diverse microbial communities take up residence in the host and during reproductive life are unexplored at a population level. Because microbial abundance and diversity might differ in pregnancy, we sought to generate comparative metagenomic signatures across gestational age strata. DNA was isolated from the vagina (introitus, posterior fornix, midvagina) and the V5V3 region of bacterial 16S rRNA genes were sequenced (454FLX Titanium platform). Sixty-eight samples from 24 healthy gravidae (18 to 40 confirmed weeks) were compared with 301 non-pregnant controls (60 subjects). Generated sequence data were quality filtered, taxonomically binned, normalized, and organized by phylogeny and into operational taxonomic units (OTU); principal coordinates analysis (PCoA) of the resultant beta diversity measures were used for visualization and analysis in association with sample clinical metadata. Altogether, 1.4 gigabytes of data containing >2.5 million reads (averaging 6,837 sequences/sample of 493 nt in length) were generated for computational analyses. Although gravidae were not excluded by virtue of a posterior fornix pH >4.5 at the time of screening, unique vaginal microbiome signature encompassing several specific OTUs and higher-level clades was nevertheless observed and confirmed using a combination of phylogenetic, non-phylogenetic, supervised, and unsupervised approaches. Both overall diversity and richness were reduced in pregnancy, with dominance of Lactobacillus species (L. *iners crispatus*, *jensenii* and *johnsonii*, and the orders Lactobacillales (and *Lactobacillaceae* family), Clostridiales, Bacteroidales, and Actinomycetales. This intergroup comparison using rigorous standardized sampling protocols and analytical methodologies provides robust initial evidence that the vaginal microbial 16S rRNA gene catalogue uniquely differs in pregnancy, with variance of taxa across vaginal subsite and gestational age.

Editor: Adam J. Ratner, Columbia University, United States of America

Funding: This work was funded by the National Institutes of Health (NIH Director New Innovator Award (DP2120OD001500-01 K.A.), NICHD/NIDDK #R01DK080558-01 (K.A.), UH3 DK083990I (J.V.), NIH 1R01HG005969 (C.H.), and the NIH Common Fund Human Microbiome-HMP (K.A., J.V., J.P., C.H., (NIHU54HG004969 D.G.)), and the Burroughs Welcome Fund (Burroughs Welcome Fund Preterm Birth Initiative (K.A. and J.V.)). The funders had no role in study design, data collection and analysis, decision to publish, or preparation of the manuscript.

Competing Interests: The authors have declared that no competing interests exist.

* E-mail: aagaardt@bcm.edu

Introduction

To date, the dominant paradigm in Western medicine considers microbes as "foreign" and has led to the prevailing view that elimination of predominant pathogens will result in amelioration of disease. Such a view is seemingly in contrast to longstanding observations that humans serve as host to co-evolving microbes residing in highly plethoric communities. Indeed, microbiota are present from the time of birth, with up to 10-fold the number of microorganisms to adult human cells and a collective genome (the "metagenome") which exceeds our human genome in terms of gene content by more than 100-fold [1]. Moreover, we appreciate that the human microbiota are a metabolically and antigenically vibrant and diverse community which may function as *mutualists* (symbiotically beneficial), *commensals* (of neither harm nor benefit), or *pathogens* (of host detriment) [2–4].

Current major national research efforts (*i.e.,* the NIH Road Map initiative known as the Human Microbiome Project (HMP)) will enable sequence-based comprehensive characterization of the adult human microbiota and theoretically allow for cataloguing of the microbiota into core guilds, which can be thereafter interrogated for their associations with disease states [1]. Understanding the processes that govern the structure and dynamics of these human microbial communities is essential for gaining a complete understanding of human development and physiology [5–9]. However, questions pertaining to how and when diverse microbial communities reside in the host (and how they differ during an individual's lifetime) are under-explored at a population-wide level. In other words, while we may soon know *what* constitutes the adult human microbial guild, we will neither know *how* it is established nor whether it is dynamic during intervals in reproductive life when the next generation's microbial community is being established.

Primate fetal development is thought to occur within an intrauterine microbiota-free environment, and yet within a short interval following birth the human microbiome is colonizes and "differentiates" until the adult complement of 90 trillion or so microbiota is achieved [1,4,7,9]. Based on a relative paucity of data, it is proposed that the naïve neonatal microbiome is first established with rupture of the amniotic membranes, with further microbiota being introduced as the fetus traverses the vaginal birth canal. By the time of delivery, the neonate has been exposed to the maternal vaginal microbial ecosystem [9–12]. Passage through the vaginal canal is an integral part of this process, as mode of delivery alters the neonatal microbiome [7–12]. However, since a comprehensive characterization of the vaginal microbiome signature in pregnancy has not yet been undertaken, conclusions regarding mechanisms of neonatal colonization are likely premature [13–17]. Since the infant is exposed to several environmental sources of bacteria in the early neonatal interval (maternal vaginal canal and feces, swallowing and breathing, skin to skin contact, maternal breastmilk, etc.) it is important to discern the relative potential contribution of the maternal vaginal community to the neonate.

Established 16S ribosomal RNA (rRNA) gene sequence-based methodologies have enabled primary cataloguing of the bacterial composition of the human microbiome [18–21]. While multiple studies during the past several years have launched the era of human metagenomics, few reports have examined microbiomes outside of the gastrointestinal tract in more than a few individuals [22–26], and none have systematically examined the vaginal microbiome throughout pregnancy. Here, we use cultivation-independent, molecular-phylogenetic techniques to characterize the first comparative bacterial assemblages across gestational age strata and in a rigorous clinical study. By interrogating the "healthy" human microbiome in pregnancy and in paralleled comparison with non-pregnancy, we reasoned that the ensuing metagenomic profile would optimally reveal the comparative diversity and richness of microbial species. In this manuscript, we describe our intergroup comparison using rigorous standardized sampling protocols and analyses methodologies in order to provide robust initial evidence that the vaginal microbial gene catalogue uniquely differs in pregnancy, with variance of molecular phylogeny (species richness and diversity) across both subsite and gestational age.

Results

Subject Characteristics

Subject characteristics are as outlined in **Table 1**. As anticipated by our use of a parallel protocol design, comparable age, pregravid (or nongravid) BMI, race and ethnicity, and tobacco use were observed among both pregnant and non-pregnant subjects. Although the enrollment percentages by virtue of race and ethnicity were distinct in the two cohorts, these distinctions did not reach statistical significance ($p>0.05$, both independent samples t-test and ANOVA). As further anticipated among pregnant subjects *per se*, significant variance in medication use with respect to vitamins and antacids were similarly observed. Recalling prospective subject enrollment, the majority of gravidae had uncomplicated pregnancy outcomes: mean gestational age at delivery exceeded 39 weeks and included appropriately grown infants (mean 3265 grams), with 2/24 (8%) of subjects delivering <37 weeks (34 5/7, and 36 6/7 weeks). The cesarean delivery rate was consistent with the regional population (33%). A single fetal comorbidity was observed among the cohort after enrollment (fetal gastroschisis), and a total of 3 subjects manifest 4 comorbidites common to the obstetrical population (**Table 1**). Among the non-pregnant cohort, all subjects were not menstruating at the time of sampling per study protocol, and 58.8% would be anticipated to be anovulatory secondary to use of contraceptive (30 of 51 subjects).

Pregnancy Structures the Vaginal Microbial Community

Other investigators have employed cultivation-independent, molecular phylogenetic approaches to characterizing maternal microbial communities [9]. In order to leverage our data from the HMP, we developed a parallel sampling strategy in our 24 pregnant subjects with equivalent stringent screening criteria [27]. Both subject cohorts were sampled in a uniform and highly consistent manner by a single obstetrician (K.A.) from three distinct vaginal sites (vaginal introitus, posterior fornix, and midvagina); all samples employed in this report were extracted in a single laboratory. With sequencing on the 454 Titanium FLX platform, our approach yielded robust bacterial 16S V5V3 enriched data sets for subsequent analysis (**Table 2**). Because we could not reliably measure nor control for differences in the sampled area or volume, we focused our analyses within these microbial communities on shifts in bacterial community structure and diversity which occur solely by virtue of pregnancy.

As demonstrated in **Figure 1**, microbiome sequence surveys with 16S rRNA pyrosequencing reveal primary structuring by virtue of pregnancy (green versus blue). Given our aim to ultimately describe the vaginal taxa that contribute to a unique community structure in pregnancy, we applied both phylogenetic (UniFrac) and non-phylogenetic (Canberra, Chord, and Ochiai) methods. As a quality filtering step, each dataset was minimally (left plots, **Figure 1**) or modestly (right plots) preprocessed and applied to a flowgram clustering step for removal of chimeric sequences (which occur as a byproduct of the PCR-amplification and pyrosequencing) [28]. This robust analysis approach reveals evident significant community clustering structured predominately by pregnancy regardless of method (phylogenetic or non-phylogenetic) or potential noise for noisy or denoised datasets (**Figure 1**).

When beta diversity metrics were considered by virtue of the vaginal subsite, pregnancy persisted as the primary arbitrator of community microbial structure (**Figure 2A**). However, when assessing alpha diversity (community richness and Shannon diversity index) with respect to either proximity to the uterus (posterior fornix, which is just posterior to the cervix, versus the vaginal introitus or midvagina) or gestational age (interval in weeks), variable differentiation was observed (**Figure 2B**). When within community (alpha diversity) variance was analyzed with OTU-based methodologic approaches [29–33] the vaginal microbiome in pregnancy was equally rich but less diverse than non-pregnant communities (**Figure 2B**, left panel). This result was true in proximity to the cervix and throughout the latter midtrimester as the cusp of fetal viability (24–32 weeks) is reached. However, when taxa were analyzed with phylogenetic applications uniformly projected to the genus level, vaginal introitus and midvaginal communities preterm were both less diverse and less rich than either late preterm (>32 weeks) pregnancy or non-pregnant communities. At the posterior fornix, phylogenetic projections suggested unchanged communities among pregnant and non-pregnant (**Figure 2B**, right panel).

Given our distinct observations with respect to intracommunity richness and diversity by virtue of taxonomic and phylogenetic

Table 1. Subject Characteristics and Pregnancy Outcomes.

Subject Demographics	**Pregnant n = 24**	**Non-Pregnant n = 60**
Subject age (mean, years)	31.4 (5.8)	26.9
BMI (kg/m2) at sampling	30.4 (7.3)	23.9
BMI (kg/m2) prepregnancy	27.6 (7.6)	NA
Ethnicity/Race		
Hispanic	4 (16%)	11 (18%)
Non-Hispanic	15 (63%)	35 (58%)
Black or African American	3 (13%)	3 (5%)
Other	2 (8%)	11 (18%)
Tobacco Use		
Yes	2	4
No	22	56
Medication (Category)		
Vitamins or supplements	24	7
Endocrine metabolic agents	1	7
Antacids/H2 antagonists	11	2
GI medication (Antiemetics and stool softeners)	3	2
Pregnancy Outcomes		
Mean gestational age (weeks)	39 weeks 2 days	
Preterm birth <37 weeks (rate)	8% (34 5/7, 36 6/7 weeks)	
Mean birthweight (grams)	3265 g	
Mean Apgar score		
1 minute	8	
5 minute	9	
Cesarean delivery (rate)	33%	
Vaginal delivery (rate)	67%	
Comorbidities		
(3 subjects with 4 comorbidities)	GDMA1, fetal gastroschisis, mild preeclampsia, severe preeclampsia	

Characteristics of both pregnant (gravidae) and non-pregnant subject cohorts, alongside pregnancy outcomes of the gravid cohort. GDMA1, gestational diabetes mellitus White classification A1 (diet controlled). There were no significant differences with respect to Race/Ethnicity among gravidae and non-pregnant subjects ($p>0.05$ by independent samples t-test and ANOVA).

approaches, we were concerned with the potential for inflated OTU counts due to sequencing artifacts and/or binning errors. We therefore measured alpha diversity in both pregnant and non-pregnant vaginal communities with AbundantOTU (**Figure 3**, left panels) and QIIME denoising (in addition to chimera and singleton removal) (**Figure 3**, right panels). Although these denoising pipelines did, in fact, differ in their assessments of sequencing noise (and thus resulting OTU counts), both resulted in significant and robust differences among the pregnant (black lines) and non-pregnant (red lines) vaginal microbial communities. They also agreed in that community diversity in pregnancy is both significantly less rich (rarefaction metrics for microbial richness or taxa quantification, **Figure 3A**) [34] and less diverse (Renyi metrics for microbial diversity or number of distinct taxa, **Figure 3B**) [34–37]. Taken together, these data are consistent with the notion that community aggregates to genus level projections (**Figure 2B**, right panel) are unchanged in the posterior fornix, but the more robust analysis of genus and sub-

Table 2. V5V3 Sequence Metrics.

Subject cohort	Number of Subjects	Number of Samples	Total Sequences	Average Sequence Length	Average Sequences/Sample
Pregnant	24	68	670,921	498 nt	9,867
Non-pregnant	60	301	1,852,039	491 nt	6,153
Combined	84	369	2,522,960	493 nt	6,837

Sequence metrics for the pregnant (gravidae), non-pregnant, and combined cohorts. Pregnant subject's samples were of comparable average sequence length, but retained a higher average number of sequence reads per sample.

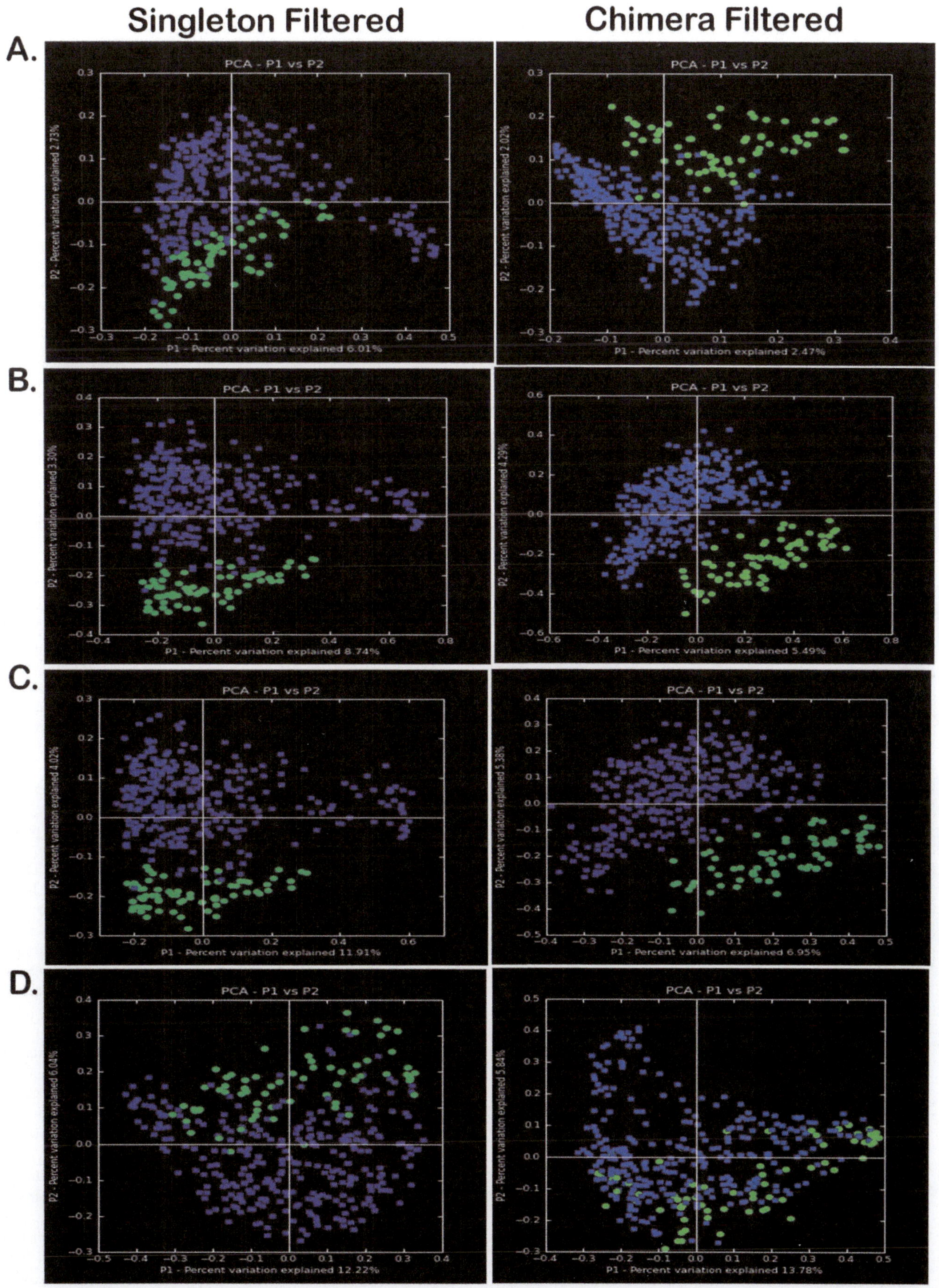
Singleton Filtered
Chimera Filtered
A.
B.
C.
D.
PCA - P1 vs P2
P1 - Percent variation explained 6.01%
P2 - Percent variation explained 2.73%
P1 - Percent variation explained 2.47%
P2 - Percent variation explained 2.02%
P1 - Percent variation explained 8.74%
P2 - Percent variation explained 3.30%
P1 - Percent variation explained 5.49%
P2 - Percent variation explained 4.29%
P1 - Percent variation explained 11.91%
P2 - Percent variation explained 4.02%
P1 - Percent variation explained 6.95%
P2 - Percent variation explained 5.38%
P1 - Percent variation explained 12.22%
P2 - Percent variation explained 6.04%
P1 - Percent variation explained 13.78%
P2 - Percent variation explained 5.84%

Figure 1. Beta diversity metrics of bacterial 16S rRNA genes reveal distinctly clustered vaginal microbiome communities structured by pregnancy. Datasets were minimally filtered for removal of singletons (left panels) or filtered for chimeras (right panels; QIIME ChimeraSlayer). Beta diversity microbiome community clustering is observed for non-phylogenetic methods ((A) normalized Canberra), binary non-phylogenetic methods ((B) binary Chord, (C) binary Ochiai), and phylogenetic beta diversity metrics ((D) unweighted UniFrac). In each panel, each point corresponds to a vaginal sample from either a pregnant (green) or non-pregnant (blue) subject. The percentage of variation explained by the plotted principal coordinates is indicated on the axes.

genus OTU based projections reveal significantly less diverse and less rich community structure in pregnancy (**Figure 3**).

Taxa Contributing to the Unique Vaginal Microbiome in Pregnancy

Given the robust evidence that the structure of the vaginal microbiome significantly differs in pregnancy (**Figures 1** and **2**), and further evidence suggesting that the pregnant microbial community is less diverse and rich (**Figure 3**), we next sought to identify which bacterial taxa were contributing to an altered community in pregnancy. OTU tables and representative sequences generated from AbundantOTU (**Figure 4A**) and QIIME denoised (**Figure 4B**) datasets were employed in generating global phylogenetic trees. Additionally, the RDP classifier was used to assign taxonomy and label each element in the OTU table, and the resultant alignment enabled character-based construction of a phylogenetic tree from the filtered NAST alignment using FastTree [38]. Clear family (internal cluster with annotations in figure legend) and order level (middle circle) differences in the pregnancy microbiome are visible in both datasets (**Figure 4**), with an overall predominance of the order Lactobacillales (and *Lactobacillaceae* family), followed by Clostridiales, Bacteroidales, and Actinomycetales. Distinctions between pregnant and non-pregnant communities by unweighted phylogenetic clustering are observed and annotated in the outermost circle (**Figure 4**).

Finally, to identify the specific taxa differentially present or abundant in the vaginal microbiome in pregnancy, we employed two complementary supervised machine learning approaches. The first, the random forest algorithm paired with Boruta feature selection, was validated on classification of subjects by their vaginal microbiome with a pi statistic (improvement over random) exceeding 0.8 at OTU abundance thresholds of 100 or 500 for the AbundantOTU data set (**Figures S1** and **S2**, **Tables S1** and **S2**) and likewise above 0.6 for the QIIME denoised data set (**Figure S1, Tables S1** and **S2**). Pregnancy status was, in fact, the single feature best predicted by the microbiome (as opposed to BMI, vaginal site of sampling, and ethnicity), and sub-sites were not well distinguished (pi statistic of discrimination <0.2 in all cases; **Tables S1** and **S2**, **Figure S3** with binned taxonomy). Using Boruta feature selection during RF analysis, 12 specific taxa were identified as discriminating between pregnant and non-pregnant cohorts at the genus level (at $\geq 80\%$ bootstrap cutoff, **Figure S1**). In order to identify a taxonomic biomarker with high stringency, we additionally employed the LEfSe method [39], which confirmed the differential abundance of 29 and 27 clades (AbundantOTU and QIIME data, respectively, using default significance and LDA thresholds) at all taxonomic levels between pregnant and non-pregnant microbiomes (**Figure 5**). Supporting our findings in Figures 3 and 4, these comprised primarily specific *Lactobacillus* OTUs and potentially the Bifidobacteriaceae and Streptococcaceae. Our findings suggest that the composition of the microbial community in pregnancy represents a relatively diminished profile of species richness and diversity (**Figures 2, 3, 4, 5**), culminating in the specific reduction of many usual vaginal community members and enrichment for a targeted set of *Lactobacillus* species (**Figure 5**). We present the species-level data in Table S3. Briefly, OTUs generated from AbundantOTU and QIIME were aligned to the Greengenes database with the top OTUs identified by LEfSe (**Figure 5A**). OTU sequences selected by Boruta feature selection are deposited in GenBank and are presented in Figure S2 and Table S3. Top OTUs identified by LEfSe are marked with an asterics. For example, AbundantOTU generated consensus 2/33 aligned to *Lactobacillus iners*, consensus 102/84/70/98/67/44/29/46 to *Lactobacillus crispatus*, and consensus 28 to *Lactobacillus jensenii*. On QIIME, 882 aligned to *Lactobacillus johnsonii*, and 322 to *Lactobacillus crispatus*.

Discussion

With a robust sampling, sequencing, and analysis approach, we generated the first comprehensive catalogue of the vaginal microbiome in pregnancy across subsite and gestational age. When compared to the non-pregnant vaginal microbiota, the community is uniquely and distinctly structured during pregnancy (**Figure 1** and **2A**), in ways that cannot be attributed to alterations in BMI (**Tables S1** and **S2**), to subject race or ethnicity, nor to readily identifiable clinical confounders. Interrogations of discrete contributors to community diversity revealed that the vaginal microbial community varied in pregnancy by gestational age and proximity to the cervix, but was less diverse and less rich overall (**Figures 2B** and **3**). To our knowledge, this structured molecular study of gravidae is unique in terms of stringency of a parallel clinical approach, sample acquisition from subjects, depth and robustness of analysis, and notable findings. In sum, the vaginal microbiome is distinctly structured by a state of health in most women's lifetime, *i.e.*, pregnancy.

Others have taken similar but limited approaches to interrogating the vaginal microbiota in pregnancy. Dominguez-Bello utilized 16S 454-generated molecular signatures to generate vaginal profiles in a limited sample set of 9 subjects at term with delivery (including non-laboring and active laboring mothers) from a remote population of Amerindians [9]. In this small sample set, the dominant vaginal taxa varied from mother to mother, also with notable variance in *Lactobacillus* spp. However, these investigations did not include parallel sampling of both non-pregnant and pregnant subjects, nor from multiple vaginal subsites [9]. However, this study was remarkable for its parallel acquisition of neonatal microbial community sampling. As supported by other studies [4–12], the infant gut microbiome largely reflects the maternal mode of delivery, although it bears mention that in several studies women were delivered by cesarean for obstetrical indications in active and advanced labor thereby revealing a potential bias by virtue of infant handling in cesarean and vaginal birth and not solely a reflection of fetal descent via the birth canal [9–12].

Our study suggests that although human adults have highly differentiated bacterial communities that are relatively stable [3–6,13–17], in such prevalent and healthful states as pregnancy the vaginal community in particular shifts naturally in its structure with respect to diversity and richness. Indeed, Ravel *et al* have previously reported that the vaginal microbiome in healthy, reproductive-aged women occupies states dominated by *Lactoba-*

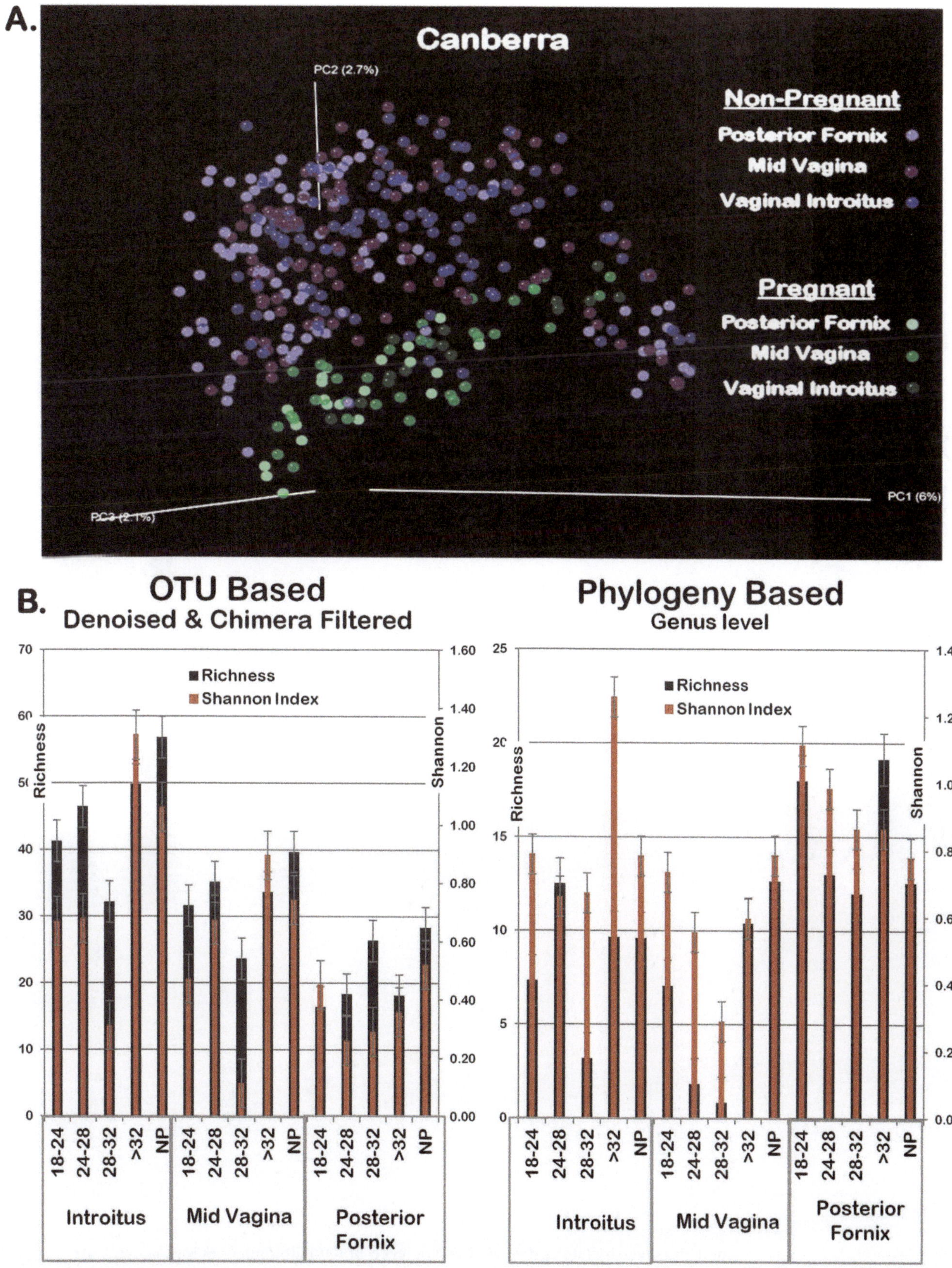

Figure 2. Subclassification of microbial community structure by vaginal subsite and week of gestation. (A) Pregnancy clusters vaginal microbial communities, while site of vaginal sampling minimally contributes to within cluster formation. Canberra beta diversity metric with PCoA plot clustering. Each dots represents one sample from the distinct vaginal subsites (mid vagina, posterior fornix, and vaginal introitus) of individual

subjects from pregnant (green shades) and non-pregnant (blue and purple shades). (B) Among gravid subjects, microbial community richness and diversity (Shannon indices) vary by week of gestation and proximity to the uterus. Community richness and Shannon diversity indices by gestational age and vaginal sampling site against normalized abundance values from both OTU and phylogeny based analysis charted by vaginal site (posterior fornix, mid vagina, introitus) and gestational age. Richness - Black; Diversity – Dark Red; Left panel designates OTU based; Right panel designates Phylogeny based. Gestational age interval shown in weeks, or designated as non-pregnant (NP). Error bars denote variance (standard error of the mean, s.e.m.). In each of the gestational age intervals, an equivalent number of gravidae were sampled and compared (n = 6 per strata).

cillus iners and *Lactobacillus crispatus* specifically in association with low vaginal pH [40]. Others speculated [41] that vaginal community changes with pregnancy, but ours is the first direct evidence as such. Indeed, our findings suggest that at least among reproductive aged women, the vaginal microbiome remains a dynamic community in adult reproductive life, and that terminal differentiation does not occur *per se* [41]. Moreover, we observe persistent relative prevalence (but not sole nor absolute predominance) of *Lactobacillus* (**Figure 4**, AbundantOTU). However, across the entirety of our study population, less diversity and richness occurred in measured variance throughout weeks of gestation and in proximity to the uterus (posterior fornix), leading us to speculate on variances within the cusp of preterm viability. Of interest, in subjects closer to term OTU-based projections suggest that the non-pregnant community structure may return to some extent in the latter weeks of gestation. Our study is potentially limited by employing a cross-sectional comparison in gravidae (**Figure 2**), relative to a limited number of non-pregnant subjects with multiple samplings (18/301 specimens represented thrice sampling, see methods). Alternatively, when we compared only first samplings of non-pregnant women to gravidae, we still observed consistent cluster separation (data not shown). The most

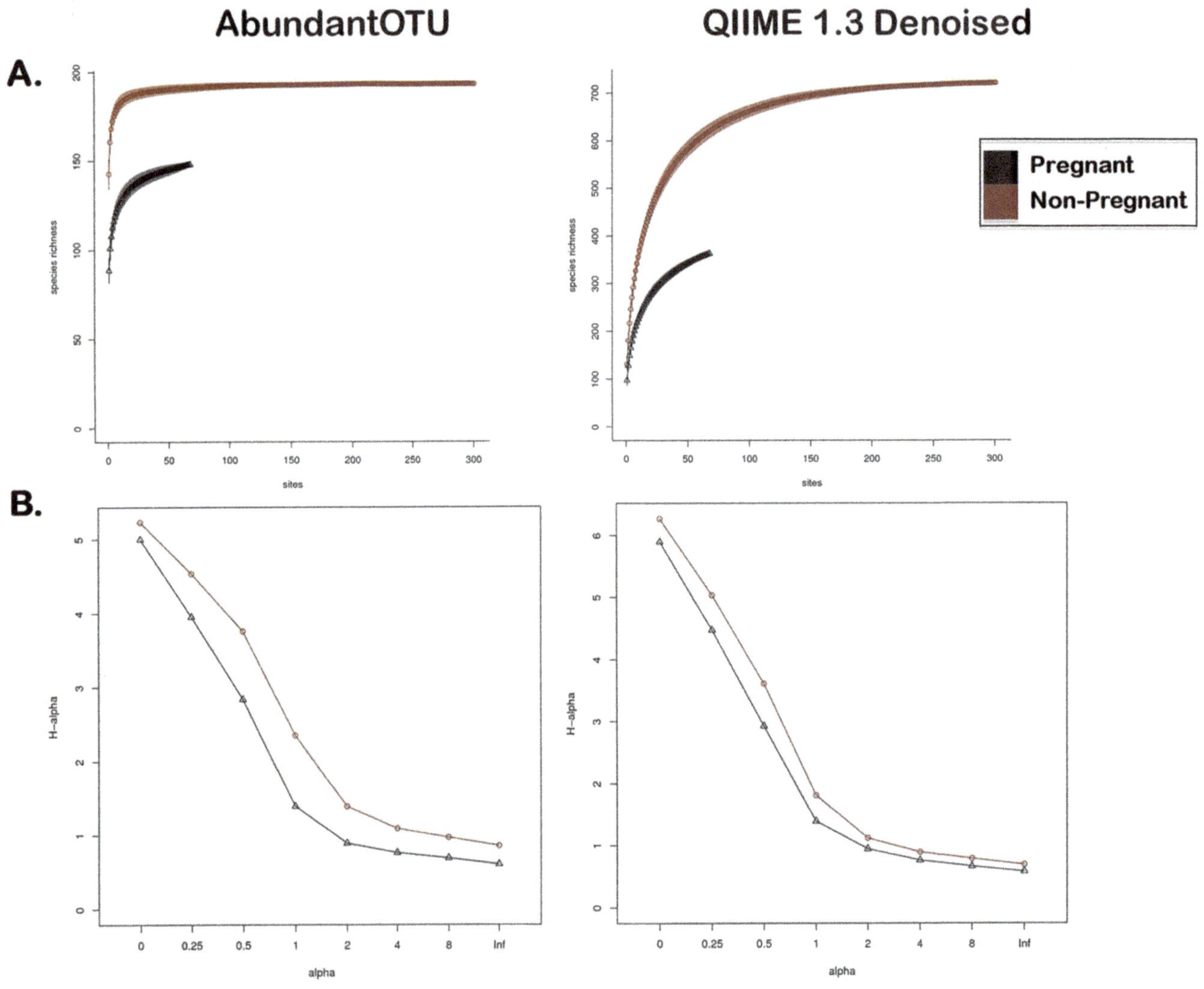

Figure 3. Measures of within community diversity (alpha diversity) at two levels of data filtering. Black lines indicate pregnant cohort, with red lines indicating non-pregnant cohort. Data sets were subjected AbundantOTU (left panels), or denoising and chimera slaying with removal of singletons and chimeras (right panels). (A) Rarefaction alpha diversity metrics note significantly lower richness in the pregnant data set, while (B) Renyi alpha diversity metrics indicates significantly less diversity among pregnant vaginal communities following denoising. Significance is denoted by the absence of curves crossing over at any point following denoising and chimera slaying (right panels).

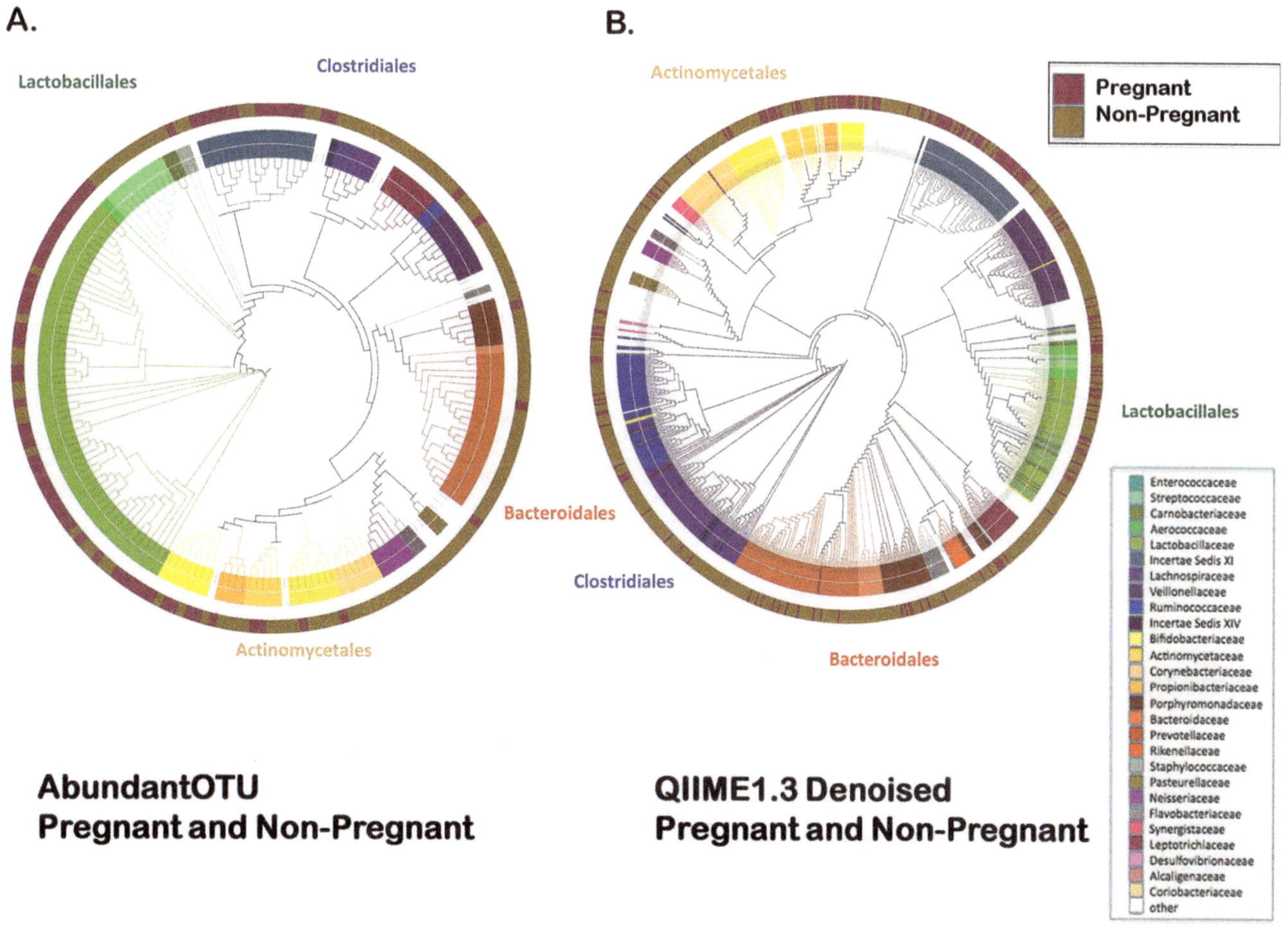

Figure 4. Global phylogenetic trees after AbundantOTU and QIIME denoising. Global phylogenetic trees show the distribution of taxonomy among all the pregnant and non-pregnant subject samples. The internal cluster dendrograms are colored by taxa Family level projections (annotated in figure legend), while the mid-circle is colored by the majority origins of OTUs from pregnant or non-pregnant subject samples (pregnant-magenta; non-pregnant-brown). Outermost circle using text to indicates OTU projection to Order level (Bacteroidales-red, Actinomycetales-yellow, Lactobacillales-green, Clostridiales-blue). OTU tables and representative sequences generated from AbundantOTU (**Figure 4A**) and QIIME denoised (**Figure 4B**) datasets were employed in generating these global phylogenetic trees.

robust method to formally address the structure shifts in pregnancy would be to employ a longitudinal approach whereby each subject is sampled at ongoing weekly intervals across pregnancy. However, that is outside the scope of this initial study.

We opted to employ a parallel sampling strategy by stringent inclusion and exclusion criteria to the Human Microbiome Project. While this enabled us to make true comparisons to a large, robust, and unparalleled dataset of non-pregnant subjects, it similarly opened the possibility that we were sampling an unperturbed but not "normal" population. However, it bears mention that our outcomes among gravidae were entirely what might be anticipated in a health pregnant population (**Table 2**), and did not differ significantly among subjects. As with all large human cohorts, our study is prone to both alpha error and induced bias. We attempted to minimize error and bias with a single physician performing all subject sampling among both cohorts, and all samples being extracted from primary specimens within a single laboratory utilizing a common and rigorously tested protocol (HMP).

It remains a distinct possibility that our significant observed community clustering with the vaginal microbiome being evidentially structured by pregnancy reflects a secondary trait in our pregnant population, but not gravid condition itself. Of note, we did not exclude gravid subjects by virtue of posterior fornix vaginal pH. In contrast, in the non-gravid HMP cohort subjects were excluded at the time of screening if the posterior fornix pH exceeded (see Methods). Given that <10% of screen failures (and <4% of the entire potential cohort) met such pH criteria for exclusion, we feel that this is an exceedingly unlikely potential confounder or bias. However, it cannot be formally excluded as such. To this end, in the recent publication of Ravel *et al.*, [40], the authors reported that while the pH and Nugent scores of each community demonstrated strong correlation between high pH and high Nugent scores and the highest pH values were associated with community states not dominated by species of Lactobacillus. It is of importance to note that these investigators employed self-sampling in their study. Nevertheless, the investigators also reported that elevated pH and high Nugent scores were observed in some communities with high proportions of lactobacillus species and that this was most true in communities which contained decreasing proportions of *L. iners*. However, this was not universally true, leading the authors to summize that these metrics cannot be predicted with absolute certainty solely on the basis of the proportion of Lactobacillus in

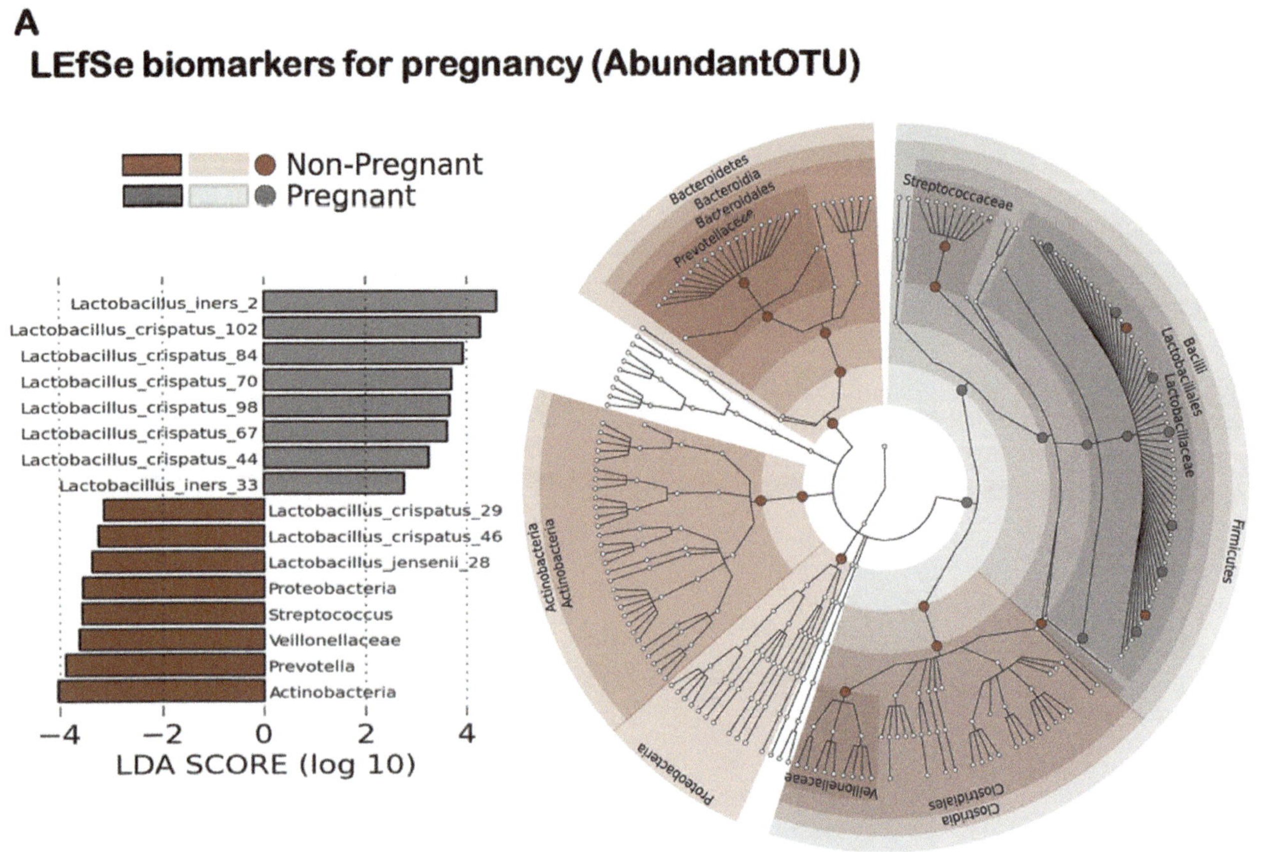
A
LEfSe biomarkers for pregnancy (AbundantOTU)
Non-Pregnant
Pregnant
Lactobacillus_iners_2
Lactobacillus_crispatus_102
Lactobacillus_crispatus_84
Lactobacillus_crispatus_70
Lactobacillus_crispatus_98
Lactobacillus_crispatus_67
Lactobacillus_crispatus_44
Lactobacillus_iners_33
Lactobacillus_crispatus_29
Lactobacillus_crispatus_46
Lactobacillus_jensenii_28
Proteobacteria
Streptococcus
Veillonellaceae
Prevotella
Actinobacteria
−4
−2
0
2
4
LDA SCORE (log 10)
Bacteroidetes
Bacteroidia
Bacteroidales
Prevotellaceae
Streptococcaceae
Bacilli
Lactobacillales
Lactobacillaceae
Firmicutes
Actinobacteria
Actinobacteria
Proteobacteria
Veillonellaceae
Clostridiales
Clostridia

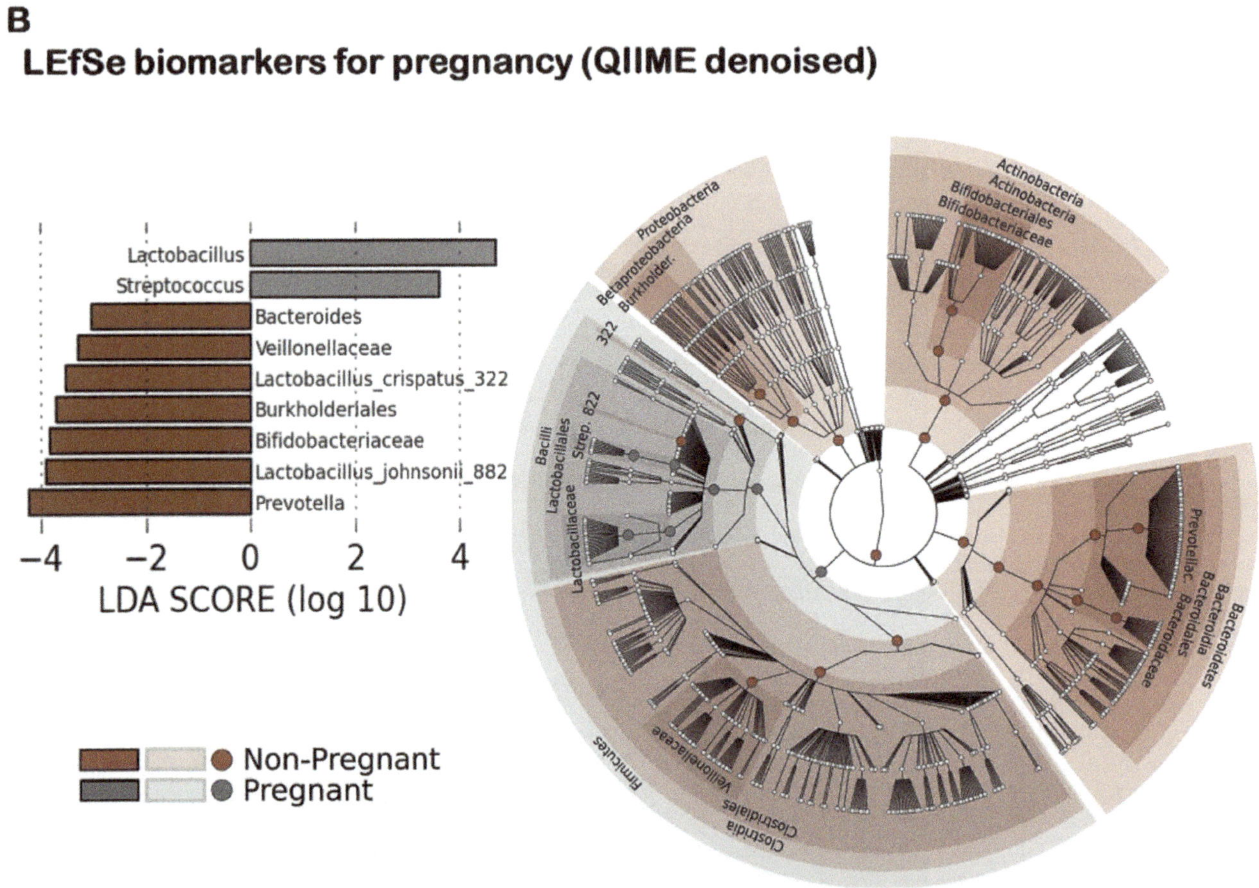
B
LEfSe biomarkers for pregnancy (QIIME denoised)
Lactobacillus
Streptococcus
Bacteroides
Veillonellaceae
Lactobacillus_crispatus_322
Burkholderiales
Bifidobacteriaceae
Lactobacillus_johnsonii_882
Prevotella
−4
−2
0
2
4
LDA SCORE (log 10)
Non-Pregnant
Pregnant
Proteobacteria
Betaproteobacteria
Betaproteobacteria
Burkholder.
Actinobacteria
Actinobacteria
Bifidobacteriales
Bifidobacteriaceae
322
Bacilli
Lactobacillales
Strep.
822
Lactobacillaceae
Prevotellac.
Bacteroidetes
Bacteroidia
Bacteroidales
Bacteroidaceae
Firmicutes
Veillonellaceae
Clostridiales
Clostridia

Figure 5. Metagenomics-based discovery of bacterial taxa contributing to differentiation of vaginal communities in pregnancy. Bacterial taxa were selected as significantly differentially abundant between pregnant and non-pregnant communities (regardless of sampling site) by the LDA Effect Size (LEfSe) algorithm, (left panel) sorted by degree of difference (listing only taxonomic leaves) and (right panel) overlayed on a complete taxonomy. Taxa are again reduced in diversity during pregnancy, with several specific *Lactobacillus* OTUs detected uniquely among pregnant individuals. OTU tables and representative sequences generated from AbundantOTU (**Figure 5A**) and QIIME denoised (**Figure 5B**) datasets were employed in LEfSe analyses.

a community [40]. We concur with these investigators summary statements, and note that when comparing our population of pregnant subjects to the non-pregnant HMP cohort we observed community discrimination by virtue of Lactobacillus species, namely *L. iners*, *L. crispatus*, *L. jensenii* and *L. johnsonii*. If our community distinctions were the result of an incidental inclusion of pregnanc subjects with a posterior fornix pH >4.5, then we would anticipate potentially seeing a decreasing proportion L. iners in the community structure. However, the opposite holds true (**Figure 5**) making confounding by version of pH unlikely.

As the number and robustness of computational approaches to analysis of metagenomics data increases, investigators are faced with distinct methodologic approaches to analyzing community profiles. In any emerging field of study, optimal measures of data analysis are not evident to investigators at the forefront and different methodologic approaches may yield variance in significance of findings [28,42–46]. With this in mind, we employed a diverse and robust set of bioinformatic tools in analysis of our datasets (see Methods). For community cluster distinction (beta metrics), we analyzed taxa by nonphylogenetic and phylogenetic methods. Regardless of distance metric or phylogenetic analysis, the vaginal microbiome distinctly clustered by virtue of pregnancy (**Figures 1, 2A,** and **4**). As AbundantOTU uses a consensus alignment algorithm, thus tending to concentrate on OTUs of greater abundance. Detection of rare species is difficult to differentiate from sequencing error. QIIME denoiser preempts this difficulty with a pre-filter to reduce the needs of all-on-all comparison; each additional unclustered read is compared to the most abundant clusters to discern sequencing error probability from detection of rare bacterial species retained in the OTU table. Regardless of methodology, our results from AbundantOTU and QIIME denoising are strikingly similar in terms of differentiating OTU identified (**Figure 4**). This finding is further evident at the species level, as detailed in **Figure 5**.

Similarly, for measurement of within community diversity (alpha) we employed variations in data filtering ranging from minimal removal with scant trimmed reads to well-described modest "denoising" and slaying of chimeras. We persistently observed less richness and diversity in pregnant communities when compared with parallel non-pregnant subject cohorts, regardless of computational pipeline, tool employed, or means of data projection (**Figures 2B** and **3**). The limited results analyses at the genus level, and the number of significantly differential OTUs (**Figure 5**), both suggest that it is subgenus taxa that most strongly contribute to observed alterations in community structure in pregnancy. This was supported by two complementary rigorous denoising approaches to our dataset. AbundantOTU and QIIME. The former resulted in diminished OTU estimates (868 versus 1,121 OTUs), but both agreed in the predominance of lactobacilli (**Figure 4**) irrespective of gravid condition and clades differential during pregnancy (**Figure 5**). With QIIME denoising, a broader set of taxon differences and less absolute predominance of *Lactobacillus* could be observed. While each denoising pipeline has its own strengths and limitations, one undeniable observation persists: the vaginal microbiome community is structured by pregnancy and varies with respect to richness, diversity, and specific microbial members.

Employing such robust analyses methods, we were able to detect species which are discriminately and specifically relative enriched in pregnancy (albeit in the face of overall diminished community richness and diversity). These include *Lactobacillus iners*, *Lactobacillus crispatus*, *Lactobacillus jensenii* and *Lactobacillus johnsonii.* Although it is outside the scope of this initial manuscript to delve deeply into the species differentiation and clinical implications, these findings are of probable biologic significance nevertheless. For example, *L. johnsonii* encodes enzymes and transporters essential for the release bile salt hydrolase and is primarily found in the upper GI tract [47]. In addition, the capacity for production of bacteriocins is a broad trait of the lactic acid bacteria and *L. johnsonii* production of Lactacin F both limits other lactobacillus as well as Enterococcus species in the GI tract. It's notable increased dominance in the vagina in pregnancy may be important for establishing the neonatal upper GI microbiota upon delivery, or preserving the integrity of the community to reduce risk of ascending infection or preterm birth.

The vaginal microbiome signature in pregnancy is thus distinct from non-pregnant, and this distinction comprises both from lesser diversity and, to a lesser degree, from the absence and occasionally presence of unique taxa. Our reporting by gestational age and vaginal subsite now lays the foundations for further interrogations into microbial variance, including such presumed pathogen-related perinatal morbidities as preterm birth. Moreover, it lends to the growing understanding of the remarkable dynamic nature of our metagenome and its role in vertical transmission of the microbiota through subsequent generations.

Materials and Methods

Subjects

The intent of this study was to compare the vaginal microbiome in pregnancy from healthy individuals whose core microbiomes were likely to be minimally perturbed by virtue of infectious comorbidities and exogenous exposures. We took advantage of our role as clinical investigators with the Human Microbiome Project (K.A., J.P., and J.V., Baylor College of Medicine) [27] and employed parallel recruitment of non-gravid (non-pregnant) and gravid subjects using rigorous standardized sampling protocols. Subjects were recruited from the general population with general media and institutional advertisements, in addition to institutional study enrollment web sites and approach during previously scheduled clinical appointments. An initial telephone query included a general health questionnaire to screen interested individuals who would be evaluated by a list of inclusion/exclusion criteria including a pre-pregnancy or current body mass index range of 18–35, history of cancer, compromised immune status, history of specified chronic diseases, or medication exposure within the last six months (*e.g.*, antibiotics, corticosteroids, cytokines, large doses of probiotics, etc.). Major dietary changes and history of moderate-high alcohol intake excluded individuals. Medications and dietary components that potentially might alter the human microbiome intentionally such as antibiotics and probiotics served as additional important exclusion criteria.

Females were required to have a regular 21–35 day menstrual cycle or a history of regular cycles prior to beginning contraception or at the time of conception, and the use of specific contraceptives such as the combination hormone vaginal ring was exclusionary in non-gravid subjects. For gravidae, inclusion criteria additionally detailed best obstetrical dating for gestational age, defined as reliable last menstrual period consistent with <12 week sonogram or <10 week sonographic dating with unreliable last menstrual period. At the time of enrollment, all gravid subjects had presumed normal singleton gestations and were without known maternal comorbidities (such as type II DM, gestational diabetes type A2, hypertensive disorders, or chronic medical conditions). However, given the prospective nature of the study with early second trimester enrollment, some subjects developed these comorbidities (**Table 1**). Individuals who remained interested in the study and met study criteria were invited for a screening visit in the research clinic (non-gravid), or approached at their next scheduled clinic visit (gravid).

At the screening visit, the study and associated risks were explained and subjects documented informed consent by signing a study consent form. Investigators determined final study eligibility by taking a medical history and performing a review of systems, documenting concomitant medications, measuring vital signs, collecting urine for pregnancy testing (non-gravid) and either obtaining blood for serum HIV, HBV and HCV testing or reviewing previously obtained serologies (all required to be negative). In addition, all subjects underwent targeted physical exams with attention to body site-specific exclusion criteria (oral cavity, skin and nasal cavity), and among non-gravid subjects measuring vaginal pH in females (pH>4.5 at posterior fornix was exclusionary). Gravid subjects did not undergo vaginal pH measurements as per study protocol limitations but were queried as to signs and symptoms of bacterial vaginosis and excluded if reported. Individuals who passed all screening criteria were eligible for enrollment and were sampled the first time within 30 days of the screening visit (non-gravid), or at the time of the prenatal visit (gravid). Although exclusion of non-gravidae for posterior fornix pH>4.5 at the time of screening was exceedingly rare (notably with only 22 of 254 screen failures in the HMP occurring by virtue of vaginal irritation or posterior fornix pH>4.5; 554 potential subjects were screened to ultimately enroll 300 in total for the HMP [http://www.hmpdacc.org/micro_analysis/microbiome_sampling.php]), it remains a formal possibility that our cohort comparison may be biased as posterior fornix pH was not an exclusionary criteria in our pregnant cohort.

For the non-gravid (non-pregnant) comparison cohort, of the 150 females from the HMP JumpStart initiative [18,19] all individuals were designated for study enrollment *a priori*. Enrollment criteria were established at approximately 20% minority (racial and ethnic) subjects. While in the index study protocol, two ethnically, racially, and socieconomically diverse U.S. cities (Houston, Texas and St. Louis, Missouri) served as geographically distinct study sites for clinical sampling for this comparison. However, only subjects sampled at Baylor College of Medicine were considered in the comparison cohort. This served to limit potential additional variables, including personnel doing the sampling (all gravid and non-gravid subjects were sampled by one investigator [K.A.]), site of microbial DNA extraction, and regional variability. To assure compliance of screening and sampling measures, a common set of human study protocols and consent forms were created, reviewed independently and subsequently approved at Baylor College of Medicine. An age range of 18–40 years was established to minimize variability due to childhood growth and development, aging, and hormonal influences during adolescence and menopause. Each of the enrolled non-gravid subjects agreed to primary and repeat (second) sampling so as many individuals as possible were sampled twice. A subset of non-gravid subjects was designated to undergo three samplings at each body site, at *a priori* designated intervals.

Subject Compliance/Protection and Informed Consent

Specimen collection procedures involved minimal physical risk to subjects. As defined in 45 US Code of Federal Regulations (CFR) 46.102 (i), "Minimal risk" infers that the "probability and magnitude of harm or discomfort anticipated in the research are not greater in and of themselves than those ordinarily encountered in daily life or during the performance of routine physical or psychological examinations or tests." Potential risks of participation as discussed with subjects included those associated with biologic sample collection and repository, and the unintentional release of protected health information; the protocol and informed consent form described precautions taken to reduce these risks. The study sites developed informed consent documents using NHGRI guidelines for genomics studies, which address the ethical, legal and social implications of such research. The Institutional Review Board reviewed and approved the protocol, informed consent and other study documents. Additional protections for subjects included Certificates of Confidentiality intended to protect against the forced disclosure of identifiable research data, coding of genomic and metagenomic specimens and sequence data, and use of controlled access databases for medical data and human genome sequence data. As part of the consent process, we informed subjects about data protection, including coding, use of controlled-access databases for human genomic data and controlled-access repositories for extracted metagenomic nucleic acids. If a subject withdrew consent after providing specimens, remaining specimens and extracted nucleic acids were to be destroyed; however, any metagenomic sequence data that was already published in open access databases could not be retracted.

Clinical Metadata Collection

EMMES and the Data Analysis and Coordination Center for the Human Microbiome Project (www.hmpdacc.org) established an internet data entry system for investigators to enter coded non-gravid subject information; gravid subject information was entered and maintained on a local database with secured protection but not released to the DACC. Clinical data elements included gender, race, ethnicity, age, place of birth, occupation, body mass index, vital signs, vaginal pH (non-gravid), date of last menstrual period and first sonogram, tobacco use, and both dental and health insurance status. A medical history, comprehensive obstetrical data, and targeted physical examination findings alongside medication history were also recorded for each subject.

Specialized Reagents and Instrumentation for Human Sampling

The Catch-All swab from Epicentre Technologies was selected as the swab of choice for sampling based on preliminary evaluations done in conjunction with HMP. The Catch-All swab was used for collection of all vaginal samples. Prior to vaginal sampling in non-gravid subjects, a digital display pH meter with accuracy to >0.01 was employed for precision measure of vaginal pH (Waterproof BigDisplay pH Spear, Oakton pH meter, Vernon Hills, IL.).

Subject Sampling

Inclusion criteria included stringent requirements as otherwise detailed [27]. Healthy, young adults (ages 18–40) who retained the ability to provide informed consent and were willing and available to provide samples during the study interval were targeted for enrollment. In non-gravid subjects, specimen types included anterior nares, oral cavity (9 samples), peripheral blood, skin (4 samples), stool, and vaginal (3 samples per female) samples [27]; in gravidae, only vaginal samples at the posterior fornix, midvagina, and vaginal introitus were collected. All subjects were required to have a minimum of 24 teeth with no more than 8 missing teeth. Three vaginal specimens were collected in a systematic and uniform fashion from the vaginal introitus, the posterior fornix, and the midpoint of the vagina. All specimens were collected using sterile Catch-All™ specimen collection swabs by applying the swab to a single site, swirling it 6 times, and then withdrawing from the site without contamination. The protocol for collection was as follows: in non-gravid subjects, the pH was measured at the vaginal introitus (Oakton pH meter, Vernon Hills, IL) and among both cohorts the vaginal introitus specimen was collected first. A clear, small or medium Pederson speculum (manufacturer) was thereafter introduced in the absence of lubrication, and turned to an approximate 45 degree angle to enable placement of the Oakton pH meter at the posterior fornix for determination of posterior fornix pH (non-gravid subjects). In both gravid and non-gravid, the posterior fornix sample was collected, then the speculum was slowly withdrawn to the midpoint of the vagina and the midpoint specimen was obtained. Subjects were not menstruating on the days of specimen collection and were abstaining from sexual intercourse, douching, tampon usage or vaginal creams for the preceding 48 hours.

Specimen Processing

In this study protocol, vaginal specimens were coded, stored and processed for nucleic acid extraction at a single laboratory (J.P.) using a common protocol to reduce variability between samples. Genomic DNA was isolated on standard protocol with PowerSoil DNA Isolation Kit (MoBio) per HMP modifications (http://www.hmpdacc.org/doc/HMP_MDG_454_16S_Protocol_V4_2_102109.pdf).

16S Metagenomic 454 Sequencing Data Generation

Extracted DNA samples were used for 16S rRNA sequence-based survey. High Fidelity PCR reactions were performed in 96 well plates. 16 μL of master mix composed of 13.85 μL RNAse/DNAse free water, 2 μL 10X AccuPrime PCR Buffer II, and 0.15 μL AccuPrime™ Taq DNA Polymerase (Invitrogen) were mixed into individual wells in the 96 well reaction plate. The covered plate was spun and centrifuged at 2000 rpm to collect sample at the bottom of the wells. For the initial reaction, a 2 μL sample of DNA was added to the reaction wells. Barcoded primers (2 μM) from the primer plate were added to corresponding wells in the 96 well PCR plate. Two different PCRs were set up separately with a set of barcoded primers targeting the V3V1 region and V3V5 regions; this analysis examined V3V5 regions. The V3V5 regions of the 16S rRNA gene were amplified by PCR using bar-coded universal primers 354F and 926R (V3-V5) containing the A and B sequencing adaptors (454 Life Sciences, Branford, CT) obtained from Invitrogen. Primer sequences are as follows:B-354F(5′-cctatcccctgtgtgccttggcagtctcaGCCTACGG-GAGGCAGCAG-3′; B adaptor in lowercase letters); A-926R (5′ccatctcatccctgcgtgtctccgactcagNNNNNCCGTCAATTCMTT-TRAGT; A adaptor in lowercase letters, and N represents a bar code that is unique for each sample). Cycling conditions were 95°C for 2 min, followed by 30 cycles at 94°C for 20 s, 50°C for V3V5 primer sets. PCR products were cleaned using AmPure Beads Agencourt (Beckman Coulter, Beverly, MA) using 1.8× volume beads. Beads were eluted with 25 μL 1× low TE, pH 8.0 and transferred to a new 96-well plate. PCR products were quantified using Quant-IT dsDNA high sensitivity assay (Invitrogen) according to the manufacturer's specifications. All samples were diluted according to the sample that had the lowest concentration. Equal volumes of each (5–10 μL) sample were pooled and then concentrated using MinElute columns (Qiagen, Valencia, CA). DNA pool emulsion PCR amplification and 454 sequencing were performed at the BCM Human Genome Sequencing Center (HGSC) in Houston, TX. Sequencing was performed using the 454/Roche B sequencing primer kit in the Roche Genome Sequencer GS-FLX Titanium platform. Samples were combined in a single region of the picotiter plate such that approximately 20,000 to 40,000 sequences were obtained from each group with each primer set. Samples were isolated and quality-filtered from each multiplexed Standard Flowgram Format (SFF) file.

In the gravid sample cohort, three subjects did not have valid sequences from a total of 4 body sites (subject 14 did not have valid sequences from midvagina nor posterior fornix, and subjects 6 and 9 did not have from the midvagina and posterior fornix, respectively). In the non-pregnant cohort, 29 subjects were sequenced for all three vaginal subsites 1 time (29 of 301 samples), 21 subjects for all three subsites twice (42 of 301 samples), 6 subjects thrice (18 of 301 samples), and 2 subjects for all three subsites four times (8 of 301 samples).

Denoising Data Sets

As a quality filtering step, each sample was preprocessed to remove sequences with length less than 200 nucleotides and sequences with minimum average quality less than 20. If they could be identified, reverse primers were also removed from sequences. In a first pass denoising, singletons and chimeric sequences (which occur as a byproduct of the PCR-amplification) were identified by ChimeraSlayer [28] and removed from the representative sequence file generated by the output of the minimal filtering pipeline and subsequent OTU table.

For second pass denoising, two programs were employed: AbundantOTU [29] and the QIIME denoising pipeline. AbundantOTU is a robust and fast OTU picking approach, which uses a consensus alignment algorithm to infer consensus sequences from full-length 16S pyrosequences. Since it relies on sequence redundancy, sequencing errors will have less effect on the OTU picking process compared with other clustering based methods. Rare species, which tend to cause inflated species diversity estimations [30], are not included. In our case, 87% of all input sequences were assigned to 194 consensus sequences. For QIIME denoising, the sequences from all samples were processed through QIIME's 454 dataset denoising pipeline (version 1.3), whereby the sff files from 301 non-pregnant samples and the sff file from the 68 pregnant samples were grouped into one complete dataset for denoising [31], resulting in the production of an OTU table for all 369 samples in a single run on a local cluster utilizing 75 processing cores [32,42,43]. Chimeric sequences in the representative sequence sets picked from denoised fasta files were detected by ChimeraSlayer and subsequently removed from the OTU table. In order to account for artifacts that may arise from multiple sequencing runs and separate denoising runs, all samples were denoised together in all alpha diversity, supervised and unsupervised learning, and feature selection pipelines.

16S Metagenomic 454 Sequencing Analyses

RDP pipeline. Unique reads were classified with the MSU RDP classifier v2.2 [44], maintained at the Ribosomal Database Project (RDP 10 database, version 6), and normalized data were produced from the relative abundance of taxa present in each sample based on a naïve Bayesian classifier [28,43–45]. Output sequences were classified as domain, phylum, family and genus, depending on the depth of reliable classifier assignments. Combined sample reports included the total counts and normalized (relative abundance) data.

Creating input data. QIIME was utilized to produce OTU tables from the quality-filtered sequences as outlined above [28–32,42–44]. The generated OTU tables combined with the clinical metadata comprised the data matrix used as input for alpha diversity (biodiversity within a group of samples), beta diversity (biodiversity between groups of samples) [3], and machine learning pipelines (randomForest). Quality filtered sequences were analyzed using three standard microbiome analysis techniques: operational taxonomic unit (OTU) generation, phylogenetic tree construction, and taxonomic binning of classified sequences.

Alpha diversity. A custom in-house pipeline has been written to calculate alpha diversity measurements and plots using R packages [46]: BiodiversityR [48] and vegan [49]. This pipeline was designed to utilize the output of the OTU picking step (OTU table and meta data file) as the only necessary input. Analyzing the biodiversity within a group of samples provides insight into the differences in species richness, evenness, abundance; we employed the methods of Chao1 [35] and Shannon [36] as we were analyzing groups of samples. Plots were generated and exported for species richness, rank-abundance, and Renyi's diversity indices [34] whereby species richness plots depict the number of species (OTUs) on the y-axis and the number of sites (samples) on the x-axis such that rank-abundance curves rank and list the most abundant OTUs from left to right on the x-axis and top to bottom on the y-axis. The width of the curve on the on the horizontal axis of a species richness plot serves as an indicator of richness: a wide curve indicates higher species richness and a narrow curve indicates the opposite. The shape of the rank-abundance curve is an indicator of species evenness: a horizontal curve indicates a completely evenly distributed system, whereas a steep curve indicates a less even distribution of species [48]. Rank abundance has been demonstrated to successfully show the extent that tag pyrosequencing illuminates the rare biosphere of the human gut [37]. Based on prior evidence, we assumed that Renyi diversity profiles are helpful in analyzing the differences in diversity and evenness between multiple subsets of samples [34,37,50]. If the curves from one sample set contain greater y-axis values, then it can be concluded that this sample set has greater diversity. However, if the two curves intersect at any point, the sample sets are said to be non-comparable, which may reflect important ecological processes [51].

Beta diversity. Beta diversity analysis incorporated 9 binary non-phylogenetic (binary chi-square, binary chord, binary Euclidean, binary Hamming, binary Jaccard, binary Lennon, binary Ochiai, binary Pearson, and binary Sörensen-Dice), 14 non-phylogenetic (Bray-Curtis, Canberra, Chi-square, Chord, Euclidean, Gower, Hellinger, Kulczynski, Manhattan, Morisita-Horn, Pearson, Soergel, Spearman rank, and species profile), and 6 phylogenetic (UniFrac G metric, UniFrac full tree, unweighted UniFrac, unweighted UniFrac full tree, weighted normalized UniFrac, and weighted UniFrac) beta diversity metrics [51–54]. Each beta diversity distance metric calculated was systematically displayed for the top 3 principal coordinates (PCoA (Principal Coordinates Analysis)) for both normalized and non-normalized OTU tables in both 2D and 3D formats for further analysis [55].

Phylogenetic analysis. Phylogenetic trees were produced as necessary input for the phylogenetic beta diversity metrics (*i.e.* UniFrac) in order to determine whether derived microbiome communities were significantly different as an estimate of the degree of divergence between different representative sequences [53–55]. Phylogenetic trees were produced in Newick format employing interactive tree of life (iTOL) interface [56].

Machine learning (randomForest modeling with Boruta feature selection and LEfSe). In order to supplement the data obtained from the taxonomically binned reports, alpha diversity, beta diversity, and phylogenetic data sets, a machine learning pipeline was written to study the patterns that can be detected within the sub-groups of various microbiomes [33,38,39,57–63]. Machine learning algorithms are useful in determining the strength of meta data clusters (bagging, binning, etc.) as well as listing the most important variables involved in discriminating two groups of samples (feature selection). The algorithm randomForest enabled classification of groups of samples by constructing a classification tree, randomly sampling the predictors, choosing the best splitting variables, and predicting new data by combining the predictions from all trees in order to estimate the error rate ["out-of-bag" (OOB)] and list the highest performing variables. We secondarily applied the R package Boruta to explicitly perform feature selection [33,39,58–60]. Linear discriminate analysis effect size (LEfSe) is a novel method developed to support for high dimensional class comparisons in metagenomics analysis [61]. LEfSe combines the standard tests for statistical significance (Kruskal-Wallis test and pairwise Wilcoxon test) with linear discriminate analysis for feature selection. In addition to detecting significant features, it also ranks features by effect size, which put features explain most of the biological difference at top. Pregnant and Non-pregnant were indicated as two classes with no subclass indicated. Alpha value for the factorial Kruskal-Wallis test is 0.05. Threshold on the logarithmic LDA score for discriminative features is 2.0.

Supporting Information

Figure S1 Accuracy of phenotypic predictions from microbiome composition by randomForest. OTU abundances using increasingly many minimal row contributions were used by RF machine learning to predict clinical metadata including pregnancy status, vaginal sampling subsite, subject body mass index, and race/ethnicity. Pregnancy was extremely well predicted by OTU features (Scott's pi >0.8, as compared to a random baseline of 0), exceeding the (still high) accuracy of other metadata predictions.

Figure S2 Supervised (machine) learning with definition by randomForest and confirmation by Boruta feature selection enables visualization of bacterial taxa contributing to clustering of vaginal communities in pregnancy. Bacterial taxa (leftmost column) were defined by randomForest (Table S1) and confirmed by Boruta feature selection. Taxa are sorted first by Mann-Whitney U score, followed by the largest disparity in medians for each group. Taxa represent the lowest taxonomic depth (Genus) that are labeled by RDP Classifier (at $\geq$80% bootstrap cut off). Boxes represent the first quartile, median, and third quartile of the distribution of OTUs for each sample group. Empty circles represent outliers that are 1.5-fold greater than the respective interquartile ranges.

Figure S3 LEfSe analysis of binned taxonomy at discrete vaginal subsites. Bacterial taxa were selected as significantly differentially abundant between pregnant and non-pregnant communities by virtue of discrete sampling site and displayed by LDA Effect Size (LEfSe) algorithm. Taxa level projections are defined by pregnancy at each subsite, with specific *Lactobacillus* species detected consistently among pregnant individuals.

Table S1 Estimated error rate of the randomForest simulation by virtue of potentially contributable clinical metadata (pregnancy, BMI, vaginal sampling site, and ethnicity) following abundant OTU pipeline for denoising of dataset. Top row header: Minimal row contribution cut off sum for each OTU to determine the best performing data set (*i.e.*, contains the most discriminative features with least amount of noise). When describing estimated error rate per minimal row contribution, pregnancy was retained as the only significant clinical metadata category in the model simulation that had an acceptable level of estimated error (<10%, in bold face type).

Table S2 Estimated error rate of the randomForest simulation by virtue of potentially contributable clinical meta data (pregnancy, BMI, vaginal sampling site, and ethnicity) following QIIME pipeline for denoising of dataset. Top row header: Minimal row contribution cut off sum for each OTU to determine the best performing data set (*i.e.*, contains the most discriminative features with least amount of noise). When describing estimated error rate per minimal row contribution, pregnancy was retained as the only significant clinical metadata category in the model simulation that had an acceptable level of estimated error (<10%, in bold face type).

Table S3 The OTUs selected by Boruta feature selection are assigned to the species level by realigned to Greengenes database and BLAST with NCBI microbial database. Top OTUs identified by LEfSe are marked with an asterics.

Author Contributions

Conceived and designed the experiments: KA JV. Performed the experiments: KA JP JV. Analyzed the data: KA KR JM NS TM CC S. Raza AM DG CH JV. Contributed reagents/materials/analysis tools: KA KR JM NS AM CH JV. Wrote the paper: KA KR JM CC DG CH JV. Critical review of the manuscript: IVdV. Clinical recruitment of subjects: S. Rosenbaum.

References

1. Turnbaugh PJ, Ley RE, Hamady M, Fraser-Liggett CM, Knight R, et al. (2007) The Human Microbiome Project. Nature 449: 804–810.
2. Qin L, Li R, Raes J, Arumugam M, Burgdorf KS, et al. (2010) A human gut microbial gene catalogue established by metagenomic sequencing. Nature 464: 59–65.
3. Costello EK, Lauber CL, Hamady M, Fierer N, Gordon JI, et al. (2009) Bacterial community variation in human body habitats across space and time. Science 326: 1694–1697.
4. Ley RE, Peterson DA, Gordon JI (2006) Ecological and evolutionary forces shaping microbial diversity in the human intestine. Cell 124: 837–848.
5. Ley RE, Hamady M, Lozupone C, Turnbaugh PJ, Ramey RR, et al. (2008) Evolution of mammals and their gut microbes. Science 320: 1647–1651.
6. Ley RE, Lozupone CA, Hamady M, Knight R, Gordon JI (2008) Worlds within worlds: evolution of the vertebrate gut microbiota. Nat Rev Microbiol 6: 776–788.
7. Penders J, Thijs C, Vink C, Stelma FF, Snijders B, et al. (2006) Factors influencing the composition of the intestinal microbiota in early infancy. Pediatrics 118: 511–521.
8. Turnbaugh PJ, Hamady M, Yatsunenko T, Cantarel BL, Duncan A, et al. (2009) A core gut microbiome in obese and lean twins. Nature 457: 480–487.
9. Dominguez-Bello MG, Costello EK, Contreras M, Magris M, Hidalgo G, et al. (2010) Delivery mode shapes the acquisition and structure of the initial microbiota across multiple body habitats in newborns. Proc Natl Acad Sci 107: 11971–11975.
10. Favier C, de Vos W, Akkermans A (2003) Development of bacterial and bifidobacterial communities in feces of newborn babies. Anaerobe 9: 219–229.
11. Mackie RI, Sghir A, Gaskins HR (1999) Developmental microbial ecology of the neonatal gastrointestinal tract. Am J Clin Nutr 69: 1035S–1045S.
12. Gronlund MM, Lehtonen OP, Eerola E, Kero P (1999) Fecal microflora in healthy infants born by different methods of delivery: Permanent changes in intestinal flora after cesarean delivery. J Pediatr Gastr Nutr 28: 19–25.
13. Forney LJ, Foster JA, Ledger W (2006) The vaginal flora of healthy women is not always dominated by Lactobacillus species. J Infect Dis 194: 1468–1469; author reply 1469–1470.
14. Brown CJ, Wong M, Davis CC, Kanti A, Zhou X, et al. (2007) Preliminary characterization of the normal microbiota of the human vulva using cultivation-independent methods. J Med Microbiol 56: 271–276.
15. Forney LJ, Gajer P, Williams CJ, Schneider GM, Koenig SS, et al. (2010) Comparison of self-collected and physician-collected vaginal swabs for microbiome analysis. J Clin Microbiol 48: 1741–1748.
16. Kim TK, Thomas SM, Ho M, Sharma S, Reich CI, et al. (2009) Heterogeneity of vaginal microbial communities within individuals. J Clin Microbiol 47: 1181–1189.
17. Xhou X, Brotman RM, Gajer P, Abdo Z, Schuette U, et al. (2010) Recent Advances in Understanding the Microbiology of the Female Reproductive Tract and the Causes of Premature Birth. Infect Dis Obstet Gynecol 2010: 737425.
18. Zoetendal EG, Akkermans AD, DeVos WM (1998) Temperature gradient gel electrophoresis analysis of 16SrFNA from human fecal cells reveals stable and host-specific communities of active bacteria. Appl Environ Microbiol 64: 3854–3859.
19. Eckburg PB, Bik EM, Bernstein CN, Purdom E, Dethlefsen L, et al. (2005) Diversity of the human intestinal microbial flora. Science 308: 1635–1638.
20. Favier CF, Vaughan EE, DeVos WM, Akkermans AD (2002) Molecular monitoring of succession of bacterial communities in human neonates. Appl Environ Microbiol 68: 219–226.
21. Hooper LV, Wong MH, Thelin A, Hansson L, Falk PG, et al. (2001) Molecular analysis of commensal host-microbial relationships in the intestine. Science 291: 881–884.
22. Reinhardt C, Reigstad CS, Backhed F (2009) Intestinal microbiota during infancy and its implications for obesity. J Pediatr Gastr Nutr 48: 249–256.
23. Penders J, Thijs C, Vink C, Stelma FF, Snijders B, et al. (2006) Factors influencing the composition of the intestinal microbiota in early infancy. Pediatrics 118: 511–521.
24. Qin J, Li R, Raes J, Arumugam M, Burgdorf KS, et al. (2010) A human gut microbial gene catalogue established by metagenomic sequencing. Nature 464: 59–65.
25. Gill SR, Pop M, Deboy RT, Eckburg PB, Turnbaugh PJ, et al. (2006) Metagenomic analysis of the human distal gut microbiome. Science 312: 1355–1359.
26. Kurokawa K, Itoh T, Kuwahara T, Oshima K, Toh H, et al. (2006) Comparative metagenomics revealed commonly enriched gene set in human gut microbiomes. DNA Res 14: 169–181.
27. Aagaard KJ (2011) A comprehensive strategy for sampling the human microbiome. Sci Transl Med (In submission).
28. Haas B, Gevers D, Earl AM, Feldgarden M, Ward DV, et al. (2011) Chimeric 16S rRNA sequence formation and detection in Sanger and 454-pyrosequenced PCR amplicons. Genome Res 21: 494–504.
29. Ye Y (2010) Identification and quantification of abundant species from pyrosequences of 16S rRNA by consensus alignment. Proceedings (IEEE Int Conf Bioinformatics Biomed) 2010: 153–157.
30. Quince C, Lanzen A, Curtis TP, Davenport RJ, Hall N, et al. (2009) Accurate determination of microbial diversity from 454 pyrosequencing data. Nat Methods 6: 639–641.
31. Reeder J, Knight R (2010) Rapidly denoising pyrosequencing amplicon reads by exploiting rank-abundance distributions. Nat Methods 7: 668–669.
32. Caporaso JG, Kuczynski J, Stombaugh J, Bittinger K, Bushman FD, et al. (2010) QIIME allows analysis of high-throughput community sequencing data. Nat Methods 7: 335–336.
33. Knights D, Costello EK, Knight R (2011) Supervised classification of human microbiota. FEMS Microbiol Rev 35: 343–359.
34. Renyi A (1961) On measures of entropy and information. In: Neyman J, editor. Proceedings of the Fourth Berkeley Symposium in Mathematical Statistics. Berkeley: University of California Press. pp. 547–561.
35. Chao A (1984) Nonparametric estimation of the number of classes in a population. Scand J Stat 11: 265–270.
36. Shannon CE (1948) A mathematical theory of communication. Bell System Technical Journal 27: 379–423 and 623–656.

37. Dethlefsen L, Huse S, Sogin ML, Relman DA (2008) The pervasive effects of an antibiotic on the human gut microbiota, as revealed by deep 16S rRNA sequencing. PLoS Biol 6: e280.
38. Riehle KP (2011) The Genboree microbiome toolset for the analysis of 16S rRNA microbial sequences. BMC Bioinformatics (in press).
39. Charlson ES, Chen J, Custers-Allen R, Bittinger K, Li H, et al. (2010) Disordered Microbial Communities in the Upper Respiratory Tract of Cigarette Smokers. PLoS ONE 5: e15216.
40. Ravel J, Gajer P, Abdo Z, Schneider GM, Koenig SS, et al. (2011) Microbes and health sackler colloquium. vaginal microbiome of reproductive age women. Proc Natl Acad Sci 108: Suppl 1: 4680–4687.
41. Dominguez-Bello MG, Blaser MJ, Ley RE, Knight R (2011) Development of the human gastrointestinal microbiota and insights from high throughput sequencing. Gastroenterology 140: 1713–1719.
42. Hamady M, Knight R (2009) Microbial community profiling for human microbiome projects: Tools, techniques, and challenges. Genome Res 19: 1141–1152.
43. Wang Q, Garrity GM, Tiedje JM, Cole JR (2007) Naive Bayesian classifier for rapid assignment of rRNA sequences into the new bacterial taxonomy. Appl Environ Microbio 73: 5261–5267.
44. Cole JR, Wang Q, Cardenas E, Fish J, Chai B, et al. (2009) The Ribosomal Database Project: improved alignments and new tools for rRNA analysis. Nucleic Acids Res 37: D141–145.
45. Whittaker RH (1972) Evolution and measurement of species diversity. Taxon 21: 213–251.
46. R Development Core Team (2008) R: A language and environment for statistical computing. Vienna Austria R Foundation for Statistical Computing 1: 2673.
47. Pridmore RD, Berger B, Desiere F, Vilanova D, Barretto C, et al. (2004) The genome sequence of the probiotic intestinal bacterium Lactobacillus johnsonii NCC 533. Proc Natl Acad Sci U S A 101: 2512–2517.
48. Kindt R, Coe R (2005) Tree diversity analysis. A manual and software for common statistical methods for ecological and biodiversity studies. Nairobi: World Agroforestry Centre (ICRAF) ISBN 92-9059-179-X.
49. Oksanen J (2007) Vegan: community ecology package. R package version 1(6).
50. Tóthmérész B (1995) Comparison of Different Methods for Diversity Ordering. J Veg Sci 6: 283–290.
51. Ciccarelli FD, Doerks T, von Mering C, Creevey CJ, Snel B, et al. (2006) Toward automatic reconstruction of a highly resolved tree of life. Science 311: 283–287.
52. Gavin DG, Oswald WW, Wahl ER, Williams JW (2003) A statistical approach to evaluating distance metrics and analog assignments for pollen records. Quatern Res 60: 356–367.
53. Oswald WW, Brubaker LB, Hu FS, Gavin DG (2003) Pollen-Vegetation Calibration for Tundra Communities in the Arctic Foothills, Northern Alaska. J Ecol 91. 1022 1033.
54. Lozupone C, Knight R (2005) UniFrac: a new phylogenetic method for comparing microbial communities. Appl Environ Microbio 71: 8228–8235.
55. Lozupone CA, Hamady M, Kelley ST, Knight R (2007) Quantitative and qualitative beta diversity measures lead to different insights into factors that structure microbial communities. Appl Environ Microbio 73: 1576–1585.
56. Fierer N, Lauber CL, Zhou N, McDonald D, Costello EK, et al. (2010) Forensic identification using skin bacterial communities. Proc Natl Acad Sci USA 107: 6477–6481.
57. Letunic I, Bork P (2007) Interactive Tree Of Life (iTOL): an online tool for phylogenetic tree display and annotation. Bioinformatics 23: 127–128.
58. Knights D, Kuczynski J, Koren O, Ley RE, Field D, et al. (2010) Supervised classification of microbiota mitigates mislabeling errors. ISME J 5: 570–573.
59. Manly B (1986) Multivariate statistical method: A primer. Chapman & Hall Ltd, London UK, UK.
60. Kursa MB, Rudnicki WR (2010) Feature selection with the Boruta package. J Stat Softw 36: 1–13.
61. Segata N, Izard J, Waldron L, Gevers D, Miropolsky L, et al. (2011) Metagenomic biomarker discovery and explanation. Genome Biol 12: R60.
62. Li M, Wang B, Zhang M, Rantalainen M, Wang S, et al. (2008) Symbiotic gut microbes modulate human metabolic phenotypes. Proc Natl Acad Sci USA 105: 2117–2122.
63. Price MN, Dehal PS, Arkin AP (2009) FastTree: computing large minimum evolution trees with profiles instead of a distance matrix. Mol Biol Evol 26: 1641–1650.

15

Soil Bacterial Communities Respond to Climate Changes in a Temperate Steppe

Ximei Zhang[1,2], Guangming Zhang[2], Quansheng Chen[2], Xingguo Han[1,2*]

1 State Key Laboratory of Forest and Soil Ecology, Institute of Applied Ecology, Chinese Academy of Sciences, Shenyang, China, **2** State Key Laboratory of Vegetation and Environmental Change, Institute of Botany, Chinese Academy of Sciences, Beijing, China

Abstract

Climate warming and shifting precipitation regimes are affecting biodiversity and ecosystem functioning. Most studies have focused on the influence of warming and altered precipitation on macro-organisms, whereas the responses of soil microbial communities have been neglected. We studied the changes in the abundance, richness, and composition of the entire bacterial kingdom and 16 dominant bacterial phyla/classes in response to increased precipitation, warming, and their combination, by conducting a 5-year experiment in a steppe ecosystem in Inner Mongolia, China. Watering had a greater effect than warming on almost all the bacterial groups as indicated by changes in all the three attributes (abundance, richness, and composition). The 16 phyla/classes responded differentially to the experimental treatments, with Acidobacteria and Gamma-proteobacteria being the most sensitive. Stepwise regression analyses further revealed that climate changes altered the abundance and richness of bacterial groups primarily through direct routes (e.g., increasing soil water content), and changed the community composition through both direct and indirect routes (e.g., reducing soil total nitrogen content and increasing soil pH). The diverse responses of various bacterial groups could imply some potential shift in their ecosystem functions under climate changes; meanwhile, the indirect routes that are important in altering bacterial composition suggest that specific strategies (e.g., adding NH_4NO_3 to maintain soil nitrogen content and pH) could be adopted to maintain soil microbial composition under climate changes.

Editor: Hauke Smidt, Wageningen University, The Netherlands

Funding: This work was supported by the National Natural Science Foundation (30830026), the State Key Basic Research Development Program (973 Program) (2009CB825103) and Postdoctoral Science Foundation (2012T50158, 2011M500440) of China. The funders had no role in study design, data collection and analysis, decision to publish, or preparation of the manuscript.

Competing Interests: The authors have declared that no competing interests exist.

* E-mail: xghan@ibcas.ac.cn

Introduction

The rising concentration of atmospheric greenhouse gases has elevated the global mean temperature by 0.76°C since 1850, and the temperature is predicted to continue increasing by 1.8–4.0°C by the end of this century [1]. Precipitation patterns are also expected to shift with climate warming [2,3]. These climate changes will have profound influences on biodiversity and ecosystem functioning, and subsequently create feedbacks to the climate changes. For example, climate warming reduced plant richness and community coverage in a temperate steppe ecosystem, causing net carbon loss and a positive feedback to climate warming [4,5]. Previous studies have mainly focused on the response of higher organisms to climate changes. Although soil microbial communities are among the most abundant and diverse groups of organisms on Earth and are responsible for numerous key ecosystem processes such as carbon and nitrogen cycling [6–8], their response to climate changes, especially to the combination of climate warming and altered precipitation, has not been comprehensively explored [9].

Less attention has been paid to soil microbial communities primarily because of their extreme complexity and lack of accurate methods to quantify their composition and function. Specifically, they are composed of diverse taxonomic groups (e.g., phyla and classes) with different ecological niches and ecosystem functions. Furthermore, more than 99.9% of them are uncultivable and, hence, we have relatively limited knowledge about their niches and functions [7,10–12]. The traditional PCR-dependent cloning and sequencing technology allows us to acquire tens or hundreds of gene sequences from a soil sample, and thus, is a useful tool for the investigation of some special microbial groups (e.g., the ammonia-oxidizing bacteria [13]) with limited phylogenetic diversity. However, this technology is very labor-intensive and, hence, is not effective for the study of complex microbial groups such as the entire soil bacterial community. Fortunately, the current metagenomic approaches that exploit next-generation sequencing technologies can facilitate investigation of the microbial composition of hundreds of samples simultaneously, accurately, and rapidly [14,15].

Climate changes may affect soil microbial communities through both direct and indirect routes. We define a direct route as one in which a climate variable (e.g. surface air temperature) alters a synonymous soil physicochemical variable (e.g. soil temperature). Specifically, climate warming may directly affect soil microbial communities by elevating soil temperature, and indirectly affect them by changing other soil physicochemical features such as soil pH, water availability, and carbon and nitrogen content [16,17]. Similarly, shifting precipitation regime may directly affect soil microbial communities through altering soil water content, and indirectly affect them through changing other physicochemical features. It should be noted that climate warming and precipitation may have interactive effects. Specifically, warming can decrease soil water content, and precipitation can decrease soil temperature. Warming and precipitation may also have interactive effects on soil microbial communities. To gain a mechanistic understanding of the response of soil microbial communities to climate changes, elucidating direct and indirect effects is essential.

The semiarid temperate steppe in northern China is an important part of the Eurasian grassland biome and is sensitive to climate changes [18,19]. According to the climate history of our study site, the air temperature has increased by 2.4°C over the past several decades (1953–2008), and summer precipitation has also been predicted to increase in the future [1,20–22]. To comprehensively examine the response of soil microbial communities to climate changes in the steppe ecosystem, a field manipulative experiment of watering and warming (mimicking increased precipitation and elevated temperature, respectively) with four treatments (control, watering, warming, and watering plus warming) was conducted since 2005. The pyrosequence technology targeting bacterial 16S rRNA gene was used to measure the structure of soil bacterial communities [14]. The specific questions of this study were as follows: (1) Whether and how will the abundance, richness, and composition of the entire soil bacterial community and the dominant phyla/classes respond to climate warming, increased precipitation, and their combination? (2) What are the relative contributions of the direct and indirect effects to the changes in bacterial abundance, richness, and composition?

Materials and Methods

Study Site and Experimental Design

This study was part of a long-term experiment conducted at the Duolun Restoration Ecology Station of Institute of Botany, Chinese Academy of Sciences, approximately 30 km from Duolun County (42°02′N, 116°17′E), Inner Mongolia Autonomous Region of China. This experiment commenced from 2005. As the study site and experimental design have been described earlier [4,5,19], we provided only a brief description here. The experimental site was a typical temperate zone characterized by a semiarid continental monsoon climate. Mean annual temperature was 2.1°C with monthly mean temperature ranging from 18.9°C in July to −17.5°C in January. Mean annual precipitation was about 385.5 mm with 80% precipitation occurred from June to September. Soil was chestnut soil (Chinese classification), Calcis-orthic Aridisol in the US Soil Taxonomy classification, with sand, silt, and clay being 62.7%, 20.3%, and 17.0%, respectively. Mean soil bulk density was 1.31 g/cm^3 and pH was 6.84. This temperate steppe was dominated by perennials, including *Stipa krylovii*, *Artemisia frigida*, *Potentilla acaulis*, *Cleistogenes squarrosa*, *Allium bidentatum*, and *Agropyron cristatum*.

The experiment used a paired, nested design with watering as the primary factor and warming as the secondary factor [4,19]. There were three pairs of 10 m×15 m plots; one plot in each pair was randomly assigned to watering treatment and the other to the control. At each watering plot, six sprinklers were arranged evenly into two rows with a distance of 5 m between two sprinklers. Each sprinkler covered a circular area with a diameter of 3 m, so the six sprinklers covered the 10 m×15 m plot. In July and August, 15 mm of water was added weekly to the watering plots. Therefore, a total amount of 120 mm water (approximately 30% of mean annual precipitation at the study site) was added each year. Within each 10 m×15 m plot, four 3 m×4 m sub-plots were randomly assigned as the warming and control subplots (each with two replicates). The warming subplots were heated continuously since April 28 of 2005 using 165 cm×15 cm MSR-2420 infrared radiators (Kalglo Electronics, Bethlehem, PA, USA) suspended 2.5 m above the ground. The effect of infrared radiators on soil temperature was spatially uniform within the plot [23]. In the control subplot, one 'dummy' heater with the same shape and size as the infrared radiator was suspended 2.5 m high to mimic the shading effect of the heater. Overall, there were six replicates for each treatment (control, watering, warming, and watering plus warming).

Measurement of Soil Physiochemical Indexes

Soil temperature at the depth of 10 cm was measured with a CR1000 datalogger (Campbell Scientific, Logan, UT, USA) at 1-h intervals from June 4 of 2005. Soil samples were taken on 22 August of 2010. Four soil cores (10 cm deep, 3.5 cm diameter) were collected from each sub-plot (3 m×4 m) at random and thoroughly mixed, part of which was frozen for DNA extraction and the left part was used to measure soil water content, total carbon (TC) content, total nitrogen (TN) content and soil pH. Soil water content was determined as the weight loss after drying for 24 hr at 105°C. Soil TC and TN contents were quantified with the potassium dichromate-vitriol oxidization method and the Kjeldahl acid-digestion method, respectively. Soil pH was measured in 1:2.5 (*W*/*V*) suspensions of soil in distilled water.

Measurement of Bacterial Abundance and Community Structure

For each soil sample, we extracted DNA from 0.5 g of mixed soil using the Fast DNA SPIN kit for soil according to the manufacturer's instructions (Qbiogene, Carlsbad, CA, USA); however, we used 350 μL instead of 50 μL DNA elution solution to elute the DNA in the tenth step of the procedure. The DNA solution was then stored at −20°C until analysis.

The content of bacterial 16S rRNA gene was measured using real-time PCR to represent the abundance of the entire bacterial community [16]. The standard curve was generated using a 10-fold serial dilution of a plasmid containing a copy of the *Escherichia coli* 16S rRNA gene. The 20 μL PCR reaction mixtures contained 10 μL SYBR Premix (TaKaRa Biotechnology Co., Ltd., China), 0.4 μL each of 10 μmol/L forward and reverse primers (Eub338:5′-ACT CCT ACG GGA GGC AGC AG-3′; Eub518:5′-ATT ACC GCG GCT GCT GG-3′), 0.4 μL Rox II, 2 μL BSA (10 mg/mL), and 5.8 μL sterile and DNA-free water. The amount of standard and soil DNA samples added per reaction was 1.0 μL (1.2–5.1 ng). The reaction was conducted with a Roche LightCycler^{TM} Real-time PCR system using the following program: 95°C for 1 min followed by 40 cycles of 95°C for 5 sec, 55°C for 15 sec and 72°C for 15 sec. Melting curve and gel electrophoresis analyses were conducted to confirm that the amplified products were of the appropriate size. The bacterial 16S rRNA gene copy number was calculated using a regression equation that related the cycle threshold (Ct) value to the known number of copies in the standards. For each soil sample, the qPCR reactions were repeated three times. We added BSA to the PCR reaction mixtures to reduce the inhibitory effects of co-extracted polyphenolic compounds in the soil. Additionally, three rounds of PCR were conducted after adding known amounts of standard plasmid with the soil DNA extract to estimate the possible inhibitory effects of co-extracted polyphenolic compounds. The inhibitory effects were found to be negligible.

The method of 454 pyrosequences was used to measure the bacterial community structure of each soil sample. The primers 27F (5′-AGA GTT TGA TCC TGG CTC AG-3′) and 338R (5′-TGC TGC CTC CCG TAG GAG T-3′) were used to amplify the fragment of 16S rRNA gene. To measure all 24 samples in a run, a unique 10-mer tag for each sample was added to the 5′-end of the primer 338R [14]. Each 20 µl PCR mixture contained 4 µl FastPfu Buffer (5×; Transgen), 2 µl of 2.5mM dNTPs, 0.4 µl of each primer (5 µM), 0.8 µl of DNA template, and 0.4 µl of FastPfu Polymerase (Transgen). The PCR protocol was as follows: 95°C for 2 min (denature); 25 cycles of 95°C for 30 sec (denature), 55°C for 30 sec (anneal), 72°C for 30 sec (elongate); and 72°C for 5 min (elongate). Three replicates of PCR were performed for each sample, after which the products were combined and purified by agarose gel electrophoresis, recovered, and quantified with Pico-Green using a TBS-380 Mini-Fluorometer. Equal molar concentrations of PCR products for each sample were then pooled and sequenced in a Roche 454 Genome Sequencer FLX Titanium system at Shanghai Majorbio Bio-pharm Technology Co., Ltd. The sequence reads for all samples have been deposited in the National Center for Biotechnology Information Sequence Reads Archive (accession no. SRA057669).

Data Analysis

The pyrosequence reads were analyzed with the Mothur software [24]. These reads were first assigned to samples according to their tags, after which reads <150 bp in length and with ambiguous characters were removed. The first 150 bp (corresponds to the V3 region of 16S rRNA gene) of the remaining reads were aligned to the Silva database to determine their taxonomic classification and non-bacterial reads were further removed [25]. To minimize the influence of unequal sampling on the following calculated indexes, we randomly selected 3,478 reads for each sample. All these sequences (3,478×24) were clustered into OTUs (operational taxonomic units) with larger than 97% similarity. The OTU number of 3,478 reads for each sample was used to represent the OTU richness of the entire bacterial community. For each pair of samples, the measure of weighted UniFrac was calculated to represent the compositional variation of the entire bacterial community [26]. The Bray-Curtis distance basing on OTU abundance was also calculated to represent the compositional variation of the entire bacterial community. Briefly, we first calculated the difference in the number of reads for each OTU within this pair of samples, and then calculated the sum of the absolute values of these differences for all OTUs. Finally, we divided the sum by 6,956 (3,478×2) to represent the Bray-Curtis distance.

For each of the 16 dominant bacterial phyla/classes, (its sequence number in all 3,478 sequences)/3,478 was calculated to represent its relative abundance. To compare the OTU richness of each phylum/class among different samples, we used a different calculation approach from that for the entire bacterial community, because there was different number of sequences for the same phylum/class among different samples. For each phylum/class, we randomly selected a certain number of representative sequences from a sample and calculated the OTU number represented by these sequences, repeated this process 1,000 times and used the mean OTU number to represent its richness. Although the number of 16S rDNA sequence for some phyla/classes might be very low and the OTU richness could not be estimated quantitatively and accurately from these sequences, this calculation method would still be effective in comparing qualitatively the difference in OTU richness among different samples. For each of the 16 phyla/classes, we also calculated the Bray-Curtis distance basing on the abundance of OTUs to represent the compositional variation between each pair of samples.

Two-way analysis of variance (ANOVA) for a blocked split-plot design was used to determine the main and interactive effects of watering and warming on the abundance and OTU richness of the entire bacterial community as well as the relative abundance and OTU richness of each of the 16 dominant phyla/classes [27]. Before doing two-way ANOVA, the method of Levene test was used to analyse whether these indexes were homogeneous among different treatments and found that all of them were homogenous statistically (P>0.05). To visualize the relative difference in community composition among different treatments, we used non-metric multidimensional scaling (NMDS) plots with the software of Primer. Permutational multivariate analysis of variance (PERMANOVA) was further used to reveal the effects of experimental treatments on the composition of the entire bacterial community and that of each of the 16 phyla/classes [28]. For the abundance (or relative abundance), richness, or composition exhibiting significant (P<0.05) or marginally significant (P<0.10) response to experimental treatments, we used stepwise regression analysis to identify the factor that could effectively explain the changes in the abundance, richness, and composition from the five potential soil physiochemical indexes of soil water content, temperature, TC content, TN content, and pH. Specifically, principal coordinate analyses were first used to determine the difference in community composition among different treatments [29], and the first principal coordinate was subsequently used to represent the dependent variable in the stepwise regression analyses. Two-way ANOVA and stepwise regression analyses were conducted with SPSS software (SPSS 13.0 for WINDOWS).

Results

Bacterial Abundance

The experimental treatments did not alter the abundance of the entire bacterial community significantly (P>0.10; Fig. 1a). However, the treatments significantly (P<0.05) or marginally significantly (P<0.10; Fig. 2) altered the relative abundance of some phyla/classes. Watering alone decreased the relative abundance of Acidobacteria and Delta-proteobacteria (Fig. 2a, o), but increased the relative abundance of Proteobacteria, Beta- and Gamma-proteobacteria (Fig. 2i, n, p). Warming alone decreased the relative abundance of Delta-proteobacteria (Fig. 2o), but increased the relative abundance of Nitrospirae and Rubrobacteridae (Fig. 2 g, l). Watering and warming had interactive effects. Warming alone increased the relative abundance of Rubrobacteridae, whereas combined warming and watering had a negligible effect (Fig. 2l). Watering or warming alone decreased the relative abundance of Delta-proteobacteria, while the decline in relative abundance was small under concurrent watering and warming (Fig. 2o).

OTU Richness

The experimental treatments did not significantly alter the OTU richness of the bacterial community (P>0.10; Fig. 1b). However, significant (P<0.05) or marginally significant (P<0. 10; Fig. 3) changes in OTU richness occurred with some phyla/classes. Watering alone decreased the richness of Acidobacteria, Chloroflexi, Nitrospirae, Planctomycetes, and Actinobacteridae (Fig. 3a, d, g, h, k), but increased the richness of Firmicutes and Gamma-proteobacteria (Fig. 3e, p). Warming alone decreased the OTU richness of Bacteroidetes, Nitrospirae, Actinobacteridae, and Beta-proteobacteria (Fig. 3c, g, k, n), but increased that of Acidobacteria (Fig. 3a). Watering and warming had interactive effects on richness. Whereas warming alone decreased the richness of Bacteroidetes, joint watering and warming increased the OTU richness (Fig. 3c). Watering or warming alone decreased the OTU richness of Actinobacteridae, but the negative effect was negligible when watering and warming occurred concurrently (Fig. 3 k).

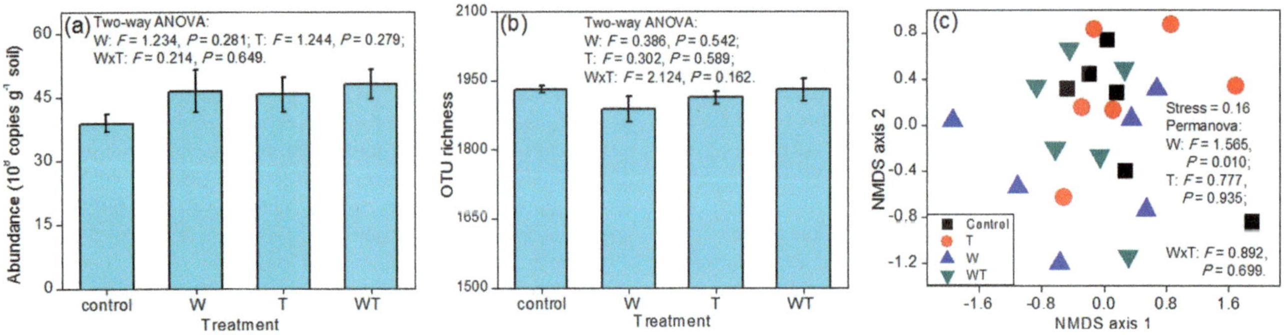

Figure 1. Effects of treatments on the abundance, richness, and composition of the entire soil bacterial community. The effects of block on abundance and richness were non-significant (P>0.05), and were not shown in the figure. The compositional variation is represented with the measure of weighted UniFrac, with the r^2 values between ordination distance and distance in the original space being 0.52 and 0.10 for axis 1 and axis 2, respectively (c). W and T represent watering and warming, respectively. The bars represent one standard error.

Bacterial Composition

Watering altered soil bacterial composition significantly (P<0.05; Fig. 1c). An NMDS plot of weighted UniFrac values showed that watered plots shifted relative to un-watered plots (Fig. 1c). However, there was a non-significant effect of warming and a non-significant interactive effect of watering and warming

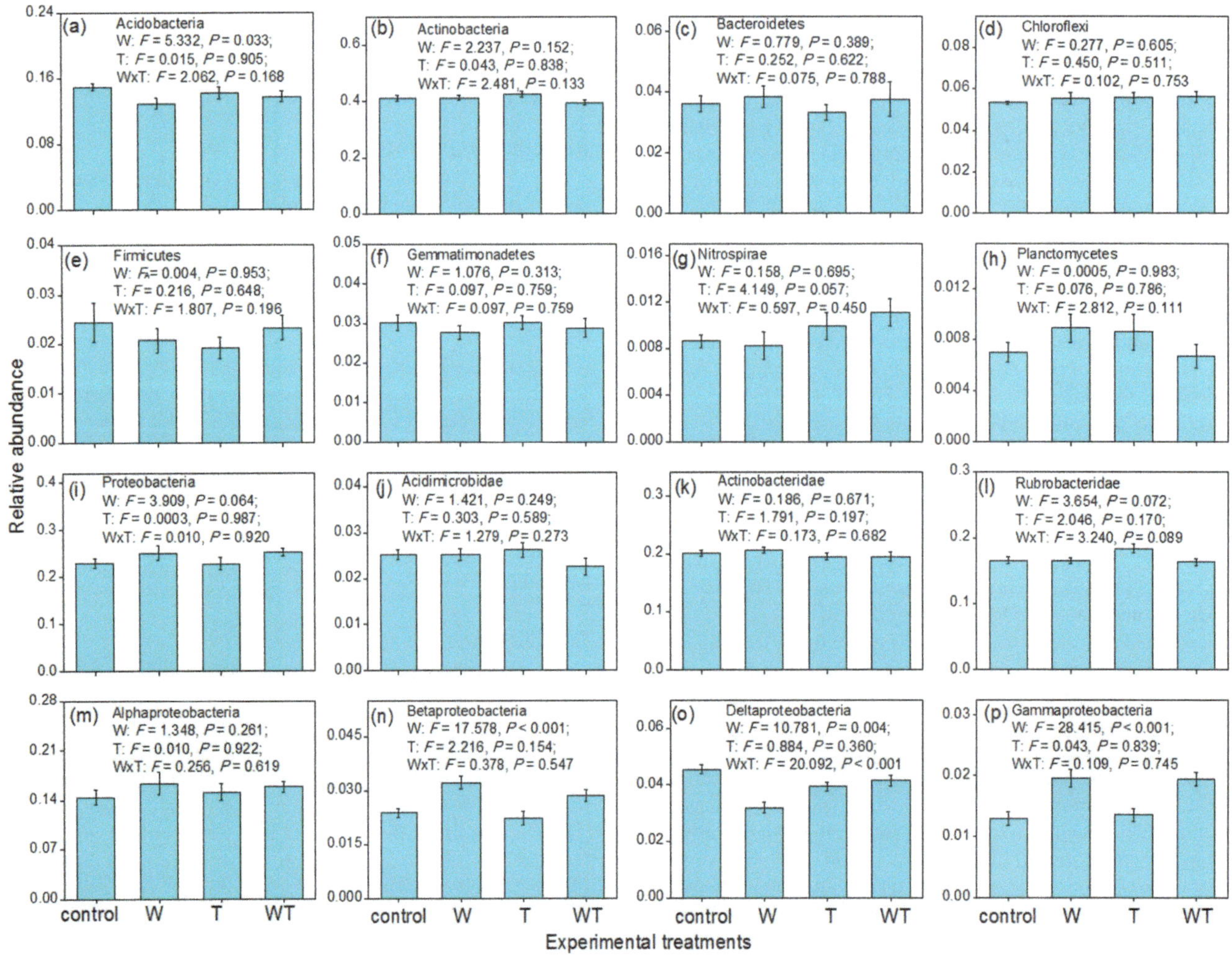

Figure 2. Effects of watering, warming, and their combination on the relative abundance of the 16 bacterial phyla/classes. Two-way ANOVA for a blocked split-plot design was used to test the effects of experimental treatments. The effects of block were non-significant (P>0.05), and were not shown in the figure. W and T represent watering and warming, respectively. The bars represent one standard error.

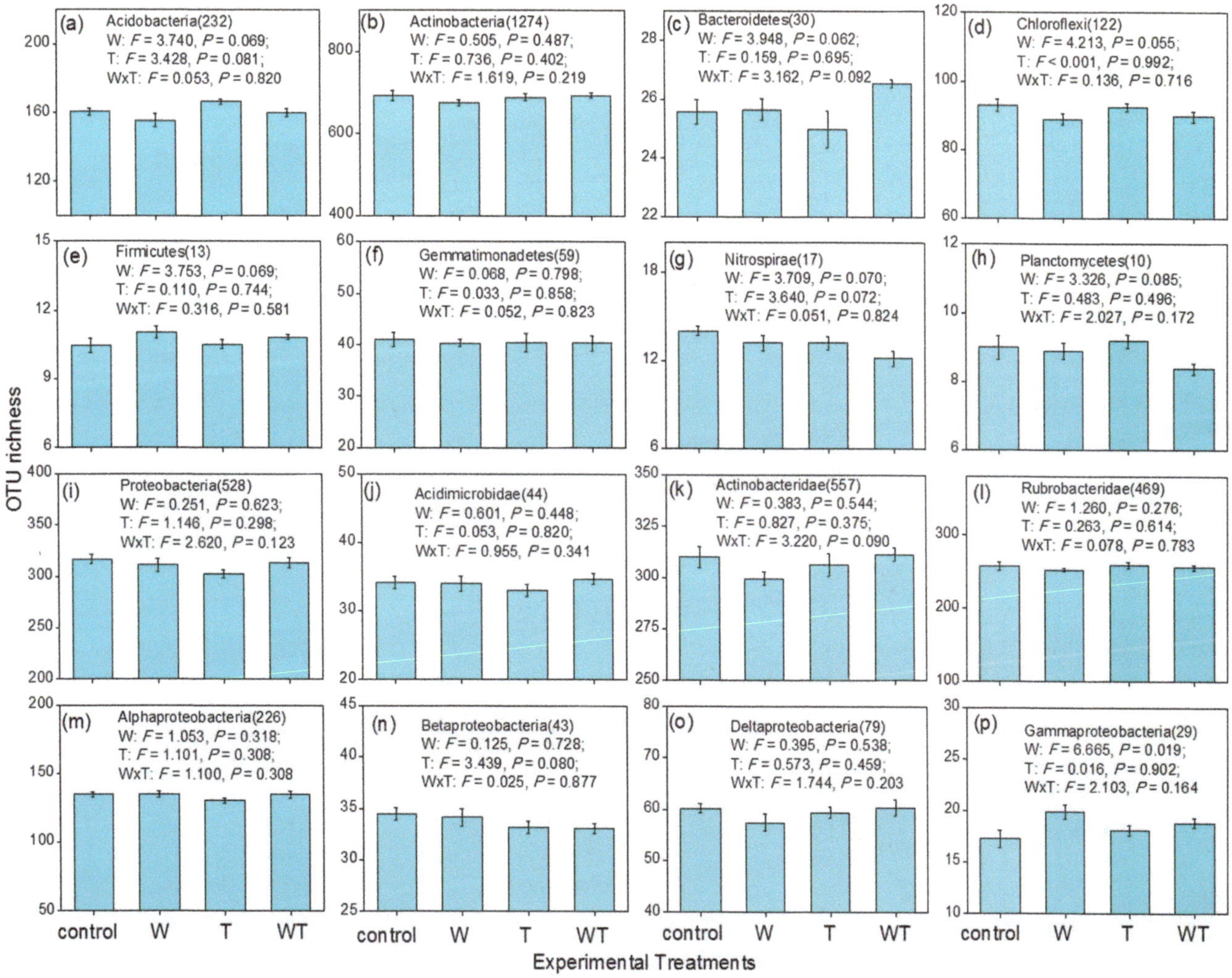

Figure 3. Effects of watering, warming, and their combination on OTU richness of the 16 bacterial phyla/classes. Two-way ANOVA for a blocked split-plot design was used to test the effects of experimental treatments. The effects of block were non-significant ($P>0.05$), and were not shown in the figure. W and T represent watering and warming, respectively. The bars represent one standard error. The number in the brackets following the phylum/class name (e.g., 232 in Acidobacteria(232) in Fig. 3a) represent the sampled sequence number from which OTU richness calculated.

on bacterial composition ($P>0.05$; Fig. 1c). When we quantified bacterial compositional variation with the Bray-Curtis distance based on the abundance of OTUs, watering also showed a significant effect (PERMANOVA: watering, $F = 1.124$, $P = 0.002$; warming, $F = 0.963$, $P = 0.873$; watering × warming, $F = 0.991$, $P = 0.572$) and the NMDS plot showed a similar pattern (Figure not shown). Among the 16 dominant phyla/classes, watering only altered the composition of six groups, namely Acidobacteria, Bacteroidetes, Proteobacteria, Acidimicrobidae, Delta-proteobacteria, and Gamma-proteobacteria ($P<0.10$; Fig. 4; Table 1). Warming had non-significant effects and there were non-significant interactions between watering and warming on the composition of the 16 groups ($P>0.10$; Table 1).

Variables Correlated with Bacterial Abundance, Richness, and Composition

Stepwise regression analyses revealed that the relative abundance of the five groups of Acidobacteria, Rubrobacteridae, Beta-proteobacteria, Delta-proteobacteria, and Gamma-proteobacteria were correlated with soil water content, and the relative abundance of Proteobacteria was correlated with soil pH (Table 2). Although experimental treatments also altered the relative abundance of Nitrospirae ($P<0.10$; Fig. 2g), stepwise regression analyses revealed that none of the five soil physiochemical variables that we examined (water content, temperature, TC content, TN content, and pH) could explain the changes in its relative abundance.

Although experimental treatments altered the OTU richness of nine bacterial groups ($P<0.10$; Fig. 3), stepwise regression analyses revealed that none of the five soil variables could explain the changes in OTU richness of six groups, namely Chloroflexi, Firmicutes, Nitrospirae, Planctomycetes, Actinobacteridae, and Beta-proteobacteria (Table 2). Only the OTU richness of Acidobacteria was correlated with soil temperature, and the richness of Bacteroidetes and Gamma-proteobacteria was correlated with soil water content.

Experimental treatments altered the composition of seven bacterial groups (including the entire bacterial community;

Table 1. The effects of watering (W), warming (T), and their combination on the composition of the 16 dominant bacterial phyla/classes revealed by PERMANOVA.

	Acidobacteria		Actinobacteria		Bacteroidetes		Chloroflexi	
	F	*P*	*F*	*P*	*F*	*P*	*F*	*P*
W	1.390	<0.001*	1.037	0.213	1.134	0.067*	0.961	0.672
T	0.983	0.579	0.975	0.674	0.950	0.717	0.954	0.707
W×T	1.019	0.371	0.995	0.496	1.045	0.283	0.932	0.797
	Firmicutes		**Gemmatimonadetes**		**Nitrospirae**		**Planctomycetes**	
	F	*P*	*F*	*P*	*F*	*P*	*F*	*P*
W	1.106	0.260	1.084	0.228	0.917	0.631	0.882	0.852
T	0.833	0.835	0.800	0.966	1.175	0.207	0.922	0.744
W×T	0.875	0.767	0.923	0.746	0.882	0.701	0.863	0.893
	Proteobacteria		**Acidimicrobidae**		**Actinobacteridae**		**Rubrobacteridae**	
	F	*P*	*F*	*P*	*F*	*P*	*F*	*P*
W	1.217	0.004*	1.412	0.003*	1.024	0.342	0.956	0.729
T	0.950	0.786	0.923	0.709	1.059	0.200	0.883	0.963
W×T	0.991	0.530	0.817	0.926	1.047	0.240	0.965	0.682
	Alpha-proteobacteria		**Beta-proteobacteria**		**Delta-proteobacteria**		**Gamma-proteobacteria**	
	F	*P*	*F*	*P*	*F*	*P*	*F*	*P*
W	1.125	0.102	1.121	0.171	1.253	0.010*	1.725	0.003*
T	0.851	0.961	1.103	0.213	1.080	0.198	1.093	0.301
W×T	0.957	0.649	1.079	0.270	1.086	0.180	0.637	0.979

*denotes $P<0.10$.

$P<0.10$; Figs. 1 and 4; Table 1), but stepwise regression analyses revealed that none of the five soil variables could explain the compositional variation of Acidimicrobidae and Gamma-proteobacteria (Table 2). Variation in composition of the entire bacterial community was correlated with both soil TN content and water content. The composition of Acidobacteria was correlated with both soil water content and soil pH. On the other hand, the composition of Bacteroidetes, Proteobacteria, and Deltaproteobacteria was correlated with soil water content, pH, and temperature, respectively (Table 2).

Discussion

Although we detected the response of soil microbial communities to experimental treatments at only one time point, the observed results were the 5-year effect of climate changes. In other words, there were accumulative effects. In addition, the response of microbial communities shifted with the intensity of experimental treatments. As the marginally significant response ($0.05<P<0.10$) might become significant ($P<0.05$) with the increasing experimental duration or intensity, we used a lower threshold for significance ($P<0.10$) than the traditional threshold ($P<0.05$) in this study.

Overall, our results demonstrated that watering had a greater effect than warming on soil microbial communities. There were three types of evidence. First, watering significantly altered the composition of the entire bacterial community, whereas warming had no significant effect (Fig. 1c). Second, for each of the three community attributes (abundance, richness and composition), watering affected some of the 16 bacterial phyla/classes, while warming affected only a small number or none of these bacterial groups (Figs. 2, 3 and 4; Table 1). Finally, the abundance and richness of seven bacterial groups were correlated with changes in soil water content, whereas soil temperature was correlated with the abundance and richness of only one group (Table 2). This conclusion is consistent with previous findings that water played a dominant role in the responses of soil/microbial respiration, plant community composition, and ecosystem carbon fluxes to simulated climate changes in this semiarid steppe ecosystem [4,5,19]. These phenomena might be due to the fact that this ecosystem was constrained more by water availability than by temperature.

Our results showed that these bacterial phyla/classes responded differently to the experimental treatments in terms of abundance, richness, and composition. In addition, for each bacterial group, its abundance, richness, and composition changed non-synchronously (Figs. 1, 2, 3 and 4; Table 1). For example, while the richness of the Chloroflexi phylum changed significantly in response to watering, its abundance and composition changed non-significantly. On the whole, Acidobacteria and Gamma-proteobacteria were most sensitive among the 16 bacterial phyla/classes, because these two groups exhibited changes in all three community attributes (abundance, richness, and composition) in response to the experimental treatments. Actinobacteria, Gemma-

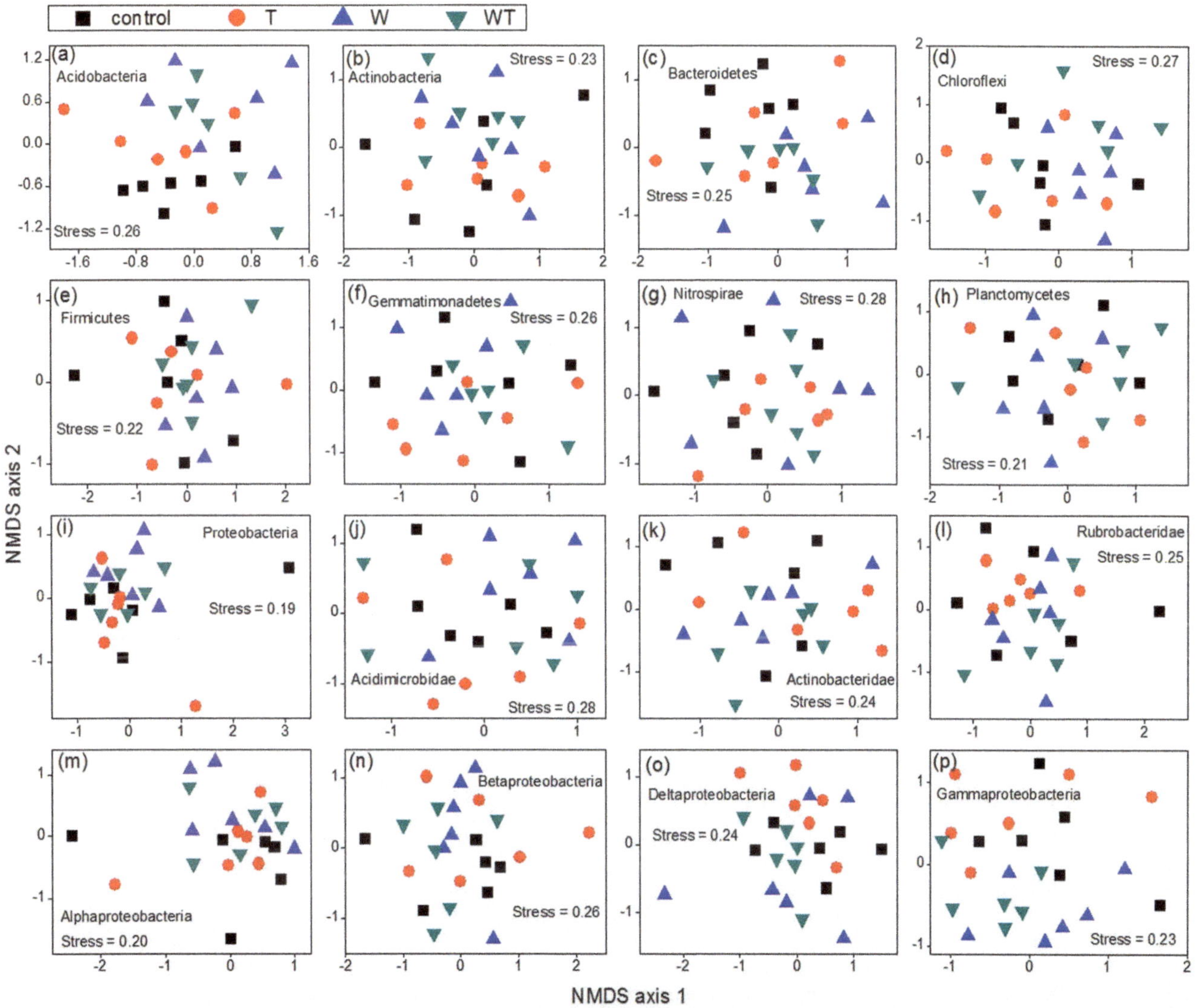

Figure 4. Non-metric multidimensional scaling (NMDS) plots of the 16 bacterial phyla/classes. These plots show the effects of watering (W), warming (T), and their combination on community composition. The compositional variation is represented with Bray-Curtis distance based on the abundance of OTUs.

timonadetes, and Alpha-proteobacteria were the most insensitive to experimental treatments, with no apparent changes in abundance, richness, or composition.

Although warming did not change the abundance, OTU richness, and composition of soil bacterial communities ($P>0.10$; Fig. 1), it caused significant reduction in microbial respiration [4]. Furthermore, warming significantly altered soil enzyme activities [27]. These results suggested that climate warming reduced soil respiration primarily through decreasing microbial activities, rather than altering their community structure. In contrast, watering altered microbial respiration [4], bacterial composition (Fig. 1C), and soil enzyme activities simultaneously [27], implying that increased precipitation altered microbial respiration through both shift in their composition and changes in their activities. However, the effect of climate changes on soil microbial activities still needs be investigated in future studies.

Besides soil respiration, soil microbial communities drove many other ecosystem functions, and the changes in their abundance, richness, and composition could guide the potential functional shifts. For example, a large proportion of members in the Chloroflexi phylum could acquire energy and fix CO_2 through photosynthesis [30–32]; thus, the decreased Chloroflexi OTU richness under watering (Fig. 3d) implies that increased precipitation might reduce their carbon sink function. Beta-proteobacteria have been found to play a key role in transforming ammonium into nitrite [13], which is available source of nitrogen to various types of plants. Therefore, increased Beta-proteobacteria abundance under watering (Fig. 2n) suggests that the ammonium-transforming function might be stimulated under increased precipitation. In contrast, decreased Beta-proteobacteria OTU richness under warming (Fig. 3n) implies that this function might be restrained under climate warming. A large proportion of the members of the Nitrospirae phylum could transform nitrite into nitrate [33,34]; thus, increased Nitrospirae abundance under warming (Fig. 2 g) suggests that this nitrogen-cycling function might be stimulated under climate warming. As there are many plant pathogenic bacteria in the Firmicutes phylum [35], the increased OTU richness under watering (Fig. 3e) implies that the

Table 2. Variables responsible for the changes in the abundance, richness, and composition of various bacterial groups.

Index	Group	Results	r^2	*F*	*P*
	Acidobacteria	y = 0.166–0.320 (water content)	0.318	10.278	0.004
	Proteobacteria	y = −0.510+0.103 (soil pH)	0.508	22.706	<0.001
abundance	Rubrobacteridae	y = 0.190–0.248 (water content)	0.197	5.411	0.030
	Betaproteobacteria	y = 0.015+0.136 (water content)	0.441	17.361	<0.001
	Deltaproteobacteria	y = 0.049–0.116 (water content)	0.234	6.730	0.017
	Gammaproteobacteria	y = 0.007+0.117 (water content)	0.567	28.763	<0.001
	Acidobacteria	y = 60.525+6.677 (soil temperature)	0.239	6.902	0.015
richness	Bacteroidetes	y = 24.184+18.237 (water content)	0.192	5.218	0.032
	Gammaproteobacteria	y = 14.768+44.975 (water content)	0.462	18.878	<0.001
	Bacteria	y = 0.878–357.383 (TN content)	0.415	15.612	0.001
		y = 0.914–297.430 (TN content) −2.213 (water content)	0.648	19.349	<0.001
	Acidobacteria	y = 0.326–3.938 (water content)	0.671	44.860	<0.001
composition		y = 1.700–3.130 (water content) −0.199 (soil pH)	0.746	30.838	<0.001
	Bacteroidetes	y = 0.273–3.298 (water content)	0.298	9.330	0.006
	Proteobacteria	y = −2.268+0.313 (soil pH)	0.261	7.757	0.011
	Deltaproteobacteria	y = −2.191+0.146 (soil temperature)	0.233	6.689	0.017

health of plants might be threatened under increased precipitation. Nevertheless, as soil microbial communities are very complex, we still have very limited knowledge about the ecosystem functions of most of the bacterial phyla/classes [36], not to mention the functional shifts under climate changes.

Among the three attributes of all the 17 bacterial groups that we examined, 24 changes in abundance, richness, or composition were at least marginally significant ($P<0.10$; Figs 1, 2, 3 and 4; Table 1), but possible explanatory physicochemical variables–water content, temperature, carbon content, nitrogen content or pH–were identified for only 14 of the changes (Table 2). The absence of correlated variables for the remaining 10 changes can be explained as follows: first, there are other important soil properties (e.g., phosphorus content, trace element contents) that were not measured. Second, although the effects of experimental treatments on the bacterial groups have lasted for 5 years, most of the soil properties were measured only at the time of sampling. A more comprehensive experimental design may reveal additional variables linked to the observed bacterial responses.

For the changes in bacterial abundance or richness that were correlated with a soil physicochemical variable, most of the changes were direct climate effects. For example, the five bacterial groups with changes in relative abundance were correlated with soil water content (direct factor), while only the Proteobacteria were linked to an indirect factor, soil pH (Fig. 2; Table 2). Soil pH was found to be significantly increased by watering treatment in this steppe ecosystem (unpublished data). Similarly, the three bacterial groups with changes in richness were correlated with the direct factors, soil water content or soil temperature (Fig. 3; Table 2). In contrast, changes in bacterial group composition that were correlated with soil physicochemical factors were linked to both direct and indirect factors. Specifically, the compositional variation of the entire bacterial community was explained by both water content (direct factor) and TN content (indirect factor), and that of Acidobacteria was explained by both water content (direct factor) and soil pH (indirect factor; Table 2). Furthermore, the compositional variation of Bacteroidetes and Proteobacteria were explained by water content (direct factor) and soil pH (indirect factor), respectively (Table 2). Although the composition of Deltaproteobacteria community changed in response to watering (Table 1), it was linked to the indirect factor, soil temperature (Table 2). Overall, climate changes not only altered the abundance and richness of soil bacterial communities primarily through direct routes, but also altered their composition through both direct and indirect routes. Abundance and richness of a bacterial group are primarily determined by the environmental capacity (e.g., the total amount of energy), and hence, the direct effects of watering (soil water content) and warming (soil temperature) are logical. In contrast, besides environmental capacity, the composition of a bacterial group is also partly determined by environmental heterogeneity, and thus, watering/warming will also affect bacterial composition through alteration of indirect soil properties [16]. For example, under increased watering, the nearly neutral soil became alkaline; meanwhile, stimulated growth of plant community in this semiarid steppe transformed most of the soil nitrogen to aboveground plant biomass, resulting in a reduction in soil TN content (unpublished data). Therefore, to maintain soil bacterial composition under climate changes, management strategies may be required to address the changes in the indirect factors such as soil TN content and pH, in addition to alterations in the direct factors.

Although climate warming had small influence on the abundance, richness and composition of the entire soil bacterial community in the semiarid temperate steppe, it was found to have much larger effect in a tall grass prairie in central Oklahoma [6,9]. This difference is due to two potential causes. First, the experimental intensity and duration are different. In other words, climate warming may have larger influence with the increasing experimental duration or intensity. Second, water/precipitation is found to be important in affecting soil bacterial communities in both ecosystems [9], and the mean annual precipitation in the steppe (~386 mm) is smaller than that in the prairie (~900 mm). We propose that climate warming primarily affects soil bacterial community through decreasing soil water content (drying the soil), and that the bacterial communities in the semiarid steppe are more adaptive to the dried soil environment than those in the prairie due to the long-term evolutionary process. Thus, soil bacterial communities in the steppe are more insensitive to climate warming. However, these hypotheses need further test. Anyway, the effect of precipitation/water should be taken into account with care in studying the influence of climate warming on soil microbial communities.

Acknowledgments

We thank Linghao Li, Shiqiang Wan, Wenming Bai and many others for conducting and managing the long-term field experiment. We thank Qiuping Hu, Guo Yu and many others in Shanghai Majorbio Bio-pharm Technology Co., Ltd. for their help in pyrosequencing.

Author Contributions

Conceived and designed the experiments: XZ XH. Performed the experiments: XZ. Analyzed the data: XZ GZ QC. Wrote the paper: XZ.

References

1. IPCC (2007) IPCC WGI Fourth Assessment Report. Climatic Change 2007: the Physical Science Basis. Intergovernmental Panel on Climate Change, Geneva.
2. Dore MHI (2005) Climatic change and changes in global precipitation patterns: what do we know? Environment International 31: 1167–1181.
3. Groisman PY, Knight RW, Easterling DR, Karl TR, Hegerl GC, et al. (2005) Trends in intense precipitation in the climate record. Journal of Climate 18: 1326–1350.
4. Liu WX, Zhang Z, Wan SQ (2009) Predominant role of water in regulating soil and microbial respiration and their responses to climate change in a semiarid grassland. Global Change Biology 15, 184–195.
5. Yang HJ, Wu MY, Liu WX, Zhang Z, Zhang NL, et al. (2011) Community structure and composition in response to climate change in a temperate steppe. Global Change Biology 17: 452–465.
6. Zhou JZ, Xue K, Xie JP, Deng Y, Wu LY, et al. (2012) Microbial mediation of carbon-cycling feedbacks to climate warming. Nature Climate Change 2: 106–110.
7. Torsvik V, Øvreås L, Thingstad TF (2002) Prokaryotic diversity-magnitude, dynamics, and controlling factors. Science 296: 1064–1066.
8. Fierer N, Strickland MS, Liptzin D, Bradford MA, Cleveland CC (2009) Global patterns in belowground communities. Ecology Letters 12: 1238–1249.
9. Sheik CS, Beasley WH, Elshahed MS, Zhou XH, Luo YQ, et al. (2011) Effect of warming and drought on grassland microbial communities. The ISME Journal 5: 1692–1700.
10. Lennon JT, Aanderud ZT, Lehmkuhl BK, Jr. Schoolmaster DR (2012) Mapping the niche space of soil microorganisms using taxonomy and traits. Ecology 93: 1867–1879.
11. Fierer N, Bradford MA, Jackson RB (2007) Toward an ecological classification of soil bacteria. Ecology 88: 1354–1364.
12. Schloss PD, Handelsman J (2005) Metagenomics for studying unculturable microorganisms: Cutting the Gordian Knot. Genome Biology 6: 229.1–229.4.
13. Martiny JBH, Eisen JA, Penn K, Allison SD, Horner-Devine MC (2011) Drivers of bacterial beta-diversity depend on spatial scale. Proceedings of the National Academy of Sciences 108: 7850–7854.
14. Hamady M, Walker J, Harris J, Gold N, Knight R (2008) Error-correcting barcoded primers for pyrosequencing hundreds of samples in multiplex. Nature Methods 5: 235–237.
15. Rousk J, Bååth E, Brookes PC, Lauber CL, Lozupone C, et al. (2010) Soil bacterial and fungal communities across a pH gradient in an arable soil. The ISME Journal 4: 1340–1351.
16. Zhang XM, Han XG (2012) Nitrogen deposition alters soil chemical properties and bacterial communities in the Inner Mongolia grassland. Journal of Environmental Sciences 24: 1483–1491.
17. Fierer N, Jackson RB (2006) The diversity and biogeography of soil bacterial communities. Proceedings of the National Academy of Sciences USA 103: 626–631.
18. Christensen L, Coughenour MB, Ellis JE, Chen ZZ (2004) Vulnerability of the Asian typical steppe to grazing and climatic change. Climatic Change 63: 351–368.
19. Niu SL, Wu MY, Han Y, Xia JY, Li LH, et al. (2008) Water mediated responses of ecosystem carbon fluxes to climatic change in a temperate steppe. New Phytologist 177: 209–219.
20. Wan SQ, Xia JY, Liu WX, Niu SL (2009) Photosynthetic overcompensation under nocturnal warming enhances grassland carbon sequestration. Ecology 90: 2700–2710.
21. Sun Y, Ding Y (2010) A projection of future changes in summer precipitation and monsoon in East Asia. Science China (Earth Science), 53: 284–300.
22. Niu SL, Xing XR, Zhang Z, Xia JY, Zhou XH, et al. (2011) Water-use efficiency in response to climate change: from leaf to ecosystem in a temperate steppe. Global Change Biology 17: 1073–1082.
23. Wan SQ, Luo Y, Wallace LL (2002) Changes in microclimate induced by experimental warming and clipping in tall-grass prairie. Global Change Biology 8: 754–768.
24. Schloss PD, Westcott SL, Ryabin T, Hall JR, Hartmann M, et al. (2009) Introducing mothur: open-source, platform-independent, community-supported software for describing and comparing microbial communities. Applied and Environmental Microbiology 75: 7537–7541.
25. Pruesse E, Quast C, Knittel K, Fuchs BM, Ludwig W, et al. (2007) SILVA: a comprehensive online resource for quality checked and aligned ribosomal RNA sequence data compatible with ARB. Nucleic Acids Research 35: 7188–7196.
26. Lozupone CA, Hamady M, Kelley ST, Knight R (2007) Quantitative and qualitative beta diversity measures lead to different insights into factors that structure microbial communities. Applied and Environmental Microbiology, 73: 1576–1585.
27. Zhou XQ, Chen CR, Wang YF, Xu ZH, Han HY, et al. (2013) Warming and increased precipitation have differential effects on soil extracellular enzyme activities in a temperate grassland. Science of the Total Environment 444: 552–558.
28. Anderson MJ (2005) PERMANOVA: a FORTRAN computer program for permutational multivariate analysis of variance. Department of Statistics, University of Auckland, New Zealand.
29. Anderson MJ (2003) PCO: a FORTRAN computer program for principal coordinate analysis. Department of Statistics, University of Auckland, New Zealand.
30. Klappenbach JA, Pierson BK (2004) Phylogenetic and physiological characterization of a filamentous anoxygenic photoautotrophic bacterium 'Candidatus Chlorothrix halophila'. Gen. Nov., sp. Nov., recovered from hypersaline microbial mats. Archives of Microbiology 181: 17–25.
31. Hanada S, Hiraishi A, Shimada K, Matsuura K (1995) "Chloroflexus aggregans sp. nov., a Filamentous Phototrophic Bacterium Which Forms Dense Cell Aggregates by Active Gliding Movement". International Journal of Systematic Bacteriology 45: 676–681.
32. Pierson BK, Castenholz RW (1974) "A phototrophic gliding filamentous bacterium of hot springs, Chloroflexus aurantiacus, gen. And sp. Nov". Archives of Microbiology 100: 5–24.
33. Watson SW, Bock E, Valois FW, Waterbury JB, Schlosser U (1994) "Nitrospira marina gen. nov. sp. nov.: a chemolithotrophic nitrite-oxidizing bacterium." Archives of Microbiology 144: 62–9.
34. Ehrich S, Behrens D, Lebedeva E, Ludwig W, Bock E (1995) "A new obligately chemolithoautotrophic, nitrite-oxidizing bacterium, Nitrospira moscoviensis sp. nov. and its phylogenetic relationship." Archives of Microbiology 164: 16–23.
35. Hogenhout SA, Loria R (2008) Virulence mechanisms of Gram-positive plant pathogenic Bacteria. Current Opinion in Plant Biology 11: 449–456.
36. Quaiser A, Ochsenreiter T, Lanz C, Schuster SC, Treusch AH, et al. (2003) "Acidobacteria form a coherent but highly diverse group within the bacterial domain: evidence from environmental genomics". Molecular Microbiology 50: 563–75.

16

Analysis of the Cystic Fibrosis Lung Microbiota via Serial Illumina Sequencing of Bacterial 16S rRNA Hypervariable Regions

Heather Maughan[1*9], Pauline W. Wang[29], Julio Diaz Caballero[1], Pauline Fung[2], Yunchen Gong[2], Sylva L. Donaldson[2], Lijie Yuan[2], Shaf Keshavjee[3], Yu Zhang[3,4], Yvonne C. W. Yau[5], Valerie J. Waters[5], D. Elizabeth Tullis[6], David M. Hwang[3,4], David S. Guttman[1,2*]

1 Department of Cell and Systems Biology, University of Toronto, Toronto, Ontario, Canada, **2** Centre for the Analysis of Genome Evolution and Function, University of Toronto, Toronto, Ontario, Canada, **3** Latner Thoracic Surgery Laboratories, University Health Network, University of Toronto, Toronto, Ontario, Canada, **4** Department of Pathology, University Health Network, University of Toronto, Toronto, Ontario, Canada, **5** Research Institute, The Hospital for Sick Children, Toronto, Ontario, Canada, **6** Adult Cystic Fibrosis Clinic, St. Michael's Hospital, Toronto, ON, Canada

Abstract

The characterization of bacterial communities using DNA sequencing has revolutionized our ability to study microbes in nature and discover the ways in which microbial communities affect ecosystem functioning and human health. Here we describe Serial Illumina Sequencing (SI-Seq): a method for deep sequencing of the bacterial 16S rRNA gene using next-generation sequencing technology. SI-Seq serially sequences portions of the V5, V6 and V7 hypervariable regions from barcoded 16S rRNA amplicons using an Illumina short-read genome analyzer. SI-Seq obtains taxonomic resolution similar to 454 pyrosequencing for a fraction of the cost, and can produce hundreds of thousands of reads per sample even with very high multiplexing. We validated SI-Seq using single species and mock community controls, and via a comparison to cystic fibrosis lung microbiota sequenced using 454 FLX Titanium. Our control runs show that SI-Seq has a dynamic range of at least five orders of magnitude, can classify >96% of sequences to the genus level, and performs just as well as 454 and paired-end Illumina methods in estimation of standard microbial ecology diversity measurements. We illustrate the utility of SI-Seq in a pilot sample of central airway secretion samples from cystic fibrosis patients.

Editor: Neeraj Vij, Johns Hopkins School of Medicine, United States of America

Funding: This work was supported by: the Canadian Institutes of Health Research, Human Microbiome Emerging Team Grant (CMF108027) (http://www.cihr-irsc.gc.ca/e/193.html); the National Sanitarium Association (http://nationalsanitarium.ca/); and Cystic Fibrosis Canada (www.cysticfibrosis.ca/). The funders had no role in study design, data collection and analysis, decision to publish, or preparation of the manuscript.

Competing Interests: The authors have declared that no competing interests exist.

* E-mail: heather.maughan@utoronto.ca; david.guttman@utoronto.ca

9 These authors contributed equally to this work.

Introduction

Microbes inhabit all environments on earth and contribute to ecosystem function in ways that are only vaguely understood. Decades of culture-based study have provided a solid foundation for our understanding of microbial physiology, genetics, and evolution, but investigations of microbial ecology have been notoriously difficult due to the small size and vast diversity of microbes. Recent developments of culture-independent methods have illuminated the microbial ecosystems that are found in terrestrial and aquatic environments, as well as those intimately associated with organisms. Of particular interest are microbial ecosystems within the human body (the human microbiome), as they are expected to profoundly impact human health and disease. For example, bacterial communities within human intestines are being studied for their effects on atopic and bowel diseases [1,2], and altered structure of bacterial communities in the vagina is associated with susceptibility to a plethora of diseases including bacterial vaginosis and sexually transmitted infections [3].

The development of next-generation sequencing technology has enabled the large-scale investigation of microbial communities and stimulated broad interest in microbiota research. Similar to earlier studies that relied on Sanger sequencing, most of the next-generation microbiota studies have focused on a portion of the taxonomically informative 16S rRNA gene. 454 pyrosequencing was the first and is still the dominant next-generation sequencing technology applied to microbiota research because it can produce reads of sufficient length for appreciable taxonomic resolution. This need for adequate read length has generally hampered the use of Illumina sequencing by synthesis technologies that produce shorter reads despite the fact that the vastly higher throughput of the Illumina systems would enable much deeper and cost-effective sampling. Several groups have circumvented this problem by overlapping Illumina reads produced by paired-end sequencing (sequencing reads produced from opposite ends of the same template) to provide longer sequence albeit at the cost of decreased taxonomic resolution due to the unavoidable sequencing of the two highly conserved primer-binding regions [4,5,6,7,8,9].

Here we propose 16S rRNA gene-based community analysis by Serial Illumina Sequencing (SI-Seq), which provides a taxonomic resolution similar to 454 pyrosequencing with the throughput and cost-effectiveness of Illumina sequencing. SI-Seq is able to get the best of both technologies by adapting Illumina sequencing technology to read three hypervariable regions of the 16S rRNA gene. Here we describe the development of SI-Seq, its validation using single species controls and a mock community, and its comparison to 454 FLX 16S rRNA sequencing of 56 sputum samples collected from Cystic Fibrosis patients.

Materials and Methods

Control communities

Genomic DNA was extracted from overnight cultures of five bacterial species (*Bacillus subtilis*, *Pseudomonas aeruginosa*, *Staphylococcus aureus*, *Erwinia amylovora*, and *Burkholderia cepacia*) using the Gentra Puregene Cell kit (Qiagen). Genomic DNAs were quantified using the Qubit Fluorometer (Life Technologies) and mixed in abundances that differed 10-fold.

CF sample collection and processing

Ethics Statement. Protocols for specimen collection, distribution and use were approved by the University Health Network (UHN) and St. Michael's Hospital Research Ethics Boards.

Central airway secretion samples were aspirated intraoperatively using sterile syringes at the time of lung explantation, from consenting patients with an established diagnosis of CF undergoing transplantation in the Toronto Lung Transplant Program between July 2008 and December 2010. Sputum samples were voluntarily produced by consenting CF patients attending the CF clinic at St. Michael's Hospital. Samples were flash frozen in liquid nitrogen and stored in cryovials at −80°C until use. Whole genomic DNA was extracted using the DNeasy Blood & Tissue Kit, with lysozyme pretreatment (Qiagen, Toronto, ON).

16S rRNA amplicons

Primers for amplification and sequencing were designed using consensus nucleotides at each position of the Greengenes 16S rRNA alignment while also considering the high melting temperatures (>70°C) required for Illumina technology (Table 1). Primers were assessed for their matching to reference sequences using the "Probe Match" function of the Ribosomal Database Project (http://rdp.cme.msu.edu/probematch/search.jsp). Multiplexing versions of amplification primers V5+791 and V7-1104 had 8-mer barcodes attached to their 5′ ends. Eight barcodes were used for initial SI-Seq development, whereas a subsequent set of 96 barcodes was designed using the barcrawl program [10] for amplification of the CF sputum templates (Table S1).

Initially each 25ul PCR consisted of 1X Taq buffer (Invitrogen or Fermentas), 1.5 mM $MgCl_2$, 0.2 mM each primer, 0.2 mM dNTPs, 50 ng DNA and 0.4 U Taq polymerase (Native, Invitrogen); later reactions were also performed using the Kapa2G Robust Hotstart Ready Mix (Kapa Biosystems), which provided more reliable amplification. Minus-template control reactions were always included. Amplification conditions were: 2 minutes at 94°C; 30 cycles of 30 seconds at 94°C, 30 seconds at 56°C or 58°C, and 30 seconds at 72°C; followed by 5 minutes final extension at 72°C. All reactions were prepared in a sterile PCR hood.

Four independent 25ul PCR reactions were generated for each patient template. Replicate reactions were then pooled and concentrated using purification columns (Fermentas). To ensure removal of primers and any nonspecific amplicons, reactions were then run on 1% low melting temperature agarose gels and amplicons were extracted from agarose using Agarase (Fermentas) or purification columns (Fermentas), according to the manufacturer's instructions. Amplicons were quantified using the Qubit dsDNA BR assay kit (Life Technologies) and equal quantities of product from each patient were pooled before preparation for Illumina sequencing.

DNA library construction and Illumina sequencing

Libraries were constructed based on Illumina's paired-end library sample preparation protocol with the exception of starting at the 'A' base overhang addition, omitting the fragmentation and end repair steps to maintain the 16S rRNA gene amplicons.

Libraries were loaded onto the Illumina cBot cluster station for cluster generation according to the manufacturer's instructions. Sequencing was performed on an Illumina GA-IIx, and the sequencing recipe used is available for download in xml format from the University of Toronto Centre for the Analysis of Genome Evolution & Function (CAGEF) website (www.cagef.utoronto.ca). First strand sequencing began with the Illumina sequencing primer for eight cycles to read the barcode, followed by a denaturation chemistry and hybridization with a mix of the V5 forward and V7 reverse primers at 5 μM final concentration each. After 36 cycles of sequencing, another round of denaturation was performed and a mixture of V6 forward and V6 reverse primers (5 μM each) was hybridized for another 36 cycles of sequencing. The second strand was then synthesized using the paired-end protocol and the same procedures were carried out to generate the barcoded reads and reads for V5, V7 and V6 forward and V6 reverse. Data were pipelined using the standard Illumina computation pipeline. Library construction, sequencing, and data pipelining were performed at CAGEF.

454 pyrosequencing

Barcoded amplicons from 56 CF sputum samples were pooled and then split in half for sequencing using SI-Seq (described above) or 454 pyrosequencing; therefore, the same amplicons generated using primers V5+791 and V7-1104 (Table 1) were sequenced by

Table 1. Primers.

Primer Name	Primer Sequence (5′-3′)	% Match*
V5+791	GGGKAKCRAACVGGATTAGATACCCBGGTAGTCCWNRCHSTAAACGWTG	70.4
V6+953 (V6a)	AAGCRGHGGADYRTGYGGYTYAATTCGANGMWAMGCGMRRAACCTTACC	67.3
V6-976 (V6b)	CTCACRRCACGAGCTGACGACRRCCATGCASCACCT	88.5
V7-1104	GGSCRTRMKGAYTTGACGTCRYCCCCDCCTTCCTCC	79.7

*RDP Database match (% Bacteria), including perfect match to sequences of 'Good' quality and at least 1200nt

both technologies. For 454 sequencing, amplicons were submitted to The Centre for Applied Genomics (Toronto, ON) for adaptor ligation and one half plate of single read pyrosequencing on a 454 FLX Titanium sequencer. The average raw 454 read length was 386nt (range 40 to 677); QIIME quality trimming (see below) resulted in an average read length of 415nt (range 200 to 573).

Structuring local version of reference databases

Sequence reads were simulated using JAligner [11] and in-house Java scripts as follows. Sequencing primers (Table 1 or published [4,5,12,13]) were aligned to each taxon in the database (Bergey, Silva, or Greengenes) using JAligner with the EDNA-FULL matrix, and gap opening and extension penalties of 15 and 1, respectively. DNA fragments downstream from the aligned primer (SI-Seq), or starting at the 5′ end of the aligned primer (454 and Illumina Paired-End), were then extracted and concatenated. The proportion of each read set that could be assigned to the genus level was determined by classifying the simulated reads using the RDP Classifier [14] (trained with the same read structure) and choosing the taxonomic classification with bootstrap support greater than or equal to 0.7.

Preparing sequence reads for classification

Sequence data in FASTQ format were subjected to quality control analysis and prepared for taxonomic classification using an in-house Java application made available on CAGEF website (Figure S1). Reads were assigned to patient samples based on the barcode data by concatenating the two barcodes sequences obtained from each of the two individual paired-end reads and comparing the sequences to a reference barcode dataset. Barcode sequences were required to match a reference barcode sequence at 13 of 16 sites to be retained. Read quality was checked using the FASTQ data. Multiple quality-filtering measures were tested with the controls and those with >6 bases with a Phred quality score below 20 were discarded.

Since SI-Seq produces reads in two possible orientations it was necessary to concatenate the two paired-end reads in both orientations (i.e. Read1::Read2 and Read2::Read1) to ensure that one of resulting concatenated sequences was reconstructed in the proper orientation to match the reference database. Paired reads were trimmed to discard the leading 8-bp barcode and then concatenated in both orientations. The concatenated reads were then individually mapped by UBLAST [15] to a Greengenes dataset (reconstructed with simulated reads structured in V5::V6a::V7::V6b orientation) using an identity cutoff of 0.8. The concatenated read that provided the best UBLAST result was retained as the "correct orientation".

OTU clustering and taxonomic classification

SI-Seq reads. Operational Taxonomic Units (OTUs) were identified using the OTU_PIPE function within USEARCH [15]; this step also identifies and excludes chimeric reads. The appropriate identity cutoff for OTU clustering was determined empirically using the single species controls, as follows. We determined the average pairwise nucleotide identity of SI-Seq structured sequences from all members of the genera *Pseudomonas* (89% average pairwise identity) and *Bacillus* (85%) in the Bergey database. We clustered starting at the average of these two identities (87%) and repeated the process using identities of 97%, 95%, 93%, 90%, 85%, 83%, and 80%. The consensus sequences for each cluster at each identity cutoff were then classified using the RDP Classifier [14] and the clustering identity that most closely recapitulated the single species input with a low level of false positive classification was chosen. Because 87% was found to be the average nucleotide identity within *Pseudomonas* and *Bacillus*, and also had the optimal empirical results with the single species controls, we performed all clustering of SI-Seq data using an identity cutoff of 87%. Seed sequences from each cluster from each community were classified by further clustering with the SI-Seq structured Silva database using USEARCH, with parameters values chosen to ensure the best hit was chosen (--maxaccepts = 0 and --maxrejects = 0).

Additional classifications were performed on each read (i.e., not clustered into OTUs) using the RDP Classifier [14]. Reference databases included three versions of the Bergey database: the SI-Seq structured database, the V5 454 simulated dataset, and the full-length database. RDP assignments with greater than 0.8 bootstrap support were incorporated into OTU tables using an in-house PERL script.

454 reads. 454 pyrosequencing data files were de-barcoded and quality filtered using the QIIME pipeline [16] with default parameters except for: a minimum quality score of 20, a maximum primer mismatch of 15, and a maximum of 2 errors in the barcode. A separate pipeline was run using a minimum quality score of 25 and a maximum primer mismatch of 5. The resultant reads were analyzed using the QIIME analysis pipeline with default parameters, unless otherwise indicated. Read trimming and filtering were also done separately using Mothur [17] with the following parameters: minlength = 200, maxlength = 300, maxambig = 0, maxhomop = 10, qwindowsize = 25, qwindowaverage = 30, bdiffs = 1, pdiffs = 10.

For both read sets (454 and SI-Seq) questionable OTUs (i.e., below the abundance cutoff) were removed by manual inspection of OTU tables. Abundance cutoffs were determined by run-specific single species controls, as described in the Results section and shown in Tables S2 & S3.

Community ecology analyses

OTU tables for 454 and SI-Seq sequences were analyzed using QIIME and the R package VEGAN [16,18]. Rarefaction analyses for assessment of community coverage were performed in QIIME using 100 rarefactions for each sequence sampling step. The alpha diversity metrics Chao1, observed species, and Phylogenetic Diversity (PD) whole tree were estimated in QIIME during rarefaction. The beta diversity metrics Bray-Curtis (BC), unweighted UniFrac (UF), and weighted UniFrac (WUF) were estimated in QIIME on OTU tables rarefied to the lowest number of sequences per sample (454: 478 sequences; SI-Seq: 38,120 sequences).

The reference tree used for estimating phylogenetic based metrics with the SI-Seq data was inferred using full-length Silva database hits. Full length sequences were aligned using gap opening and extension costs of 30 and 10, respectively, and alignments were trimmed to remove gaps at the beginning and end of the alignment that were due differences in sequence length. Phylogenetic inference was performed using the Neighbor-Joining algorithm with 1000 bootstrap pseudoreplicates. Alignment and phylogenetic inference were performed in CLC Genomics Workbench version 4.9 (Århus, Denmark).

Principal components analyses of 454 and SI-Seq data were performed separately in QIIME [16], using BC, UF, or WUF dissimilarity matrices. Procrustes analysis was used to compare the communities in the 454 and SI-Seq datasets in QIIME. Separate analyses were performed on the BC, UF, and WUF derived principal components. 1000 Monte Carlo simulations were performed to assess significance of the Procrustes matrix correlations. Mantel tests with 1000 permutations were also performed in QIIME to assess significance.

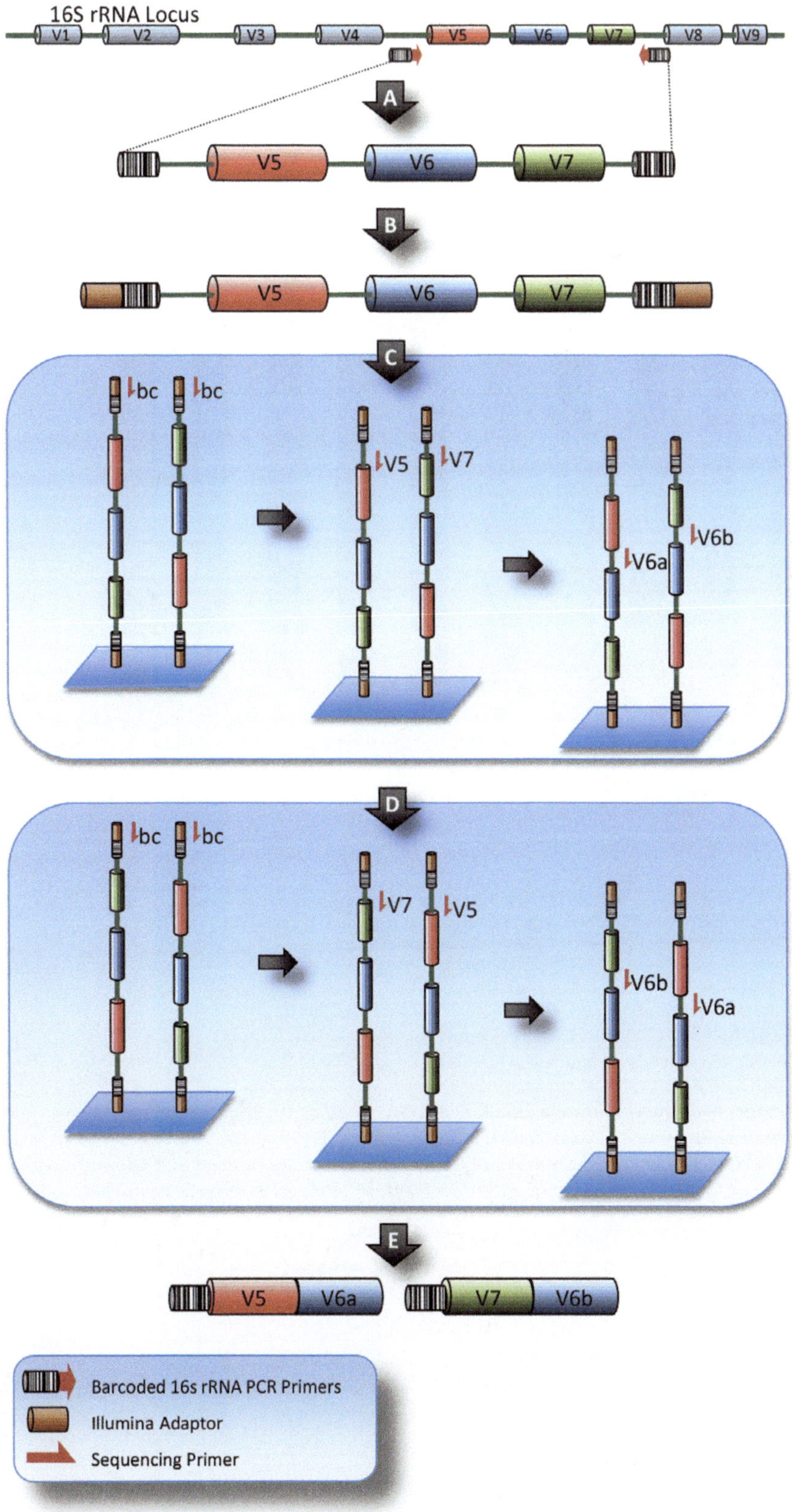

Figure 1. Serial Illumina Sequencing (SI-Seq). (A) Schematic of the 16S rRNA locus with the hypervariable regions indicated as cylinders. The V5 through V7 hypervariable regions are PCR amplified with using universal primers tagged with 8-mer barcodes on their 5′ ends. (B) Ligation of Illumina paired-end adaptors to the 16S rRNA amplicon. (C) 5′ attachment to of each template onto the Illumina flowcell followed by cluster generation (not shown) and serial sequencing of each template using three sequencing primer sets: Illumina sequencing primers to read barcodes (bc); a mix of V5 and V7 primers; and a mix of V6a and V6b primers. Note that each template can bind in either the V5:V6:V7 or the V7:V6:V5 orientation. (D) Fixed templates are flipped on the flowcell and regenerated for pair-end sequencing. Serial sequencing as described in (C) is repeated. (E) SI-Seq data structure showing the barcode, the V5 read and the V6a read, or the barcode, the V7 read and the V6b read.

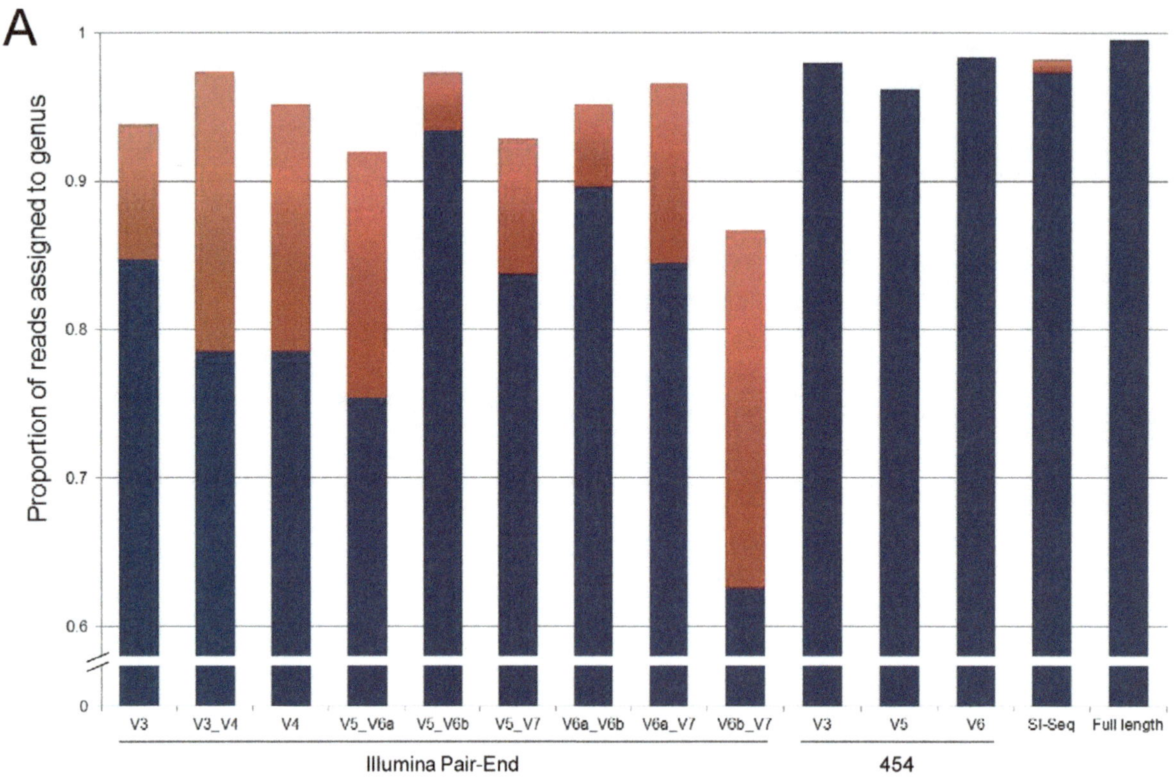

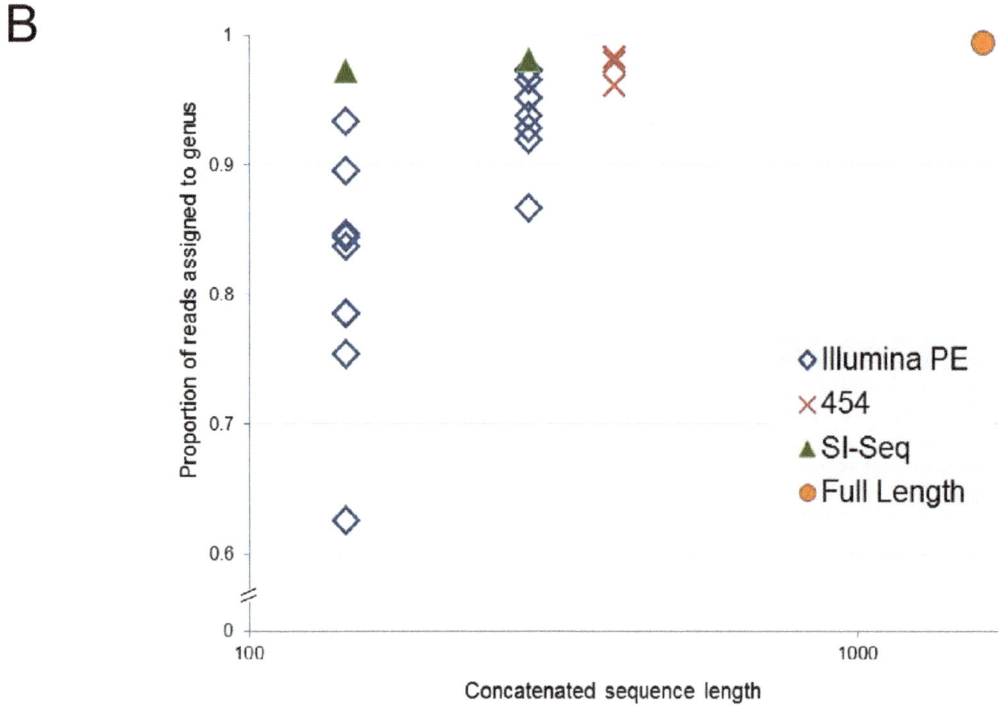

Figure 2. ***In silico*** **simulation of next-generation 16S rRNA sequences based on Bergey database.** (A) The proportion of simulated sequences that can be assigned to the genus level using the RDP Classifier for various next-generation 16S rRNA gene sequencing approaches. For the Illumina paired-end data, the blue bars indicate 72nt paired-end reads, while the red bars indicate 144nt paired-end reads (similar to the expected 150 bp Illumina MiSeq reads). For the SI-Seq data, the blue bar indicates 36nt reads, while the red bar indicates 72nt reads. 454 data corresponds to 400nt reads. Full length corresponds to sequencing of the entire 16S rRNA gene. (B) The proportion of reads assigned to the genus level as a function of total concatenated sequence length.

Species Accumulation curves were calculated in VEGAN [18], using rarefied OTU tables. Patient samples were randomly drawn and the cumulative number of OTUs was calculated. This was repeated 100 times. An additional curve was calculated after all OTUs with a combined abundance lower than 0.1% were removed from the dataset.

Results

Description of SI-Seq method

We sequenced the V5 through V7 region of the 16S rRNA locus to balance taxonomic resolution with fragment size restrictions imposed by Illumina sequencing technology. PCR amplicons of ~300 bp were generated by amplification of the 16S rRNA gene using bacterial universal primers with 5′ barcodes (Table 1 & Table S1). These amplification products were pooled and libraries were prepared for SI-Seq Illumina sequencing as described in the Materials and Methods. The resulting SI-Seq library consisted of templates with 16S rRNA V5–V7 amplicons flanked by the barcoded universal 16S rRNA PCR primers flanked by the standard Illumina paired-end adaptors (Figure 1). It is important to recognize that each individual SI-Seq template has the potential to bind to the flowcell in either the V5–V7 orientation or the V7–V5 orientation (Figure 1).

As discussed in the Methods section, the SI-Seq library was serially sequenced using multiple template-specific primer sets. The first sequencing step used the standard Illumina sequencing primer to sequence off the Illumina adaptors and into the barcode attached at the 5′ end of the 16S rRNA PCR primers. After eight cycles, this sequencing primer was removed by denaturation and replaced with a mix of V5 and V7 sequencing primers. These primers hybridize in conserved 16S rRNA regions just outside of the appropriate hypervariable regions and sequence into the hypervariable regions. Sequencing was performed with these primers for 36 cycles and followed by another round of denaturation and the addition of two distinct V6 sequencing primers [V6+953 (V6a) and V6-976 (V6b)]. The V6 sequencing primers hybridize to two conserved regions on opposite sides of the V6 hypervariable region and sequence into the V6 region from both orientations. After another round of sequencing, the two V6 primers were removed by denaturation and the clusters were flipped and regenerated using the Illumina pair-end sequencing protocol. After pair-end cluster generation the entire process was repeated by serially sequencing with the Illumina barcode primer, the V5 and V7 primers, and finally the two V6 primers (Figure 1). Overall, a total of six sequencing steps were performed with 144 nucleotides read from the 16S rRNA locus (4 sequencing steps of 36 cycles each) and the eight nucleotide barcode read twice. The

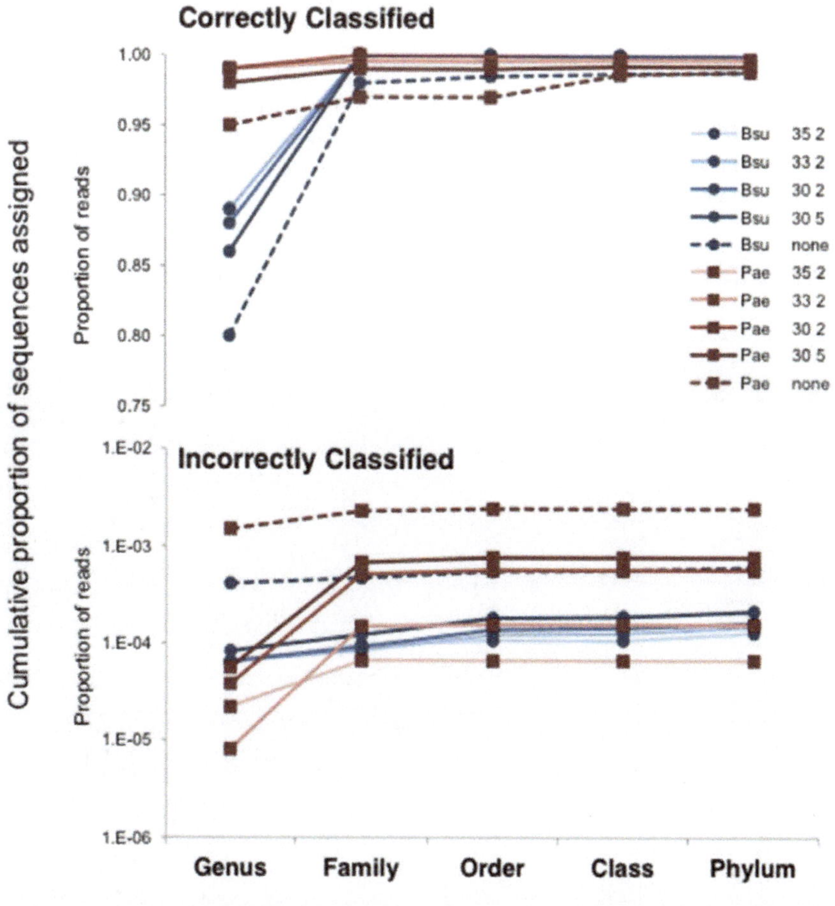

Figure 3. The effect of quality filtering on taxonomic classification. Cumulative distributions are shown for the proportion of sequences assigned to a particular level of taxonomy. Correctly classified sequences are shown in the top panel; incorrectly classified sequences are shown in the bottom panel. Bsu = *B. subtilis*; Pae = *P. aeruginosa*. The numbers following Bsu and Pae indicate the quality filtering applied. The first number is the Phred score cutoff and the second number is the number of sites allowed with a quality score below the Phred cutoff. For example, 35 2 indicates that sequences were only retained if they had two or fewer sites with a Phred score lower than 35. "none" indicates no quality filtering was performed.

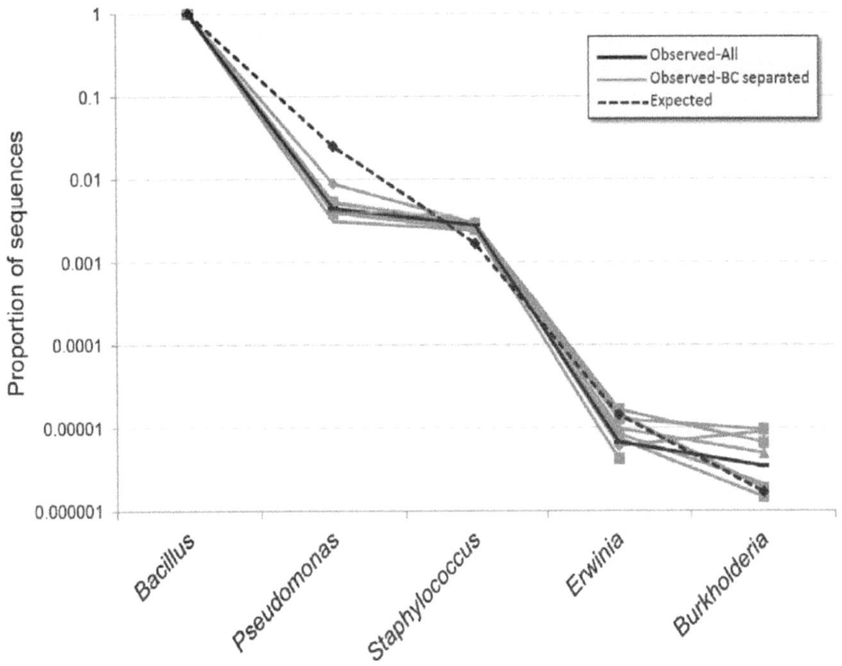

Figure 4. Observed and expected abundances of taxa in mock community control. Observed abundances are given as totals and as barcode-separated datasets. Expected abundances have been corrected for the number of rRNA operons present in each species' genome.

primers used and specific number of cycles performed can be easily altered to fit specific needs.

SI-Seq data is generated in two possible orientations depending on how the template binds to the flowcell. The first orientation is barcode:V5:V6a for read one of the pair and barcode:V7:V6b for read two, while the second orientation is the reverse with barcode:V7:V6b for read one and barcode:V5:V6a for read two (Figure 1). These two orientations can be distinguished by a rapid UBLAST analysis [15] as described in the Materials and Methods.

To compare SI-Seq to other commonly used 16S rRNA sequencing protocols (e.g., 454 and Illumina Paired-End), we simulated reads of varying lengths using published primer sequences [4,5,12,13]. From these simulated reads we calculated the proportion of reads that could be assigned to the genus level using the RDP Classifier [14]. As shown in Figure 2, SI-Seq and 454 were similar in their ability to produce a high proportion of reads that could be assigned to the genus level. These data indicate that SI-Seq is in theory comparable to 454 in its ability to produce reads with enough sequence variation for good taxonomic

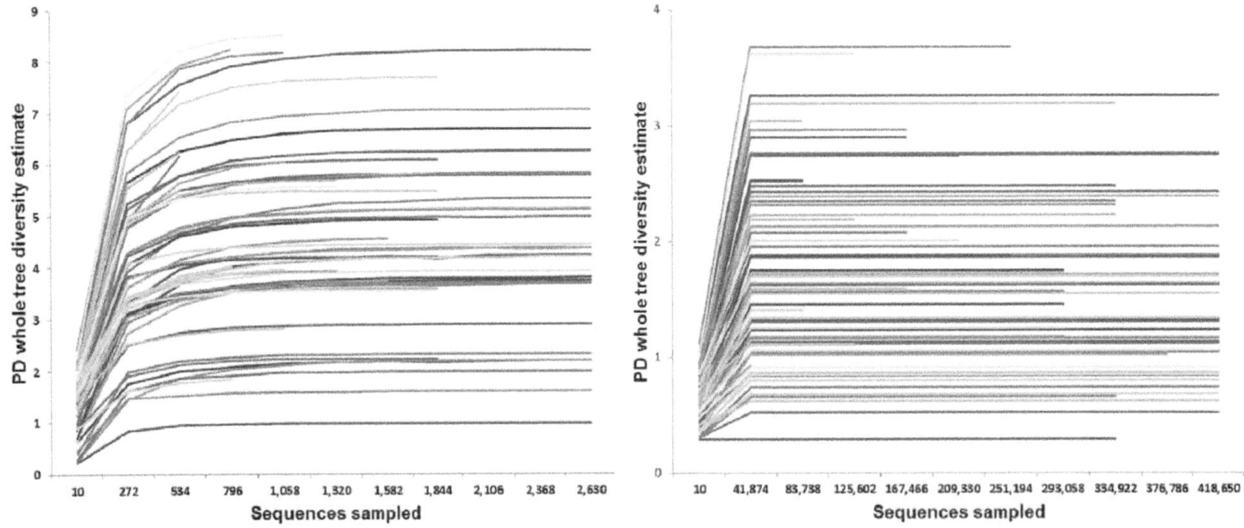

Figure 5. Rarefaction curves from 100 resamplings of each patient's community at different sequencing depths. The y-axis indicates the average Phylogenetic Diversity (PD) whole tree diversity estimate for each sample at each sequencing depth. Data shown are from the 454 OTU table (left), and SI-Seq OTU table (right). Each line corresponds to one sample listed in Table 3.

assignment, yet being Illumina-based, produces over 100-fold more reads at similar cost.

Validation with single species controls

Amplicons from either *Bacillus subtilis* or *Pseudomonas aeruginosa* genomic DNA were sequenced separately by SI-Seq and the resulting reads were used to determine the overall quality of the SI-Seq data and how quality filtering influenced the accuracy of taxonomic classification.

Filtering the sequences to obtain the optimal dataset is a balance between retaining too many sequences of poor quality that may give erroneous taxonomic classifications in order to favor dataset size (type I error) versus discarding too many moderate quality sequences in order to favor dataset quality (type II error). We filtered sequences for quality using four different cutoffs and then classified the retained sequences using the RDP Classifier [14] (Table S2); RDP reference sequences were trimmed to retain only those regions of the 16S rRNA locus recovered by the SI-Seq protocol.

We calculated the rate of taxonomic classification error by calculating the number of SI-Seq sequences whose taxonomy was inconsistent with *B. subtilis* or *P. aeruginosa*. Across several SI-Seq development runs, this average rate of classification error was ~0.03%. Figure 3 shows the cumulative distributions of sequence proportions assigned to each taxonomic level, for both the correct and incorrect classifications. These data highlight the usefulness of quality filtering: the stricter the quality filter, the larger the proportion of sequences assigned to the genus level. This was particularly apparent for *B. subtilis*. Also evident is that the overwhelming majority of the single species datasets are comprised of sequences that are classified correctly.

Because every Illumina run varies in quality, we included a single species control as one barcoded sample to empirically determine the rate of classification error. This value is then used as a lower, run-specific threshold for elimination of low abundance OTUs.

Validation with a mock community

A mock community control was sequenced to determine whether PCR amplification introduced taxonomic biases and to identify the dynamic range of SI-Seq taxon detection. Our control community consisted of genomic DNA from *Bacillus subtilis*, *Pseudomonas aeruginosa*, *Staphylococcus aureus*, *Erwinia amylovora*, and *Burkholderia cepacia* that were mixed at abundances that differed from each other by an order of magnitude. As shown in Figure 4, the observed taxon abundances were nearly identical to those expected. Furthermore, we were able to detect taxon abundances that ranged over five orders of magnitude. The taxon of lowest abundance, *B. cepacia*, had only 120 pg of total genomic DNA added to the original pool, of which less than 1,000th of which would be 16S rRNA; thus, the reliable detection of *Burkholderia* in all separate barcoded samples verifies the repeatability and sensitivity of SI-Seq.

Comparison of SI-Seq to 454 FLX Titanium

To further validate SI-Seq, we sequenced 56 sputum samples collected from CF patients using both SI-Seq and 454 FLX Titanium. Samples for both platforms were taken from the same PCR stock to ensure no PCR bias. Sequence reads from 454 and SI-Seq were filtered for quality and de-barcoded either using QIIME (454) or a pipeline developed in-house (SI-Seq) (as described in Materials and Methods). The range of sequences obtained per sample differed substantially between the two sequencing approaches: between 478 and 11,405 quality reads were obtained using 454, whereas between 38,120 and 859,764 quality reads were obtained using SI-Seq (Table 2). The average length for the 454 reads was 415nt, with 33% of the reads shorter than the average. A rarefaction sampling analysis indicated that the number of reads obtained for each patient sample saturated the community diversity in all cases for the SI-Seq data and in most cases for the 454 data (Figure 5). Both rarefaction analyses suggest that ~1,000 sequences would be sufficient for characterizing the majority of bacteria present in these CF lung communities. The saturation of community diversity with such low read numbers indicates that a much higher degree of multiplexing would further reduce per patient sequencing costs without reducing community information.

Table 2. Cystic Fibrosis samples.

		Number of Sequences					
	Sample	Time 1		Time 2		Time 3	
Patient	type	SI-Seq	454	SI-Seq	454	SI-Seq	454
0002	Sputum	319,749	11,405	339,814	7,842		
0004	Sputum	474,226	8,467	443,729	3,941		
0005	Sputum	547,415	7,973	472,506	3,703		
0006	Sputum	859,784	8,008	824,181	8,471		
0007	Sputum	484,648	4,299	538,229	6,587	613,251	5,527
0028	Sputum	383,201	1,650	347,991	3,179		
0035	Sputum	488,906	1,620	363,620	2,008		
0037	Sputum	286,778	2,069	205,463	581		
0052	Sputum	86,380	570	103,260	1,876	100,212	1,089
0060	Sputum	567,023	1,877	799,027	4,760	438,495	4,721
0064	Sputum	418,650	4,075	241,436	4,162	495,277	4,138
0068	Sputum	437,179	3,189	328,141	3,248		
0073	Sputum	360,834	5,418	468,731	6,364		
0077	Sputum	517,309	6,827	576,069	5,161	488,426	6,848
0082	Sputum	554,580	7,614	692,440	8,516		
0083	Sputum	315,906	5,127	509,827	6,394	139,260	580
0089	Sputum	288,846	938	167,530	1,230		
0092	Sputum	38,120	478	81,624	526	315,702	2,088
0096	Sputum	378,450	960	371,370	1,400	299,837	1,489
0104	Sputum	349,007	1,855	157,340	1,276		
0105	Sputum	46,065	1,170	159,400	980		
0106	Sputum	354,331	1,999	190,479	683		
0107	Sputum	183,695	947	221,825	1,392		
0109	Sputum	77,311	1,176	97,820	1,892		
S1	Airway	11,426,004					
S2	Airway	7,037,404					
S3	Airway	1,385,423					
S4	Airway	1,448,413					
S5	Airway	7,191,912					
S6	Airway	2,149,307					
S7	Airway	3,271,206					
S8	Airway	3,380,628					
S9	Airway	6,546,059					
S10	Airway	2,495,002					

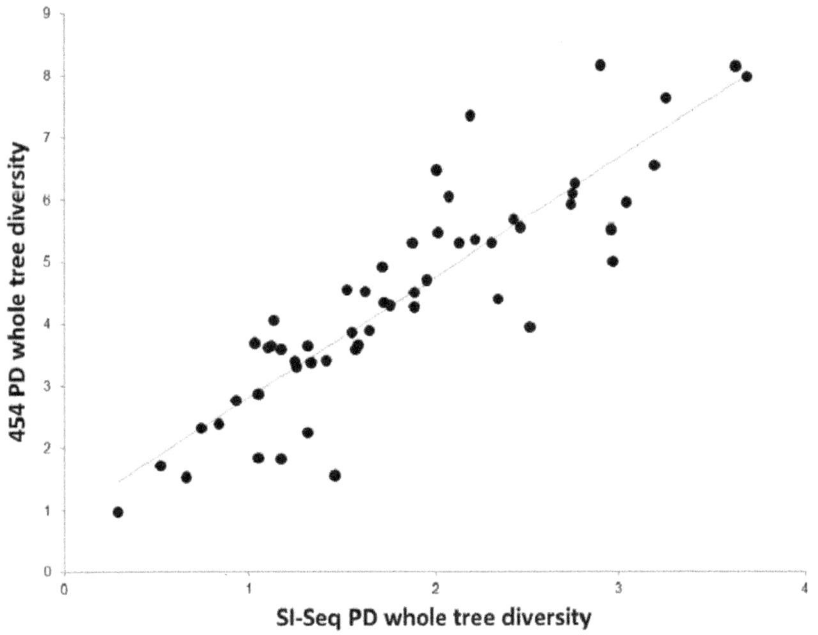

Figure 6. Comparison of phylogenetic diversity (PD). Whole tree PD diversity measurements from the 454 and SI-Seq analysis using rarefied OTU tables with 478 reads per patient for 454 and 38,120 reads per patient for SI-Seq. Very strong correlation is observed between the two approaches.

Table 3. Taxonomic Classification Comparison.

			Proportion of Sequences Classified to Each Taxonomic Level					
Platform	Template[1]	Ref.DB[2]	Domain	Phylum	Class	Order	Family	Genus
454	PAO1	FL	0	0	0.29	0	0.55	0.16
454	PAO1*	FL	0	0	0.17	0	0.20	0.63
454	PAO1	V5	0	0	0.25	0	0.58	0.17
454	PAO1*	V5	0	0	0.14	0	0.20	0.65
454	PAO1	SI-Seq	0.76	0.18	0.06	0	0	0
SI-Seq	PAO1	FL	0.91	0.09	0	0	0	0
SI-Seq	PAO1	V5	0.56	0.37	0.07	0	0	0
SI-Seq	PAO1	SI-Seq	0	0	0.03	0	0.06	0.91
454	CFS	FL	0.01	0	0.14	0.02	0.27	0.56
454	CFS*	FL	0.01	0	0.06	0	0.08	0.85
454	CFS	V5	0.01	0	0.13	0.02	0.28	0.57
454	CFS*	V5	0.01	0	0.05	0	0.08	0.86
454	CFS	SI-Seq	0.82	0.13	0.04	0	0	0.01
454	CFS-SS	SI-Seq	0.14	0.05	0.10	0.04	0.06	0.61
454	CFS-HQ	V5	0.01	0	0.13	0.02	0.28	0.56
SI-Seq	CFS	FL	0.91	0.06	0.01	0.01	0	0.01
SI-Seq	CFS	V5	0.62	0.27	0.07	0.02	0	0.01
SI-Seq	CFS	SI-Seq	0.01	0	0.01	0	0.02	0.95

[1]Template sequenced: PAO1, *P. aeruginosa* PAO1; CFS, CF sputum community; CFS-SS, CF sputum community with 454 reads trimmed to correspond to the SI-Seq data format; CFS-HQ, CF sputum community using only high quality 454 reads (Q>25). *indicates reads were trimmed using Mothur [17] instead of QIIME [16].
[2]Reference Database: FL, full length database sequences; V5, all database sequences trimmed to 400 bp starting at our V5 primer; SI-Seq, database sequences formatted to correspond to the SI-Seq read structure.

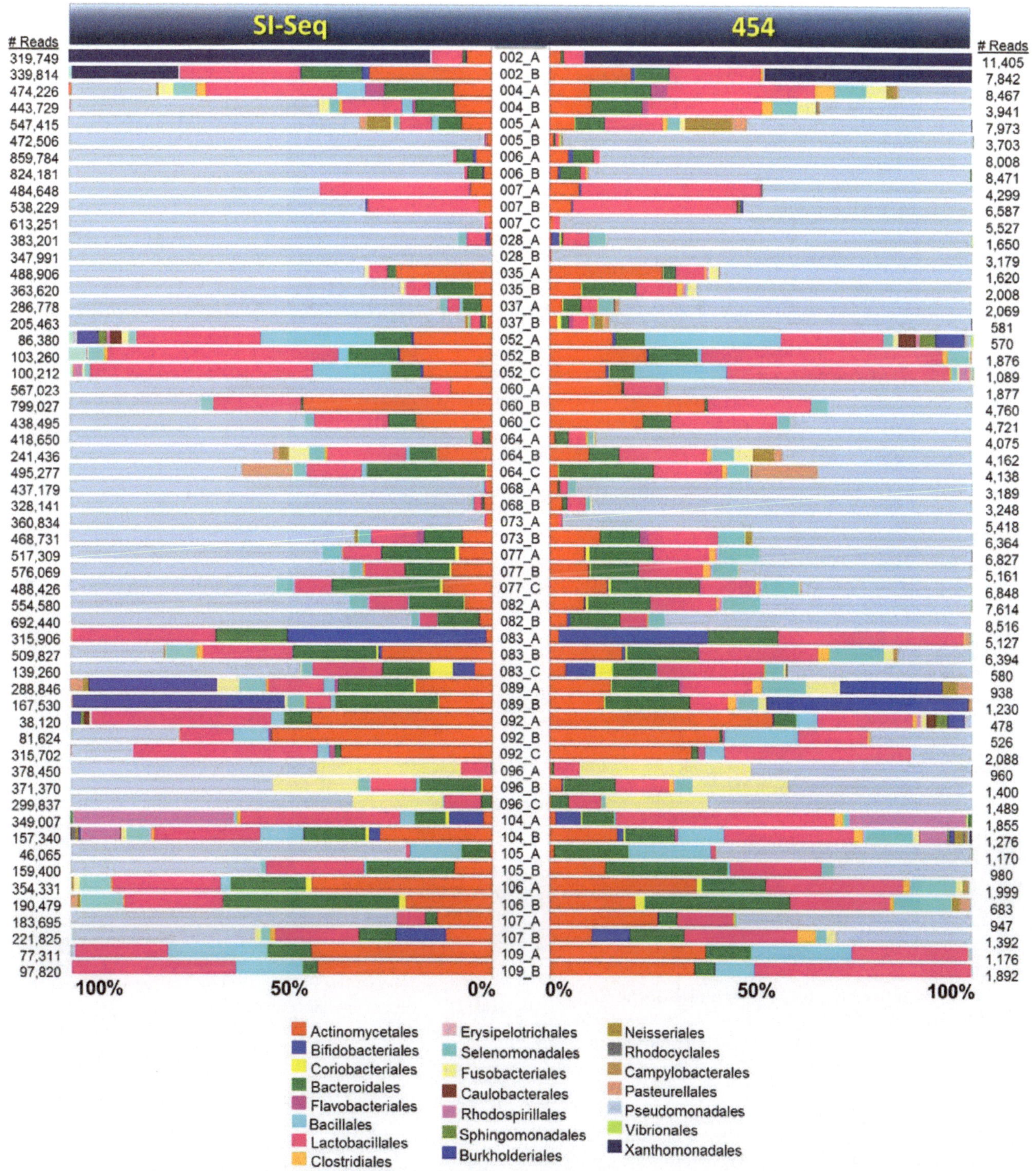

Figure 7. Relative proportions of bacterial orders found in each sample as assessed by SI-Seq and 454 sequencing approaches. Note that the relative taxonomic structure of the communities inferred for each sample are nearly identical using the two methods, although SI-Seq generates ~100-times more data, thereby permitting a higher level of resolution and larger dynamic range. Data is based on OTUs collapsed to the order level using the MEGAN ver4 software package [35]. Sample IDs are presented in the center of the figure. Relative abundance of bacterial orders for each sample are presented in mirror order for SI-Seq on the left and 454 on the right. The color code for each bacterial order is presented on the bottom. The number of reads (sequences) for each sample are presented on either side.

To compare within sample diversity estimates from 454 and SI-Seq communities, we calculated alpha diversity (Phylogenetic Diversity) on rarefied OTU tables (454: 478 reads per patient; SI-Seq: 38,120 reads per patient). It is important to use rarefied data to normalize the number of sequences per sample; this is because samples with greater sequencing depth are expected to have higher diversity due to sequence coverage alone. Alpha diversity metrics from 454 and SI-Seq data were significantly correlated (Spearman's rho: 0.90; $P<2.2\times10^{-16}$; Figure 6).

We used the RDP Classifier to determine the number of reads from each approach that could be classified down to the genus level. Quite strikingly, an average of only 56% of 454 reads were classified to the genus level, while 95% of SI-Seq reads carried this resolution (Table 3). Trimming the 454 reads to include only those regions sequenced by SI-Seq increased this percentage to 61%. Separate quality filtering and processing of raw reads using Mothur [17] increased the percentage of reads assigned at the genus level to 85%. These results support previously published findings showing that the 454 platform has a substantially higher error rate than the Illumina platform, particularly when generating only very short Illumina reads as is done in the SI-Seq protocol [19,20,21].

To determine whether the high proportion of genus level assignments for the SI-Seq data were the result of incorrect high-confidence classifications due to the short SI-Seq read length, we compared the classifications of the single species control data generated by 454 and SI-Seq runs (Table 3). We also used multiple versions of the reference database to ensure that the database structure did not bias the results, including full length reference sequences, reference sequences trimmed to 400 bp starting at our V5 primer, and SI-Seq formatted reference sequences. When classifying each dataset with the appropriately formatted reference database, this analysis remarkably showed that 91% of reads from the single species control (*P. aeruginosa* PAO1) were correctly classified with SI-Seq data compared to only 16% of the 454 data.

We compared the 454 and SI-Seq community profiles visually and statistically to verify their similarities. We first compared the per-sample distribution of bacterial orders obtained from the 454 and SI-Seq OTU tables (Figure 7). Despite differing sequencing methods, OTU clustering thresholds, and taxonomy classification approaches, the composition of each community for each sample was remarkably similar between the 454 and SI-Seq datasets. We next tested for statistical significance using a Procrustes analyses to determine whether the OTU dissimilarity matrices obtained from each approach were statistically indistinguishable. We applied a variety of distance metrics and found that the match between community profiles generated by 454 and SI-Seq were highly significant regardless of the dissimilarity matrix used (Table 4; Figure S2).

Application of SI-Seq to study the ecology of the CF lung

To demonstrate the utility of SI-Seq, we performed additional analyses on the CF sputum samples to begin our investigation of CF lung ecology, although a more in depth analysis of this subject will be completed in the future with a larger patient cohort.

We used a Species Accumulation curve to determine if the bacterial diversity observed in our patient cohort represented the overall bacterial diversity present in CF lungs. This approach measures how many new OTUs are identified as additional samples are cumulatively added to the analysis. As shown in Figure 8A, the number of OTUs increased quickly between 0 and 20 samples, and began to plateau by the end of our sampling; this indicated that we have largely saturated the diversity found in this environment. Many of the OTUs detected with increased sputum sampling were rare, as reduction of the dataset to include only those OTUs with a total abundance greater than 0.001 resulted in a plateau around 25 sputum samples (Figure 8B).

We also performed SI-Seq on ten samples extracted from mucus plugs obstructing the airways of lungs surgically removed from CF lung transplant patients, and combined these results with those from the 56 sputa discussed above. Clustering of community profiles from these two sample types indicated substantial overlap in community composition of sputum and lung airway samples (Figure 9A). The heatmap in Figure 9A shows three types of patient samples: (1) those with only *P. aeruginosa*; (2) those with predominantly *P. aeruginosa* with additional minor constituents; and (3) those with little or no *P. aeruginosa*, but either a diversity of other bacterial species or a community dominated by *Streptococcus* or *Burkholderia*. The ten airway samples were evenly divided among these three groups. Despite this similarity, a principal components analysis using weighted UniFrac dissimilarities showed a separation of the sputum and lung airway samples (Figure 9B). This suggests that when phylogenetic distance is taken into account (as was done with the weighted UniFrac but not the Euclidean distance based heatmap), sputum and airway samples slightly differ in their microbiota.

We also compared our SI-Seq data with clinical culture data and found complete agreement (data not shown). There were no samples in which a particular bacterial species was cultured but not detected using SI-Seq.

Table 4. Procrustes comparisons of 454 and SI-Seq PCA plots.

Dissimilarity matrix used for PCA plot	$M^{2\dagger}$	P value*
Bray-Curtis	0.257	<0.001
Unweighted UniFrac	0.649	<0.001
Weighted UniFrac	0.325	<0.001

†Goodness of fit between 454 and SI-Seq PCA plots
*Based on 1000 random permutations of SI-Seq matrix

Discussion

Targeted sequencing of taxonomically informative regions of the 16S rRNA gene has been instrumental in describing and comparing microbial communities and ecosystems. Associating changes in these microbiota with disease etiology or ecosystem function requires a method that fulfills the following criteria: good taxonomic resolution; deep enough sequencing to ensure community saturation; and a cost that permits multiple biological replicates and fine temporal or spatial sampling. Our side-by-side comparison of 454 and SI-Seq profiling of single species controls and CF sputum samples revealed the superiority of SI-Seq regarding each of these criteria; below we discuss these in turn.

SI-Seq profiling permits as good or better taxonomic resolution as 454 profiling. Analysis of CF communities using the RDP Classifier found that a maximum of 85% of the 454 reads were classified to the genus level, compared to 95% of SI-Seq reads. This result seems counter-intuitive since the 454 reads are substantially longer than SI-Seq reads. However, highly conserved regions are included in the 454 reads and these regions do not segregate for sufficient or appropriate genetic variation to discriminate genera. SI-Seq passes over these highly conserved regions and reads only hypervariable regions that permit greater taxonomic resolution.

An additional factor affecting taxonomic resolution is the error profiles of the two platforms. It is well established that the 454 platform has an approximately 10-fold higher error rate compared to the Illumina platform [19,20], and most Illumina sequencing errors are generated in later cycles [21]. Because SI-Seq uses only very short reads, it generates data with an extremely low error rate. This conclusion is supported by our classifications of CF

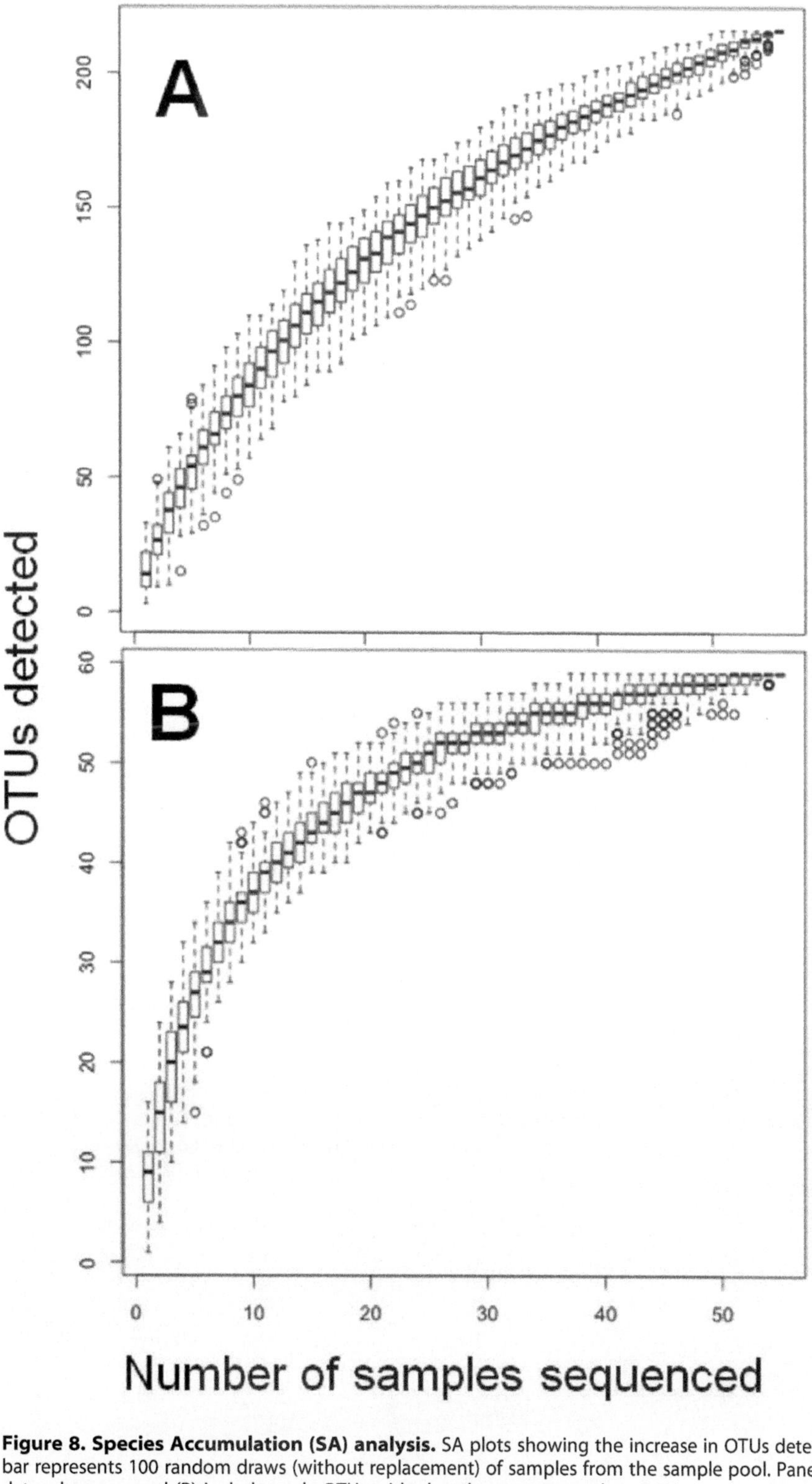

Figure 8. Species Accumulation (SA) analysis. SA plots showing the increase in OTUs detected with the addition of each patient sample. Each bar represents 100 random draws (without replacement) of samples from the sample pool. Panel (A) shows the curve obtained using all of the OTU data whereas panel (B) includes only OTUs with abundances greater than 0.1%.

microbiome reads trimmed to match the SI-Seq format, which allowed direct comparisons of 454 and SI-Seq on real samples using the same data format and the same reference database; this trimming only increased the percent of reads assigned to genus by 5% (from 56% to 61%). The substantial difference that remains between 454 and SI-Seq (61% vs. 95% of reads classified to the genus level, respectively) can only be explained by sequencing error (Table 3).

Our analyses of both control and real samples show that bacterial communities profiled using the two approaches were significantly correlated, even though different parameters were used for OTU clustering, different reference databases were used

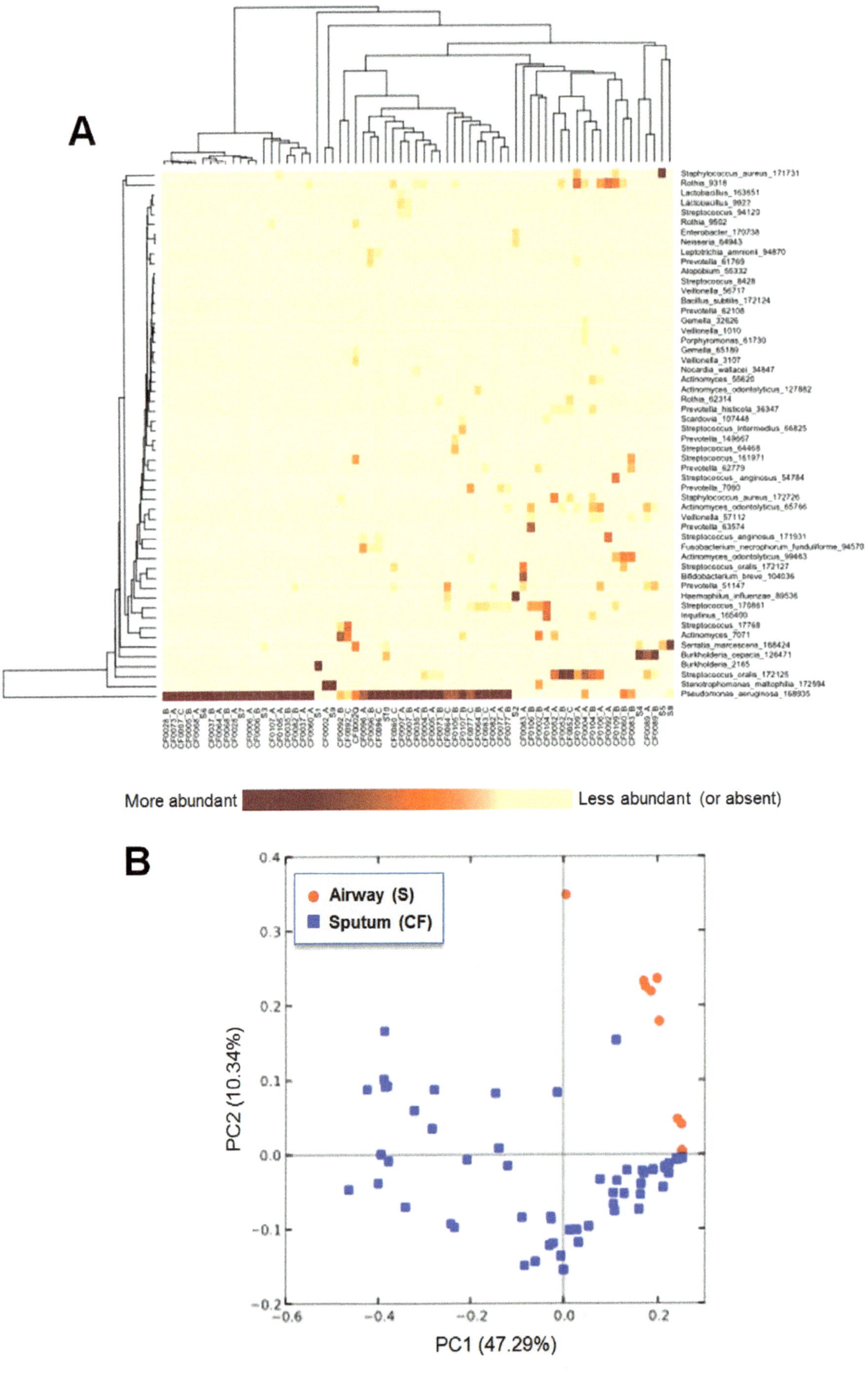
A
More abundant
Less abundant (or absent)
B
Airway (S)
Sputum (CF)
PC2 (10.34%)
PC1 (47.29%)
0.4
0.3
0.2
0.1
0.0
-0.1
-0.2
-0.6
-0.4
-0.2
0.0
0.2

Figure 9. Comparison of bacterial communities in CF sputum and lung airways. (A) Heatmap showing samples from sputum (CF prefix) and lung airways (S prefix) often cluster together. Sample suffixes 'A', 'B', and 'C' indicate the time point for longitudinal samples. For better visualization, OTUs were only included if they had a total abundance greater than 10%. (B) Principal coordinates analysis plot highlighting the sputum (blue squares) and lung airway (red circles) samples. This plot was obtained using the weighted UniFrac estimates for community dissimilarity.

for taxonomic classification, and different reference trees were used for phylogeny-based diversity metrics. The convergence of these sequencing methods and analysis pipelines on a similar picture of the CF lung community speaks to SI-Seq being a robust method.

In addition to superior taxonomic resolution, SI-Seq provides a major cost advantage relative to 454 by producing over 100-fold more data per dollar. Clearly though, it is a unique and more complicated protocol, which requires the careful addition of multiple sequencing primers during the sequencing run and additional data processing steps to deal with the structured reads. Nevertheless, the incredibly high throughput of the Illumina platform provides a compelling case for using SI-Seq. Assuming that the Illumina GA-IIx and MiSeq platforms can sequence 40M and 15M templates per channel per run, respectively; SI-Seq should produce over 400K or 150K reads per sample when using 96-fold multiplexing per channel. Importantly, pooling of SI-Seq samples is performed after PCR, but prior to sample preparation, cluster generation and sequencing; therefore, only one pooled-sample needs to be carried through the sample preparation protocol for all of the sequencing. The net result is that the cost of performing high-multiplex SI-Seq is in the range of tens of dollars per sample, while still obtaining as many as 400K sequences per sample. Multiplexing can be increased for simpler communities (such as the CF lung) if lower depth coverage is needed, or if amplicons are run on machines with higher throughput, such as the Illumina HiSeq.

Paired-end Illumina sequencing of the 16S rRNA gene has been proposed by a number of authors [4,5,6,7,8,9], particularly since the Illumina MiSeq can currently produce read lengths of 150nt, and are expected to reach 250nt within the next couple of years. While these approaches have great merit, they have some of the same issues as 454, namely: they are forced to sequence through conserved regions of the 16S rRNA gene; they have a higher error rate since the reads are longer and error rates increase dramatically near the end of long Illumina reads [21] (with the caveat that some Illumina-based methods overlap the ends of the two paired-reads to reduce the overall error); and they cost more per reaction and take longer to run due to the fixed cost and time accrued per Illumina sequencing cycle. SI-Seq, as described here, runs in 144 total cycles, while Illumina-based microbiome approaches need at least 144nt paired-end runs (288 total cycles) to get taxonomic resolution similar to 454 and SI-Seq (Figure 2). So while paired-end Illumina approaches are initially simpler to apply than SI-Seq, they cost at least twice as much per sample and take twice as long to run.

While this study focuses on the interrogation of the V5, V6 and V7 hypervariable regions of the 16S rRNA locus, it should be clear that any region or locus can be analyzed by SI-Seq if appropriate PCR and sequencing primers can be designed. Additionally, the methods can be easily adapted to sequence even larger numbers of regions or provide longer reads for each region. In general, SI-Seq has many potential applications, including targeted re-sequencing of amplicons for SNP detection and multilocus sequence analysis.

Our study of the CF lung microbiota indicated that SI-Seq produces results consistent with laboratory culture and the published literature [22,23,24,25,26,27,28,29,30,31,32,33]. This very preliminary analysis reveals some distinction between samples collected from the CF sputum and those collected from airway mucus plugs from explanted lungs. Sampling the microbiota present in mucus plugs is important for determining whether some taxa are truly present in the lower airways, or present in the more commonly sampled sputa because of contamination from upper airway bacteria during sputum expectoration (reviewed in [34]). However, our results show that anaerobes are indeed present within the lower airways, as mucus plugs from one patient's airways contained *Veillonella*, *Parvimonas*, and *Streptococcus anginosus*, The DNA from these anaerobes was in low abundance, and likely to only be detectable using a method like SI-Seq that offers a great depth of coverage; nevertheless, its presence suggests anaerobic bacteria are within the lower airways of CF patients, and not merely oropharangeal contaminants of sputum. Further analysis will assess the role of these anaerobes in relation to other bacteria present, and the clinical status, antibiotic therapy, and genotype on a larger sample of CF patients.

Although here we describe the use of SI-Seq for characterizing relatively simple CF communities, additional work shows that the approach easily generates sufficient sequence coverage to thoroughly examine more complex communities such as mouse or human feces, house dust, and saliva (unpublished). We believe SI-Seq will be generally applicable to any microbiota. It can be readily adapted by designing primers to target different regions of the 16S rRNA gene, or different loci altogether. This flexibility also opens the possibility of using SI-Seq to study eukaryotic microbes and Archaea.

Supporting Information

Figure S1 Flow chart of SI-Seq data analysis pipeline. Beginning with raw FASTQ reads, the SI-Seq analysis pipeline filters reads based on quality, checks read orientation and corrects orientation if needed, and uses barcode sequences to parse read data into separate output FASTA files.

Figure S2 Procrustes plot comparison of 454 and SI-Seq community data. Unweighted UniFrac dissimilarity data were used to generate a Procrustes plot as described in the Materials and Methods, and discussed in the main text and Table 4.

Table S1 Barcode file. Barcode sequences designed for SI-Seq development and testing, and for multiplexing CF samples are listed. Barcodes tested with SI-Seq are indicated as such in the 'Sample' column.

Table S2 Quality filtering and incorrect taxonomic classification. For the single species controls used in the development of SI-Seq, varying quality cut-off thresholds were used to filter the raw FASTQ data. These results were used to identify the best quality cut-off for maximizing the number of retained reads while minimizing misclassification of reads.

Table S3 Example of *B. subtilis* classifications. Taxonomy classifications of each read were done using RDP Classifier as discussed in the Materials and Methods. Prior to classification, the reads were filtered based on quality score to exclude any read with more than five sites having a Phred score less than 30 (Table S2).

Acknowledgments

The authors would like to thank Paul Chartrand for technical assistance in sample collection.

Author Contributions

Conceived and designed the experiments: HM PWW SK YCWY VJW DET DMH DSG. Performed the experiments: HM PWW PF LY YZ. Analyzed the data: HM JDC YG SLD. Contributed reagents/materials/ analysis tools: SK YCWY VJW DET DMH DSG. Wrote the paper: HM DSG.

References

1. Cho I, Blaser MJ (2012) The human microbiome: at the interface of health and disease. Nat Rev Genet 13: 260–270.
2. Robinson CJ, Bohannan BJ, Young VB (2010) From structure to function: the ecology of host-associated microbial communities. Microbiol Mol Biol Rev 74: 453–476.
3. Wilson BA, Thomas SM, Ho M (2011) The Human Vaginal Microbiome. In: Nelson KE, editor. Metagenomics of the Human Body. New York: Springer. pp. 91–115.
4. Bartram AK, Lynch MD, Stearns JC, Moreno-Hagelsieb G, Neufeld JD (2011) Generation of Multimillion-Sequence 16S rRNA Gene Libraries from Complex Microbial Communities by Assembling Paired-End Illumina Reads. Appl Environ Microbiol 77: 3846–3852.
5. Caporaso JG, Lauber CL, Walters WA, Berg-Lyons D, Lozupone CA, et al. (2011) Global patterns of 16S rRNA diversity at a depth of millions of sequences per sample. Proc Natl Acad Sci U S A 108 Suppl 1: 4516–4522.
6. Degnan PH, Ochman H (2012) Illumina-based analysis of microbial community diversity. ISME J 6: 183–194.
7. Gloor GB, Hummelen R, Macklaim JM, Dickson RJ, Fernandes AD, et al. (2010) Microbiome profiling by illumina sequencing of combinatorial sequence-tagged PCR products. PLoS One 5: e15406.
8. Lazarevic V, Whiteson K, Huse S, Hernandez D, Farinelli L, et al. (2009) Metagenomic study of the oral microbiota by Illumina high-throughput sequencing. J Microbiol Methods 79: 266–271.
9. Zhou HW, Li DF, Tam NF, Jiang XT, Zhang H, et al. (2011) BIPES, a cost-effective high-throughput method for assessing microbial diversity. ISME J 5: 741–749.
10. Frank DN (2009) BARCRAWL and BARTAB: software tools for the design and implementation of barcoded primers for highly multiplexed DNA sequencing. BMC Bioinformatics 10: 362.
11. Moustafa A (2010) JAligner: Open source Java implementation of Smith-Waterman. JAligner website. Available: http://jaligner.sourceforge.net. Accessed 2012 June 11.
12. Dethlefsen L, Huse S, Sogin ML, Relman DA (2008) The pervasive effects of an antibiotic on the human gut microbiota, as revealed by deep 16S rRNA sequencing. PLoS Biol 6: e280.
13. Sogin ML, Morrison HG, Huber JA, Mark Welch D, Huse SM, et al. (2006) Microbial diversity in the deep sea and the underexplored "rare biosphere". Proc Natl Acad Sci U S A 103: 12115–12120.
14. Wang Q, Garrity GM, Tiedje JM, Cole JR (2007) Naive Bayesian classifier for rapid assignment of rRNA sequences into the new bacterial taxonomy. Appl Environ Microbiol 73: 5261–5267.
15. Edgar RC (2010) Search and clustering orders of magnitude faster than BLAST. Bioinformatics 26: 2460–2461.
16. Caporaso JG, Kuczynski J, Stombaugh J, Bittinger K, Bushman FD, et al. (2010) QIIME allows analysis of high-throughput community sequencing data. Nat Methods 7: 335–336.
17. Schloss PD, Westcott SL, Ryabin T, Hall JR, Hartmann M, et al. (2009) Introducing mothur: open-source, platform-independent, community-supported software for describing and comparing microbial communities. Appl Environ Microbiol 75: 7537–7541.
18. Oksanen J, Blanchet FG, Kindt R, Legendre P, Minchin PR, et al. (2011) vegan: Community Ecology Package.
19. Glenn TC (2011) Field guide to next-generation DNA sequencers. Mol Ecol Resour 11: 759–769.
20. Luo C, Tsementzi D, Kyrpides N, Read T, Konstantinidis KT (2012) Direct comparisons of Illumina vs. Roche 454 sequencing technologies on the same microbial community DNA sample. PLoS One 7: e30087.
21. Minoche AE, Dohm JC, Himmelbauer H (2011) Evaluation of genomic high-throughput sequencing data generated on Illumina HiSeq and Genome Analyzer systems. Genome Biol 12: R112.
22. Armougom F, Bittar F, Stremler N, Rolain JM, Robert C, et al. (2009) Microbial diversity in the sputum of a cystic fibrosis patient studied with 16S rDNA pyrosequencing. Eur J Clin Microbiol Infect Dis 28: 1151–1154.
23. Bittar F, Richet H, Dubus JC, Reynaud-Gaubert M, Stremler N, et al. (2008) Molecular detection of multiple emerging pathogens in sputa from cystic fibrosis patients. PLoS One 3: e2908.
24. Cox MJ, Allgaier M, Taylor B, Baek MS, Huang YJ, et al. (2010) Airway microbiota and pathogen abundance in age-stratified cystic fibrosis patients. PLoS One 5: e11044.
25. Guss AM, Roeselers G, Newton IL, Young CR, Klepac-Ceraj V, et al. (2011) Phylogenetic and metabolic diversity of bacteria associated with cystic fibrosis. ISME J 5: 20–29.
26. Han MK, Huang YJ, Lipuma JJ, Boushey HA, Boucher RC, et al. (2012) Significance of the microbiome in obstructive lung disease. Thorax 67: 456–63.
27. Harris JK, De Groote MA, Sagel SD, Zemanick ET, Kapsner R, et al. (2007) Molecular identification of bacteria in bronchoalveolar lavage fluid from children with cystic fibrosis. Proc Natl Acad Sci U S A 104: 20529–20533.
28. Klepac-Ceraj V, Lemon KP, Martin TR, Allgaier M, Kembel SW, et al. (2010) Relationship between cystic fibrosis respiratory tract bacterial communities and age, genotype, antibiotics and Pseudomonas aeruginosa. Environ Microbiol 12: 1293–1303.
29. Rogers GB, Carroll MP, Serisier DJ, Hockey PM, Jones G, et al. (2006) Use of 16S rRNA gene profiling by terminal restriction fragment length polymorphism analysis to compare bacterial communities in sputum and mouthwash samples from patients with cystic fibrosis. J Clin Microbiol 44: 2601–2604.
30. Sibley CD, Parkins MD, Rabin HR, Duan K, Norgaard JC, et al. (2008) A polymicrobial perspective of pulmonary infections exposes an enigmatic pathogen in cystic fibrosis patients. Proc Natl Acad Sci U S A 105: 15070–15075.
31. Sibley CD, Surette MG (2011) The polymicrobial nature of airway infections in cystic fibrosis: Cangene Gold Medal Lecture. Can J Microbiol 57: 69–77.
32. van der Gast CJ, Walker AW, Stressmann FA, Rogers GB, Scott P, et al. (2011) Partitioning core and satellite taxa from within cystic fibrosis lung bacterial communities. ISME J 5: 780–791.
33. Han MK, Huang YJ, Lipuma JJ, Boushey HA, Boucher RC, et al. (2012) Significance of the microbiome in obstructive lung disease. Thorax 67: 456–463.
34. Lipuma JJ (2010) The changing microbial epidemiology in cystic fibrosis. Clin Microbiol Rev 23: 299–323.
35. Huson DH, Auch AF, Qi J, Schuster SC (2007) MEGAN analysis of metagenomic data. Genome Res 17: 377–386.

17

Resource Availability and Spatial Heterogeneity Control Bacterial Community Response to Nutrient Enrichment in Lakes

KathiJo Jankowski*, Daniel E. Schindler, M. Claire Horner-Devine

University of Washington, School of Aquatic and Fisheries Sciences, Seattle, Washington, United States of America

Abstract

The diversity and composition of ecological communities often co-vary with ecosystem productivity. However, the relative importance of productivity, or resource abundance, versus the spatial distribution of resources in shaping those ecological patterns is not well understood, particularly for the bacterial communities that underlie most important ecosystem functions. Increasing ecosystem productivity in lakes has been shown to influence the composition and ecology of bacterial communities, but existing work has only evaluated the effect of increasing resource supply and not heterogeneity in how those resources are distributed. We quantified how bacterial communities varied with the trophic status of lakes and whether community responses differed in surface and deep habitats in response to heterogeneity in nutrient resources. Using ARISA fingerprinting, we found that bacterial communities were more abundant, richer, and more distinct among habitats as lake trophic state and vertical heterogeneity in nutrients increased, and that spatial resource variation produced habitat specific responses of bacteria in response to increased productivity. Furthermore, changes in communities in high nutrient lakes were not produced by turnover in community composition but from additional taxa augmenting core bacterial communities found in lower productivity lakes. These data suggests that bacterial community responses to nutrient enrichment in lakes vary spatially and are likely influenced disproportionately by rare taxa.

Editor: Martin Heil, Centro de Investigación y de Estudios Avanzados, Mexico

Funding: University of Washington Keeler Professorship to Daniel Schindler, Seattle ARCS Foundation and EPA STAR Fellowship to KathiJo Jankowski. The funders had no role in study design, data collection and analysis, decision to publish, or preparation of the manuscript.

Competing Interests: The authors have declared that no competing interests exist.

* E-mail: kathijo@u.washington.edu

Introduction

Ecosystem productivity is an important driver of the diversity and composition of ecological communities. Much attention has been given to understanding how communities change with increased productivity, due to the desire to understand how species and their threats are distributed globally [1] and the widespread increase in nutrient enrichment and primary productivity of many ecosystems [2]. Productive ecosystems often support high species richness [3], as evidenced by diversity hotspots in ecosystems such as marine upwelling zones [4] and tend to host distinct communities from low productivity ecosystems. Productivity is thought to promote changes in species richness and composition due to the increased energy available to support the coexistence of multiple species and trophic levels [5,6], as well as by promoting shifts to species that dominate in productive environments. However, productivity is not always a good predictor of species richness [7], and the mechanisms behind observed richness and compositional changes in response to increased ecosystem productivity remain obscure [8].

Spatial or temporal heterogeneity in resource availability can also facilitate the coexistence of species in many environments [9,10], and is commonly used to explain why species richness varies with ecosystem productivity [11–13]. Yet, the relative importance of resource availability and heterogeneity in influencing patterns of species richness and composition in productive ecosystems remains unclear for many ecological communities [8,14–16], especially for prokaryotes. Bacteria are a fundamental component of food webs and provide the foundation for overall ecosystem functioning, yet we know relatively little about how bacterial communities respond to increases in productivity in most ecosystems [17–19]. In addition, bacteria have unique characteristics, such as metabolic flexibility and dormancy that might make their response to productivity and resource heterogeneity unique. In addition, bacteria can acquire new functional capacities through the exchange of genetic material [20], thus, taxonomic richness may be unresponsive to changes in productivity [21].

Lakes vary widely in productivity and the heterogeneity of resource distribution in response to variation in nutrient loading from human and watershed sources [15,22]. Increased primary production, or trophic status, in lakes is associated with changes in species richness and composition of many ecological communities, including bacteria [23,24]. The richness of macroorganisms often declines at richer trophic state, due to the dominance of phytoplankton that are less palatable or toxic to consumers [25], declines in littoral productivity [26], and changes to the physical and chemical characteristics of the lake environment [22]. Therefore, changes associated with increased lake trophic status often negatively impact diversity of lake communities, change their composition, and lead to the dominance of a few species through homogenization of food resources and reduction in habitat availability [15,27]. Bacterial communities are known to shift in response to increased lake trophic status [28], but the fundamental mechanisms and importance of resource distribution in mediating those changes have not been fully explored.

Nutrient enrichment in lakes tends to magnify the vertical differences in physical and chemical characteristics such as nitrogen (N), phosphorus (P) and dissolved oxygen (DO) among lake strata [22]. However, existing studies of the response of bacterial communities to eutrophication have only evaluated the responses of surface communities or the integrated water column rather than habitat-specific responses [28,29]. In stratified lakes, the surface layer (epilimnion) is typically warm, nutrient-poor, and productive, whereas the deep layer (hypolimnion) is cooler, richer in nutrients, and often low in dissolved oxygen (DO). These differences may be especially important when considering how the response of lake bacteria may differ from eukaryotic communities since vertical differences in physical and chemical conditions are known to structure bacterial communities in stratified lakes [30–32]. For example, low DO in the hypolimnion promotes the use of diverse energy pathways by bacteria such as denitrification and sulfate reduction that are not energetically advantageous in the oxic epilimnion, and therefore, could promote higher diversity of bacterial communities in the entire water column in response to increased trophic status. Therefore, bacterial communities are likely less similar among lake strata in high productivity (eutrophic) than in low productivity (oligotrophic) lakes.

In addition, although several studies have observed changes in bacterial communities with increased lake trophic status [24,29,33], few have identified which type of bacterial taxa are responsible for driving shifts in overall composition [34,35]. For example, while there is increasing evidence that some taxa flourish in high productivity lakes [36], it is unclear whether taxonomic changes result from a complete turnover in the community [37], an increase in the relative abundance of a few key taxa [38], or the increased presence of previously rare or novel taxa that augment a core community of taxa present in low nutrient lakes. For example, a study that evaluated how dominant, common and rare taxa responded to another important disturbance in lakes, lake mixing, found that shifts in the bacterial community were driven by the increased dominance of a few taxa [38].

We evaluated how bacterial abundance, taxonomic richness, and composition changed among and within lakes along a gradient of increasing trophic state. In particular, we quantified the amount of variation in the bacterial response that was explained by trophic state, resource heterogeneity, and their combination. Second, we evaluated whether communities associated with different lake habitats (specifically the epi-, meta- and hypolimnion) responded differently to increased trophic status than communities assessed in the surface layer or integrated at the whole-lake scale. Finally, we evaluated which taxa were responsible for changes in community characteristics; specifically, we asked whether patterns were driven by turnover in the community or by additional taxa augmenting a core community present across all lakes. Thus, in this study we were able to address whether changes in the observed number of taxa and composition of bacterial communities followed the same patterns as eukaryotic communities in response to productivity in lakes and whether the distribution or abundance of resources was more important in shaping those patterns.

Methods

We sampled 21 lakes in the Puget Sound region of western Washington (USA) and southern British Columbia (Canada, Figure 1) that spanned a large gradient of anthropogenic nutrient loading and productivity [39]. We sampled during the summer-stratified period of July and August 2008. Therefore, our samples reflected the communities that had developed following two to three months of stratified conditions within the water column [31]. As previously described [39], the lakes included in this study were physically similar. Twenty of the 21 lakes were monomictic, and one lake was too shallow to develop thermal stratification. No permissions were required to access 17 of these lakes since they were accessible via a public boat launch. We obtained permission from the University of British Columbia to access the remaining four lakes, which were on the property of their Malcolm Knapp Research Forest. No endangered or protected species were involved in this research.

All bacterial community samples and measurements of lake environmental characteristics were collected over the deepest point in each lake. Water samples for nutrient and chlorophyll *a* analyses were collected from the epilimnion (surface), metalimnion (thermocline depth), and hypolimnion (within 3 m of the lake bottom) with a van Dorn bottle. Total N (TN) was determined using the perchloric acid digestion method [40] followed by analysis with automated colorimetry on a Lachat autoanalyzer (Lachat Instruments, Loveland, CO, USA). Total P (TP) concentration was determined colorimetrically after persulfate digestion and reaction with molybdate and stannous chloride [40]. Water samples for inorganic N and P determination were pre-filtered through a 0.2 mm Supor filters (Supor-200, Pall Gelman, East Hills, NY) and then analyzed colorimetrically using the same methods as above without a pre-digestion step. Chlorophyll *a* concentration was determined fluorometrically (Turner Designs, Sunnyvale, California) and used as a surrogate for algal community biomass. Temperature, dissolved oxygen (DO), and pH measurements were taken at 1-m depth intervals with a YSI sonde 6600 (YSI Integrated Systems & Services, Yellow Springs, OH, USA). Other physical lake data such as mean and maximum depth, lake area, and drainage area were obtained from the King County Water and Land Resources Division and the Washington Department of Ecology.

Two water samples for bacterial community analysis were collected from the epilimnion, metalimnion and hypolimnion of each lake with a Van Dorn Bottle. Two 300-mL samples were pooled and bacteria collected on 0.2-μm filters (Supor-200, Pall Gelman, East Hills, NY). Filters were frozen immediately and stored at −80°C until further processing. DNA was extracted from replicate filters using the Qiagen DNEasy Blood and Tissue Mini-kit (Qiagen, Valencia, CA). Samples for bacterial cell enumeration were preserved with 2% formalin, filtered onto a 0.2 μm black polycarbonate filter, stained with 4′, 6-diamidino-2-phenylindole (DAPI), and viewed with a Nikon Eclipse 80i digital microscope at 1000× magnification.

Bacterial community composition and observed richness were assessed using automated ribosomal intergenic spacer analysis [41]. ARISA generates fingerprints of the microbial community based on the length heterogeneity in the intergenic spacer region between the 16S and 23S rRNA genes, which varies among organisms. ARISA has similar limitations as other PCR-based fingerprinting approaches [41] and tends to only survey dominant taxa in a community, thus our assessment of bacterial community composition is really a comparison of the community of dominant taxa among lakes. However, ARISA has been shown to give a robust, high-resolution view of bacterial assemblages in aquatic ecosystems [42,43], to generate results that are consistent with more high resolution techniques [42,44], and can represent species-level taxonomic resolution (98–99% sequence similarity;

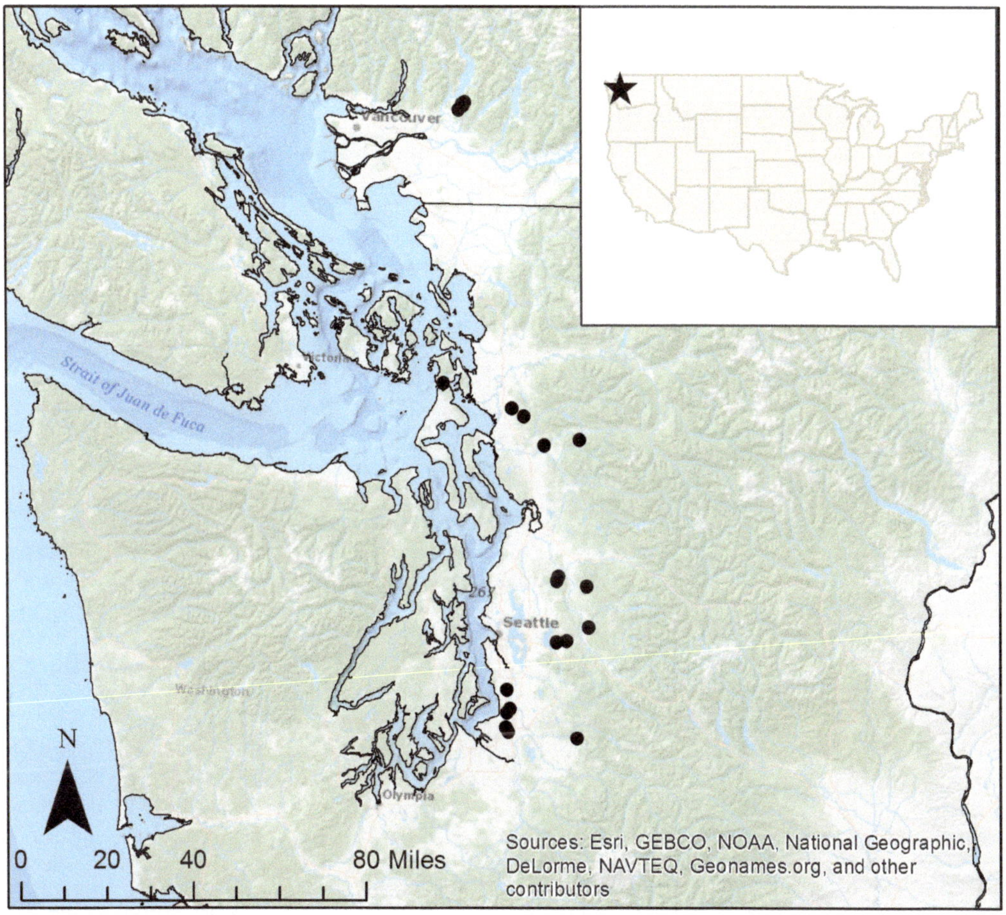

Figure 1. Map of study sites. Lakes included in the study were located in the Puget South Basin in Washington state, USA and British Columbia, Canada. Lakes are indicated by black points.

[42]). The 16S-23S intergenic region was amplified using the polymerase chain reaction (PCR) from the total extracted DNA using 6-FAM-labelled universal 1406-F primer (5′ TGYACACACCGCCCGT-3′) and bacterial specific primer 23S-R(5′-GGGTTBCCCCATTCRG-3′) [41,45]. PCRs were conducted on a Mastercycler gradient thermocycler (Eppendorf, New York). PCR products were pooled, quantified, and analyzed on a MegaBACE 1000 automated capillary sequencer (GE Healthcare Corporation, New Jersey). Operational taxonomic units (OTUs) were generated by binning ARISA fragments into successively larger length bins based on their size and eliminating fragments that were <150 and >1300 bp [42]. We used peak area to estimate relative abundance of OTUs in our samples [43], which we considered to be the ratio of the peak area of an OTU in a sample to the total peak area of the sample. We also converted the peak area matrix to presence-absence to assess the composition of bacterial communities in the ARISA profiles by occurrence patterns. We calculated observed richness from ARISA profiles by summing the number of OTUs observed in each sample, hereafter referred to as profile richness. We found no differences in bacterial community patterns using peak height vs. peak area.

Statistical Analyses

We used a principal components analysis (PCA) to summarize physical and chemical variation related to trophic status among lakes. We found that lakes varied little in relevant physical characteristics (lake area and mean depth), and thus we used the first axis of the resulting PCA as a multivariate proxy for increasing lake trophic state (Figure 2A). Although we did not measure primary productivity directly, other studies have found good agreement between primary productivity and chlorophyll *a* and nutrient concentrations in lakes with similar concentrations as lakes in this study [46].

To quantify vertical heterogeneity in chemical and physical variables within each lake (e.g., TN, TP, temperature, and DO), we used the standard deviation of measurements among lake strata (Table S1). We then performed a PCA that included only the standard deviations of these physical and chemical variables to establish a gradient of resource heterogeneity among lakes.

To compare the influence of increasing trophic state ("trophic status"), depth variation in resource availability ("resource heterogeneity"), and their combination (trophic status and resource heterogeneity) on the bacterial community, we then took the scores from the first principal component (PC 1) of each PCA and regressed them against metrics of bacterial abundance, ARISA profile richness, and an index of community similarity among ARISA profiles (see below for description). The combined trophic status and resource heterogeneity model contained two predictor variables: PC 1 of the trophic status PCA, and PC1 of the resource heterogeneity PCA. All variables were transformed to meet the assumptions of normality prior to the PCA. We evaluated the support for each of the three candidate models describing the

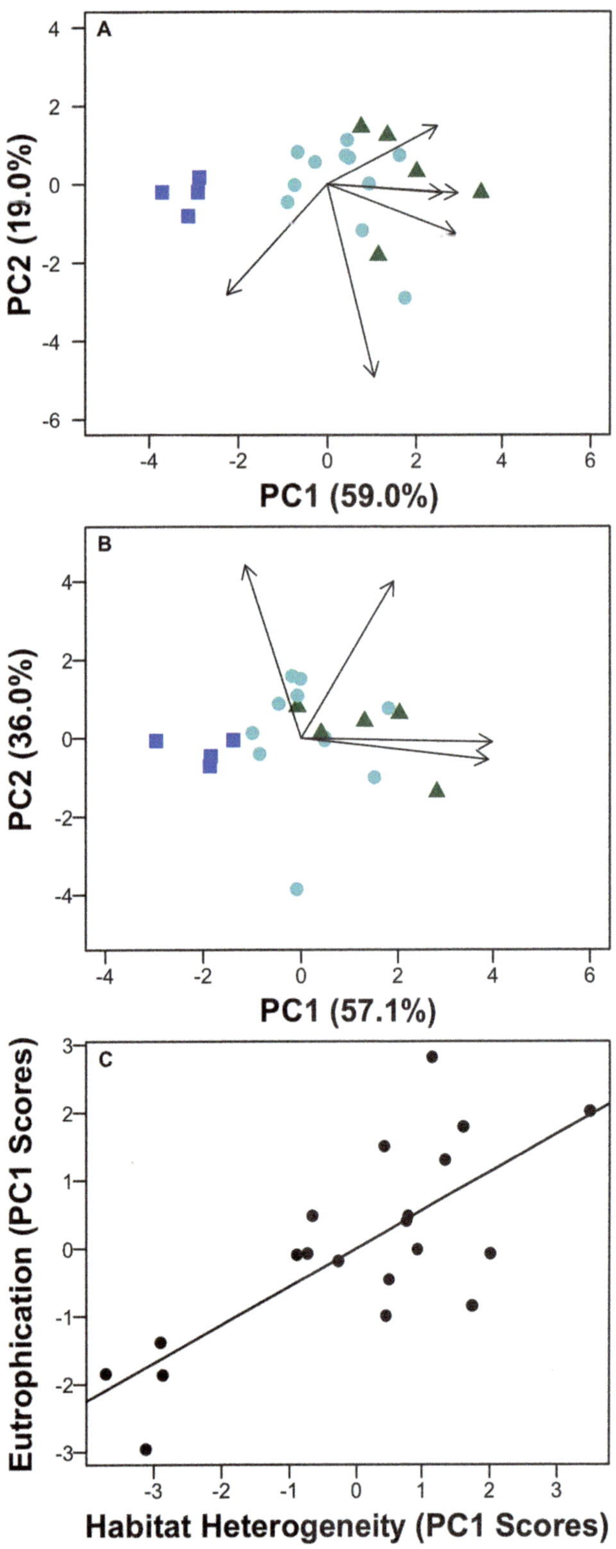

Figure 2. Principal component analyses (PCAs) showing environmental variation across lakes in this study. Panel A shows PCA results based on trophic state variables, panel B shows PCA results based on Heterogeneity variables (standard deviation among depths in variables) and panel C shows the correlation between the trophic state PC 1 scores and heterogeneity PC 1 scores ($R^2 = 0.57$). Arrows show significant variables ($p < 0.05$) and values in parentheses show percentages of total environmental variation among lakes explained by each axis. 'EpiTN' = epilimnetic TN, 'EpiTP' = epilimnetic TP, 'HypoTP' = Hypolimnetic TP, 'HypoDO' = hypolimnetic DO, and DO = epilimnetic DO. '.std' indicates that the standard deviation of measurements of the specified variable among layers. Triangles = eutrophic, stars = mesotrophic, and squares = oligotrophic lakes.

relationships between environmental conditions and the bacteria community attributes using Akaike's Information Criteria adjusted for small sample sizes [47]. The model with the lowest AIC_c was considered the best model, and models within 2 AIC_c units of one another were considered to be equally good [47]. In addition, we calculated AIC weights (w_i) for each individual model, which estimates a probability that model *i* is the best model given the set of models we considered. Finally, to evaluate overall importance of the individual variables, trophic status and resource heterogeneity, we calculated w_i for each term across the three models we compared (Table 1).

Bacterial community similarity among samples was assessed using Sorensen's coefficient for occurrence data [48] and the Chao-Sorensen abundance estimator for relative abundance data [49]. We assessed the overall similarity of communities within a lake by using an average dissimilarity value among ARISA profiles from all two-layer comparisons. We used a constrained analysis of principal coordinates (CAP) to evaluate if changes in community composition were associated with increasing trophic state [50] since it allowed us to use the Chao-Sorensen similarity index.

Finally, we investigated whether changes in the bacterial community with increased lake trophic status were realized by shifts in "widespread" or "narrowly distributed" taxa. We assessed how the relative contribution of widespread OTUs (taxa observed in the majority of lakes) changed with trophic state and habitat heterogeneity. We considered widespread taxa to be those that were observed in 90% of lakes in our study (but see Table S2 for evaluation of different thresholds). We then assessed whether the occurrence and relative abundance of these taxa changed across the lake trophic gradient and with increasing heterogeneity (i.e., PC1 of trophic status and resource heterogeneity PCAs). All analyses were done in R Version 2.14.0 [51] using the vegan package [52].

Results

Lake Characteristics

Productivity-related variables such as TP, TN, and chlorophyll a explained a substantial portion of the environmental variation among lakes in this study (59%; Figure 2, Table S1). Epilimnetic TP concentrations ranged from 4.6 $\mu g\ L^{-1}$ to over 30 $\mu g\ L^{-1}$, and chlorophyll *a* ranged from 0.23 to 10.2 $\mu g\ L^{-1}$, thus the lakes ranged from oligotrophic to eutrophic [22]. Environmental conditions did not change similarly in each layer with increased trophic state; for example, the epilimnion was less variable among lakes than either the metalimnion or hypolimnion in most environmental characteristics (Table S3). As a result, conditions within the water column were more heterogeneous as trophic state increased ($R^2 = 0.57$, Figure 2C). We observed the most significant differences in TN and TP concentrations among layers as trophic state increased (Figures 2B and C). TN and TP were correlated with the availability of NH_4 ($r = 0.71$) and PO_4 ($r = 0.86$), respectively. Thus, nutrient availability was variable within the water column. Finally, the percent change in nutrient concentrations was greater in the hypolimnion than in either the epi- or metalimnion (Table S3).

Table 1. Comparison of models evaluating the effects of trophic state, heterogeneity, and their combination on bacterial communities.

	Model	n	k	R^2	AIC_c	ΔAIC_c	w_i*	w_i of T⁺	w_i of H⁺⁺
ABUNDANCE									
	Trophic State	21	3	0.46	612.2	9.1	0.01	0.34	0.83
	Heterogeneity	21	3	0.66	603.1	0.0	0.50		
	PC1 T+PC1 H	21	4	0.69	603.3	0.0	0.33		
RICHNESS									
	Trophic State	17	3	0.30	129.3	2.9	0.16	0.33	0.84
	Heterogeneity	17	3	0.41	126.4	0.0	0.67		
	PC1 T+PC1 H	17	4	0.40	129.2	2.8	0.17		
DISSIMILARITY									
	Trophic State	17	3	0.12	−23.8	0.9	0.35	0.46	0.65
	Heterogeneity	17	3	0.17	−24.7	0.0	0.54		
	PC1 T+PC1 H	17	4	0.12	−21.5	3.1	0.11		
WIDESPREAD									
TAXA	Trophic State	18	3	0.29	−36.9	1.5	0.26	0.44	0.74
	Heterogeneity	18	3	0.35	−38.4	0.0	0.56		
	PC1 T+PC1 H	18	4	0.35	−36.1	2.3	0.18		

*AIC_c weight,
⁺Trophic State,
⁺⁺Heterogeneity.
Relationship of trophic state, heterogeneity, and their combination with abundance, richness, dissimilarity of bacterial communities among lake habitats ("Dissimilarity"), and the proportional abundance of common taxa. 'n' = sample size and 'k' = number of parameters in each model. 'w_i' refers to the AIC_c weight calculated for each model and the weights for the individual terms 'T' and 'H' across all models.

Did Lake Bacterial Communities Shift in Response to Increasing Trophic State?

Bacterial communities at the whole-lake scale shifted significantly in association with increasing lake trophic state (Figure 3, Figure S1). Average bacterial abundance ($R^2 = 0.46$, Table 1, Figure 2A) and ARISA profile richness increased linearly with our proxy for lake eutrophication ($R^2 = 0.30$, Table 1, Figures 3A & E) and ranged from 66 to 106 OTUs per lake. Our CAP model showed that bacterial community composition shifted with increased lake trophic status and shifts were strongly associated with increasing chlorophyll *a* ($r = 0.99$) and epilimnetic TN ($r = 0.76$) concentration (Figure S1). The CAP model explained 29% of the total variation in community composition among lakes. The first CAP axis captured the majority of that explained variation (71.4%), indicating that community composition changed in response to increased lake trophic state.

Did the Responses of the Bacterial Community Vary among Habitats in Lakes?

Bacterial communities associated with surface and deep habitats displayed different patterns of abundance, and the richness and composition of ARISA profiles changed significantly as lake trophic state increased (Figure 3, Figure S1). Average bacterial abundance increased with trophic state ($R^2 = 0.46$), was highest in the metalimnion (ANOVA; $F = 5.6$, $p = 0.006$), but increased significantly in all layers across the trophic gradient (Figure 3). The richness of ARISA profiles also varied significantly among layers (ANOVA, $F = 10.6$, $p < 0.001$), but only increased notably in the hypolimnion in response to trophic state ($R^2 = 0.15$, Figure 3H). Profile richness was highest on average in the hypolimnion (55 ± 7 SD), which also had the largest range of observed richness, ranging from 37 OTUs in Gwendoline Lake, an oligotrophic lake, to 71 OTUs found in more nutrient-rich Geneva Lake. Therefore, increases in the profile richness in the hypolimnion accounted for the increases we observed in overall lake richness ($R^2 = 0.39$).

When all lake communities were considered together, there were significant, but small, compositional differences among epi-, meta- and hypolimnetic communities (ANOSIM, $R = 0.16$, $p = 0.001$), and surface and deep communities shared the fewest taxa (data not shown). Furthermore, surface communities were significantly less variable than deep communities across the trophic gradient (Homogeneity of dispersion, $p < 0.001$), and surface and deep communities within a given lake tended to become less similar to one another as trophic state increased ($R^2 = 0.12$). However, heterogeneity in nutrient concentrations among strata explained slightly more of that variation than trophic state alone ($R^2 = 0.17$, Figure 4, Table 1).

In all cases, the heterogeneity model or the trophic status plus heterogeneity model explained more variation in bacterial communities among lakes than the trophic status model alone (Table 1). We found that greater vertical heterogeneity of nutrient availability (Figure 2) was strongly related to increased abundance, ARISA profile richness, and decreasing similarity of communities among lake strata (Table 1, Figure 4). Total abundance and observed richness were both more strongly related to increases in the heterogeneity of P and N than increases in their concentrations alone or to differences in temperature and DO among strata, which other studies have shown to be associated with heterogeneity in bacterial community composition (Table S1; [31,32]). Therefore, although we saw an increase in observed richness with increased trophic state ($R^2 = 0.30$), observed richness was more closely linked to greater heterogeneity of nutrients within the water column ($R^2 = 0.41$). Overall, the AIC_c shows that the heteroge-

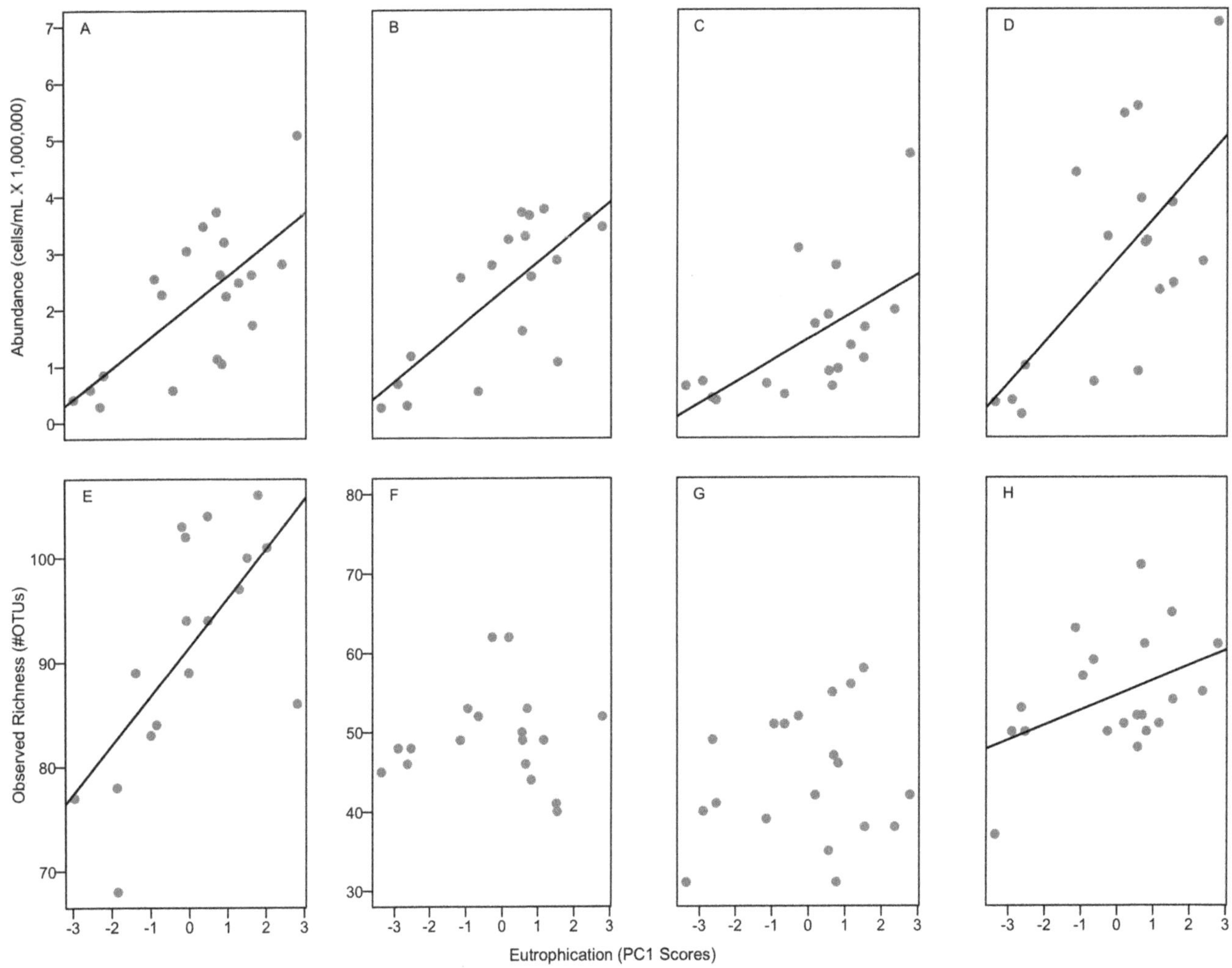

Figure 3. Relationships of increasing trophic state (PC 1 scores) with abundance and richness. A) Whole-lake average abundance ($R^2 = 0.46$, $p < 0.001$), B) Epilimnetic abundance ($R^2 = 0.52$, $p < 0.001$), C) Metalimnetic abundance ($R^2 = 0.33$, $p = 0.007$), D) Hypolimnetic abundance ($R^2 = 0.40$, $p = 0.003$), E) Whole-lake average richness ($R^2 = 0.30$, $p = 0.003$), F) Epilimnetic richness ($R^2 = -0.06$, $p = 0.98$), G) Metalimnetic richness ($R^2 = -0.03$, $p = 0.49$), and H) Hypolimnetic richness ($R^2 = 0.15$, $p = 0.05$).

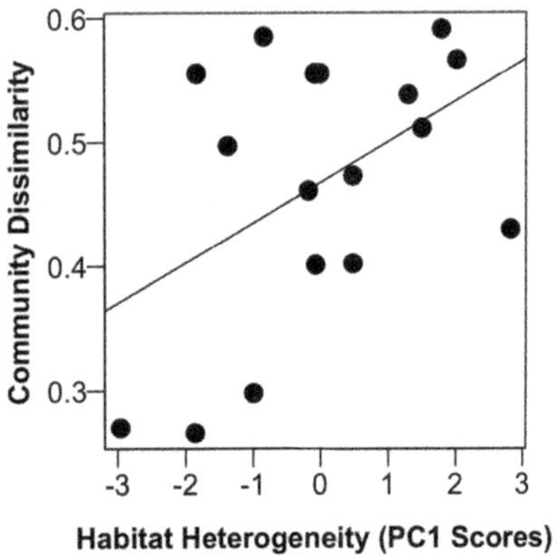

Figure 4. Bacterial communities were less similar among lake strata as chemical heterogeneity increased ($R^2 = 0.17$, $p = 0.05$). Similarity of communities was based on the average Chao Sorensen Dissimilarity from comparisons of relative abundance of taxa among three depth strata.

neity model had the most support, and that, in fact, the heterogeneity term had the most weight across models (Table 1). Thus, as measured here, bacterial communities exhibited habitat-specific responses to lake eutrophication, and spatial variation in resource availability often influenced bacterial community composition more than simple increases in nutrient concentration and productivity (Figures 4 and 5).

Which Taxa Accounted for changes in Community Composition with Eutrophication?

We observed a total of 221 OTUs across all lakes and found that some of those taxa were widespread among lakes. We observed that while a core community of 11 OTUs was present and detected using ARISA in ~90% of lakes in this study ("widespread taxa"), and while still present across lakes, made up a decreasing proportion of both the number ($R^2 = 0.23$) and relative abundance of taxa in lakes as lake trophic state increased ($R^2 = 0.35$, Figure 5). Specifically, while these widespread taxa comprised 35–40% of the relative abundance of the community in the more homogenous oligotrophic lakes, they were less than 15% in the more heterogeneous eutrophic lakes. These trends were

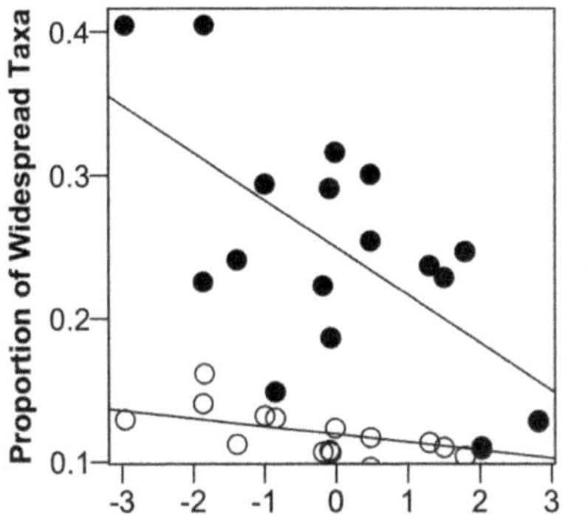

Figure 5. Relationship of habitat heterogeneity with the proportional representation of widespread taxa in each lake. Heterogeneity was measured as the PC 1 scores of 'heterogeneity' model. Filled circles show the total proportion of relative abundance made up by widespread taxa ($R^2=0.35$, $p=0.007$) and open circles show widespread taxa as a proportion of the total number of taxa observed ($R^2=0.23$, $p=0.03$).

robust to using different thresholds to define "widespread" (e.g., present in 70–90% of lakes), and strengthened as we considered thresholds up to 90% (Table S2). We only observed four OTUs that were present in >90% of lakes, which likely were ubiquitous taxa that would be present regardless of lake trophic state [36]. In addition, we found that the increasing representation of previously low abundance or new taxa in response to trophic status was explained most by increasing heterogeneity among habitats (Table 1). Although the AIC_c values for all three models were very close, the AIC value and variable w_i's suggest that heterogeneity explained the declining contribution of the widespread OTUs we observed to the overall lake community (Table 1). This suggests that either novel, previously low abundance, (i.e., below the detection limit of ARISA), or dormant taxa [35] increased in their relative importance as nutrient status of lakes increased, and that chemical heterogeneity in the lake environment most likely facilitated the increased prevalence of these taxa in the lake community.

Discussion

We found that the abundance and richness of bacterial communities increased as lake nutrient status increased, in parallel to what has been reported in other studies [28,33,45]. However, we found even stronger relationships between the overall abundance and number of OTUs observed and the spatial heterogeneity in nutrient conditions among lake layers as trophic status increased. Additionally, because environmental conditions among lake strata diverged and communities associated with these habitats responded differently to lake trophic status (Figure 3), bacterial communities as measured by ARISA fingerprinting were much less similar among lake strata as lake trophic state increased (Figure 4, Table S3). Furthermore, our analyses demonstrated that a core bacterial community of dominant bacteria did not change systematically in high nutrient lakes but rather that increasing habitat heterogeneity, specifically due to large changes in conditions in the deepest layer of lakes, provided additional habitat for previously low abundance or absent taxa that became detectable in more eutrophic conditions. Our study shows that the response of bacterial communities to increased productivity in lakes may differ from that of other lake organisms as a result of spatial heterogeneity in resources specifically affecting the ecology of bacteria and supports the idea that the recruitment of rare or dormant bacterial taxa in lake communities may facilitate much of this response (e.g., [35]).

Effects of Trophic State on Lake Bacterial Communities

We observed significant increases in bacterial abundance and changes in the richness and composition of ARISA profiles at the whole-lake scale in response to increases in lake trophic status. As expected, abundance was positively related to increases in nutrient availability (TN and TP) and chlorophyll *a* (Table 1) across the entire range of ecosystem productivity we observed. High abundances of bacteria are often correlated with high rates of bacterial productivity [46], which can be nutrient-limited in oligotrophic lakes. We also observed increased ARISA profile richness as trophic state increased (Figure 3E), suggesting that more taxa coexisted as dominant types (and thus able to be detected by ARISA) as resource availability increased. As has been observed for macroorganisms in lakes [23], other studies have shown lake bacterial richness to exhibit a range of linear and unimodal responses to productivity depending on the spatial and taxonomic resolution of the study [17,18,29]. We likely observed a linear increase in observed richness since our lakes represented a relatively modest gradient in nutrient loading and chlorophyll *a* compared to the global distribution of lake productivity [22], thus, we may have only captured the initial upward slope of a unimodal response across the global range of lake trophic states. Alternatively, is also possible that we observed a consistently increasing number of taxa rather than a unimodal trend due an effect of increased sampling; in other words, because we took discrete samples from each lake habitat and ran a separate ARISA analysis on each sample rather than on one integrated sample from each lake, we were able to detect more taxa than previous studies [23,29].

Bacterial Community Response to Resource Heterogeneity

Bacterial community shifts were most strongly associated with changes in the heterogeneity of the lake environment in all cases (Table 1). Specifically, we observed more OTUs and decreased similarity among bacterial communities in surface and deep habitats in response to heterogeneity in N and P concentrations (Figures 2B and 3). Environmental conditions within lakes changed to different degrees among the epi-, meta- and hypolimnion (Table S3), which translated to habitat specific responses of bacterial communities in each layer across lakes (Figure 3). For instance, although bacterial abundance increased in all layers, we only observed increases in richness in the hypolimnion with increased trophic status (Figure 3H), which also had the largest range in nutrient concentrations across the trophic gradient (Figure S1). Interestingly, communities in the surface layer varied less across the trophic gradient than communities in the hypolimnion, suggesting that studies that evaluate trends in bacterial communities in response to eutrophication miss an important aspect of the bacterial response.

While our study design does not allow us to definitely conclude whether patterns in bacterial communities were related more to increased trophic state or to habitat heterogeneity because the two were correlated (Figure 2C), we found strong support for our heterogeneity model (Table 1) and patterns suggesting that both were important for bacterial communities. For instance, we observed lower richness (66 OTUs) and higher similarity of communities among lake strata (Sorensen similarity = 0.86) in the only lake in our study that did not stratify into discrete habitats.

Although this lake had similar nutrient concentrations as eutrophic lakes, environmental conditions were more homogeneous within the water column in this lake suggesting that heterogeneity in resource availability influenced bacterial richness more than nutrient concentration alone in all lakes in our study.

While changes in bacterial communities in response to increasing lake trophic status have been widely observed [28,29,45] as have differences in communities among lake strata [30,31,53] previous studies have not linked these observations to evaluate the combined effects of lake resource availability and resource heterogeneity on the richness and composition of lake bacterial communities. Thus, while similar communities may inhabit eutrophic lakes [36], communities become increasingly distinct from one another as surface and deep habitat conditions diverge and biotic interactions change [54]. Furthermore, these findings contrast with studies of other lake organisms, such as zooplankton and fish, which find increased productivity reduces diversity as a result of homogenization of food resources and loss of habitat [15,26,27]. Thus, our study demonstrates that the type of heterogeneity that influences communities varies among macro- and microorganisms and that the response to a large-scale environmental change is expressed differently among habitats within ecosystems.

In addition, our results suggest that changes in bacterial communities with increased trophic status may be more strongly related to vertical differences in nutrient concentrations than dissolved oxygen, temperature, and light, which are typically thought to structure differences among communities in different strata within lakes [24,30,31,53]. We tested this by regressing axis 1 and 2 of our environmental PCA (Figure S1), which represented differences among layers in nutrient concentrations and DO/ temperature, respectively. We found that in all cases axis 2 (variation in DO and temperature) did not have a strong relationship with any bacterial response measure. Few studies have tested for this and those that have, have often focused on a single lake or small set of lakes [30,53]. Thus, these results suggest that increased vertical differences in nutrient concentrations may be more important in structuring the bacterial community than vertical differences in DO and temperature as the trophic state of lakes increases.

Were Widespread or Narrowly Distributed Taxa Responsible for Shifts in Community Composition?

We found that habitat heterogeneity played an important role in shaping the bacterial response to increasing lake trophic status through enabling the increased contribution of new or previously low abundance taxa in lake communities. Furthermore, we found that increased resource heterogeneity simultaneously allowed for the retention of a core group of taxa that were widespread among lakes while also providing new habitat and resources for these previously unobserved or low abundance taxa (Figure 5). Thus, our results support the general notion of the importance of rare taxa in microbial communities [35,55] and suggest that bacterially-relevant habitat heterogeneity may be an important mechanism driving the bacterial response to increased lake ecosystem productivity. Additionally, this pattern may be common to many types of planktonic communities [34]. For instance, a study of the response of a lake phytoplankton community to eutrophication and recovery found that there were phytoplankton that were consistently present through time, but that temporally rare taxa were most responsive to changes in lake nutrient status and drove changes in community composition [56].

While the use of ARISA allowed us to screen the bacterial community in a large number of lakes and has been shown to have species-level taxonomic resolution, as with so many methods used to sample bacterial communities, there are limitations associated with using this approach [57]. For example, ARISA underestimates the total richness of the bacterial community and has biases such as preferential amplification of abundant organisms [41]. For example, many of the additional taxa that we observed in more eutrophic lakes could have been present at low abundances in oligotrophic lakes, and therefore below the detection limit of ARISA. However, increased detection of these taxa in eutrophic lakes suggests that they are at higher abundances and thus may be more functionally important in those communities. In addition, other studies have shown that ARISA captures similar patterns in diversity among communities as more high-resolution techniques such as clone libraries [58]. Further, although sequencing and clone library techniques would have allowed us to identify specific taxa, the higher costs associated with those techniques would have limited our ability to sample the entire trophic gradient in our study. Finally, our results are comparable to other studies using similar techniques [24,33,35], and recent studies of bacterial responses to nutrient additions that have used high throughput sequencing techniques suggest that using a more thorough sampling approach likely would not likely reveal a different trend in how richness and composition respond to increased ecosystem productivity (e.g., [19,44]). Thus, the fact that we observed such striking trends using this approach suggests that higher resolution sampling would have only strengthened observed patterns.

In summary, we showed that bacterial community composition changed and was richer and more heterogeneous within lakes as trophic status increased. In contrast to trends in macroorganisms whose diversity is often negatively associated with increases in lake productivity [15,23,27], we showed that the high degree of heterogeneity in bacterial resources in eutrophic lakes promoted higher richness as a result of differentiation of bacterial taxa among lake habitats. We found that eutrophication alters the drivers of bacterial community differences within lakes from physical and redox related variables to changes in nutrient availability. Furthermore, our results suggest that rare or dormant taxa may be most responsible for changes in bacterial communities with increased lake trophic state [35,59]. This "seed bank" of taxa has increasingly been recognized to be important in responding to changes in many types of ecosystems [60], and understanding the role of rare and dormant taxa an important frontier for understanding the processes that regulate how microbial communities respond to ecosystem change in general.

Acknowledgments

The authors would like to acknowledge M. Dyen, H. Bekris, and A. Coogan for extensive help with field data collection and laboratory analyses. We are grateful to R. Lange, J. Griffiths, J. Armstrong and two anonymous reviewers for helpful reviews on the manuscript.

Author Contributions

Conceived and designed the experiments: KJJ DES MCHD. Analyzed the data: KJJ. Contributed reagents/materials/analysis tools: DES MCHD. Wrote the paper: KJJ DES MCHD.

References

1. Reid WV, Mooney HA, Cropper A, Capistrano D, Carpenter SR, et al. (2005) Ecosystems and human well-being: synthesis. Washington, D.C.: Island Press.
2. Carpenter SR, Caraco NF, Correll DL, Howarth RW, Sharpley AN, et al. (1998) Nonpoint pollution of surface waters with phosphorus and nitrogen. Ecological Applications 8: 559–568.
3. Rozenzweig ML (1995) Species diversity in space and time. Cambridge, U.K.: Cambridge University Press.
4. Ribalet F, Marchetti A, Hubbard KA, Brown K, Durkin CA, et al. (2010) Unveiling a phytoplankton hotspot at a narrow boundary between coastal and offshore waters. Proceedings of the National Academy of Sciences of the United States of America 107: 16571–16576.
5. Wright DH (1983) Species-energy theory - An extension of species-area theory. Oikos 41: 496–506.
6. Mittelbach GG, Steiner CF, Scheiner SM, Gross KL, Reynolds HL, et al. (2001) What is the observed relationship between species richness and productivity? Ecology 82: 2381–2396.
7. Adler PB, Seabloom EW, Borer ET, Hillebrand H, Hautier Y, et al. (2011) Productivity Is a Poor Predictor of Plant Species Richness. Science 333: 1750–1753.
8. Cardinale BJ, Hillebrand H, Harpole WS, Gross K, Ptacnik R (2009) Separating the influence of resource 'availability' from resource 'imbalance' on productivity-diversity relationships. Ecology Letters 12: 475–487.
9. Hutchinson GE (1961) The paradox of the plankton. American Naturalist 95: 137–145.
10. Tilman D (1982) Resource Competition and Community Structure. Princeton, New Jersey: Princeton University Press.
11. Chase JM, Leibold MA (2002) Spatial scale dictates the productivity-biodiversity relationship. Nature 416: 427–430.
12. Shurin JB, Winder M, Adrian R, Keller W, Matthews B, et al. (2010) Environmental stability and lake zooplankton diversity - contrasting effects of chemical and thermal variability. Ecology Letters 13: 453–463.
13. Reich PB, Frelich LE, Voldseth RA, Bakken P, Adair EC (2012) Understorey diversity in southern boreal forests is regulated by productivity and its indirect impacts on resource availability and heterogeneity. Journal of Ecology 100: 539–545.
14. Stevens MHH, Carson WP (2002) Resource quantity, not resource heterogeneity, maintains plant diversity. Ecology Letters 5: 420–426.
15. Barnett A, Beisner BE (2007) Zooplankton biodiversity and lake trophic state: Explanations invoking resource abundance and distribution. Ecology 88: 1675–1686.
16. Eilts JA, Mittelbach GG, Reynolds HL, Gross KL (2011) Resource Heterogeneity, Soil Fertility, and Species Diversity: Effects of Clonal Species on Plant Communities. American Naturalist 177: 574–588.
17. Horner-Devine MC, Leibold MA, Smith VH, Bohannan BJM (2003) Bacterial diversity patterns along a gradient of primary productivity. Ecology Letters 6: 613–622.
18. Smith VH (2007) Microbial diversity-productivity relationships in aquatic ecosystems. Fems Microbiology Ecology 62: 181–186.
19. Bowen JL, Ward BB, Morrison HG, Hobbie JE, Valiela I, et al. (2011) Microbial community composition in sediments resists perturbation by nutrient enrichment. Isme Journal 5: 1540–1548.
20. Madigan MT, Martinko JM, Stahl DA, Clark DP (2010) Brock biology of microorganisms. San Francisco: Benjamin Cummings.
21. Green JL, Bohannan BJM, Whitaker RJ (2008) Microbial biogeography: From taxonomy to traits. Science 320: 1039–1043.
22. Wetzel RC (2001) Limnology: lake and river ecosystems. San Diego: Academic Press.
23. Dodson SI, Arnott SE, Cottingham KL (2000) The relationship in lake communities between primary productivity and species richness. Ecology 81: 2662–2679.
24. Longmuir A, Shurin JB, Clasen JL (2007) Independent gradients of producer, consumer, and microbial diversity in Lake Plankton. Ecology 88: 1663–1674.
25. Smith VH (2003) Eutrophication of freshwater and coastal marine ecosystems - A global problem. Environmental Science and Pollution Research 10: 126–139.
26. Vadeboncoeur Y, Jeppesen E, Vander Zanden MJ, Schierup HH, Christoffersen K, et al. (2003) From Greenland to green lakes: Cultural eutrophication and the loss of benthic pathways in lakes. Limnology and Oceanography 48: 1408–1418.
27. Vonlanthen P, Bittner D, Hudson AG, Young KA, Muller R, et al. (2012) Eutrophication causes speciation reversal in whitefish adaptive radiations. Nature 482: 357–U1500.
28. Kolmonen E, Haukka K, Rantala-Ylinen A, Rajaniemi-Wacklin P, Lepisto L, et al. (2011) Bacterioplankton community composition in 67 Finnish lakes differs according to trophic status. Aquatic Microbial Ecology 62: 241–U249.
29. Korhonen JJ, Wang JJ, Soininen J (2011) Productivity-Diversity Relationships in Lake Plankton Communities. Plos One 6.
30. De Wever A, Muylaert K, Van der Gucht K, Pirlot S, Cocquyt C, et al. (2005) Bacterial community composition in Lake Tanganyika: Vertical and horizontal heterogeneity. Applied and Environmental Microbiology 71: 5029–5037.
31. Shade A, Jones SE, McMahon KD (2008) The influence of habitat heterogeneity on freshwater bacterial community composition and dynamics. Environmental Microbiology 10: 1057–1067.
32. Garcia SL, Salka I, Grossart HP, Warnecke F (2013) Depth-discrete profiles of bacterial communities reveal pronounced spatio-temporal dynamics related to lake stratification. Environ Microbiol Rep 5: 549–555.
33. Lindstrom ES (2000) Bacterioplankton community composition in five lakes differing in trophic status and humic content. Microbial Ecology 40: 104–113.
34. Galand PE, Casamayor EO, Kirchman DL, Lovejoy C, Karl DM (2009) Ecology of the rare microbial biosphere of the Arctic Ocean.. Proceedings of the National Academy of Sciences of the United States of America 106: 22427–22432.
35. Jones SE, Lennon JT (2010) Dormancy contributes to the maintenance of microbial diversity. Proceedings of the National Academy of Sciences of the United States of America 107: 5881–5886.
36. Newton RJ, Jones SE, Eiler A, McMahon KD, Bertilsson S (2011) A Guide to the Natural History of Freshwater Lake Bacteria. Microbiology and Molecular Biology Reviews 75: 14–49.
37. Bell T, Bonsall MB, Buckling A, Whiteley AS, Goodall T, et al. (2010) Protists have divergent effects on bacterial diversity along a productivity gradient. Biology Letters 6: 639–642.
38. Shade A, Chiu CY, McMahon KD (2010) Seasonal and Episodic Lake Mixing Stimulate Differential Planktonic Bacterial Dynamics. Microbial Ecology 59: 546–554.
39. Jankowski KJ, Schindler DE, Holtgrieve GW (2012) Assessing non-point source nitrogen loading and nitrogen fixation in lakes using δ15N and nutrient stoichiometry. Limnology and Oceanography 57: 671–683.
40. Eaton A, Greenberg AE, Rice EW (2005) Standard Methods for the Examination of Water and Wastewater: American Public Health Association.
41. Fisher MM, Triplett EW (1999) Automated approach for ribosomal intergenic spacer analysis of microbial diversity and its application to freshwater bacterial communities. Applied and Environmental Microbiology 65: 4630–4636.
42. Brown MV, Schwalbach MS, Hewson I, Fuhrman JA (2005) Coupling 16S-ITS rDNA clone libraries and automated ribosomal intergenic spacer analysis to show marine microbial diversity: development and application to a time series. Environmental Microbiology 7: 1466–1479.
43. Yannarell AC, Triplett EW (2005) Geographic and environmental sources of variation in lake bacterial community composition. Applied and Environmental Microbiology 71: 227–239.
44. Logue JB, Langenheder S, Andersson AF, Bertilsson S, Drakare S, et al. (2012) Freshwater bacterioplankton richness in oligotrophic lakes depends on nutrient availability rather than on species-area relationships. ISME J 6: 1127–1136.
45. Yannarell AC, Kent AD, Lauster GH, Kratz TK, Triplett EW (2003) Temporal patterns in bacterial communities in three temperate lakes of different trophic status. Microbial Ecology 46: 391–405.
46. Pace ML, Cole JJ (1994) Comparative and experimental approaches to top-down and bottom-up regulation of bacteria. Microbial Ecology 28: 181–193.
47. Burnham KP, Anderson DR (2002) Model selection and multi-model inference: A practical information-theoretic approach. New York: Springer-Verlag.
48. Legendre P, Legendre L (1998) Numerical Ecology. Amsterdam: Elsevier Science B.V.
49. Chao A, Chazdon RL, Colwell RK, Shen TJ (2005) A new statistical approach for assessing similarity of species composition with incidence and abundance data. Ecology Letters 8: 148–159.
50. Legendre P, Anderson MJ (1999) Distance-based redundancy analysis: Testing multispecies responses in multifactorial ecological experiments. Ecological Monographs 69: 1–24.
51. R Core Team (2012) R: A language and environment for statistical computing. Vienna, Austria: R Foundation for Statistical Computing.
52. Oksanen J, Guillaume Blanchet F, Kindt R, Legendre P, Minchin PR, et al. (2012) vegan: Community Ecology Package. R package version 2.0–5 ed.
53. Dorigo U, Fontvieille D, Humbert JF (2006) Spatial variability in the abundance and composition of the free-living bacterioplankton community in the pelagic zone of Lake Bourget (France). FEMS Microbiol Ecol 58: 109–119.
54. Kent AD, Jones SE, Yannarell AC, Graham JM, Lauster GH, et al. (2004) Annual patterns in bacterioplankton community variability in a humic lake. Microbial Ecology 48: 550–560.
55. Sogin ML, Morrison HG, Huber JA, Mark Welch D, Huse SM, et al. (2006) Microbial diversity in the deep sea and the underexplored "rare biosphere".

Proceedings of the National Academy of Sciences of the United States of America 103: 12115–12120.
56. Schindler DE, Chang GC, Lubetkin S, Abella SEB, Edmonson WT (2003) Rarity and functional importance in a phytoplankton community.. In: Kareiva P, Levin SA, editors. The importance of species: perspectives on expendability and triage. Princeton, New Jersey: Princeton University Press.
57. Bent SJ, Forney LJ (2008) The tragedy of the uncommon: understanding limitations in the analysis of microbial diversity. ISME J 2: 689–695.
58. Kovacs A, Yacoby K, Gophna U (2010) A systematic assessment of automated ribosomal intergenic spacer analysis (ARISA) as a tool for estimating bacterial richness. Res Microbiol 161: 192–197.
59. Lyons KG, Brigham CA, Traut BH, Schwartz MW (2005) Rare species and ecosystem functioning. Conservation Biology 19: 1019–1024.
60. Lennon JT, Jones SE (2011) Microbial seed banks: the ecological and evolutionary implications of dormancy. Nature Reviews Microbiology 9: 119–130.

18

How Does Conversion of Natural Tropical Rainforest Ecosystems Affect Soil Bacterial and Fungal Communities in the Nile River Watershed of Uganda?

Peter O. Alele[1,2,3,7]*, Douglas Sheil[4,5,6,7], Yann Surget-Groba[1], Shi Lingling[1,2], Charles H. Cannon[1,8]

1 Key Laboratory of Tropical Forest Ecology, Xishuangbanna Tropical Botanical Garden (XTBG), Chinese Academy of Sciences, Kunming, Yunnan, P. R. China, **2** University of the Chinese Academy of Sciences, Beijing, P. R. China, **3** Great Nile Conservation Centre (GNCC), Lira, Uganda, **4** Department of Ecology and Natural Resource Management, Norwegian University of Life Sciences, Ås, Norway, **5** Center for International Forestry Research (CIFOR), Bogor, Indonesia, **6** Department of Ecology and Natural Resource Management, School of Environment, Science and Engineering, Southern Cross University, Lismore, New South Wales, Australia, **7** Institute of Tropical Forest Conservation (ITFC), Mbarara University of Science and Technology (MUST), Kabale, Uganda, **8** Texas Tech University, Lubbock, Texas, United States of America

Abstract

Uganda's forests are globally important for their conservation values but are under pressure from increasing human population and consumption. In this study, we examine how conversion of natural forest affects soil bacterial and fungal communities. Comparisons in paired natural forest and human-converted sites among four locations indicated that natural forest soils consistently had higher pH, organic carbon, nitrogen, and calcium, although variation among sites was large. Despite these differences, no effect on the diversity of dominant taxa for either bacterial or fungal communities was detected, using polymerase chain reaction-denaturing gradient gel electrophoresis (PCR-DGGE). Composition of fungal communities did generally appear different in converted sites, but surprisingly, we did not observe a consistent pattern among sites. The spatial distribution of some taxa and community composition was associated with soil pH, organic carbon, phosphorus and sodium, suggesting that changes in soil communities were nuanced and require more robust metagenomic methods to understand the various components of the community. Given the close geographic proximity of the paired sampling sites, the similarity between natural and converted sites might be due to continued dispersal between treatments. Fungal communities showed greater environmental differentiation than bacterial communities, particularly according to soil pH. We detected biotic homogenization in converted ecosystems and substantial contribution of β-diversity to total diversity, indicating considerable geographic structure in soil biota in these forest communities. Overall, our results suggest that soil microbial communities are relatively resilient to forest conversion and despite a substantial and consistent change in the soil environment, the effects of conversion differed widely among sites. The substantial difference in soil chemistry, with generally lower nutrient quantity in converted sites, does bring into question, how long this resilience will last.

Editor: Morag McDonald, Bangor University, United Kingdom

Funding: This work was funded by Xishuangbanna Tropical Botanical Garden (XTBG) and the Chinese Academy of Sciences. The funders had no role in study design, data collection and analysis, decision to publish, or preparation of the manuscript.

Competing Interests: The authors have declared that no competing interests exist.

* Email: alelepeter@gmail.com

Introduction

Tropical rainforests (TRF) possess most of the world's terrestrial biodiversity and deforestation is the leading cause of biodiversity loss [1,2]. Due to their high biodiversity and endemism, the tropical rainforests in Uganda's Nile river watershed are among the world's most important for their conservation values. But these areas are under pressure. The United Nations Population Division [3] predicts that the population of the Nile Basin states will increase by 57% from 2010 to 2030, reaching 647 million people. This rapid population growth, high levels of poverty and prevalent civil insecurity continue to exert severe pressure on natural resources in the region. Uganda in particular has one of the world's highest population growth rates (3.2% per year) [4]. Most of this growing population (nearly 80%) is dependent on agriculture leading to large scale and continuing conversion of natural habitats [5].

Soil communities form the foundation of any ecosystem, in terms of nutrient cycling and availability, so understanding how land conversion affects these communities is an important first step. The effect of land use change on soil microbial communities has been studied in South American and Southeast Asian forests [6,7], but not in the biodiversity hotspots of the Nile river watershed. There is considerable global concern about the loss of biodiversity and the consequences for human well-being [8]. Microorganisms in particular play a vital role in many ecological processes and environmental services [9]: these roles are not always apparent or well characterized but if all microbes died the world would rapidly become buried in undecomposed dead material. Due to their significance in maintaining ecosystem function and productivity [9,10], our study offers a vital exploratory appraisal of microbial community dynamics in natural TRF and human-converted sites. We don't know if there are reasons to be concerned unless we look. Developing such knowledge is critical at this point, because populations in the Nile river watershed are highly dependent on forests for basic requirements such as food and fuel wood, with the environment contributing between 40–60% of the gross domestic product (GDP) of the Nile riparian states [11].

Because of widespread loss of biodiversity, focus from species conservation within particular habitats has been shifted to conservation of communities [12,13]. It is therefore important to explore and understand how composition and diversity changes across spatial scales in a given context [14–16]. Changes in ecosystems caused by conversion to intensive management can lead to biotic homogenization, the increase in community similarity over time and/or space and an implied loss of rare and vulnerable taxa when examined at larger scales [17–19]. Because microorganisms are the most diverse organisms on earth with most taxa and respective functions and behaviors as yet unknown, determining their sensitivities and biogeography remains a major challenge. But in the longer term such knowledge will help us better understand the sustainability of land-use systems and associated environmental values.

This study was therefore necessary as a first step in exploring these relationships, and to enhance understanding so as to contribute to the informed and appropriate stewardship of the region's natural resources. Our objective was to establish how forest conversion and soil factors affect soil bacterial and fungal diversity and community composition in the tropical rain-forests in the Nile river watershed of Uganda. We chose four forest sites found within protected areas, with paired treatments within each forest; (1) natural and (2) converted ecosystem sites. The natural forest ecosystem at each site had suffered minimal human disturbance, while converted areas had been transformed to cropland. These matched sites found in different locations and environmental conditions each experienced different land use histories, conservation circumstances and individual challenges for management.

In each matched set of natural and converted sites, we compared soil physical and chemical properties and microbial community diversity and composition using standard PCR-based genotyping techniques. We then calculated community similarity indices between sites. This approach would allow us to examine both environmental and biotic changes in the soil community associated with conversion. Disturbances of sufficient magnitude or duration may alter an ecosystem and force a different regime of predominant processes and structures that favor some populations over others [20].

We tested the null hypotheses that there was no difference in soil properties, band-types, and diversity between treatments. The influence of soil properties on microbial community diversity was measured by discriminant analysis and canonical correspondence analysis (CCA), [21,22]. Because additive partitioning of diversity provides a useful framework for quantifying the spatial patterns of diversity across hierarchical spatial scales [23,24], we partitioned total diversity (γ) in each ecosystem type (natural and converted) into additive components representing within-community diversity (α) and between-community diversity (β). Our objective was to identify the most important sources of total diversity so as to propose conservation measures for microbial communities in the TRF ecosystems of the Nile river watershed of Uganda.

Methods

Site description

We selected four tropical rainforest (TRF) sites because of their relative size, biodiversity, socio-economic and scientific importance (Fig 1). Mabira forest is located between the highly populated and urbanized Kampala city on the western side; the extensive and mechanized Lugazi sugar and tea plantations on the Eastern; and Lake Victoria on the southern side. Budongo forest is located next to the extensive Kinyala sugar plantations on one side and a densely populated mainly subsistence population scattered around it. Maramagambo and Kaniyo Pabidi are located within Queen Elizabeth and Murchison Falls national parks (NP) respectively. These two NP forests had perhaps the best protection due to presence of Uganda Wildlife Authority (UWA) personnel. However, Maramagambo's location starting on the steep slopes of the rift valley subjected it to frequent storms with strong runoff flow that swept away most of its top soil (Table 1).

Soil sampling design

We collected 400 core soil samples within 40 plots (1000 m^2 each) in four TRF sites (Fig 1). We sampled five plots from each site of the natural TRF and five plots from the converted TRF. We established the plots at least 100 m from the ecosystem edge and 500 m apart and collected 10 evenly placed core subsamples of top soil (0–15 cm) from each plot and homogenized them into one sample per plot. We then derived a 500 g composite sample from the mixture, sieved and packed it for physical and chemical analyses and DNA extraction.

Sample preparation

We sieved 100 g of the soil on-site through a 4 mm mesh, transported it to the laboratory on ice, and stored in a freezer at −40°C prior to nucleic acid extraction and analysis. We kept the rest of the soil for drying and physical and chemical analysis. We performed DNA extractions from 1 g of soil using the Ultra Clean soil DNA kit (Mo Bio Labs, Solana Beach, CA, USA) following the manufacturer's protocol. The purified DNA was detected by agarose gel electrophoresis, and the DNA was amplified by polymerase chain reaction (PCR).

Soil property analyses

We measured the soil pH in 2.5:1 water to soil suspension using a pH meter (10 g soil+ 25 ml of distilled water, shaken for 30 min and read on a calibrated pH meter). We then used the Walkley and Black method [25] to analyze soil organic carbon (SOC) and the Kjeldahl method [26] to determine soil nitrogen. We measured the soil phosphorus by the Bray and Kurtz no. 1 method [27]. The photoelectric flame photometer was used to determine the soil potassium, sodium and calcium after extraction with neutral ammonium acetate. We used the atomic absorption spectrometer to measure the soil magnesium after extraction with neutral ammonium acetate. The Bouyoucos hydrometer method adopted from Gee and Bauder [28] was used to determine soil texture. The soil copper and iron were then determined using the atomic absorption spectrometer after extraction with EDTA.

PCR amplification and DGGE analysis

Polymerase chain reaction-denaturing gradient gel electrophoresis (PCR-DGGE) method has been used extensively in microbial ecology and is a robust and cost effective method for exploratory classification of microbial communities [29]. Following soil DNA extraction, we performed a PCR for each DNA extraction to

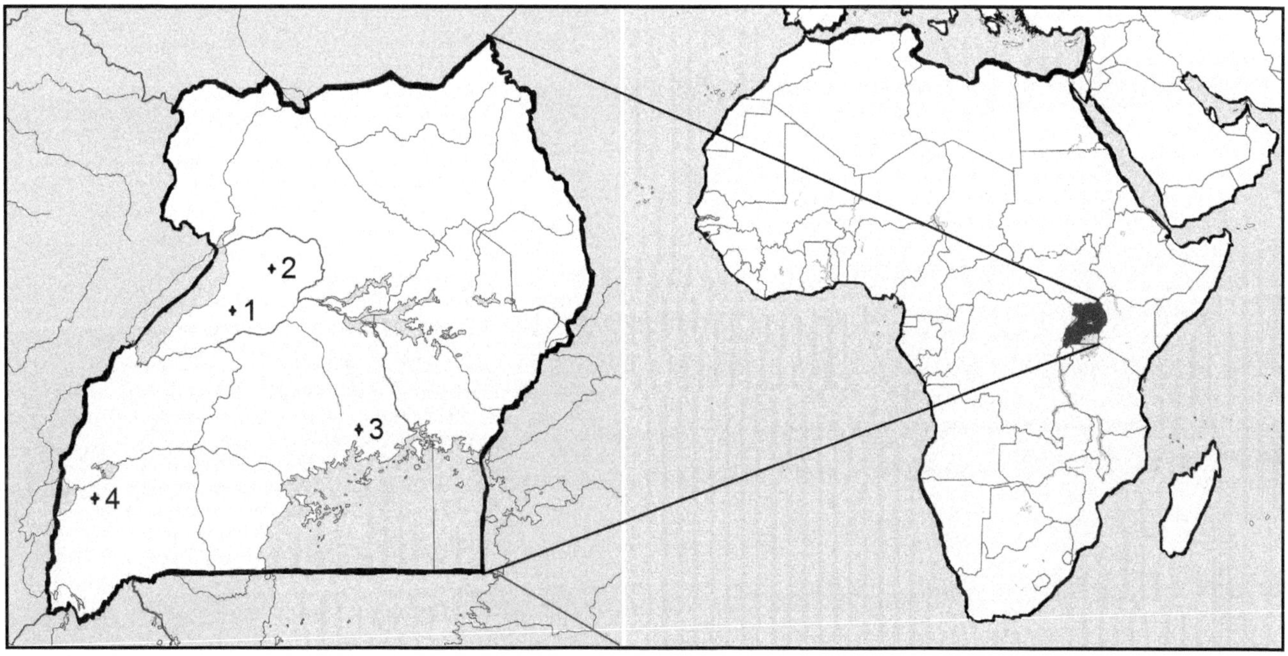

Figure 1. Map of Uganda showing the distribution of sampling sites; Budongo forest (1), Kaniyo Pabidi (2), Mabira forest (3), and Maramagambo forest (4).

amplify the 16S rRNA genes for bacteria and 18S rRNA genes for fungi using universal primers (Table 2).

PCR reactions had a final volume of 25 µl containing a final concentration of 1× TaKaRa ExTaq PCR buffer with $MgCl_2$, 300 pM of primers for bacteria. We then added 200 µM dNTPs, 2.5 U ExTaq DNA polymerase (TaKaRa Bio, Otsu, Japan) and milliQ H_2O to complete the volume, BSA was also added for the fungal community analysis. We performed PCR cycles with an initial denaturing temperature of 95 °C for 5 min, followed by 35 cycles of 95 °C for 30 sec, annealing temperature of 50 °C for 30 sec, extension of 72 °C for 1 min; and a final extension of 72 °C for 10 min. We checked the product of the PCR-rounds and quantified by agarose gel-electrophoresis.

We then performed 16S rRNA and 18S rRNA-DGGE analysis using a universal mutation detection system (Dcode Bio-Rad, Richmond, CA, USA) with a 6% and 8% acrylamide gel for bacteria and fungi respectively containing a gradient of 40–60% denaturant (100% denaturant contains 7 mol urea and 40% formamide). We applied 100 ng of PCR samples to the DGGE gel. DGGE was performed in 1 × TAE Buffer (40 mol Tris/acetate, pH 8; 1 mol ethylene diaminetetra acetic acid) at 60 °C and a constant voltage of 60 V for 16 hours. After staining with SYBR Green1, we recorded the DGGE gels as digital images and analyzed the DNA band numbers using image-processing software after subtracting background noise.

Data analysis

We used the Rolling disk method with Quantity One (Bio-Rad laboratories Inc.), which normalizes the band pattern from electrophoresis for identification of each band. We then converted the band patterns into binary data based on the presence or absence of each band for part of our analysis. The DGGE fingerprints were interpreted in terms of band richness (number of predominant DGGE bands/population). The pixel intensity of each band was detected by Quantity One software and is expressed as relative abundance (P_i) [30]. Shannon Index (H') and Simpson index (D), the most widely used diversity indices were then calculated using the richness and relative abundance data following the equations:

$$H' = -\sum_{i=1}^{R} P_i InP_i \quad (1)$$

$$D = \sum_{i=1}^{R} (P_i)^2 \quad (2)$$

Where R, the richness, is number of different bands each data set contains, $P_i = \frac{n_i}{N}$ and n_i is the abundance of the *ith* band and N the total abundance of all bands in the sample.

Band-type data of the DGGE fingerprints was then used to derive the alpha diversity (bands per sample and ecosystem type), beta diversity (total bands per site) [31]. Jaccard's similarity indices [32] between converted and natural TRF sites were determined using the equation:

Jaccard's Similarity Index $= A/(A+B+C)$

Where,

A = Total number of bands present in both converted (C) and natural (N) ecosystem samples (plots) (also β-diversity)

B = Number of bands present in C but not in N

C = Number of bands present in N but not in C

We determined the influence of site factors as revealed by soil physicochemical properties on the variation of soil microbial communities by applying discriminant analysis using Statistical Package for the Social Sciences (SPSS). This was done to assess the relative importance of each predictor variable (pH, SOC, N, P, K, Na, Ca, Mg, and soil texture). We also used the Mann-Whitney

Table 1. Summary of study site description.

Forest Site	Location	Size (km^2)	Altitude (masl)	Geology	Forest type	Habitat type	Ecosystem description
Budongo	31°N 35° E 1°S 45° N	793	700–1,270	Weathered pre-cambium with ferrallitic sandy clay loams	Ironwood forest (*Cynometra alexandri*); Mixed forest (*Maesopis*), and colonizing forest (*Entandrophragma*)	Primary forest	Consists of a medium altitude moist semi-deciduous forest with areas of savanna and woodland. Converted areas consisted of deforested agricultural land being cultivated and planted with maize, beans, sweet potatoes and cassava. This land has existed as agricultural land for at least 15years.
Mabira	33° 0.00′ E 0° 30.00′ N	300	1,070–1,340	Ferrallitic soils with mainly sandy clay loams	Mixed forest	Secondary forest, heavily influenced by humans	The forest is surrounded by a densely populated area. Converted areas were actively cultivated and used to grow maize, groundnuts, beans, yams, cassava, sweet potatoes and a few scattered plants of coffee and sugarcane. The converted land had existed as agricultural land for at least 10 years.
Maramagambo	00° 33′ 00" S and 29° 53′ 00" E	1,978 (QENP)	910–1390	Ferrallitic soils with undifferentiated dark horizons	Medium altitude, moist, semi-deciduous forest	Secondary forest influenced by wildlife and humans	Forms part of the Queen Elizabeth NP (QENP) which is 1,978 km2. Converted areas consisted of cultivated and grazed areas with gardens of sweet potatoes, beans, maize, and sorghum with areas commonly grazed by cattle and goats.
Kaniyo Pabidi	Lat 1.916667 Long 31.666667	N/A	700–1,270	Freely drained ferruginous tropical soils	Mixed forest	Primary forest	Located north of Budongo forest and part of Murchison Falls N.P. Converted areas consisted of cultivated areas with crops like maize, beans, cassava and sweet potatoes.

[5,61,70,71].

Table 2. Sequences of primers used in study.

Microorganism	Primer	Sequence (5′–3′)	Reference
Bacteria	F357	CGC CCG CCG CGC GCG GCG GGC GGG GCG GGG GCA CGG GGG GCC TAC GGG AGG CAG CAG	[72]
	907R	CCG TCA ATT CMT TTG AGT TT	
Fungi	FF390	CGA TAA CGA ACG AGA CCT	[73]
	FR1GC	AIC CAT TCA ATC GGT AIT	

test to examine differences between soil properties in natural and converted ecosystems, and microbial communities in natural and converted ecosystems.

We tested the null hypothesis that diversity is uniform at all spatial scales by additive partitioning of total diversity (γ diversity). To determine contributions of α and β diversity to overall diversity across a range of spatial scales [14,23,33], an additive relationship between diversity components (i.e., $\beta = \gamma - \alpha$) was derived (Fig 2.). The scale at which diversity is maximized was therefore identified [23,34], to facilitate planning processes and management strategies to conserve natural levels of diversity accordingly [35–38].

We used PARTITION 3.0 software [39] to calculate average diversity at each scale and diversity was measured as band richness. Individual-based randomization procedure in the software was used to test whether the observed partitions of diversity within the ecosystem could have been obtained by a random allocation of lower-level samples nested among higher-level samples [34]. Null values of β_i obtained from 1,000 randomizations were used to obtain a *p*-value for the observed β_i at each hierarchical scale. Deviations of the observed diversity from the null expectation indicated a nonrandom spatial distribution of fungi or bacteria at a given scale.

Results

Soil property variations

Soil pH comparisons using a Mann-Whitney U test of significance, between five plots of natural and five plots of converted TRF ecosystems in each of the four forest sites, found significantly higher (less acidic) pH in three of the four sites at Budongo ($p = 0.0107$), Kaniyo Pabidi ($p = 0.0112$), and Mabira ($p = 0.0269$); and non-significant difference at Maramagambo ($p = 0.1706$). Percentage soil organic carbon (SOC) was significantly higher in natural than converted ecosystems in all four sites i.e. Budongo ($p = 0.0119$), Kaniyo Pabidi ($p = 0.0212$), Mabira ($p = 0.0122$) and Maramagambo ($p = 0.0119$) with combined %SOC in natural sites more than double of that in converted sites; whereas %soil nitrogen was only significantly higher in natural forests at Budongo ($p = 0.0112$) and Kaniyo Pabidi ($p = 0.0119$), and non-significant at Mabira ($p = 0.6015$) and Maramagambo ($p = 0.0947$) (Table 3).

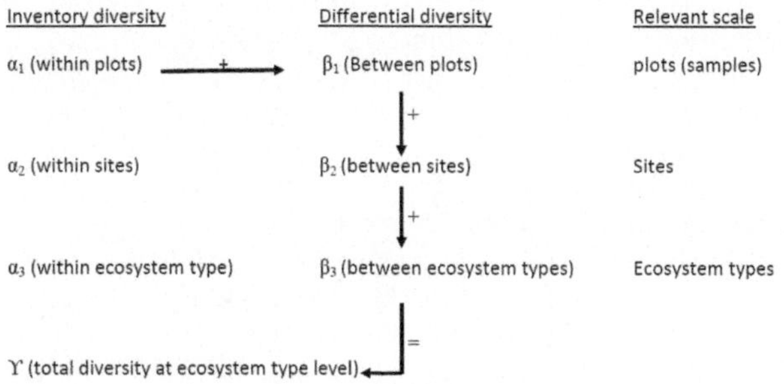

Figure 2. Illustration of hierarchical spatial scales in our additive partitioning model. The α scale is the within-level, and β scale, the between-level components. Because a diversity at a given scale is the sum of the α and β diversity at the next lower scale, the total diversity (γ) can be described by the following formula: $\alpha1+\beta1+\beta2+\beta3$ [14,22].

Ecosystem and site comparisons of microbial community diversity

Bacterial (B) communities were significantly richer ($p = 0.0304$; Mann-Whitney U) in detectable bands than fungal (F) communities in both converted (C) and natural (N) ecosystems (converted: medians; F = 36, B = 61.5; natural: medians; F = 39.5, B = 60.5). While total band richness (B+F) did not differ between natural and converted forests we observed greater fungal richness in natural than in converted forests (medians: C = 36, N = 39.5; test stat = 18.5) and more bacterial bands in converted than in natural ecosystems (medians: C = 61.5, N = 60.5; test stat = 18.5). Kaniyo Pabidi was the most diverse site overall with the highest number of bacterial and fungal bands, while Maramagambo had the least band richness (Fig 3).

Natural sites harbored more bands unique to one site than converted sites for bacteria at Kaniyo Pabidi and Maramagambo and for fungi at Maramagambo and Budongo. Mabira and Kaniyo Pabidi had higher numbers of unique bacterial bands than at Maramagambo and Budongo. There were also more unique fungal bands at Mabira and Budongo than at Maramagambo and Kaniyo Pabidi (Fig 4).

We also found that Mabira and Maramagambo had the lowest bacterial Jaccard's community similarity indices [32] between natural and converted ecosystems, whereas Budongo and Mabira had the lowest fungal community similarity between natural and converted ecosystems (Table 4). Dissimilarity between natural and converted ecosystems was nonetheless non-significant in all sites for both fungal and bacterial communities. Also, there was generally greater dissimilarity between sites of fungal communities than in bacterial communities suggesting a higher susceptibility to habitat change among fungi than bacteria (Table 4).

Ecosystem classification and importance of predictor variables

The CCA showed that, despite the relatively small amount of difference between sites, soil pH, average phosphorus, and texture (%sand) had strong influence on bacterial diversity in the TRF ecosystem (Fig 5); whereas organic carbon, sodium, pH and average phosphorus were strongly associated with fungal community variation in both natural and converted TRF ecosystems (Fig 5). The CCA also showed that bacterial communities in both Kaniyo Pabidi and Mabira were unique to bacterial communities

Table 3. Mean (Standard deviation) for diversity indices of Bacterial (B) and Fungal (F) communities and soil properties in natural (N) and converted (C) ecosystems.

	Budongo		Kaniyo Pabidi		Mabira		Maramagambo	
	Natural	**Converted**	**Natural**	**Converted**	**Natural**	**Converted**	**Natural**	**Converted**
Shannon (B)	2.49(0.68)	2.73(0.10)	3.25(0.21)	3.12(0.25)	3.06(0.18)	3.19(0.18)	3.04(0.16)	2.93(0.07)
Simpson (B)	0.89(0.10)	0.93(0.01)	0.95(0.13)	0.95(0.02)	0.95(0.01)	0.95(0.01)	0.95(0.01)	0.94(0.01)
Shannon (F)	2.44 (0.30)	2.49 (0.17)	2.83(0.19)	2.77(0.20)	2.57(0.62)	2.64(0.30)	2.10(0.50)	1.93(0.67)
Simpson (F)	0.90(0.29)	0.90(0.03)	0.93(0.01)	0.93(0.02)	0.90(0.06)	0.92(0.03)	0.84(0.08)	0.82(0.12)
pH	5.88(0.11)*	5.08(0.20)*	6.24(0.23)*	5.38(0.27)*	6.46(0.54)*	6.18(0.42)*	6.18(0.45)	5.80(0.24)
OC (%)	6.14(0.67)*	1.59(0.17)*	5.69(1.20)*	3.65(0.54)*	5.87(2.03)*	3.53(1.19)*	9.08(0.79)*	3.98(1.09)*
N (%)	0.43(0.05)*	0.11(0.01)*	0.43(0.08)*	0.25(0.03)*	0.21(0.10)	0.22(0.06)	0.28(0.04)	0.19(0.08)
Ca (Cmoles/kg)	8.00(2.56)*	4.00(0.58)*	13.54(3.36)*	6.74(0.98)*	8.14(1.19)*	6.20(0.76)*	12.12(2.13)*	7.34(1.16)*

*significant differences ($p<0.05$) between natural and converted ecosystems.

in the other sites and there was high contrast between bacterial communities of converted and natural ecosystems at Kaniyo Pabidi. Fungal communities at Maramagambo and Mabira were also unique to those in other sites and there was high contrast between fungal communities at Mabira's natural and converted ecosystems. Furthermore, the CCA showed that fungal communities responded more to soil pH levels than bacterial communities (Fig 5), with site-specific patterns showing that bacteria and fungi were grouping according to sites.

A discriminant analysis to predict whether bacterial or fungal communities were from natural or converted ecosystems found that only OC, Ca, N, and pH for bacterial communities; and OC, N, Ca, and pH for fungal communities (all ranked from most important to least important) were found to be significant predictors of soil physicochemical properties. All other variables were poor predictors in this context (Table 5).

Hierarchical scaling

We found 58 and 56 fungal bands in natural and converted forests respectively, from 17 plots of natural ecosystems and 20 plots of converted forests. There were also 92 and 88 bacterial bands in natural and converted ecosystems respectively found in 20 plots of converted ecosystems and 17 plots of natural ecosystem. All these were within four sites. β-diversity varied more than α-diversity between natural and converted ecosystems for both bacteria and fungi. We found higher bacterial and fungal β-diversity in converted ecosystems than in natural ecosystems at lower hierarchical scales (β_1); higher β-diversity in natural than converted at between-site scale (β_2), and higher β-diversity in converted than in natural ecosystems at the between-ecosystem type scale (β_3) (Fig 6).

We also found substantial contribution of observed β-diversity (β_1, β_2, and β_3) to total band richness (γ-diversity), while α-diversity of both bacteria and fungi in converted and natural ecosystems were similar. Spatial partitioning of total diversity also consistently showed that the beta components (β_1 and β_2) were always greater than expected by chance, whereas the alpha component (α_1) was always lower than expected. For both fungal and bacterial communities in natural and converted ecosystems, observed within plot diversity were substantially less than values expected from individual-based randomizations (Fig 7).

Discussion

Soil property variations and site differences

Studies in both tropical and temperate zones show that soils in converted or cropped areas normally have reduced soil aggregation, structural stability and organic matter, and an increase in bulk density when compared to forests [40,41]. Habitat conversion may also alter soil properties such as nutrient levels, and abiotic conditions and may affect associations between organisms. In our study there are some local details that may influence our results.

Both Maramagambo and Kaniyo Pabidi are located within Queen Elizabeth NP and Murchison Falls NP respectively and are protected by Uganda Wildlife Authority (UWA) personnel. They are well protected and there is little evidence of recent encroachment. There is significant wildlife populations including elephants, buffaloes, zebras and the areas are frequented by tourists. Protection by UWA and presence of dangerous animals (such as buffalos and lions) reduce damaging human activity at Kaniyo Pabidi and Maramagambo which should enhance the difference between natural and converted ecosystems. Maramagambo's location, in contrast, means the

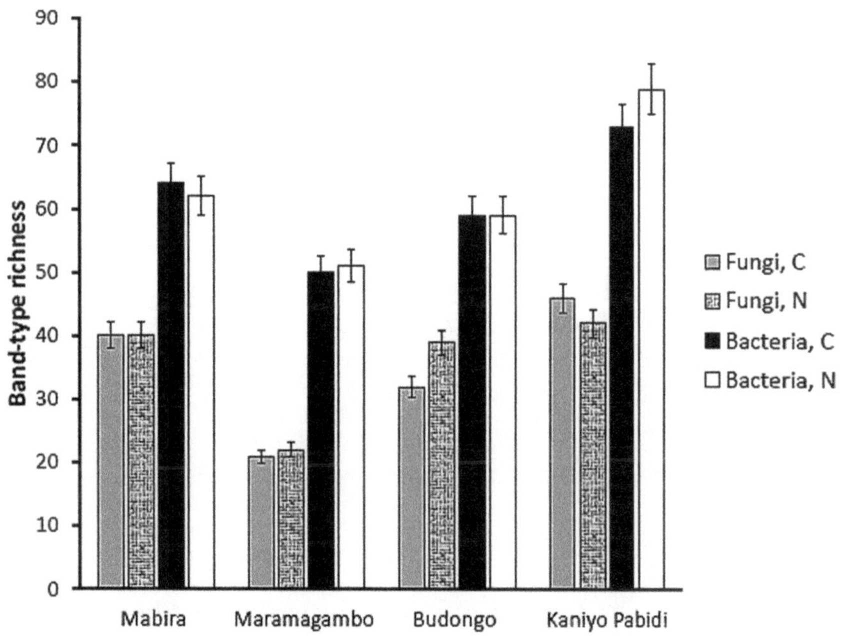

Figure 3. Band richness for fungal and bacterial communities in converted and natural ecosystems. All richness values are total bands present in five samples of each ecosystem treatment (error bars are 5% confidence interval).

forest is subjected to substantial natural disturbance from frequent storms and strong erosive runoff even within the natural forest, whereas tourist activity at Kaniyo Pabidi seemed to have little impact on soil properties. Converted areas at Kaniyo Pabidi were also sparsely populated with limited human impacts on the environment. Its sites were old and might have been cultivated for at least 20years.

In our study, Budongo is located next to a high, mainly subsistence population and resultant population activity. But even though encroachment, illegal hunting and logging in natural habitats in Budongo are not uncommon, there seems to be minimal impact of conversion on soil properties in our sample locations; whereas proximity of Mabira's natural forest to densely populated urban areas exposes it to increased human activities, likely reducing its difference with converted sites.

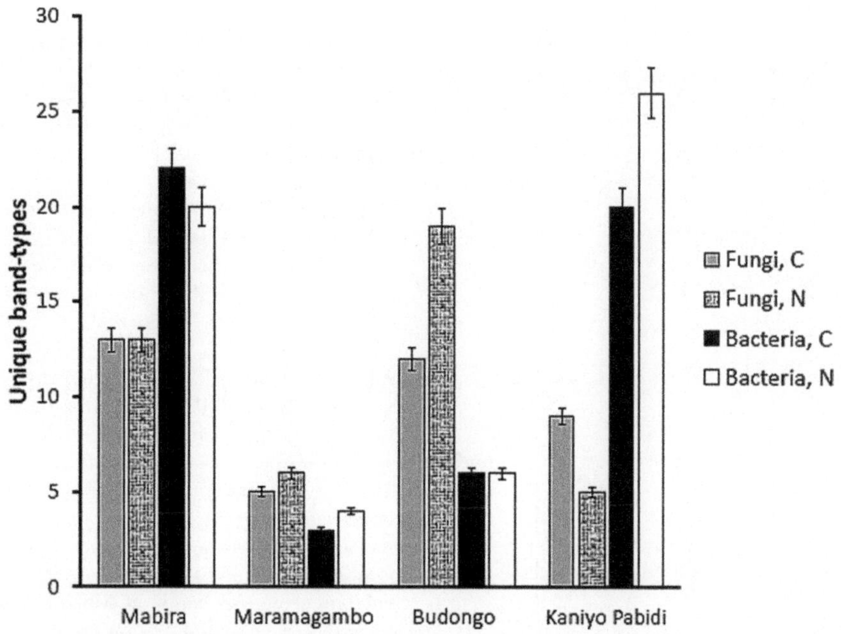

Figure 4. Bacterial and fungal bands unique to converted (C) and natural (N) ecosystems at each site (error bars are 5% confidence interval).

Table 4. Jaccard's similarity indices between bacterial and fungal communities in natural (N) and converted (C) sites of Mabira (Mb), Maramagambo (Mg), Budongo (Bd) and Kaniyo Pabidi (Kp).

			F		U		N		G		I
			Mb	*Mg*	*Bd*	*Kp*	*Mb*	*Mg*	*Bd*	*Kp*	
			N	N	N	N	C	C	C	C	F
B	*Mb*	N	1	0.713	0.769	0.732	0.671	0.685	0.696	0.768	
A	*Mg*	N	0.841	1	0.733	0.666	0.666	0.711	0.646	0.666	U
C	*Bd*	N	0.803	0.840	1	0.804	0.808	0.693	0.622	0.827	
T	*Kp*	N	0.731	0.765	0.698	1	0.774	0.670	0.794	0.785	N
E	*Mb*	C	0.667	0.788	0.739	0.741	1	0.663	0.710	0.761	
R	*Mg*	C	0.866	0.885	0.846	0.755	0.809	1	0.684	0.698	G
I	*Bd*	C	0.825	0.857	0.844	0.696	0.731	0.856	1	0.786	
A	*Kp*	C	0.805	0.869	0.716	0.683	0.759	0.776	0.733	1	I
			B	A	C	T	E	R	I	A	

Microbial community variations

Soil properties determine many aspects of soil microbial community structure [42–44]. Carbon availability [45–47], nitrogen availability [45,48,49] and soil pH [44,50,51] can all influence microbial community composition and diversity. In addition, correlation studies have shown that plant species [52–54] and soil type [45,54,55] are associated with variation in microbial communities. It has also been shown that land use indirectly affects bacterial community structure by modification of soil properties [56] but similarity between converted and natural ecosystem bacterial communities may also suggest a high number of generalists.

Nacke et al., [56] found that bacterial community composition in forests and grassland was largely determined by tree species and soil pH. Jesus et al. [6] also showed that bacterial community structure is influenced by changes linked to soil acidity and nutrient concentration. Other studies also suggest that soil pH is a major factor influencing microbial community composition [50,57–59]. This influence of soil pH has been recognized at different taxonomic levels [45,60] with most microorganisms thriving within a limited pH range. This is because acids can denature proteins and large pH changes may inhibit growth in microorganisms. Fierer and Jackson, 2006 [44] found, in contrast, that net carbon mineralization rate (an index of C availability) was the best predictor of phylum-level abundances of dominant bacterial groups, and Bisset et al., [42] found that soil microbial communities were consistent with disturbance gradients within different agricultural treatments and relatively undisturbed non-agricultural sites.

Because of widespread forest conversion in Uganda as a result of increasing population pressure, estimated at between 1.1% and 3.15% per year [61]; natural ecosystems and their inhabitant biodiversity are at risk [62]. Loss of diversity increases the likelihood of losing important functional roles and associated ecosystem processes. At landscape scale, spatial and temporal variations of microbial communities in forest soils are influenced by numerous biotic and abiotic factors. These factors may include climate, soil types, and vegetation associations [50,63,64]. Owing to this study design, many of these factors were assumed to be similar between natural and converted ecosystems. For instance, the proximity of natural and converted ecosystem sites meant that climate and geology were, we assume, similar in the two treatments. Even though there could still be a number of underlying causes of community differences, two likely influences were assumed to be soil properties [45,54,55] and vegetation types [52–54]. Clearly both of these sets of factors change when forest is converted for agriculture or range lands.

Despite substantial reductions in SOC, N, Ca and pH in converted sites in this study, differences in microbial communities were small meaning that converted sites still had sufficient SOC, N and Ca to sustain the same microbial populations. The close proximity of the matched pairs could also lead to a source-sink relationship between the natural and converted forests, with the presence of unique taxonomic groups a likely indication of habitat preference (endemism) for some taxonomic groups. It may also be an indicator of relative habitat dissimilarity. The high numbers of unique bacterial bands at Mabira and Kaniyo Pabidi and unique fungal bands at Mabira and Budongo (Fig 4) thus suggests that ecosystem alteration at these sites was sufficient to force a different regime of processes and structures enabling a new set of taxonomic groups to predominate. Mabira had high numbers of both unique bacterial and unique fungal bands that can be attributed to the extent of disturbance at its sites (Mabira

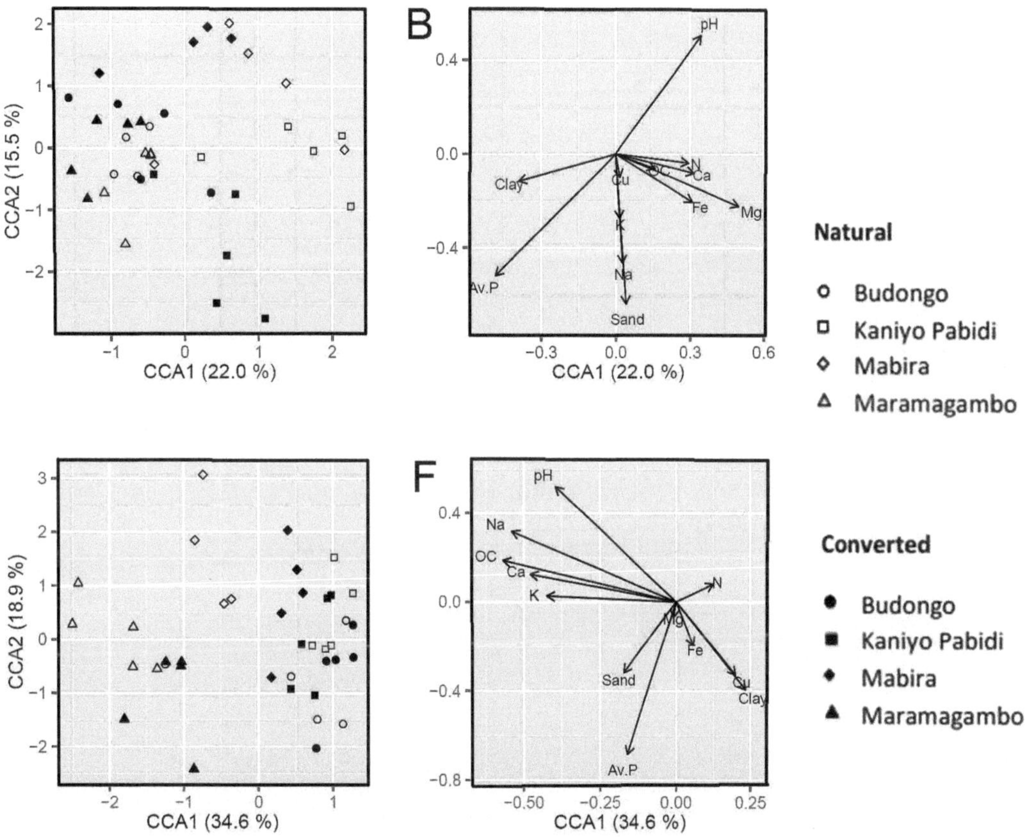

Figure 5. CCA for bacterial (B) and fungal (F) relationships using relative abundance of bands and soil physicochemical properties in natural and converted ecosystems. The symbols (left graphs) represent the similarity between each sample (plot) as defined by their diversity, and the vectors (right graphs) represent the structural matrix for soil properties and their influence on relative abundance of each band. The length of the vectors represents the relative strength of influence of the particular aspect of soil physicochemical property.

is the only peri-urban tropical rainforest site among the four selected sites).

The low numbers of unique bacterial and fungal bands at Maramagambo can be attributed to the high erosion at natural sites that reduced the contrast between the natural and converted sites. For the other sites, bacteria and fungi had different responses to ecosystem alterations. This could indicate separate influences on microbial distribution that exist when alteration is moderate. Similarity indices suggested that bacterial and fungal communities were determined by separate forces leading to distinct responses across the study locations.

Hierarchical scaling

Many studies have shown that specialist species are more negatively affected by current global changes than generalists [7,65]. The process of biotic homogenization can involve the replacement of native biota with non-natives or the introduction of generalist species [66]. In this study, the net decrease in β-diversity from natural to converted TRF ecosystem at the between-site scale (β_2) for both fungi and bacteria was an indication of biotic homogenization [18,66]. This can result from ecosystem alterations which can in-turn alter ecosystem function and reduce ecosystem resilience to disturbance [65,67].

We also showed that the β components of diversity ($\beta_1$and β_2; the average diversity between the plots and sites, respectively) were consistently higher than those expected by chance, whereas the local α_1 diversity component (α_1, the average diversity within the plots) was consistently lower than that expected (Fig 7). Such scale-dependent deviations of the observed diversity from the expected can be generally explained by aggregation at a relatively small "local" scale and, spatial differentiation of diversity at a larger "landscape" scale [33,34,68,69].

Relatively lower diversity within converted ecosystems suggests that conversion of natural TRF ecosystems results in reduced diversity for both bacteria and fungi. This is compatible with recent studies that show that conversion of TRF ecosystems threatens microbial diversity [7] and because microorganisms, like all other organisms, have habitat preferences and may be affected by land-use changes [6,64]. While we cannot be certain that such decline in diversity has led to a decline in any particular ecosystem functions or services, this is a possibility that deserves further evaluation, and we speculate that such loss of diversity will at the very least cause a reduction in functional redundancy and associated resilience.

Higher β-diversity of both bacterial and fungal communities at the between-plot scale (β_1) in converted ecosystems indicates

Table 5. Structure matrix rank showing absolute size of correlation between discriminant analysis function from most important to least important predictor variable of site factors (soil physiochemical properties) and their influence on the variation of soil microbial communities.

Bacteria		Fungi	
Predictor Variables	**Function 1**	**Predictor variables**	**Function 1**
OC	0.625*	OC	0.625*
Ca	0.471*	N	0.493*
N	0.421*	Ca	0.473*
pH	0.355*	pH	0.340*
Mg	0.298	Mg	0.281
Cu	0.197	Cu	0.221
Na	0.181	Na	0.183
K	0.169	K	0.160
Av.P	−0.077	Sand	0.064
Sand	0.070	Av.P	−0.060
Simpson	−0.065	Fe	0.024
Fe	0.029	Shannon	0.012
Shannon	−0.023	Simpson	0.011

(* = important predictor variable, with 0.30 used as the threshold).

differentiation (reduced community similarity) in converted ecosystems at this hierarchical scale. Considering the multiple land-uses and cropping systems of converted areas, this was expected. There was also substantial contribution of β-diversity to total diversity (γ). This suggests the importance of nonrandom ecological processes at the between-plot and between-site scale in determining total richness and community composition [14,34]. Differences between the observed and expected diversity components could be due to ecological processes that lead to a non-random dispersion of individuals. These processes could include intra-specific aggregation, habitat selection, and limited dispersal capacity [33].

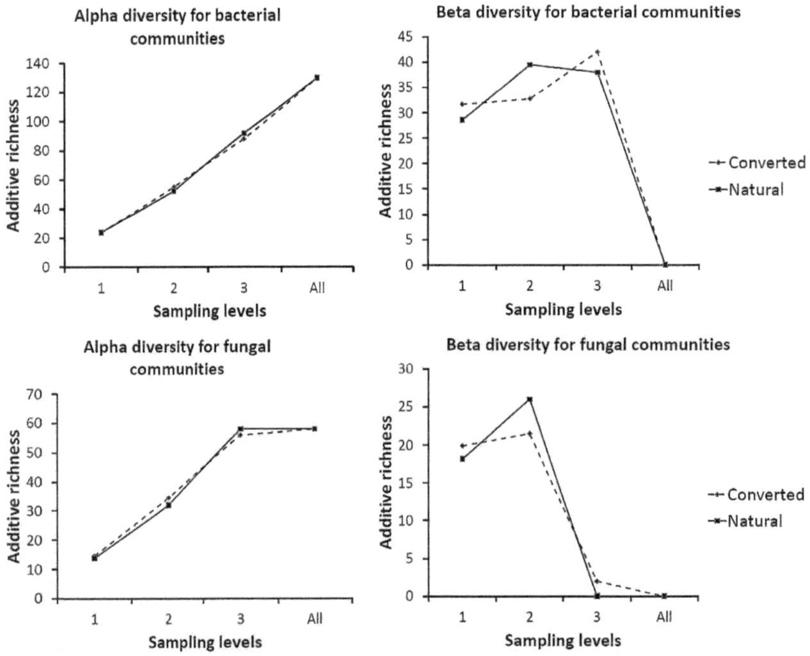

Figure 6. Additive partitioning of bacterial and fungal diversity (expressed as additive richness) across alpha and beta hierarchical spatial scales at three sampling levels (plot, site and ecosystem type) in natural and converted TRF ecosystems.

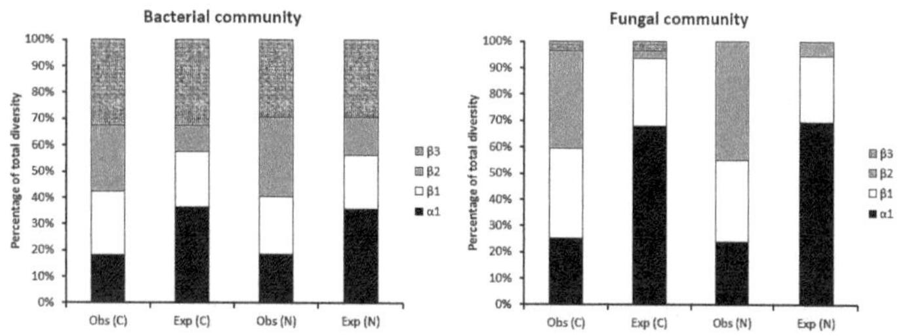

Figure 7. The additive partitioning of total bacterial and fungal community, γ -diversity into α and β components at three nested spatial scales, with each component expressing their relative contributions to γ –diversity; where γ –diversity is equal to $\alpha1+\beta1+\beta2+\beta3$. The observed (obs) partitions are compared with the expected (exp) values, as predicted by the null model based on 1000 iterations using individual-based randomization.

Conclusion

There is international concern about the threat to natural habitats in the Nile river watershed and the consequential loss of important biodiversity. Whereas aspects of microbial biogeography and influence of forest conversion in Uganda's Nile river watershed is largely unknown, this study offers an important first glimpse into indicators of spatial diversity patterns of soil fungal and bacterial communities in the Uganda's Nile river watershed. Our observations of reduced soil microbial diversity, both bacterial and fungal, in converted ecosystems though unsurprising in itself causes us some concern and would justify further work to determine the significance of the diversity lost and the wider implications.

By focusing on diversity patterns across multiple hierarchical spatial scales, we were able to identify the scale at which regional microbial diversity is maximized. We showed that there was substantial contribution of β-diversity to total ecosystem diversity (γ) which includes taxa at the between-plot, site and ecosystem scales and unique taxa, highlighting the necessity to conserve marginal habitats and ecotones. Soil microbial communities in Uganda's Nile river watershed exhibit considerable resilience to forest conversion even though SOC, N, Ca and pH were all significantly altered. This result is surprising given that these physical and chemical properties typically strongly influence microbial diversity. Additionally, the variation among sites was quite large, indicating that soil communities in this region vary considerably on a regional spatial scale. Our results do not explain this variation. Most studies suggest that biogeographic barriers play little role in the geographic structure of soil communities. Rather than a consistent general pattern of microbial community change following forest conversion we find that responses are largely site-specific and widely variable.

Acknowledgments

We confirm that our field work did not involve any endangered or protected species and that we were granted access to protected areas by Uganda Wildlife Authority (UWA) and National Forestry Authority (NFA). Our sample site coordinates included: (1) Mabira; Lat, N00°24.495'; Long, E033°02.464' (Required permission obtained from NFA); (2) Maramagambo; Lat, S00°04.421'; Long, E030°02.305' (Required permission obtained from UWA); (3) Budongo; Lat, N01°41.852'; Long, E031°29.363' (Required permission obtained from UWA/NFA); (4) Kaniyo Pabidi; Lat, N01°55.128'; Long, E031°43.116' (Required permission obtained from UWA).

We would like to thank the Institute of Tropical Forest Conservation (ITFC) of Mbarara University of Science and Technology for supporting and co-advising this research. We also thank the Genetics Lab at Faculty of Science and the soil department lab at Faculty of Agriculture both at Makerere University, Kampala, for their lab facilitation during extractions and analyses. And finally we acknowledge the ideas and insights of Masatoshi Katabuchi, Dossa Gbadamassi, and our field and lab technicians and assistants Solomon Echel, Francis Alele Ozirit and Boniface Balikuddembe.

Supporting Information

Data S1 Excel spreadsheet from Quantity One analysis of PCR-DGGE profiles of fungal and bacterial communities in natural and converted TRF ecosystems.

Author Contributions

Conceived and designed the experiments: POA CHC. Performed the experiments: POA SL. Analyzed the data: POA DS CHC. Contributed reagents/materials/analysis tools: DS YSG CHC. Contributed to the writing of the manuscript: POA DS YSG CHC.

References

1. Turner IM (1996) Species loss in fragments of tropical rain forest: a review of the evidence. J Appl Ecol 33: 200–209.
2. Joseph Wright S, Muller-Landau HC (2006) The Future of Tropical Forest Species. Biotropica 38: 287–301.
3. United Nations (2013) World Population Prospects: The 2012 Revision. Highlights and Advance Tables. New York.
4. United Nations D of E and SAPD (2007) World Population Prospects: The 2006 Revision.
5. Kayanja FIB, Byarugaba D (2001) Disappearing forests of Uganda: The way forward. Curr Sci 81: 936–947.
6. Jesus da CE, Marsh TL, Tiedje, M J, de S Moreira FM (2009) Changes in land use alter the structure of bacterial communities in Western Amazon soils. ISME J 3: 1004–1011. Available: http://www.ncbi.nlm.nih.gov/pubmed/19440233. Accessed 15 November 2012.
7. Rodrigues JLM, Pellizari VH, Mueller R, Baek K, Jesus EDC, et al. (2012) Conversion of the Amazon rainforest to agriculture results in biotic homogenization of soil bacterial communities. Proc Natl Acad Sci U S A. Available: http://www.ncbi.nlm.nih.gov/pubmed/23271810. Accessed 29 December 2012.

8. Cardinale BJ, Duffy JE, Gonzalez A, Hooper DU, Perrings C, et al. (2012) Biodiversity loss and its impact on humanity. Nature 486: 59–67. Available: http://www.ncbi.nlm.nih.gov/pubmed/22678280. Accessed 25 May 2014.
9. Venail PA, Vives MJ (2013) Positive effects of bacterial diversity on ecosystem functioning driven by complementarity effects in a bioremediation context. PLoS One 8: e72561. Available: http://www.pubmedcentral.nih.gov/articlerender.fcgi?artid = 3762786&tool = pmcentrez&rendertype = abstract. Accessed 24 June 2014.
10. Torsvik V, Øvreås L (2002) Microbial diversity and function in soil: from genes to ecosystems. Curr Opin Microbiol 5: 240–245. Available: http://www.ncbi.nlm.nih.gov/pubmed/12057676.
11. Nile Basin Initiative (2012) State of the River Nile Basin. Entebbe (Uganda): Nile Basin Initiative Secretariat.
12. Olson DM, Dinerstein E, Powell GVN, Wikramanayake ED (2002) Conservation Biology for the Biodiversity Crisis. Conserv Biol 16: 1–3.
13. Ricklefs RE (2004) A comprehensive framework for global patterns in biodiversity. Ecol Lett 7: 1–15. Available: http://doi.wiley.com/10.1046/j.1461-0248.2003.00554.x. Accessed 21 January 2014.
14. Gering JC, Crist TO, Veech J a. (2003) Additive Partitioning of Species Diversity across Multiple Spatial Scales: Implications for Regional Conservation of Biodiversity. Conserv Biol 17: 488–499. Available: http://doi.wiley.com/10.1046/j.1523-1739.2003.01465.x.
15. Whittaker RJ, Willis KJ, Field R (2001) Scale and species richness: towards a general, hierarchical theory of species diversity. J Biogeogr 28: 453–470. Available: http://doi.wiley.com/10.1046/j.1365-2699.2001.00563.x.
16. Demeny P (2003) Population Policy: A Concise Summary. Internatio. Demeny P, McNicoll G, editors New York: MacMillan Reference. doi:173.
17. Olden JD (2006) Biotic homogenization: a new research agenda for conservation biogeography. J Biogeogr 33: 2027–2039. Available: http://doi.wiley.com/10.1111/j.1365-2699.2006.01572.x. Accessed 5 May 2014.
18. Olden JD, Poff NL (2004) Ecological processes driving biotic homogenization: testing a mechanistic model using fish faunas. Ecology 85: 1867–1875.
19. Olden JD, Poff NL (2003) Toward a mechanistic understanding and prediction of biotic homogenization. Am Nat 162: 442–460. Available: http://www.ncbi.nlm.nih.gov/pubmed/14582007.
20. Folke C, Carpenter S, Walker B, Scheffer M, Elmqvist T, et al. (2004) Regime Shifts, Resilience, and Biodiversity in Ecosystem Management. Annu Rev Ecol Evol Syst 35: 557–581. Available: http://www.annualreviews.org/doi/abs/10.1146/annurev.ecolsys.35.021103.105711. Accessed 26 October 2012.
21. Ter Braak CJ F. (1986) Canonical Correspondence Analysis: A New Eigenvector Technique for Multivariate Direct Gradient Analysis. Ecol Soc Am 67: 1167–1179.
22. Legendre P, Legendre L (1998) Numerical Ecology. 2nd ed. Amsterdam: Elsevier.
23. Lande R (1996) Statistics and partitioning of species diversity, and similarity among multiple communities. Oikos, Nord Soc 76: 5–13.
24. Godfray HCJ, Lawton JH (2001) Scale and species numbers. Trends Ecol Evol 16: 400–404. Available: http://www.ncbi.nlm.nih.gov/pubmed/11403873.
25. Walkley A., Black I (1934) An examination of the Degtjareff method for determining organic carbon in soils: effect of variations in digestion conditions and of inorganic soil constituents. Soil Sci: 251–263.
26. Kjeldahl J (1883) A new method for the determination of nitrogen in organic matter. Zeitschreft fur Anal Chemie 22.
27. Bray RH, Kurtz LT (1945) Determination of total, organic, and available forms of phosphorus in soils. Soil Sci 59: 39–45.
28. Gee G., Bauder J. (1986) Particle-size analysis. p. 383–411. In A Klute (ed.) Methods of Soil Analysis, Part 1. Physical and Mineralogical Methods. Agronomy Monograph No. 9 (2ed). Am Soc Agron Sci Soc Am Madison, WI.
29. Cleary DFR, Smalla K, Mendonça-Hagler LCS, Gomes NCM (2012) Assessment of variation in bacterial composition among microhabitats in a mangrove environment using DGGE fingerprints and barcoded pyrosequencing. PLoS One 7: e29380. Available: http://www.pubmedcentral.nih.gov/articlerender.fcgi?artid = 3256149&tool = pmcentrez&rendertype = abstract. Accessed 27 October 2012.
30. Reche I, Pulido-Villena E, Morales-Baquero R, Casamayor EO (2005) Does ecosystem size determine aquatic bacterial richness? Ecology 86: 1715–1722. Available: http://www.ncbi.nlm.nih.gov/pubmed/17489473.
31. Whittaker RH (1972) Evolution and Measurement of Species Diversity. Int Assoc Plant Taxon (IAPT) 21: 213–251.
32. Jaccard P (1908) Nouvelles recherches sur la distribution florale. Bull Soc Vaud Sci Nat 44: 223–270.
33. Veech JA, Summerville KS, Crist TO, Gering JC (2002) The additive partitioning of species diversity: recent revival of an old idea. Nord Soc Oikos 99: 3–9.
34. Crist TO, Veech J a, Gering JC, Summerville KS (2003) Partitioning species diversity across landscapes and regions: a hierarchical analysis of alpha, beta, and gamma diversity. Am Nat 162: 734–743. Available: http://www.ncbi.nlm.nih.gov/pubmed/14737711.
35. Chandy S, Gibson DJ, Robertson P a. (2006) Additive partitioning of diversity across hierarchical spatial scales in a forested landscape. J Appl Ecol 43: 792–801. Available: http://doi.wiley.com/10.1111/j.1365-2664.2006.01178.x. Accessed 22 January 2014.
36. Ribeiro DB, Prado PI, Brown Jr KS, Freitas AVL (2008) Additive partitioning of butterfly diversity in a fragmented landscape: importance of scale and implications for conservation. Divers Distrib 14: 961–968. Available: http://doi.wiley.com/10.1111/j.1472-4642.2008.00505.x. Accessed 3 February 2014.
37. Wu F, Yang XJ, Yang JX (2010) Additive diversity partitioning as a guide to regional montane reserve design in Asia: an example from Yunnan Province, China. Divers Distrib 16: 1022–1033. Available: http://doi.wiley.com/10.1111/j.1472-4642.2010.00710.x. Accessed 6 February 2014.
38. Sasaki T, Katabuchi M, Kamiyama C, Shimazaki M, Nakashizuka T, et al. (2012) Diversity partitioning of moorland plant communities across hierarchical spatial scales. Biodivers Conserv. doi:10.1007/s10531-012-0265-7.
39. Veech JA, Crist TO (2009) Partition: software for hierarchical partitioning of species diversity, version 3.0. Available: http://www.users.muohio.edu/cristto/partition.htm.
40. Monkiedje A, Spiteller M, Fotio D, Sukul P (2006) The effect of land use on soil health indicators in peri-urban agriculture in the humid forest zone of southern cameroon. J Environ Qual 35: 2402–2409. Available: http://www.ncbi.nlm.nih.gov/pubmed/17071911. Accessed 9 November 2012.
41. Neris J, Jiménez C, Fuentes J, Morillas G, Tejedor M (2012) Vegetation and land-use effects on soil properties and water infiltration of Andisols in Tenerife (Canary Islands, Spain). Catena 98: 55–62. Available: http://linkinghub.elsevier.com/retrieve/pii/S0341816212001270. Accessed 28 December 2012.
42. Bissett A, Richardson AE, Baker G, Thrall PH (2011) Long-term land use effects on soil microbial community structure and function. Appl Soil Ecol 51: 66–78. Available: http://linkinghub.elsevier.com/retrieve/pii/S0929139311001922. Accessed 9 November 2012.
43. Garbisu C, Alkorta I, Epelde L (2011) Assessment of soil quality using microbial properties and attributes of ecological relevance. Appl Soil Ecol 49: 1–4. Available: http://linkinghub.elsevier.com/retrieve/pii/S0929139311001089. Accessed 9 November 2012.
44. Fierer N, Jackson RB (2006) The diversity and biogeography of soil bacterial communities. Proc Natl Acad Sci U S A 103: 626–631. Available: http://www.pubmedcentral.nih.gov/articlerender.fcgi?artid = 1334650&tool = pmcentrez&rendertype = abstract.
45. Shange RS, Ankumah RO, Ibekwe AM, Zabawa R, Dowd SE (2012) Distinct soil bacterial communities revealed under a diversely managed agroecosystem. PLoS One 7: e40338. Available: http://www.pubmedcentral.nih.gov/articlerender.fcgi?artid = 3402512&tool = pmcentrez&rendertype = abstract. Accessed 1 November 2012.
46. Wang Y, Boyd E, Crane S, Lu-Irving P, Krabbenhoft D, et al. (2011) Environmental conditions constrain the distribution and diversity of archaeal merA in Yellowstone National Park, Wyoming, U.S.A. Microb Ecol62: 739–752. Available: http://www.ncbi.nlm.nih.gov/pubmed/21713435. Accessed 9 November 2012.
47. Fierer N, Bradford MA, Jackson RB (2007) Toward an ecological classification of soil bacteria. Ecology 88: 1354–1364. Available: http://www.ncbi.nlm.nih.gov/pubmed/17601128.
48. Saiya-Cork K., Sinsabaugh R., Zak D. (2002) The effects of long term nitrogen deposition on extracellular enzyme activity in an Acer saccharum forest soil. Soil Biol Biochem 34: 1309–1315. Available: http://linkinghub.elsevier.com/retrieve/pii/S0038071702000743.
49. Steenwerth K, Jackson L, Calderon F, Scow K, Rolston D (2005) Response of microbial community composition and activity in agricultural and grassland soils after a simulated rainfall. Soil Biol Biochem 37: 2249–2262. Available: http://linkinghub.elsevier.com/retrieve/pii/S0038071705001549. Accessed 9 November 2012.
50. De Vries FT, Manning P, Tallowin JRB, Mortimer SR, Pilgrim ES, et al. (2012) Abiotic drivers and plant traits explain landscape-scale patterns in soil microbial communities. Ecol Lett 15: 1230–1239. Available: http://www.ncbi.nlm.nih.gov/pubmed/22882451. Accessed 2 November 2012.
51. Hartman WH, Richardson CJ, Vilgalys R, Bruland GL (2008) Environmental and anthropogenic controls over bacterial communities in wetland soils. Proc Natl Acad Sci U S A 105: 17842–17847. Available: http://www.pubmedcentral.nih.gov/articlerender.fcgi?artid = 2584698&tool = pmcentrez&rendertype = abstract.
52. Cadotte MW, Cardinale BJ, Oakley TH (2008) Evolutionary history and the effect of biodiversity on plant productivity. Proc Natl Acad Sci U S A 105: 17012–17017. Available: http://www.pubmedcentral.nih.gov/articlerender.fcgi?artid = 2579369&tool = pmcentrez&rendertype = abstract.
53. Garbeva P, Veen JA Van, Elsas JD Van (2004) Microbial Diversity in Soil: Selection of Microbial Populations by Plant and Soil Type and Implications for Disease Suppressiveness. Annu Rev Phytopathol 42: 243–270. doi:10.1146/annurev.phyto.42.012604.135455.
54. Schulz S, Giebler J, Chatzinotas A, Wick LY, Fetzer I, et al. (2012) Plant litter and soil type drive abundance, activity and community structure of alkB harbouring microbes in different soil compartments. ISME J doi:10.103: 1763–1774.
55. Kuramae EE, Yergeau E, Wong LC, Pijl AS, van Veen J a, et al. (2012) Soil characteristics more strongly influence soil bacterial communities than land-use type. FEMS Microbiol Ecol 79: 12–24. Available: http://www.ncbi.nlm.nih.gov/pubmed/22066695. Accessed 7 November 2012.
56. Nacke H, Thürmer A, Wollherr A, Will C, Hodac L, et al. (2011) Pyrosequencing-based assessment of bacterial community structure along different management types in German forest and grassland soils. PLoS One 6: e17000. Available: http://www.pubmedcentral.nih.gov/articlerender.

fcgi?artid = 3040199&tool = pmcentrez&rendertype = abstract. Accessed 5 November 2012.
57. Waldrop MP, Balser TC, Firestone MK (2001) Linking microbial community composition to function in a tropical soil. Soil Biol Biochem 32: 1837–1846.
58. Dinsdale E a, Edwards R a, Hall D, Angly F, Breitbart M, et al. (2008) Functional metagenomic profiling of nine biomes. Nature 452: 629–632. Available: http://www.ncbi.nlm.nih.gov/pubmed/18337718. Accessed 26 October 2012.
59. Lauber CL, Hamady M, Knight R, Fierer N (2009) Pyrosequencing-based assessment of soil pH as a predictor of soil bacterial community structure at the continental scale. Appl Environ Microbiol 75: 5111–5120. Available: http://www.pubmedcentral.nih.gov/articlerender.fcgi?artid = 2725504&tool = pmcentrez&rendertype = abstract. Accessed 19 March 2014.
60. Russo SE, Legge R, Weber K a., Brodie EL, Goldfarb KC, et al. (2012) Bacterial community structure of contrasting soils underlying Bornean rain forests: Inferences from microarray and next-generation sequencing methods. Soil Biol Biochem 55: 48–59. Available: http://linkinghub.elsevier.com/retrieve/pii/S0038071712002362. Accessed 8 November 2012.
61. Winterbottom B, Eilu G (2006) Uganda Biodiversity and Tropical Forest Assessment. Washington DC.
62. Laurance WF (1999) Reflections on the tropical deforestation crisis. Biol Conserv 91: 109–117.
63. Scheckenbach F, Hausmann K, Wylezich C, Weitere M, Arndt H (2010) Large-scale patterns in biodiversity of microbial eukaryotes from the abyssal sea floor. Proc Natl Acad Sci U S A 107: 115–120. Available: http://www.pubmedcentral.nih.gov/articlerender.fcgi?artid = 2806785&tool = pmcentrez&rendertype = abstract. Accessed 1 November 2012.
64. Martiny JBH, Bohannan BJM, Brown JH, Colwell RK, Fuhrman J a, et al. (2006) Microbial biogeography: putting microorganisms on the map. Nat Rev Microbiol 4: 102–112. Available: http://www.ncbi.nlm.nih.gov/pubmed/16415926. Accessed 27 October 2012.
65. McKinney M, Lockwood J (1999) Biotic homogenization: a few winners replacing many losers in the next mass extinction. Trends Ecol Evol 14: 450–453. Available: http://www.ncbi.nlm.nih.gov/pubmed/10511724.
66. Olden JD, Rooney TP (2006) On defining and quantifying biotic homogenization. Glob Ecol Biogeogr 15: 113–120. doi:10.1111/j.1466-822x.2006.00214.x.
67. Devictor V, Julliard R, Clavel J, Jiguet F, Lee A, et al. (2008) Functional biotic homogenization of bird communities in disturbed landscapes. Glob Ecol Biogeogr 17: 252–261. Available: http://doi.wiley.com/10.1111/j.1466-8238.2007.00364.x. Accessed 21 January 2014.
68. Summerville KS, Boulware MJ, Veech J a., Crist TO (2003) Spatial Variation in Species Diversity and Composition of Forest Lepidoptera in Eastern Deciduous Forests of North America. Conserv Biol 17: 1045–1057. Available: http://doi.wiley.com/10.1046/j.1523-1739.2003.02059.x.
69. Weiher E, Howe A (2003) Scale-dependence of environmental effects on species richness in oak savannas. J Veg Sci 14: 917–920.
70. Obua J, Agea JG, Ogwal JJ (2010) Status of forests in Uganda. Afr J Ecol 48: 853–859.
71. National Environment Management Authority (NEMA) (2008) State of the Environment Report for Uganda.
72. Amann RI, Ludwig W, Schleifer K-H (1995) Phylogenetic identification and in situ detection of individual microbial cells without cultivation. Microb Rev 59: 143–169.
73. Vainio EJ, Hantula J (2000) Direct analysis of wood-inhabiting fungi using denaturing gradient gel electrophoresis of amplified ribosomal DNA. Mycol Res 104: 927–936.

Comparison of Seasonal Soil Microbial Process in Snow-Covered Temperate Ecosystems of Northern China

Xinyue Zhang[1,2], Wei Wang[1]*, Weile Chen[1], Naili Zhang[3], Hui Zeng[2]

1 Department of Ecology, College of Urban and Environmental Sciences and the Key Laboratory for Earth Surface Processes of the Ministry of Education, Peking University, Beijing, China, **2** Key Laboratory for Cyclic Economy, School of Urban Planning and Design, Shenzhen Graduate School, Peking University, Shenzhen, China, **3** State Key Laboratory of Vegetation and Environmental Change, Institute of Botany, Chinese Academy of Sciences, Beijing, China

Abstract

More than half of the earth's terrestrial surface currently experiences seasonal snow cover and soil frost. Winter compositional and functional investigations in soil microbial community are frequently conducted in alpine tundra and boreal forest ecosystems. However, little information on winter microbial biogeochemistry is known from seasonally snow-covered temperate ecosystems. As decomposer microbes may differ in their ability/strategy to efficiently use soil organic carbon (SOC) within different phases of the year, understanding seasonal microbial process will increase our knowledge of biogeochemical cycling from the aspect of decomposition rates and corresponding nutrient dynamics. In this study, we measured soil microbial biomass, community composition and potential SOC mineralization rates in winter and summer, from six temperate ecosystems in northern China. Our results showed a clear pattern of increased microbial biomass C to nitrogen (N) ratio in most winter soils. Concurrently, a shift in soil microbial community composition occurred with higher fungal to bacterial biomass ratio and gram negative (G-) to gram positive (G+) bacterial biomass ratio in winter than in summer. Furthermore, potential SOC mineralization rate was higher in winter than in summer. Our study demonstrated a distinct transition of microbial community structure and function from winter to summer in temperate snow-covered ecosystems. Microbial N immobilization in winter may not be the major contributor for plant growth in the following spring.

Editor: Jack Anthony Gilbert, Argonne National Laboratory, United States of America

Funding: This research was funded by grants from Projects of National Natural Science Foundation of China (31222011 and 31270363), National Basic Research Program of China (2010CB950604 and 2013CB956303), the Foundation for Innovative Research Groups of National Natural Science Foundation of China (31021001) and Construction Projects between Beijing City and the University in the Central Authorities. The funders had no role in study design, data collection and analysis, decision to publish, or preparation of the manuscript.

Competing Interests: The authors have declared that no competing interests exist.

* E-mail: wangw@urban.pku.edu.cn

Introduction

More than half of the earth's terrestrial surface currently experiences seasonal snow cover and soil frost [1,2]. In these ecosystems, some soil microorganisms are likely protected by the snowpack during winter. With adequate depth and density [3,4], the heat-insulation snowpack creates a relatively warmer habitat with liquid water films around soil particles [5,6]. Other soil microorganisms may develop their physiological adaption to survive the chilly environments [7]. Therefore, despite the freezing air temperature, winter catabolic processes of the soil microbial community, detected through biogenic CO_2 production, can still make a significant contribution to annual ecosystem fluxes across a wide variety of seasonally snow-covered ecosystems [8,9,10,11,12]. However, the influence of temporal variations in soil anabolic processes (i.e. microbial biomass accumulation) on annual N cycling is not well understood in these ecosystems [13,14,15].

Recent studies have reported peak microbial biomass C (MBC) and N (MBN) in late winter, followed by a quick decline when soil temperatures rise to 0°C at alpine and arctic sites [14,16,17,18]. An enlarged N pool was retained in the microbial biomass under the snowpack and subsequently released as a nutrient pulse when soils thawed, which may lead to an increase in N availability for plants at the start of the growing season [13,14,16,19]. The microbial N immobilization observed during winter may play a crucial role in ecosystem function, since it could prevent dissolved N produced by fall litter degradation being lost from the ecosystem during a time when plants are mostly inactive [20]. Therefore, in these snow-covered ecosystems, N retained in the soil by microorganisms during winter may be vital for plant nutrient uptake in the following growing season [18,21,22,23,24,25].

Soil microbial community can adapt to changing environmental conditions on very short time scales [7], thus changes between summer and winter may be a key control on annual patterns of nutrient cycling and plant N uptake [26]. For instance, in the alpine tundra of Colorado, winter maximal microbial biomass corresponded with increased biomass in the soil fungal community [27,28]. Fungi was the primary decomposer of plant debris in the territorial ecosystems, as it could release a great number of extracellular enzymes that can digest a wide variety of substrates, even complex organic compound as lignin [29,30]. Besides, fungi can grow towards nutrient sources and force their hyphae into solid substrates [31], which help fungi to make use of any nutrient source presented in soils. Furthermore, fungi differ from bacteria in N concentrations and storage capabilities [26,32]. Based on above viewpoints, fungi dominance in winter may have profound effects on soil biogeochemical cycling in the subsequent growing season. In addition, different gram-staining groups of bacteria, categorized by their cell wall composition, were also found to have various substrate preferences and survival strategies [33]. Therefore, the relative dominance of G- to G+ bacteria may differ between winter and summer. However, it remains poorly understood about the links between the seasonal variation of microbial biomass and the relative abundances of fungi and bacteria in snow-covered temperate areas.

As various decomposer microbes differ in their ability/strategy to efficiently use soil organic matter [34,35,36,37], shifts within the community composition may affect decomposition rates. SOC decomposition is the primary pathway where plant fixed CO_2 is released back into the atmosphere [38,39,40]. Along with frequently observed maximal microbial biomass in the winter of seasonally snow-covered ecosystems, the potential SOC mineralization rate (SOCMR, indicating intrinsic substrate use efficiency) may increase, releasing more CO_2 into the atmosphere when temperatures rise. However, winter potential SOCMR are seldom conducted [41,42].

Recent studies of winter soil microbial biogeochemical processes have mainly focused on alpine and tundra ecosystems. However, the wintertime conditions in snow-covered temperate areas could differ from those in alpine and tundra areas. In comparison with alpine and tundra, temperate ecosystems experience a shallower snowpack and shorter duration of snow cover [43,44], which may modify the extent of the N pool and the seasonal dynamics of its soil microbes. Moreover, temperate soil microbial communities are sensitive to climate changes, because the soil remains close to freezing throughout the winter [45], and small changes in winter temperatures may result in large changes in the amount and timing of snow cover [2]. More importantly, many temperate ecosystems are exposed to larger atmospheric N deposition as a result of increased emissions from industrial and agricultural activities [46]. Consequently, this may induce substantial changes in microbial N dynamics and ecosystem nutrient cycling. However, so far little investigation has been conducted on the winter biogeochemical process of soil microbes in temperate ecosystems [47]. Our study aims to explore the seasonal variation of microbial biomass, community composition and SOC mineralization in temperate ecosystems. The following three questions were addressed: (1) what is the C and N retention capacity of the winter soil microbial community in seasonally snow-covered temperate ecosystems? (2) Does microbial community composition shift from summer to winter? (3) What is the microbial function (organic matter decomposition), as indicated by potential SOCMR, in winter? To answer these questions, summer and winter soil samples were collected from the top 10 cm mineral soil in six seasonally snow-covered temperate ecosystems in northern China. The MBC and MBN, microbial community composition and mineralization rate of SOC were determined. We predicted (1) increased MBN immobilization because of relatively low quality substrates in winter (decreased root exudates and increased autumn litter input) may lead to microbial immobilization of N for growth [23,48]; (2) bacterial PLFAs to dominate in summer and fungal PLFAs in winter because fungi were commonly reported with a higher C to N ratio (targeting recalcitrant substrates in winter) compared with bacteria (targeting labile substrates in summer); and (3) a higher potential SOCMR in winter than summer because fungi were reported to produce more β-1,4-glucosidase (BG) enzyme (involved in C metabolism) than bacteria [49].

Materials and Methods

Ethics Statement

The administration of the Saihanba Forestry Center gave permission for this research at each study site. We confirm that the field studies did not involve endangered or protected species.

Site information

Our study sites were situated at the Saihanba Forestry Center in Hebei Province, North China (117°12′–117°30′ E, 42°10′–42°50′ N, 1400 m a.s.l.), which is a typical forest-steppe ecotone of a temperate area. The climate in this area is semi-arid and semi-humid, with a long and cold freezing period (November-March) and relatively short growing season. Annual mean air temperature and precipitation between 1964 and 2004 were −1.4°C and 450 mm, respectively. The soils in the region are dominated by aeolian sandy soil, together with meadow and swamp soil. Soil has low nutrient content with SOC content ranging from 0.71 to 2.65% and soil total nitrogen (STN) ranging from 0.08 to 0.26%, respectively (Table 1). Soil bulk density (SBD) ranged from 0.65 to 1.06 g cm^{-3}; Soil pH varied from 5.8 to 6.5 and 6.0 to 6.6 in the summer and winter, respectively (Table 1). Soil moisture content was calculated on the gravimetric bias. Specifically, 10 g fresh soil samples were dried to a constant weight by an oven at a temperature of 105°C. All the soil and microbial properties were determined by the dry soil weight.

Primary forests were harvested via large scale industrial logging in the late 1900s and have been replaced by secondary forests and plantations. This site contains the largest area of plantation forests in China, the dominant species are *Pinus sylvestris* var. *mongolica* (Mongolia pine) and *Larix principis-rupprechtii* (Prince Rupprecht's larch). The Mongolia pine and larch herbaceous layers are similar, composed by *Radix Sanguisorbae*, *Thalictrum aquilegifolium L.*, *Agrimonia pilosa Ledeb*, and *Carex stenophylla Wahleub.* The secondary forest mainly consists of *Betula platyphylla* (birch) with an herbaceous layer of *Agrimonia pilosa Ledeb*and *Radix Sanguisorbae.* In addition, shrublands dominated by *Rosa bella Rehd. et Wils* (solitary rose) and *Malus baccata* (Siberian crabapple) and meadow grasslands are also very common. The solitary rose herbaceous layer consists of *Leymus chinensis.* The Siberian crabapple herbaceous layer is dominated by *Veronica linariifolia*, *Galium verum*, *Heteropappus hispidus*, *Trollius chinensis*, and *Bupleurum chinense.* The meadow grassland is zonal vegetation dominated by *L. chinensis.* Due to the sparse understory species of the forest sites and simple species composition of the grasslands and shrublands, we did not consider the potential impacts of vegetation diversity on microbial processes. The distribution of sampling sites was shown in Fig. 1, which was drawn by ArcMap (ArcGIS 10.0, Esri Inc., California, USA) with ancillary site information.

Study design and methods

Because of the significant effect of stand age on ecosystem C and N dynamics [50], we took forest age into consideration in the two coniferous plantation sites, namely Mongolia pine and Prince Rupprecht's larch. Three replicates were selected for each of three age classes (shown in Table 1) in Mongolia pine (sites P1, P2, and P3) and larch (sites L1, L2, and L3). We also selected three replicates for the birch stand (site BH), each of the two shrublands Siberian crabapple (site MA) and solitary rose (site RO) and one meadow grassland (site CG); 30 plot samples in total (Table 1). The area of each plot was 20 × 20 m. All sampling sites were less than 10 km away from each other (Fig. 1) to ensure similar climatic conditions.

Table 1. Site information and soil properties of the ten sampling sites.

Sites	Location	Domain species	SOC (%)	STN (%)	SBD (g cm^{-3})	ST (°C)		pH	
						JUL.	JAN.	JUL.	JAN.
P1	42°24.707′N 117°14.071′E	*Pinus sylvestris*	0.71 ± 0.05	0.08 ± 0.01	0.98 ± 0.00	19.0	−7.3	6.45 ± 0.10	6.31 ± 0.08
P2	42°24.760′N 117°14.771′E	*Pinus sylvestris*	1.26 ± 0.09	0.12 ± 0.01	0.83 ± 0.05	14.7	−12.8	6.34 ± 0.05	6.36 ± 0.04
P3	42°25.079′N 117°15.974′E	*Pinus sylvestris*	1.10 ± 0.12	0.09 ± 0.01	0.85 ± 0.03	14.9	−8.1	6.31 ± 0.06	6.30 ± 0.07
L1	42°24.332′N 117°12.933′E	*Larix principis-rupprechtii*	0.94 ± 0.04	0.09 ± 0.00	1.06 ± 0.01	15.8	−9.4	6.30 ± 0.02	6.36 ± 0.01
L2	42°24.118′N 117°12.722′E	*Larix principis-rupprechtii*	0.95 ± 0.07	0.09 ± 0.01	0.88 ± 0.01	14.5	−9.7	5.94 ± 0.09	6.02 ± 0.09
L3	42°23.911′N 117°19.052′E	*Larix principis-rupprechtii*	1.88 ± 0.17	0.18 ± 0.02	0.74 ± 0.05	13.7	−7.2	5.78 ± 0.06	6.16 ± 0.04
BH	42°23.848′N 117°19.031′E	*Betula platyphylla*	2.65 ± 0.25	0.26 ± 0.04	0.65 ± 0.00	13.7	−7.8	5.92 ± 0.10	6.16 ± 0.04
MA	42°24.729′N 117°14.132′E	*Malus baccata*	2.10 ± 0.15	0.20 ± 0.01	0.72 ± 0.02	16.1	−7.3	6.39 ± 0.07	6.63 ± 0.06
RO	42°24.107′N 117°13.866′E	*Rosa bella*	1.22 ± 0.16	0.13 ± 0.02	0.73 ± 0.07	19.4	−11.2	6.20 ± 0.07	6.46 ± 0.06
CG	42°24.717′N 117°14.107′E	*Leymus chinensis*	1.00 ± 0.04	0.09 ± 0.00	0.88 ± 0.05	19.9	−11.7	6.28 ± 0.05	6.29± 0.09

Values are presented as mean±standard errors. SOC = Soil Organic Carbon; STN = Soil totoal nitrogen; SBD = Soil Bulk Density; ST = Soil Temperature; JUL. = July; JAN. = January.

All experiments were performed on upper 10 cm mineral soils at five random locations from each plot in July 2010 and January 2011. The two months we chose had the highest (July) and lowest (January) air temperature according to the local climate records of recent decade, representing two most stable seasons of the years, namely summer and winter. This approach has been commonly applied to investigate seasonal dynamics [18,28,51]. Under this circumstance, we assumed that the two single months can represent their respective seasons, and that changes over these two seasons in the soil microbial community may be of greater magnitude than differences within the same season or from year to year [52]. In summer, soil was taken by 5.8-cm diameter soil cores, passed through a 2-mm sieve to remove plant litter and roots and homogenized. In winter, after the thickness of snow cover measured by a steel ruler, snows were swept away with a shovel before sampling. The snow thickness varied from 4-16 cm in different sampling sites (data not shown). After that, frozen soils were collected using axe or drill and the upper materials were cut

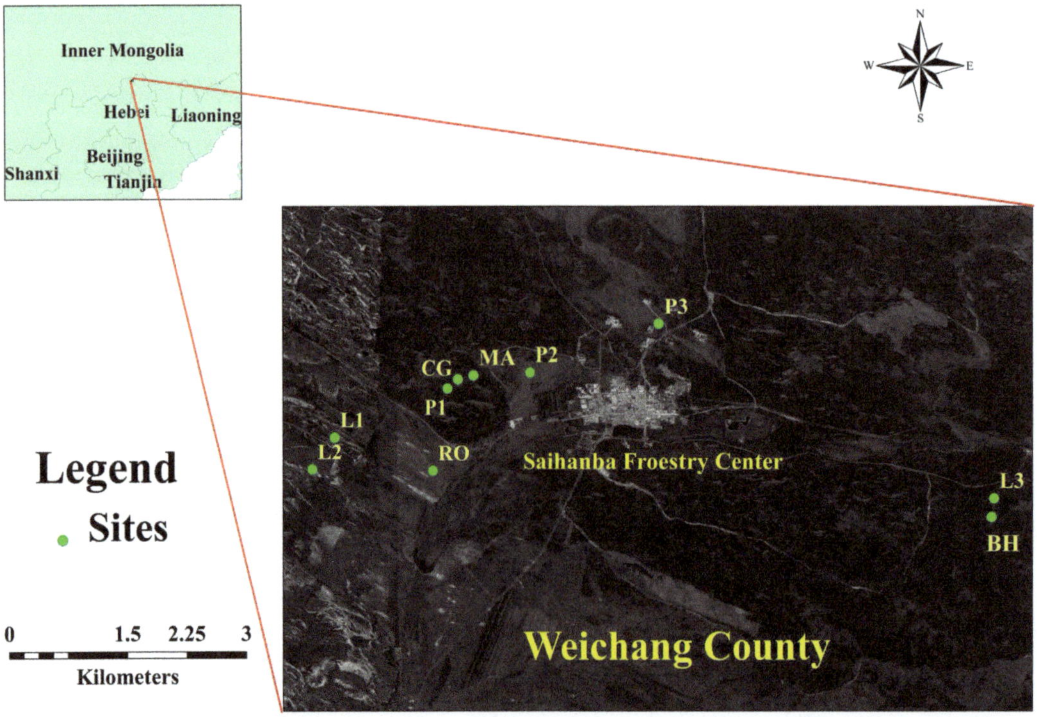

Figure 1. The distribution of sampling sites. P1 = ~15-year Mongolia pine; P2 = ~25-year Mongolia pine; P3 = ~35-year Mongolia pine; L1 = ~15-year Prince Rupprecht's larch; L2 = ~25-year larch; L3 = ~35-year larch; BH = Birch; MA = Siberian crabapple; RO = Solitary rose; CG = Meadow grassland.

by knife. The processed soils were immediately transported to the laboratory (less than 30 min away) for subsequent analysis.

In situ continuous measurements of soil temperature were conducted at 30-min intervals with StowAway loggers (Onset Comp. Corp., Bourne, MA, USA) inserted into the soil at a depth of 5 cm. Extractable ammonium and nitrate nitrogen (NH_4^+ and NO_3^-) were measured simultaneously in 2 mol/L KCl solution (1:5 w/v) [53] using a Lachat Flow Injection Analyzer (Lachat Instruments, Milwaukee, WI). The amount of extractable ammonium and nitrate nitrogen were taken as total inorganic N. SOC was measured by the potassium dichromate oxidation method with ground air-dried soils [54] while STN was measured by Kjeldahl method [55]. Because of the minor variation of SOC and STN in our study sites over a year (data not shown), we only measured the values in summer time (July) and did not take the difference between summer and winter into consideration. Soil pH was determined using a pH meter (Denver, US) on a 1:1 (w/v) air-dried soil to distilled de-ionized water slurry.

MBN and MBN

MBC and MBN were measured by the chloroform fumigation extraction (CEF) method [56]. Two replicate samples, one un-fumigated and one fumigated with alcohol-free $CHCl_3$ for 24 h and then extracted with 0.5 mol/L K_2SO_4 (1:2.5 w/v). MBC and MBN were calculated as the difference of C and N between fumigated and un-fumigated soil extraction, which was estimated using the dichromate oxidation and titration method and Kjeldahl digestion, respectively. To avoid variations in CFE values due to the choice of conversion factor (corrects for the incomplete release and extraction of microbial biomass following CEF method), no conversion factor was used for fumigation efficiency as conducted by Edwards and Jefferies (2013) [18], since it could vary across different seasons as well as among different sampling sites [57]. Soil dissolved organic N (DON) was measured from the initial un-fumigated extraction [18].

Phospholipid fatty acid (PLFA) analysis

PLFA analysis was used to determine the total microbial biomass and assess microbial community composition. In brief, PLFAs were extracted by a mixture of chloroform-methanol-phosphate buffer (1:2:0.8) [58]. Polar lipids in the initial soil extracts were separated from neutral and glycolipids by elution with 5 ml chloroform and 10 ml acetone followed by 5 ml methanol. The polar lipid fraction was used to perform mild alkaline methanolysis. The peak area of each resulting fatty acid methyl ester was recorded on the chromatogram for each sample before identification. Peaks were identified by chromatographic retention time and a standard qualitative mix that ranged from C9 to C30 using a microbial identification system (Microbial ID Inc., Newark, DE). The fatty acid 18:2ω6, 9 was recognized as the fungal biomarker [59]. Bacterial biomass was quantified as the sum of i14:0, i15:0, a15:0, 16:1ω9, 16:1ω7, i17:0, a17:0, cy17:0, 17:0, and cy19:0 [60]. G+ bacteria were marked by i15:0, a15:0, i16:0, a16:0, i17:0, and a17:0 [59]. The mono-unsaturated and cyclopropyl saturated peaks 16:1ω5, 16:1ω9, 17:1ω9, cy17:0, 18:1ω11, and cy19:0 were used as indicators for G- bacteria [59,61,62]. We also calculated the total lipids as an indicator of microbial biomass [63].

Potential SOCMR

We used a laboratory incubation experiment to measure the SOCMR through detection of CO_2 emission [41,64]. Approximately 25 g of fresh soil were placed into 250 ml glass gas tight jars and incubated at 25°C for 3 d. Respired CO_2 was captured by a connecting vial with 5 ml 1 mol/L NaOH and determined by titration with 1 mol/L HCl. Potential SOCMR were expressed as the amount of CO_2-C released per hour per gram soil dry weight. Because the winter soil temperature was below 0°C, it was unable to measure *in situ* mineralization activity using the laboratory incubation method. Thus, we conducted the potential SOCMR in winter using the same temperature as the summer (25°C, being the suitable condition for soil microbial growth in the field). Considering the condition for winter samples was quite different from the *in situ* environment, we indirectly estimated winter mineralization of SOC using the Q_{10} coefficient (increase in reaction rate per 10°C increase in temperature) obtained from our previous research in the same study sites [12].

$$SOCMR_{in-situ} = SOCMR_{25}/Q_{10}^{(25-T)/10} \quad (1)$$

Where $SOCMR_{in\text{-}situ}$ was the mineralization rate of SOC at an *in situ* temperature measured in winter, $SOCMR_{25}$ was the mineralization rate of SOC at the room temperature of 25°C, and T was the winter soil temperature at 5 cm depth.

Statistical analysis

We used the averaged value of the subsamples in each plot to conduct statistical analysis. Statistical analyses were performed using SPSS (ver. 18.0, SPSS Inc., Chicago, IL, USA). Repeated measures analysis of variance (ANOVA) was used to determine the effects of seasons and dominate vegetation types on soil N concentrations, microbial biomass, microbial community composition and potential SOCMR. Besides, ANOVA was also used to determine the effect of stand age on above soil and microbial properties. Furthermore, simple correlation analysis was conducted to explore the relationships between microbial community related parameters and soil physical and chemical properties. In all cases, differences of $P < 0.05$ were regarded as statistically significant. Results were displayed as mean ± standard error.

Besides, we analyzed the PLFAs data by principal components analysis (PCA) to determine whether the PLFA signatures of microbial community varied between summer and winter. PCA was performed on 21 different PLFAs indentified from all the samples with concentration larger than 0.005 fraction (ratio of moles individual lipid to moles total lipid biomass) using R 3.0.1 (R Core Team, 2013). Statistical differences among PLFAs data were assessed using multi-response permutation procedures (MRPP) [65,66]. MRPP is a nonparametric procedure for testing the hypothesis of no differences between two or more pre-existing groups [67]. *P* value evaluated the significant differences due to chance, where A value described within-group homogeneity compared to random expectation [67]. An A value equal to 1 was found when all items within a group are identical; when heterogeneity within groups equaled expectation by chance A = 0.

In addition, inverse Simpson index was calculated to describe the microbial diversity obtained from PLFAs data in different seasons as well as in different sampling sites. We preferred this index to other measures of alpha-diversity because it is an indication of the richness in a community with uniform evenness that would have the same level of diversity.

Results

Soil physical and chemical properties

In July, soil temperature at 5-cm depth averaged 16.5°C, and decreased to −9.5°C in January across six ecosystems (Table 1).

Table 2. Soil inorganic nitrogen (NH_4^+-N and NO_3^--N) concentrations, microbial biomass carbon to nitrogen (MBC/N) ratio and fungi to bacteria biomass (F/B) ratio of the ten sampling sites.

Sites	NH_4^+-N (mg kg^{-1})		NO_3^- -N (mg kg^{-1})		MBC/N		F/B	
	JUL.	JAN.	JUL.	JAN.	JUL.	JAN.	JUL.	JAN.
P1	1.48 ± 0.03	1.66 ± 0.21	0.04 ± 0.01	NT	15.90 ± 4.57	49.26 ± 5.18	0.024 ± 0.001	0.126 ± 0.011
P2	1.54 ± 0.00	1.92 ± 0.07	0.04 ± 0.00	NT	23.54 ± 5.07	41.58 ± 6.25	0.023 ± 0.002	0.136 ± 0.039
P3	1.60 ± 0.08	1.76 ± 0.08	0.02 ± 0.01	NT	30.68 ± 6.82	55.04 ± 5.31	0.033 ± 0.002	0.132 ± 0.011
L1	1.57 ± 0.04	1.69 ± 0.03	0.01 ± 0.00	NT	8.89 ± 2.49	31.48 ± 1.57	0.054 ± 0.013	0.115 ± 0.018
L2	1.61 ± 0.01	1.84 ± 0.06	0.01 ± 0.00	NT	35.36 ± 3.22	68.38 ± 16.33	0.083 ± 0.004	0.075 ± 0.003
L3	1.60 ± 0.03	1.78 ± 0.31	0.00 ± 0.00	NT	33.87 ± 8.95	47.83 ± 7.87	0.079 ± 0.003	0.097 ± 0.019
BH	1.73 ± 0.96	2.25 ± 0.03	0.02 ± 0.00	NT	15.34 ± 2.46	31.97 ± 4.28	0.056 ± 0.007	0.096 ± 0.006
MA	1.63 ± 0.08	2.23 ± 0.20	0.03 ± 0.01	NT	41.79 ± 13.95	33.74 ± 1.18	0.060 ± 0.004	0.076 ± 0.007
RO	1.73 ± 0.13	1.85 ± 0.10	0.06 ± 0.01	NT	9.70 ± 1.09	31.59 ± 7.83	0.073 ± 0.006	0.078 ± 0.004
CG	1.45 ± 0.03	1.67± 0.08	0.08 ± 0.01	NT	9.25 ± 3.72	33.69 ± 1.06	0.070 ± 0.008	0.066 ± 0.004

Values are presented as mean±standard errors. NT= not detectable.

Soil NH_4^+-N and NO_3^--N concentrations averaged 1.59 ± 0.02 mg kg^{-1} and 0.03 ± 0.01 mg kg^{-1} in July and 1.86 ± 0.05 mg kg^{-1} and 0.02 ± 0.01 mg kg^{-1} in January, respectively (Table 2). Although total soil inorganic nitrogen (NO_3^--N + NH_4^+-N) across all ecosystems was significantly higher in winter than in summer at a statistical level ($P < 0.001$), there was no large difference between them (1.88 ± 0.07 mg kg^{-1} in winter vs. 1.63 ± 0.03 mg kg^{-1} in summer) (Table 3; Fig. 2A; Table S1). DON was significantly higher in summer than in winter, with average value decreasing from 4.76 ± 0.52 mg kg^{-1} in summer to 0.84 ± 0.06 mg kg^{-1} in winter, respectively (Table 3; Fig. 2B; Table S1). Although DON significantly differed across different ecosystem types, no significant interaction between seasons and sampling sites was observed (Table 3; Fig. 2B).

Soil MBC and MBN

Across all the sampling sites, soil microbes showed a significantly higher MBC ($P < 0.001$) while lower MBN ($P = 0.015$) in winter than in summer (Table 3; Fig. 3). MBC consistently increased from 66.1 ± 7.4 mg kg^{-1} in summer to 112.1 ± 8.5 mg kg^{-1} in winter on average, except for site BH. MBN in most of the sites were lower in winter than in summer (4.5 ± 0.5 mg kg^{-1} in summer vs. 3.3 ± 0.3 mg kg^{-1} in winter) (Table S1). Both MBC and MBN varied significantly across different sites as well as in different seasons (Table 3). Besides, the interaction between seasons and sampling sites also showed significant effect on MBC but not MBN (Table 3). MBC significantly correlated with MBN in both summer (R^2 = 0.27, $P = 0.020$) and winter ($R^2 = 0.52$, $P < 0.001$). MBC to MBN (MBC/N) ratio varied from 9.25 ± 3.72 to 41.79 ± 13.95 in summer, and 31.48 ± 1.57 to 68.38 ± 16.33 in winter (Table 2). The MBC/N ratio consistently increased from summer to winter across all the sampling sites, except for site MA, where a slight decrease was found (Table 2). Stand age exhibited no significant influence on MBC/N ratio for both summer and winter.

Microbial community composition

Microbial biomass indicated by total PLFAs showed no significant difference between summer and winter ($P = 0.509$), but varied significantly among different sampling sites ($P < 0.001$) (Table 3; Fig. 4; Table S1). In addition, microbial community composition shifted with higher fungal biomass abundance in winter (Table 3; Fig. 5A). The bacterial community biomass was not significantly different between summer and winter ($P = 0.802$) across all the sites, as well as the G+ bacterial community biomass ($P = 0.050$). However, G- bacterial biomass significantly increased in winter ($P < 0.001$; varying from 1.98 ± 0.35 nmol PLFAs g dry soil^{-1} in summer to 3.66 ± 0.51 nmol PLFAs g dry soil^{-1} in winter). Thus the G- to G+ bacterial biomass (G−/G+) ratio also exhibited a significant increase in winter (Fig. 5B; Table S1). In summer, fungal to bacterial biomass (F/B) ratio was negatively correlated with soil pH value ($R^2 = 0.43$, $P = 0.036$), whereas, in winter, F/B ratio was positively correlated with total microbial biomass indicated by PLFAs ($R^2 = 0.78$, $P = 0.001$) (Table 4). G−/G+ ratio was negatively correlated with soil temperature only in winter ($R^2 = 0.62$, $P = 0.007$) (Table 4).

Principle component analysis (PCA) also showed the differentiation in microbial community structure between seasons (Fig. 6). The first principle component axis (PC1) alone could explain as

Table 3. The repeated measure ANOVA results of soil and microbial properties tested in this study.

ANOVA results	IN	DON	MBC	MBN	MB	F/B	G−/G+	Potential SOCMR
Season	$P < 0.001$	$P < 0.001$	$P < 0.001$	$P = 0.015$	$P = 0.509$	$P < 0.001$	$P < 0.001$	$P < 0.001$
Sites	$P = 0.044$	$P = 0.032$	$P < 0.001$	$P = 0.002$	$P < 0.001$	$P = 0.509$	$P < 0.001$	$P < 0.001$
Interaction	$P = 0.297$	$P = 0.086$	$P = 0.048$	$P = 0.121$	$P = 0.390$	$P < 0.001$	$P < 0.001$	$P = 0.017$

IN = inorganic nitrogen; DON = dissolved organic nitrogen; MB = microbial biomass; G−/G+ = gram negative to positive bacteria biomass ratio; SOCMR = soil organic carbon mineralization rate.

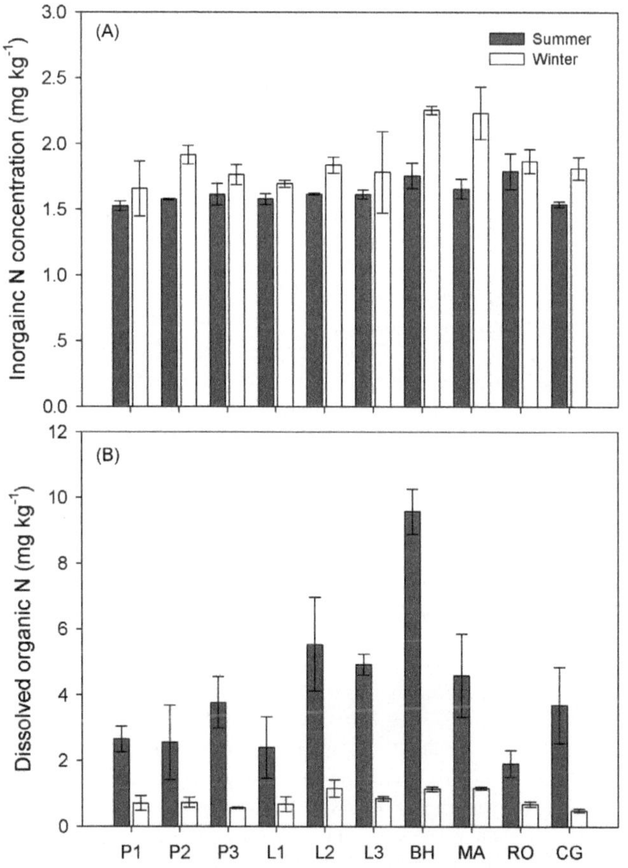

Figure 2. Inorganic nitrogen (N) (A) and dissolved organic N (B) across different sites in summer and winter. Results are presented as mean ± stander error. Support information is presented in Table S1.

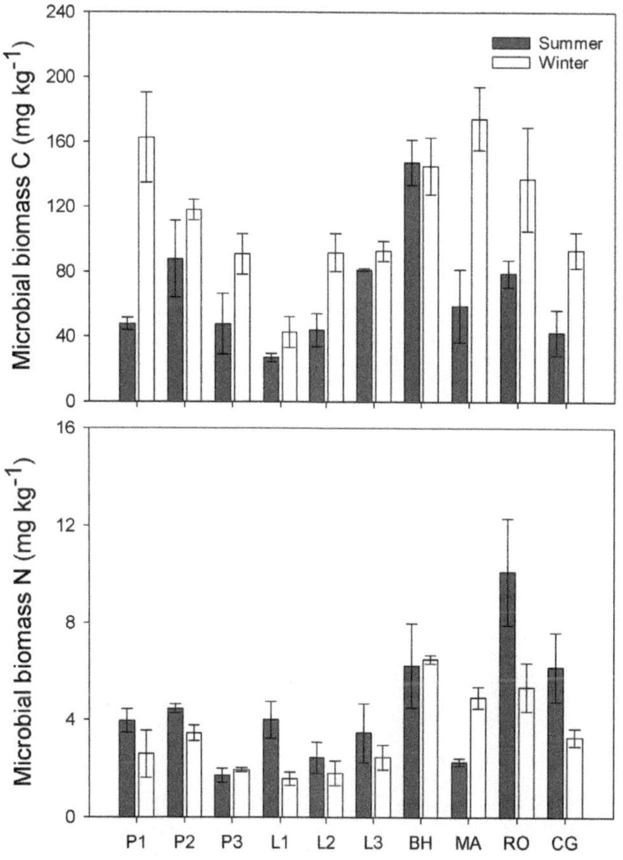

Figure 3. Microbial biomass carbon (C) (A) and nitrogen (N) (B) across different sites in summer and winter. Results are presented as mean ± stander error. Support information is presented in Table S1.

much as 96% of the variance in the PLFAs data. There was no evidence of difference in PLFAs data among different sampling sites (MRPP: $P = 0.693$, A = 0), however, significant difference of PLFAs was observed between summer and winter (MRPP: $P = 0.001$, A = 0.56). Microbial diversity was significantly lower in winter (4.76) than that in summer (8.59) indicated by inverse Simpson index ($P < 0.001$), and differed in each sampling site ($P < 0.001$). In all sampling sites, microbial diversity showed a decrease trend from summer to winter (Table S2). Stand age exerted no significant effect on microbial community composition or diversity.

Potential SOCMR

The C emission rate ranged from 5.31 ± 2.50 mg CO_2-C d^{-1} g dry $soil^{-1}$ in summer to 63.22 ± 4.02 mg CO_2–C d^{-1} g dry $soil^{-1}$ in winter under the same incubation temperature, showing a significantly higher potential SOCMR for winter microbes than summer ones (Table 3; Fig. 7; Table S1). However, when applied the Q_{10} value to model the *in situ* winter SOCMR, the average *in situ* mineralization rate was 0.26 mg CO_2–C d^{-1} g dry $soil^{-1}$, quite lower than the averaged summer ones (12.57 mg CO_2–C d^{-1} g dry $soil^{-1}$). No significant effect of stand age was observed on potential SOCMR for both summer and winter.

Discussion

Multiple studies in alpine and arctic tundra ecosystems have reported active microbial metabolism under snow-cover during winter [26,27,28,50,68,69]. However, until now little has been known about winter microbial biogeochemical processes in temperate areas. This study determined the soil microbial biomass, community composition and mineralization rate of SOC in a variety of seasonally snow-covered temperate ecosystems.

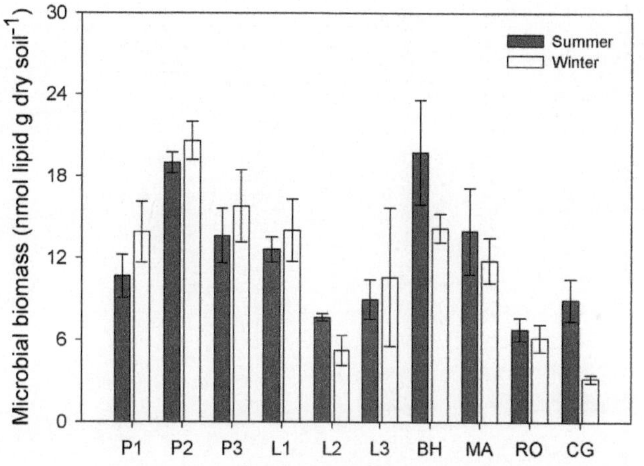

Figure 4. Total microbial biomass indicated by PLFAs across different sites in summer and winter. Results are presented as mean ± stander error. Support information is presented in Table S1.

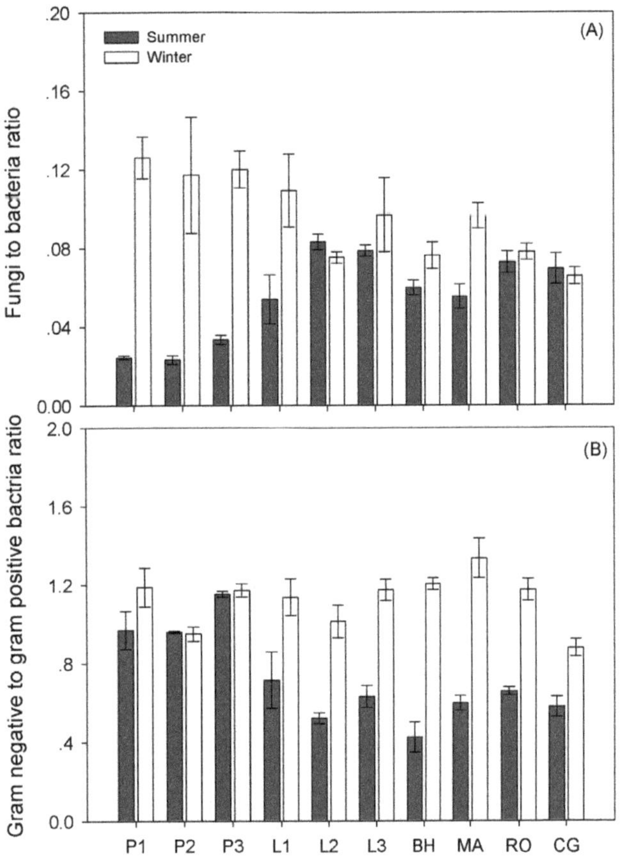

Figure 5. Soil fungal to bacterial PLFAs ratio (A) and gram negative to gram positive bacterial PLFAs ratio (B) across different sites in summer and winter. Results are presented as mean ± standard error. Support information is presented in Table S1.

It was commonly thought that relatively low quality substrates may lead to microbial immobilization of N for growth [23,48]. Therefore, we expected increased MBN immobilization because of decreased root exudates and increased autumn litter input in winter when plants were inactive. However, in our study area, there existed a clear pattern of increased MBC to MBN ratio in winter across almost all of the ecosystems, mainly through increased MBC (Fig. 3A) and reduced MBN (Fig. 3B). Non-increased microbial N pool in winter was unexpected and contrasted with the findings from alpine and tundra areas [26] and corresponding to a temperate beech forest soil [20]. Exception was only found in site MA, where MBN obviously increased from summer to winter (Fig. 3B). However, there was no soil or microbial properties measured in this study seemed to explain this exception. The lower levels of microbial N immobilization in our temperate winter soils indicate a smaller or no nitrogen pulse from winter soil microbes in the following spring. The discrepancy may be due to the differences of snow cover depth and duration. Across our studied sites, the snowpack was as thin as 4–16 cm and the duration of snow cover was short. Snowpack less than 30 cm depth could not effectively decouple soil temperatures from the atmosphere [70,71,72], and thus resulting in lower microbial N immobilization.

As total microbial PLFAs changed negligibly (Fig. 4), the microbial biomass was stable across different seasons. However, the increased fungal dominance seemed to partly explain the increased C and decreased N uptake. The positive relationship between fungal abundance and the MBC to MBN ratio has been observed in a previous global analysis [73]. Fungi were commonly reported with a higher C to N ratio (targeting recalcitrant substrates) compared with bacteria (targeting labile substrates) [74,75]. In summer, warmer temperature can increase root exudates (labile substrates) than in winter [76] and fresh plant litter input had a relatively lower C to N ratio. Therefore, we expected bacterial PLFAs to dominate. In contrast, in winter, a lot of leaf litter with a relative high C to N ratio accumulated on the soil floor and likely became a major substrate source for fungi. We did observe increased fungal PLFAs ($P < 0.001$), but no significant variation in bacterial PLFAs was found between summer and winter ($P = 0.809$). Because it is common for soil microbes to preferentially use simple organic compounds over complex polymers [77,78], winter substrate usage may be largely confined to microbial recycling of dead microbial cells and hyphae [79] or

Table 4. The correlation between soil properties (listed in the row) and microbial community parameters (listed in the line).

Correlation Matrix		MBC/N		MB		F/B		G−/G+	
		JUL.	JAN.	JUL.	JAN.	JUL.	JAN.	JUL.	JAN.
ST	R	−0.508	0.183	−0.508	0.134	−0.015	0.075	0.076	0.790**
	P	0.134	0.612	0.134	0.712	0.967	0.836	0.835	0.007
pH	R	0.138	−0.255	0.138	0.189	−0.665**	0.340	0.595	0.033
	P	0.704	0.477	0.704	0.601	0.036	0.337	0.070	0.928
SOC	R	0.515	−0.374	0.515	0.164	0.222	−0.234	−0.520	0.482
	P	0.128	0.287	0.128	0.651	0.537	0.516	0.123	0.158
STN	R	0.485	−0.408	0.485	0.145	0.226	−0.258	−0.551	0.507
	P	0.155	0.242	0.155	0.688	0.530	0.490	0.099	0.135
DON	R	0.406	0.170	0.406	−0.018	0.280	−0.340	−0.581	0.421
	P	0.244	0.639	0.244	0.961	0.433	0.336	0.078	0.225
IN	R	0.099	−0.380	0.099	0.094	0.363	−0.354	−0.448	0.361
	P	0.785	0.279	0.785	0.797	0.303	0.316	0.194	0.305

R = correlation coefficient; ** and * represents $P < 0.01$ and $P < 0.05$ respectively.

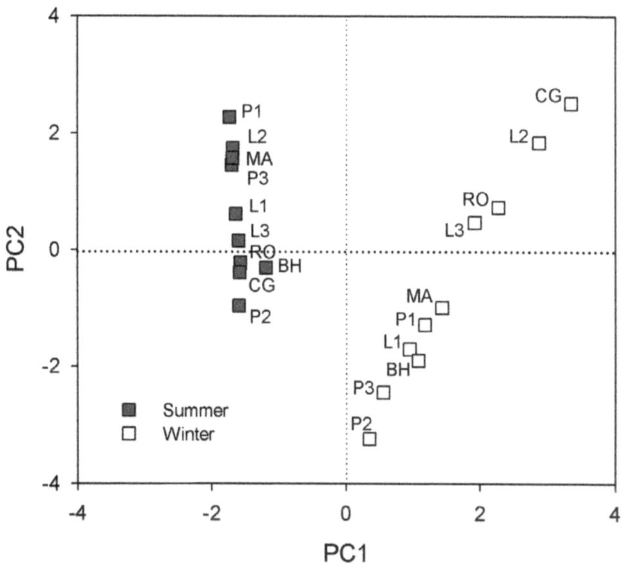

Figure 6. Principal component analysis (PCA) of PLFA signatures (mol percentages) from soil samples collected in summer and winter. PC1 and PC2 explained 96% and 2% variance of the PLFAs data, respectively.

endogenous metabolism (the breakdown of living cell constituents/storage compounds for maintenance). Therefore, the dominance of fungi in winter may be largely due to their resistance to freeze-thawing events [80], and not to substrate preference. Fungal to bacterial PLFA ratios increased in winter across all sites (Fig. 5A). This result was consistent with those obtained in alpine and tundra ecosystems [27,28,50,69]. As seasonal shifts in microbial species composition may also occur at fine taxonomic scales [81], further research into the finer microbial community composition is warranted.

Although no significant variation occurred in the common bacterial PLFAs, we observed a significant increase of G- bacteria in cold environments ($P < 0.001$); in contrast, G+ bacterial PLFAs did not change under winter conditions ($P = 0.050$), which was consistent with previous observation in late winter tussock tundra soil [82]. G- bacteria can more easily access the soil aqueous phase

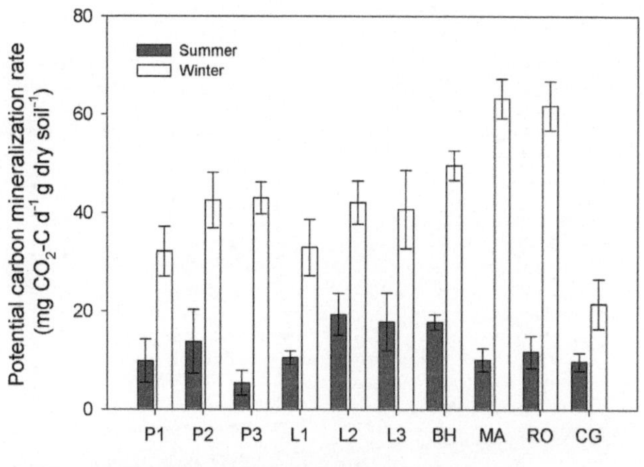

Figure 7. Potential carbon mineralization rates across different sites in summer and winter. Results are presented as mean ± standard error. Support information is presented in Table S1.

[33], thus the structure of the cell wall may help it to survive and grow in lower temperatures. Although G- bacteria seemed to take advantage over G+ at colder temperature, correlation analysis across different sampling sites showed that G- to G+ bacterial biomass ratio positively related to winter soil temperature (Table 4). Researches also suggested that G+ and G- bacteria differ in their patterns of substrate preference [83] with the former being dominant in soils with low substrate availability [60] and the latter in those with high availability of easily decomposable substrate [84]. Because the winter microbial substrate source was unknown, whether or not the increase in G- bacteria corresponds with the change of substrate preference across sites remains unclear.

A shift in microbial community composition along with fungal dominance in winter may subsequently change the enzyme production by the preference of different microbes and may ultimately influence the potential SOC mineralization rate. Fungi have been reported to produce more BG enzyme (involved in C metabolism) than bacteria [51]. As the BG enzyme was considered as an overall indicator of microbial activity [83,85], we expected a higher potential SOCMR in winter than summer. The result was consistent with our expectation, indicating without the temperature constraints substrate use efficiency was higher for winter microbes than the summer ones. However, *in situ* SOCMR was much lower in winter than in the summer. Although low winter temperature constrained the actual substrate mineralization rate, once temperature rise, the mineralization rate may experience a sharp increase, along with high C metabolism enzyme pool suggested by fungi dominance, and eventually leading to more CO_2 released back to atmosphere. The rapid increase of soil respiration from late winter to early spring has been observed in our previous field study conducted in the same ecosystems [12].

Conclusion

In summary, the trend of increased microbial C and decreased N uptake in winter dominate across all six seasonally snow-covered temperate ecosystems. Therefore, the N pool retained in the microbial biomass under the snowpack may not be major source for spring plant nutrient demand. The higher MBC to MBN ratio in winter was partly connected to the shift in microbial composition to fungal dominance in winter. Because there were significant differences in substrate use, nutrient limitation, and N storage capacity between fungi and bacteria, the changes in winter fungal to bacterial biomass ratios might substantially alter annual patterns of nitrogen cycling in seasonally snow-covered temperate ecosystems. Although *in situ* C mineralization was low due to the temperature constraints in winter, greater potential SOCMR indicated a higher intrinsic substrate use efficiency of winter microbes than summer ones. Our results suggested significant differences in microbial community structure and function between summer and winter. Considering the future changes in winter climate and N deposition in temperate areas, more detailed investigations of seasonal dynamics in soil microbial biomass, community structure and function including spring and autumn are urgently needed in the scenario of global changes.

Supporting Information

Table S1 Detailed summer (JUL.) and winter (JAN.) values of soil and microbial properties in Figure 2; 3; 4; 5B; 7. P1 = ~15-year Mongolia pine; P2 = ~25-year Mongolia pine; P3 = ~35-year Mongolia pine; L1 = ~15-year Prince Rupprecht's larch; L2 = ~25-year larch; L3 = ~35-year larch; BH = Birch; MA = Siberian crabapple; RO = Solitary rose; CG = Meadow grassland. DON = dissolved organic nitrogen,

MBC = microbial biomass carbon, MBN = microbial biomass nitrogen, MB = microbial biomass.

Table S2 The inverse Simpson index of summer (JUL.) and winter (JAN.) PLFAs data in each sampling site.

Acknowledgments

We thanked two anonymous reviewers for their very constructive and valuable comments for its early version of our manuscript.

Author Contributions

Conceived and designed the experiments: WW. Performed the experiments: XZ WC. Analyzed the data: XZ NZ WC. Contributed reagents/materials/analysis tools: HZ. Wrote the paper: XZ WW.

References

1. Zhang T, Barry RG, Armstrong RL (2004) Application of satellite remote sensing techniques to f rozen ground studies. Polar Geography 28: 63–196.
2. Brooks PD, Grogan P, Templer PH Groffman P, Öquist MG, et al. (2011) Carbon and nitrogen cycling in snow-covered environments. Geography Compass 5: 682–699.
3. Aanderud ZT, Jones SE, Schoolmater DRJ, Fierer N, Lennon JT (2013) Sensitivity of soil respiration and microbial communities to altered snowfall. Soil Biology and Biochemistry 57: 217–227.
4. Robroek BJ, Heijboer A, Jassey VE, Hefting MM, Rouwenhorst TG, et al. (2013) Snow cover manipulation effects on microbial community structure and soil chemistry in a mountain bog. Plant and soil 369: 151–164.
5. Romanovsky VE, Osterkamp TE (2000) Effects of unfrozen water on heat and mass transport processes in the active layer and permafrost. Permafrost and Periglacial Processes 11: 219–239.
6. Price BP, Sowers T (2004) Temperature dependence of metabolic rates for microbial growth, maintenance, and survival. Proceedings of the National Academy of Sciences 101: 4631–4636.
7. Schimel J, Balser TC, Wallenstein M (2007) Microbial stress-response physiology and its implications for ecosystem function. Ecology 88: 1386–1394.
8. Brooks PD, Williams MW, Schmidt SK (1996) Microbial activity under alpine snowpacks, Niwot Ridge, Colorado. Biogeochemistry 3: 93–113.
9. Grogan P, Jonasson S (2005) Temperature and substrate controls on intra-annual variation in ecosystem respiration in two sub-arctic vegetation-types. Global Change Biology 11: 465–475.
10. Groffman PM, Hardy JP, Driscoll CT, Fahey TJ (2006) Snow depth, soil freezing, and fluxes of carbon dioxide, nitrous oxide and methane in a northern hardwood forest. Global Change Biology 12: 1748–1760.
11. Monson RK, Lipson DL, Burns SP, Turnipseed AA, Delany AC, et al. (2006) Winter forest soil respiration controlled by climate and microbial community composition. Nature 439: 711–714.
12. Wang W, Peng S, Wang T, Fang J (2010) Winter soil CO_2 efflux and its contribution to annual soil respiration in different ecosystems of a forest-steppe ecotone, north China. Soil Biology and Biochemistry 42: 451–458.
13. Edwards KA, McCulloch J, Kershaw GP (2006) Soil microbial and nutrient dynamics in a wet Arctic sedge meadow in late winter and early spring. Soil Biology and Biochemistry 38: 2843–2851.
14. Buckeridge KM, Jefferies RL (2007) Vegetation loss alters soil nitrogen dynamics in an Arctic salt marsh. Journal of Ecology 95: 283–293.
15. Drotz SH, Saparrman T, Nilsson MB, Schleucher J, Öquist MG (2010) Both catabolic and anabolic heterotrophic microbial activity proceed in frozen soils. Proceedings of the National Academy of Sciences 107: 21046–21051.
16. Lipson DA, Schmidt SK, Monson RK (2000) Carbon availability and temperature control the post-snowmelt decline in alpine soil microbial biomass. Soil Biology and Biochemistry 32: 441–448.
17. Larsen KS, Grogan P, Jonasson S, Michelsen A (2007) Respiration and microbial dynamics in two subarctic ecosystems during winter and spring thaw: effects of increased snow depth. Arctic, Antarctic, and Alpine Research 39: 268–276.
18. Edwards KA, Jefferies RL (2013) Inter-annual and seasonal dynamics of soil microbial biomass and nutrients in wet and dry low-Arctic sedge meadows. Soil Biology and Biochemistry 57: 83–90.
19. Weintraub MN, Schimel JP (2005) The seasonal dynamics of amino acids and other nutrients in Alaskan Arctic tundra soils. Biogeochemistry 73: 359–380.
20. Kaiser C, Fuchslueger L, Koranda M, Gorfer M, Stange CF, et al. (2011) Plants control the seasonal dynamics of microbial N cycling in a beech forest soil by belowground C allocation. Ecology 92: 1036–1051.
21. Nielsen CB, Groffman PM, Hamburg SP, Driscoll CT, Fahey TJ, et al. (2001) Freezing effects on carbon and nitrogen cycling in northern hardwood forest soils. Soil Science Society of America Journal 65: 1723–1730.
22. Judd KE, Liken GE, Groffman PM (2007) High nitrate retention during winter in soils of the Hubbard Brook experimental forest. Ecosystems 10: 217–225.
23. Buckeridge KM, Grogan P (2008) Deepened snow alters soil microbial nutrient limitations in arctic birch hummock tundra. Applied Soil Ecology 39: 210–222.
24. Turner MM, Henry HAL (2010) Net nitrogen mineralization and leaching in response to warming and nitrogen deposition in a temperate old field: the importance of winter temperature. Oecologia 62: 227–236.
25. Larsen KS, Michelsen A, Jonasson S, Beier C, Grogan P (2012) Nitrogen uptake during fall, winter and spring differs among plant functional groups in a subarctic heath ecosystem. Ecosystems 15: 929–939.
26. Schmidt SK, Costello EK, Nemergut DR, Cleveland CC, Reed SC, et al. (2007) Biogeochemical consequences of rapid microbial turnover and seasonal succession in soil. Ecology 88: 1379–1385.
27. Lipson DA, Schadt CW, Schmidt SK (2002) Changes in soil microbial community structure and function in an alpine dry meadow following spring snow melt. Microbial Ecology 43: 307–314.
28. Schadt CW, Martin AP, Lipson DA, Schmidt SK (2003) Seasonal dynamics of previously unknown fungal lineages in tundra soils. Science 301: 1359–1361.
29. Waldrop M, Balser T, Firestone M (2000) Linking microbial community composition to function in a tropical soil. Soil Biology and Biochemistry 32: 1837–1846.
30. Sinsabaugh RL, Carreiro MM, Repert DA (2002) Allocation of extracellular enzymatic activity in relation to litter composition, N deposition, and mass loss. Biogeochemistry 60: 1–24.
31. Walder F, Niemann H, Natarajan M, Lehmann MF, Boller T, et al. (2012) Mycorrhizal networks: common goods of plants shared under unequal terms of trade. Plant physiology 159: 789–797.
32. Pokarzhevskii AD, van Straalen NM, Zaboev DP, Zaitsev AS (2003) Microbial links and element flows in nested detrital food-webs. Pedobiologia 47: 213–224.
33. McMahon SK, Wallenstein MD, Schimel JP (2009) Microbial growth in Arctic tundra soil at −2°C. Environmental Microbial Reports 1: 162–166.
34. Balser TC, Wixon DL (2009) Investigating biological control over soil carbon temperature sensitivity. Global Change Biology 15: 2935–2949.
35. Keiblinger KM, Hall EK, Wanek W, Szukics U, Hämmerle I, et al. (2010) The effect of resource quantity and resource stoichiometry on microbial carbon-use-efficiency. FEMS Microbiology Ecology 73: 430–440.
36. Lipson DA, Monson RK, Schmidt SK, Weintraub MN (2009) The trade-off between growth rate and yield in microbial communities and the consequences for under-snow soil respiration in a high elevation coniferous forest. Biogeochemistry 95: 23–35.
37. Liptzin D, Silver WL (2009) Effects of carbon additions on iron reduction and phosphorus availability in a humid tropical forest soil. Soil Biology and Biochemistry 41: 1696–1702.
38. Bond-Lamberty B, Thomson A (2010) Temperature-associated increases in the global soil respiration record. Nature 464: 579–582.
39. Wang W, Chen W, Wang S (2010) Forest soil respiration and its heterotrophic and autotrophic components: Global patterns and responses to temperature and precipitation. Soil Biology and Biochemistry 42: 1236–1244.
40. Zhou YM, Li MH, Cheng XB, Wang CG, Fan AN, et al. (2010) Soil respiration in relation to photosynthesis of quercus mongolica trees at elevated CO_2. Plos one 5:e15134.
41. Yao H, Bowman D, Shi W (2011) Seasonal variations of soil microbial biomass and activity in warm-and cool-season turfgrass systems. Soil Biology and Biochemistry 43: 1536–1543.
42. Baldrian P, Vetrovsky T, Cajthaml T, Dobiášová P, Petránková M, et al. (2013) Estimation of fungal biomass in forest litter and soil. Fungal Ecology 6: 1–11.
43. Hayhoe K, Wake CP, Huntington TG, Luo L, Schwartz MD, et al. (2007) Past and future changes in climate and hydrological indicators in the US Northeast. Climate Dynamics 28: 381–407.
44. Huntington TG, Richardson AD, McGuire KJ, Hayhoe K (2009) Climate and hydrological changes in the northeastern United States: recent trends and implications for forested and aquatic ecosystems. Canadian Journal of Forest Research 39: 199–212.
45. Henry HAL (2008) Climate change and soil freezing dynamics: historical trends and projected changes. Climatic Change 87: 421–434.
46. Galloway JN, Dentener FJ, Capone DG, Boyer EW, Howarth RW, et al. (2004) Nitrogen cycles: past, present, and future. Biogeochemistry 70: 153–226.
47. Kreyling J (2010) Winter climate change: a critical factor for temperate vegetation performance. Ecology 91: 1939–1948.
48. Pietikäinen J, Pettersson M, Bååth E (2005) Comparison of temperature effects on soil respiration and bacterial and fungal growth rates. FEMS Microbiology Ecology 52: 49–58.
49. Kramer C, Gleixner G (2006) Variable use of plant- and soil-derived carbon by microorganisms in agricultural soils. Soil Biology and Biochemistry 38: 3267–3278.
50. Yang Y, Luo Y, Finzi AC (2011) Carbon and nitrogen dynamics during forest stand development: a global synthesis. New Phytologist 190: 977–989.

51. Bjork RG, Bjorkman MP, Andersson MX, Klemedtsson L (2008) Temporal variation in soil microbial communities in Alpine tundra. Soil Biology and Biochemistry 40: 266–268.
52. Lipson DA, Schmidt SK (2004) Seasonal changes in an alpine soil bacterial community in the Colorado Rocky Mountains. Applied and environmental microbiology 70: 2867–2879.
53. Bradley RL, Fyles JW (1995) Growth of paper birch (Betula papyrifera) seedling increases soil available C and microbial acquisition of soil nutrients. Soil Biology and Biochemistry 27: 1565–1571.
54. Schumacher BA (2002) Methods for the determination of the total organic carbon (TOC) in soils and sediments. National ESD, ed. U.S. Environmental Protection Agency, Washington DC.
55. Ruzicka J (1983) Flow Injection Analysis – From Test Tube to Integrated Microconduits. Analytical Chemistry 55: 1040A–1053A.
56. Brookes PC, Landman A, Pruden G, Jenkinson DS (1985) Chloroform fumigation and the release of soil nitrogen- A rapid direct extraction method to measure microbial biomass nitrogen in soil. Soil Biology and Biochemistry 17: 837–842.
57. Ross DJ (1990) Estimation of soil microbial C by a fumigation-extraction method: influence of seasons, soils and calibration with the fumigation-incubation procedure. Soil Biology and Biochemistry 22: 295–300.
58. Bossio DA, Scow KM (1998) Impacts of carbon and flooding on soil microbial communities: phospholipid fatty acid profiles and substrate utilization patterns. Microbial Ecology 35: 265–278.
59. Zelles L (1997) Phospholipid fatty acid profiles in selected members of soil microbial communities. Chemosphere 35: 275–294.
60. Fierer N, Schimel JP, Holden PA (2003) Variations in microbial community composition through two soil depth profiles. Soil Biology and Biochemistry 35: 167–176.
61. Frostegård Å, Bååth E, Tunlio A (1993) Shifts in the structure of soil microbial communities in limed forests as revealed by phospholipid fatty acid analysis. Soil Biology and Biochemistry 25: 723–730.
62. Zogg GP, Zak DR, Ringelberg DB, White DC, MacDonald NW, et al. (1997) Compositional and functional shifts in microbial communities due to soil warming. Soil Science Society of America Journal 61: 475–481.
63. Gutknecht JL, Field CB, Balser TC (2012) Microbial communities and their responses to simulated global change fluctuate greatly over multiple years. Global Change Biology 18: 2256–2269.
64. Winkler JP, Cherry RS, Schlesinger WH (1996) The Q_{10} relationship of microbial respiration in a temperate forest soil. Soil Biology and Biochemistry 28: 1067–1072.
65. Anderson JPE (1982) Soil respiration, In Methods of Soil Analysis, Part 2. Chemical and Microbiological Properties. Soil Science Society of America, Madison: 831–871.
66. Mielke PW (1984) Meterological applications of permutation techniques based on distance functions. In: Krishnaiah PR, Sen PK (eds) Handbook of statistics, vol 4New YorkElsevier Science Publishers
67. Mielke PW, Berry KJ (2001) Permutation methods: a distance function approach. Springer, Berlin Heidelberg New York.
68. Schimel JP, Bilbrough C, Welker JM (2004) Increased snow depth affects microbial activity and nitrogen mineralization in two Arctic tundra communities. Soil Biology and Biochemistry 36: 217–227.
69. Schimel J P, Weintraub M N (2003) The implications of exoenzyme activity on microbial carbon and nitrogen limitation in soil: a theoretical model. Soil Biology and Biochemistry 35: 549–563.
70. Mariko S, Nishimura N, Mo W (2000) Winter CO_2 flux from soil and snow surfaces in a cool-temperate deciduous forest, Japan. Ecological Research 15: 363–372.
71. Uchida M, Mo W, Nakatsubo T (2005) Microbial activity and litter decomposition under snow cover in a cool-temperate broad-leaved deciduous forest. Agricultural and Forest Meteorology 134: 102–109.
72. Schimel JP, Fahnestock J, Michaelson G (2006) Cold-season production of CO_2 in arctic soils: can laboratory and field estimates be reconciled through a simple modeling approach? Arctic, Antarctic and Alpine Research 38: 249–256.
73. Fierer N, Strickland MS, Liptzin D, Bradford MA, Cleveland CC (2009) Global patterns in belowground communities. Ecology Letters 12: 1238–1249.
74. Cleveland CC, Liptzin D (2007) C: N: P stoichiometry in soil: is there a "Redfield ratio" for the microbial biomass? Biogeochemistry 85: 235–252.
75. Rovira AD (1969) Plant root exudates. The Botanical Review 35: 35–57.
76. Strickland MS, Rousk J (2010) Considering fungal: bacterial dominance in soils–Methods, controls, and ecosystem implications. Soil Biology and Biochemistry 42: 1385–1395.
77. Cardon ZG, Hungate BA, Cambardella CA, Chapin III FS, Field CB, et al. (2001) Contrasting effects of elevated CO2 on old and new soil pools. Soil Biology and Biochemistry 33: 365–373.
78. Liu LL, Kong JS, Booker FL, Giardina CP, Lee AH, et al. (2009) Enhanced litter input rather than changes in litter chemistry drive soil carbon and nitrogen cycles under elevated CO_2: a microcosm study. Global Change Biology 15: 441–453.
79. Schimel JP, Mikan C (2005) Changing microbial substrate use in Arctic tundra soils through a freeze-thaw cycle. Soil Biology and Biochemistry 37: 1411–1418.
80. Haei M, Rousk J, Ilstedt U, Öquist M, Bååth E, et al. (2011) Effects of soil frost on growth, composition and respiration of the soil microbial decomposer community. Soil Biology and Biochemistry 43: 2069–2077.
81. Wallenstein MD, McMahon S, Schimel J (2007) Bacterial and fungal community structure in Arctic tundra tussock and shrub soils. FEMS Microbiology Ecology 59: 428–435.
82. Wallenstein MD, McMahon SK, Schimel JP (2009) Seasonal variation in enzyme activities and temperature sensitivities in Arctic tundra soils. Global Change Biology 15: 1631–1639.
83. Schindlbacher A, Rodler A, Kuffner M, Kitzler B, Sessitsch A, et al. (2011) Experimental warming effects on the microbial community of a temperate mountain forest soil. Soil Biology and Biochemistry 43: 1417–1425.
84. Boer WD, Folman LB, Summerbell RC, Boddy L (2005) Living in a fungal world: impact of fungi on soil bacterial niche development. FEMS microbiology reviews 29: 795–811.
85. Sinsabaugh RL, Lauber CL, Weintraub MN, Ahmed B, Allison SD, et al. (2008) Stoichiometry of soil enzyme activity at global scale. Ecology Letters 11: 1252–1264.

20

In Vitro Continuous Fermentation Model (PolyFermS) of the Swine Proximal Colon for Simultaneous Testing on the Same Gut Microbiota

Sabine A. Tanner, Annina Zihler Berner, Eugenia Rigozzi, Franck Grattepanche, Christophe Chassard, Christophe Lacroix*

Laboratory of Food Biotechnology, Institute of Food, Nutrition and Health, ETH Zurich, Zurich, Switzerland

Abstract

In vitro gut modeling provides a useful platform for a fast and reproducible assessment of treatment-related changes. Currently, pig intestinal fermentation models are mainly batch models with important inherent limitations. In this study we developed a novel *in vitro* continuous fermentation model, mimicking the porcine proximal colon, which we validated during 54 days of fermentation. This model, based on our recent PolyFermS design, allows comparing different treatment effects on the same microbiota. It is composed of a first-stage inoculum reactor seeded with immobilized fecal swine microbiota and used to constantly inoculate (10% v/v) five second-stage reactors, with all reactors fed with fresh nutritive chyme medium and set to mimic the swine proximal colon. Reactor effluents were analyzed for metabolite concentrations and bacterial composition by HPLC and quantitative PCR, and microbial diversity was assessed by 454 pyrosequencing. The novel PolyFermS featured stable microbial composition, diversity and metabolite production, consistent with bacterial activity reported for swine proximal colon *in vivo.* The constant inoculation provided by the inoculum reactor generated reproducible microbial ecosystems in all second-stage reactors, allowing the simultaneous investigation and direct comparison of different treatments on the same porcine gut microbiota. Our data demonstrate the unique features of this novel PolyFermS design for the swine proximal colon. The model provides a tool for efficient, reproducible and cost-effective screening of environmental factors, such as dietary additives, on pig colonic fermentation.

Editor: Markus M. Heimesaat, Charité, Campus Benjamin Franklin, Germany

Funding: This study was carried out thanks to the financial support of the Commission of Technology and Innovation of Switzerland (CTI-Project Nr. 11962.1 PFLS-LS) (URL: http://www.kti.admin.ch/?lang = en). The funders had no role in study design, data collection and analysis, decision to publish, or preparation of the manuscript.

Competing Interests: The authors have declared that no competing interests exist.

* E-mail: christophe.lacroix@hest.ethz.ch

Introduction

The pig gut microbiome is a complex ecosystem, dominated by members of the phyla Firmicutes and Bacteroidetes [1]. The vast quantity and diversity of the gut bacterial community provides the host with a large set of metabolic functions and is considered to play a key role in host health and disease [2]. Diet is a principal factor shaping the gut bacterial composition and functionality [3], thereby also impacting overall animal health. This in turn largely determines productivity and efficiency of swine livestock production. Given the complexity of this interplay, it is of paramount importance to precisely evaluate the effects of specific feed ingredients or additives on composition and functionality of the gut microbiome and further elucidate the role and function of the gut bacterial ecosystem in animal health to improve productivity and efficiency of swine livestock production.

It is obvious that animal testing is one of the most prominent strategies to predict effectiveness and impact of dietary additives on the gut microbiota, but ethical concerns and costs can restrict these applications [2]. Intestinal *in vitro* models are able to partly evade these restrictions by enabling reproducible experimentation under standardized conditions, and more importantly, giving the yet host-uncoupled opportunity to investigate the complexity of gut microbiomes and the functional relatedness of specific bacterial species [4].

To date, most porcine *in vitro* models are using simple batch cultures with the aim to examine the fermentation capacity of intestinal ecosystems on a given substrate, using the cumulative gas production technique [5–8]. However, batch fermentations are limited in terms of experimental duration and the amount of substrate supply to avoid negative feedback mechanisms [9]. Batch cultures are also highly dependent on the inoculation density as it directly impacts microbial growth in these closed systems [4]. In contrast, continuous culture systems are superior in modeling the dynamic nature of the gastrointestinal tract, allowing the adaptation of various parameters, including dilution rate, retention time, pH and temperature, to meet and maintain optimal growth conditions [9]. Substrate replenishment and toxic waste removal, further, is continuous and facilitates studies on the modulation of microbial composition and activity [4]. So far, only few porcine semi-continuous or continuous intestinal fermentation models inoculated with feces or cecal content have been described, focusing on changes in gut bacterial communities tested over limited fermentation periods [10], the inhibition of *Salmonella* by medium-chain fatty acids [11] or the effect of live yeast on fermentation parameters [12]. Long-term continuous *in vitro* models of the swine gut are still lacking, likely due to the difficulty of generating highly stable free-cell suspension fermentations, inoculated with fecal extract, over a long experimental period.

To overcome possible drawbacks associated with free-cell suspension cultures, such as limited stability and washout of less dominant or slow growing bacterial species, fecal microbiota was immobilized in polysaccharide gel beads to set up stable continuous intestinal fermentation models [13,14]. The major benefits shown for immobilized fecal microbiota models include high cell density, maintenance of bacterial diversity and high stability over extended fermentation periods, tested for up to 71 days [4,15]. Besides stability requirements, reproducibility and parallel testing of treatments with the same gut microbiota are of major importance for gut *in vitro* research, but difficult to apply with classical continuous models inoculated with fecal microbiota. We have recently set up and validated a novel PolyFermS model of the child proximal colon in a two-stage system for parallel testing of treatments on the same microbiota [16]. In this model, a first-stage reactor containing immobilized fecal microbiota and operated with conditions mimicking the first section of the proximal colon was used to constantly inoculate up to four second-stage reactors, operated with conditions mimicking the remaining proximal colon. Our data demonstrated that this PolyFermS model produced reproducible and stable intestinal microbiota over a 38 days test period [16]. The microbial diversity of reactor effluents tested with the HITChip phylogenetic array was comparable to the feces of the healthy donor, whereas a high response to pH was demonstrated. However, this study used a two-stage model for upper and lower sections of the proximal colon, shown to be highly dependent on pH in the first reactor, which had to be arbitrary set at 5.5 to reach a metabolic balance for the child microbiota.

In the present study we aimed to enhance the original design of the PolyFermS model validated with child microbiota [16], and to adapt to a new model of the swine proximal colon. The novel PolyFermS model consisted of two stages; a first stage for the inoculum reactor (IR) seeded with immobilized swine fecal microbiota and used to inoculate at 10% (v/v) five parallel second-stage reactors, which were also fed with 90% fresh chyme medium. Each reactor was operated under identical conditions, selected to mimic swine proximal colon, and allowing the parallel testing of multiple treatments on the same gut microbiota. The stability of the complex intestinal microbiota in the multiple reactors was monitored over 54 days of fermentation by analyzing its composition (qPCR and 454 pyrosequencing) and metabolic activity (HPLC).

Material and Methods

Ethical statement

No specific permits were obtained for the collection of the fecal sample. The animal was not harmed during fecal sample collection and oral consent for sample collection was obtained from the owner of the farm.

Feces collection and immobilization

Feces from a healthy 5 month old sow (80 kg), raised under farming conditions and not subjected to any antibiotic treatment for the last 3 months, were collected in a sterile 50 mL Falcon tube. Anaerobiosis was maintained using anaerobic gas pack systems (Oxoid AnaeroGen TM, Oxoid AG, Basel Switzerland) during transport to the laboratory and until immobilization was performed. The entire immobilization procedure was carried out under anaerobic conditions (anaerobic chamber; Coy Laboratories, Ann Arbor, MI, USA). Briefly, a 20% (w/v) suspension of feces in pre-reduced peptone water (0.1%, pH 7) was prepared, homogenized and further immobilized in a polymer solution consisting of gellan gum (2.5%, w/v), xanthan (0.25%, w/v) and sodium citrate (0.2%, w/v) for production of 1–2 mm diameter gel beads using a two phase dispersion process as described previously [13,16].

Nutritive medium

The nutritive medium described by Macfarlane *et al.* [17] was modified for its carbohydrate and protein concentration to more closely mimic the ileal chyme of a swine (Table S1), using a similar approach as described previously [15]. For calculation of ingredient concentrations, a standard cornstarch based diet with corn (641 g kg^{-1}) as main carbohydrate and soybean meal (331 g kg^{-1}) as main N-source was used [18]. Digestibility indices of 97% for cornstarch [19] and 82.5% for soybean meal [20] were applied while considering a 2 kg/day feed intake per pig, resulting in a cornstarch:N-compound (soybean meal) ratio of 25:75. The amount of cornstarch and soy peptone supplied daily to the model was calculated by applying a scale factor of 0.09 to account for the actual volume of the proximal colon *in vivo* (approx. 2.9 L [21] compared to the proximal reactor volume of the model (260 mL). The final concentration of the two compounds in the chyme medium was estimated for a mean retention time of 9 hours in the reactors, giving a daily chyme medium supply per reactor of 693 mL. To avoid excess of nitrogen compounds, the soy peptone concentration was fixed to 13 g L^{-1} resulting in a carbohydrate (cornstarch) concentration of 4.3 g L^{-1}. Yeast extract and mucin, mimicking the contribution of endogenous secretion [13], and the non-starch polysaccharides (pectin, xylan, arabinogalactan and guar gum) were excluded from this calculation.

A volume of 0.5 mL L^{-1} of a filter-sterilized (Minisart pore size 0.2 µm, Sartorius, VWR International AG, Dietikon, Switzerland) vitamin solution described by Michel *et al.* [22] was added to the sterilized medium (20 min, 120°C). All components of the nutritive medium were purchased from Sigma-Aldrich Chemie (Buchs, Switzerland), except for soy peptone (Labo-Life Sàrl, Pully, Switzerland), yeast extract (I2CNS GmbH, Urdorf, Switzerland) and KH_2PO_4 (VWR International AG).

Experimental set-up of the PolyFermS model

The continuous fermentation was carried out for 54 days using a two-stage design with a total of six reactors (Sixfors, Ismatec, Glattbrugg, Switzerland) (Figure 1). Each reactor was aimed to simulate conditions of the swine proximal colon fermentation.

The inoculum reactor (234 mL) was seeded with 30% (v/v) swine fecal beads and connected via a peristaltic pump (Masterflex L/S, Fisher Scientific SA, Wohlen, Switzerland) to one control reactor (CR) and four test reactors (TR1-4) operated in parallel. IR was supplied with 100% fresh nutritive medium (26 mL h^{-1}) whereas the second-stage reactors CR and TR1-4 were continuously supplied with 90% (26 mL h^{-1}) fresh nutritive medium and 10% (2.9 mL h^{-1}) effluent from IR for continuous inoculation. The remaining effluent (50%) from IR was discarded. The inoculation rate of CR and TR1-4 was accurately controlled using an in-house designed distributor device equipped with valves regulated chronometrically.

Fermentation procedure

The PolyFermS model was run under conditions of the swine proximal colon with a controlled constant pH of 6.0 through

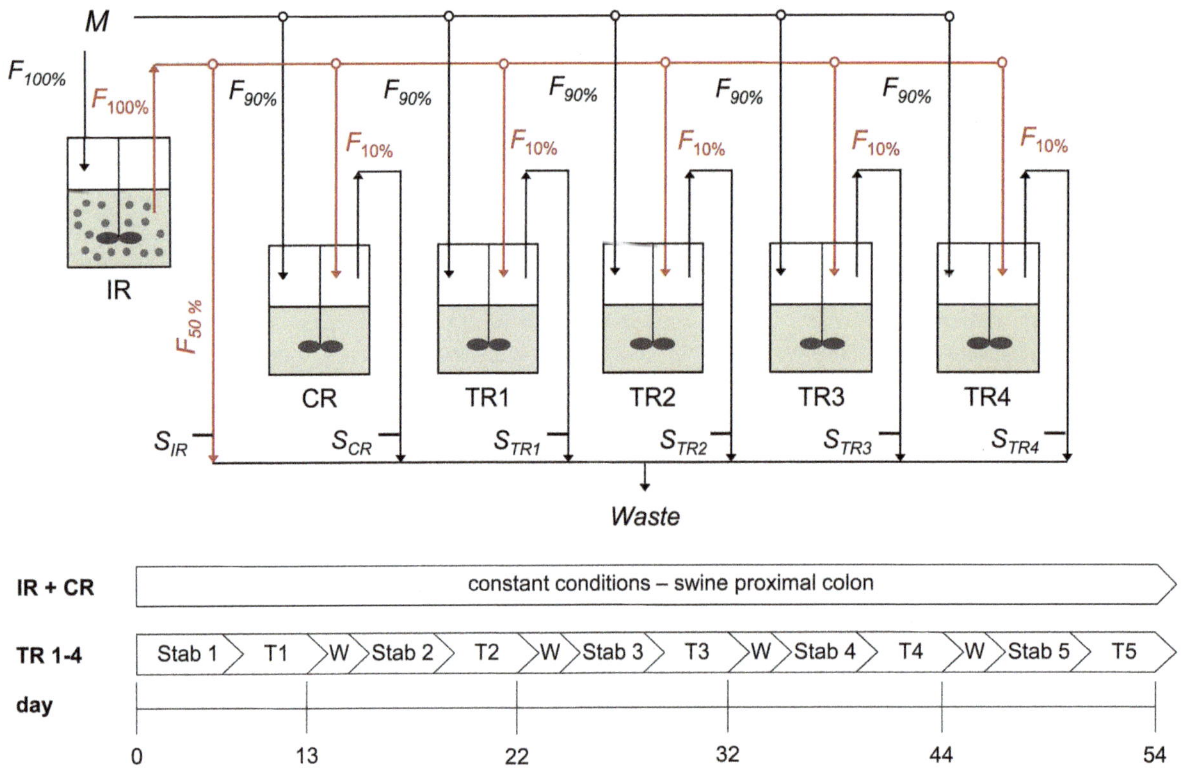

Figure 1. Experimental reactor set-up and time schedule of the swine PolyFermS model. IR: inoculum reactor, containing immobilized swine feces (30% v/v); CR: control reactor; TR1-TR4: test reactors 1-4; *M*: fresh nutritive medium supply; *S*: effluent sampling; *F*: flow rate; Stab: stabilization period; T: treatment period; W: wash period

addition of 2.5 M NaOH. The mean retention time was fixed at 9 hours, while temperature was maintained at 38°C (being in the middle of the range previously applied in porcine *in vitro* models) [7,10,11,23–25]. Anaerobic conditions were ensured by constantly flushing the headspace of the reactors with CO_2 and constant stirring was performed at 120 rpm. During the first 72 hours, IR was operated in batch mode to colonize the fecal beads and the nutritive medium was replaced by fresh medium every 12 hours. After colonization, continuous operation (26 mL h^{-1}) in IR was started, followed by a stabilization period of 5 days before connection to CR and TR1-4. The entire two-stage system was then stabilized for another 5 days.

The 54 days continuous fermentation was split into stabilization, treatment and washing periods (Figure 1). IR and CR were operated with constant conditions to assess the temporal stability of the system and were not subjected to any manipulation during the entire fermentation. In addition, CR served as a control reactor for TR1-4 that were subjected to different parallel treatment periods (data not shown).

Between two treatment periods, TR1-4 were subjected to a washing procedure with 10% chlorine to kill microbes and remove any historical effect of the previous periods. Briefly, TR1-4 were disconnected from IR, the entire medium was removed and reactors were filled with 10% freshly prepared chlorine solution. After stirring for one hour, the reactors were rinsed twice by adding sterile bidistilled water and stirring for another hour. After complete removal of water and chlorine residues, reactors were filled with sterile fresh nutritive medium and reconnected to IR. Thereafter, the system was allowed to stabilize for 3 days until reaching steady state before starting the next treatment period.

Effluent samples of all reactors were collected daily. HPLC samples were processed immediately whereas samples for DNA extraction were stored at −80°C.

HPLC analysis for metabolite determination

Short-chain fatty acids (SCFA; acetate, propionate, butyrate, valerate, formate, *iso*-butyrate and *iso*-valerate) as well as lactate concentrations in fermentation effluent samples from all reactors were determined by HPLC analysis (Thermo Fisher Scientific Inc. Accela, Wohlen, Switzerland). Briefly, effluent samples were centrifuged (14 000 g) for 10 min at 4°C. The pellet was used for DNA extraction while the supernatant was diluted 1:10 with ultrapure water and filtered directly into vials through a 4 mm HPLC filter with a 0.45 µm nylon membrane (Infochroma AG, Zug, Switzerland). The analysis was run at a flow rate of 0.4 mL min^{-1} using an Aminex HPX-87H column (Bio-Rad Laboratories AG, Reinach, Switzerland) and 10 mM H_2SO_4 as eluent. Mean metabolite concentrations were calculated from duplicate analyses and expressed in mM.

DNA extraction and qPCR analyses

Genomic DNA was extracted from effluent samples using the FastDNA SPIN Kit for soil (MP Biomedicals, Illkirch, France). DNA extracts were subjected to quantitative real-time PCR (qPCR) for enumeration of specific bacterial target groups comprising total bacteria, *Bifidobacterium* spp., *Bacteroides-Prevotella* group, *Enterobacteriaceae*, *Lactobacillus*/*Pediococcus*/*Leuconostoc* spp. and *Clostridium* Cluster IV (Table 1). In addition, the *Succinivibrio dextrinosolvens* group was quantified to verify results obtained by 454 pyrosequencing. Standard curves for each target group were

Table 1. Primers for detection of specific bacterial groups by qPCR.

Target	Primer	Sequence 5′-3′	Reference
Total 16S rRNA genes	Eub338F	ACT CCT ACG GGA GGC AGC AG	[48]
	Eub518R	ATT ACC GCG GCT GCT GG	
Bacteroides-Prevotella group	Bac303F	GAA GGT CCC CCA CAT TG	[49]
	Bfr-Fmrev	CGC KAC TTG GCT GGT TCA G	
Lactobacillus/Pediococcus/ Leuconostoc spp.	F_Lacto 05	AGC AGT AGG GAA TCT TCC A	[50]
	R_Lacto 04	CGC CAC TGG TGT TCY TCC ATA TA	
Enterobacteriaceae	Eco1457F	CAT TGA CGT TAC CCG CAG AAG AAG C	[51]
	Eco1652R	CTC TAC GAG ACT CAA GCT TGC	
Bifidobacterium spp.	xfp_fw	ATC TTC GGA CCB GAY GAG AC	[52]
	xfp_rv	CGA TVA CGT GVA CGA AGG AC	
Clostridium Cluster IV	Clep866mF	TTA ACA CAA TAA GTW ATC CAC CTG G	[49]
	Clep1240mR	ACC TTC CTC CGT TTT GTC AAC	
S. dextrinosolvens group	SucDex1F	CGTCAGCTCGTGTCGTGAGA	[53]
	SucDex1R	CCCGCTGGCAACAAAGG	

prepared as described previously [26]. All assays, were performed using the 2 x SYBR Green PCR Master Mix (Applied Biosystems, Zug, Switzerland) in a 25 µl volume and an ABI PRISM 7500-PCR sequence detection system (Applied Biosystems).

454 pyrosequencing

Selected samples were analyzed by 454 pyrosequencing for their microbial 16S rRNA based composition profile. IR and CR (days 19/20, 30/31, 42/43 and 52/53) samples were selected to assess bacterial diversity and temporal stability of the model as well as to compare the microbial composition to the fecal inoculum. To demonstrate the re-establishment of the microbiota after washing, day 25 (last day of 3rd stabilization period) was chosen as a representative day for all reactors. For IR and CR stability samples, effluent from 2 consecutive days were pooled at a ratio of 1:1. DNA was extracted with the FastDNA Spin Kit for soil (MP Biomedicals) and sent to DNAVision SA (Charleroi, Belgium) for 454 pyrosequencing analysis and subsequent taxonomic assignment of 16S rRNA gene reads. 454 pyrosequencing was performed using a 454 Life Science system combined with Titanium Chemistry (Roche) as described previously [27]. The complete 454 pyrosequencing dataset has been deposited to the National Center for Biotechnology Information (NCBI) Sequence Read Archive (SRA) under accession number SRP034540.

Statistical analysis

All statistical analyses were performed using PASW Statistics for Windows version 18.0 (SPSS Inc., Chicago). qPCR data were log10-transformed and expressed as means ± SD of the last three days of each stabilization period. To assess reproducibility of the microbial composition in CR and TR1-4 prior to a treatment period qPCR data were subjected to the non-parametric Mann-Whitney U test with exact significance and a p-value <0.05 was considered significant. The same test was used to verify results obtained from 454 pyrosequencing for *Succinivibrionaceae*. Comparisons were made between qPCR data from before (d11-37) and after (d38-54) appearance in CR. To assess temporal metabolite stability in IR and CR, linear regression of total SCFA, acetate, butyrate, propionate, *iso*-valerate, valerate and *iso*-butyrate concentrations versus time were calculated over the time period d11-54 and difference from 0 of slope coefficients was tested using the t-test ($P<0.05$).

Results

Microbial activity by HPLC

To assess the microbial activity and temporal stability of the novel PolyFermS model for the pig proximal colon during 54 days continuous fermentation, daily effluents for each reactor were analyzed using HPLC. IR and CR were operated with constant conditions and not subjected to any treatment or washing period during the entire fermentation. Therefore, the microbial activity in these two reactors was used to assess the metabolic stability of the two-stage model.

After an initial stabilization period of 10 days for reaching pseudo-steady state conditions, high and stable metabolic activity was measured in IR and CR throughout the fermentation (Figure 2). In IR the mean total SCFA concentration (d11-54) was 180.1±5.8 mM with mean values for acetate, propionate and butyrate of 102.3±7.3 mM, 45.1±4.4 mM and 20.4±1.2 mM, respectively. *Iso*-valerate was produced at 5.9±0.5 mM. In contrast, valerate and *iso*-butyrate were not detected until day 13 but were measured at concentrations of 4.5±1.3 mM and 1.9±0.6 mM over d13-54, respectively. Lactate and formate, were not detected throughout the fermentation. For CR similar mean total SCFA (174.4±7.6 mM) as well as main SCFA acetate (92.9±7.6 mM), propionate (47.1±3.8mM) and butyrate (23.1±2.4 mM) concentrations were detected. Detection of minor metabolites, *iso*-valerate, valerate and *iso*-butyrate, was delayed for 9, 14 and 13 days in CR, respectively, resulting in mean values of 5.7±1.2 mM (d11-54), 4.2±1.6 mM (d14-54) and 1.8±0.9 mM (d13-54).

To test the effect of culture time on metabolite concentrations, linear regressions of daily concentration data (d11-54) versus time were calculated. Highly significant time effects ($P<0.001$) were determined for most measured metabolites in IR and CR, except for the total SCFA concentration (Table 2). Acetate concentrations in IR and CR ($P<0.001$) and butyrate in CR ($P<0.05$) decreased

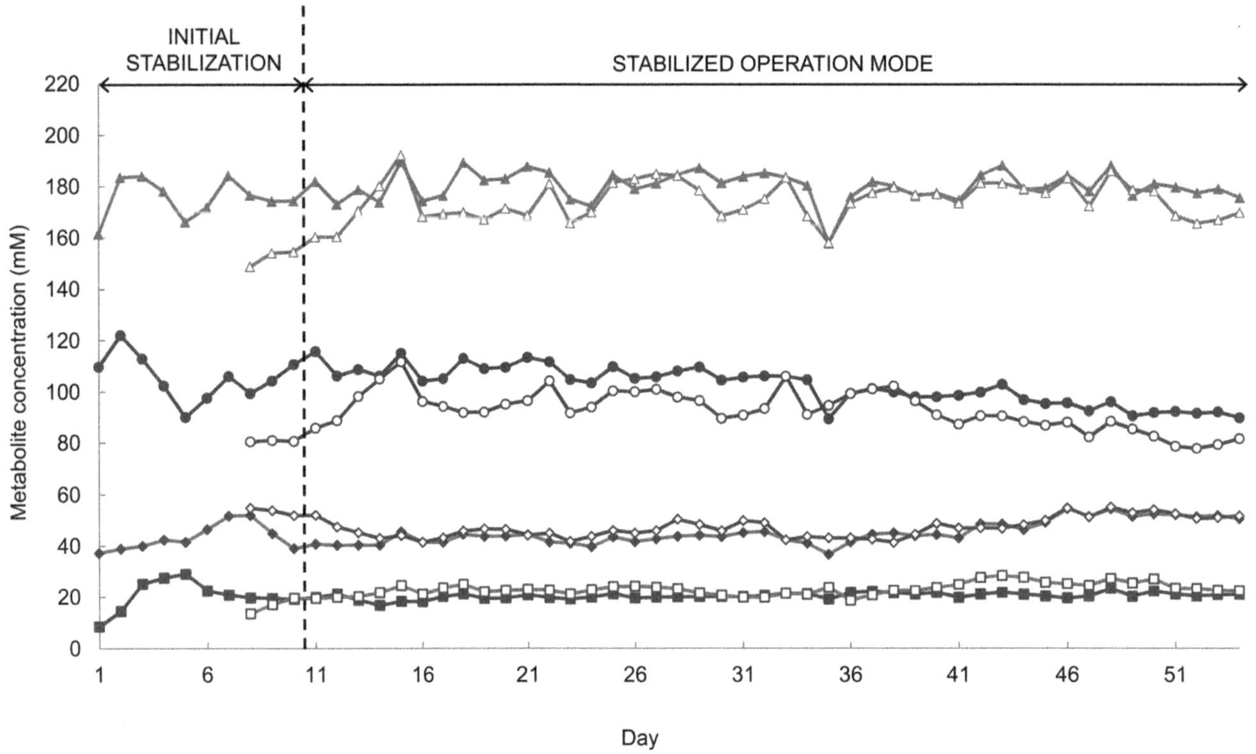

Figure 2. Daily main SCFA concentrations in fermentation effluents of IR and CR measured by HPLC. Initial stabilization: stabilization period in continuous mode to reach pseudo steady-state; closed symbol: IR; open symbol: CR; (▲, Δ) total SCFA, (●, ○) acetate, (♦, ◇) propionate, and (■, □) butyrate.

significantly over time whereas propionate, butyrate in IR, *iso*-valerate, valerate and *iso*-butyrate showed significant ($P<0.001$) concentration increases. Corresponding slope coefficients remained small (-0.493 mM/day – 0.265 mM/day) indicating a moderate time effect on tested metabolites.

Microbial composition by qPCR

Microbial composition of daily reactor effluents was assessed by analyzing the 16S rRNA gene copy numbers of bacterial target groups using qPCR. IR and CR data were used to assess the time stability of the model.

Total 16S rRNA gene copy numbers mL^{-1} effluent were approximately 1.6 $\log_{10}$ units higher in the fecal inoculum compared to the reactor effluents from IR and CR (Table 3). The fecal inoculum was dominated by the *Bacteroides-Prevotella* group and the *Lactobacillus/Pediococcus/Leuconostoc* spp. group, followed by *Clostridium* Cluster IV, *Enterobacteriaceae* and *Bifidobacterium* spp. In reactor effluents from IR and CR, stable copy numbers of the targeted bacterial groups were recorded during continuous fermentation (d11-54) after an initial stabilization as indicated by low standard deviations of the mean. Similar to the fecal inoculum, *Bacteroides-Prevotella* group was predominant in IR and CR effluents while *Bifidobacterium* spp. was the least abundant group. *Enterobac-*

Table 2. Effect of culture time on metabolite concentrations analyzed by linear regression analysis.

	Inoculum reactor (IR)			Control reactor (CR)		
	B[a]	SE B	R^2	B[a]	SE B	R^2
total SCFA	−0.030	0.069	0.005	0.092	0.090	0.024
acetate	−0.493**	0.043	0.760	−0.365**	0.072	0.376
propionate	0.265**	0.034	0.597	0.168**	0.038	0.317
butyrate	0.050**	0.012	0.292	−0.090*	0.025	0.238
iso-valerate	0.031**	0.004	0.544	0.063**	0.010	0.486
valerate	0.080**	0.009	0.637	0.093**	0.013	0.542
iso-butyrate	0.037**	0.005	0.593	0.047**	0.007	0.509

[a]slope coefficients significantly different from 0 are denoted by significance level:* $P<0.05$ ** $P<0.001$.
B: unstandardized (slope) coefficient (mM/day); SE B: Standard error of B; R^2: coefficient of determination.

teriaceae were approximately 1.8 $\log_{10}$ higher in reactor effluents compared to the fecal inoculum and displayed equally high copy numbers as for the *Clostridium* Cluster IV. In contrast, the *Lactobacillus/Pediococcus/Leuconostoc* spp. population was detected at ca. 3 $\log_{10}$ lower copy numbers in reactor effluents during the stabilized period (d11-54) compared to the fecal inoculum.

Microbial diversity by 454 Pyrosequencing

Microbial diversity and composition analyses of the fecal inoculum and selected samples from IR and CR (days 19/20, 30/31, 42/43, 52/53) were performed by 454 pyrosequencing. Sequences were aligned with the RDP classifier v 2.1 using a confidence cutoff level of 80%. After quality check the number of reads per sample was decreased from 11661±2758 to 7762±4585 and mean read length per sample was 256±2 base pairs (bp).

Relative abundance detected by 454 pyrosequencing revealed the predominance of three major phyla in all samples tested (Figure S1A). The fecal inoculum was predominated by the phylum Firmicutes whereas the Bacteroidetes phylum was most abundant in reactor effluents, followed by the Firmicutes and Proteobacteria phyla, except for samples 19/20 in IR and 30/31 in CR where the two phyla Bacteroidetes and Firmicutes were almost equally abundant. The phylum Proteobacteria was increasing from less than 1% relative abundance in the fecal inoculum to up to 29% in sample 52/53 from CR. At family level (Figure 3), the highest abundance in IR was recorded for *Prevotellaceae* (33–66%), *Lachnospiraceae* (7–17%), *Ruminococcaceae* (5–14%) and *Enterobacteriaceae* (4–8%), with unclassified reads accounting for 10–21% of total reads. The same pattern of relative abundances on family level was observed in CR, with *Prevotellaceae* (38–51%), *Lachnospiraceae* (13–18%), *Ruminococcaceae* (5–13%) and *Enterobacteriaceae* (5–9%) being the most abundant families in all samples and unclassified bacteria accounting for 10–20% of the reads. The detected families were composed of the predominant genera *Prevotella, Escherichia/Shigella, Ruminococcus, Roseburia, Blautia, Bacteroides* and *Oscillibacter* (Figure S1B). Remarkably, the family *Succinivibrionaceae* increased from 0.1% on day 19/20 in IR to up to 8% at the end of the fermentation (52/53). In CR the family *Succinivibrionaceae* even accounted for as much as 21% of total reads on day 52/53. In contrast, reads assigned to the family *Ruminococcaceae* decreased in CR from 10% on day 42/43 to 4% on day 52/53. This observation was confirmed by qPCR with specific primers targeting the 16S rRNA gene of *S. dextrinosolvens*. Similarly, the 16S rRNA gene copy numbers for the *S. dextrinosolvens* group in CR significantly increased (P<0.001) from 7.47±0.2 $\log_{10}$ copies mL^{-1} (mean ± SD; d11-37) to 9.2±0.3 $\log_{10}$ copies mL^{-1} (mean ± SD; d38-54).

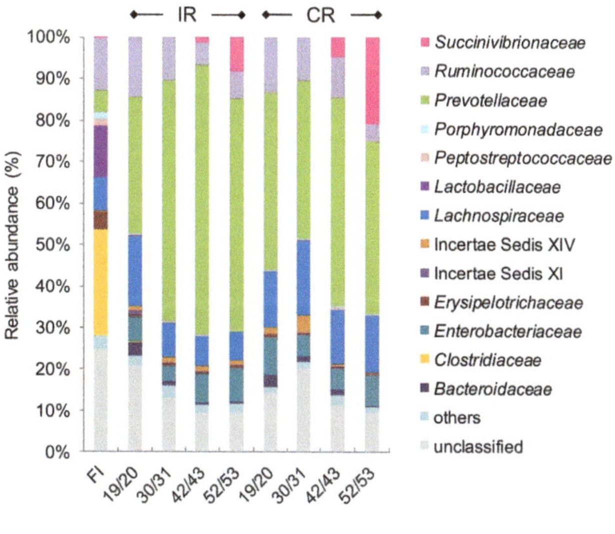

Figure 3. Microbial composition in the fecal inoculum (FI), IR and CR measured by 454 pyrosequencing. The relative abundance on family level is shown. Values <1% are summarized in the group "others".

In the fecal inoculum, the family *Clostridiaceae* (26%) was most abundant and predominantly represented by the genus *Clostridium*, followed by *Ruminococcaceae* and *Lactobacillaceae* (both 13%) and *Lachnospiraceae* (8%). The unclassified bacteria accounted for 25% of total reads. *Enterobacteriaceae* were not detected in the fecal inoculum.

Reproducibility of microbial composition and activity

The novel PolyFermS model was designed to allow reproducible testing of different treatments in parallel test reactors compared to a control reactor inoculated with the same microbiota. Consecutive multiple treatment periods could be tested by subjecting the test reactors to a washing procedure using 10% chlorine and followed by a stabilization period to regenerate comparable microbiota in CR and all test reactors (Figure 1). Therefore, to test reproducibility of control and test reactors after a washing procedure, bacterial composition (qPCR data) and activity (HPLC data) in TR1-4 after three day re-stabilization were compared with data measured in CR. No significant differences of bacterial composition of CR and TR1-4 (P<0.05) were detected using the Mann-Whitney U test (Table S2). Moreover the analysis of effluent samples of CR and TR1-4 by 454 pyrosequencing on day 25 (last day of 3rd stabilization period) showed similar microbiota

Table 3. Mean concentration ($\log_{10}$ copy numbers mL^{-1} effluent) of specific bacterial groups measured by qPCR in the fecal inoculum and effluent samples from inoculum reactor (IR) and control reactor (CR) during the stabilized period (d11-54).

	fecal inoculum	IR	CR
total 16S rRNA gene	12.2	10.6±0.2	10.6±0.2
Bifidobacterium spp.	7.8	6.6±0.7	6.3±0.5
Bacteroides-Prevotella group	11.3	10.2±0.2	10.2±0.2
Enterobacteriaceae	8.1	9.9±0.2	9.8±0.2
Lactobacillus/Pediococcus/Leuconostoc spp.	11.2	8.3±0.3	8.0±0.3
Clostridium Cluster IV	10.7	9.9±0.2	9.9±0.2

Mean values ± SD for IR and CR were calculated from daily values during the experimental stabilized period corresponding to d11-54.

composition (Figure S2). Metabolite concentrations in TR1-4 progressively re-established at similar levels to that in CR, reaching comparable values three days after restarting the system (Figures 4 and S3). High and stable bacterial concentrations were measured already after one day re-stabilization as indicated by total 16S rRNA gene copy numbers in TR1-4 over three days stabilization periods.

Discussion

The large field of research to evaluate the impact of feed ingredients and additives on the gastrointestinal health and development of pigs [28] emphasizes the need for potent porcine *in vitro* fermentation models to study feed-related impacts on gut microbial community and functionality in a fast and reproducible setting.

In this study we tested a novel porcine *in vitro* fermentation model for the proximal colon, advanced from the PolyFermS model presented by Zihler Berner *et al.* [16]. The porcine *in vitro* model was designed to generate self-contained parallel fermentations in each of the reactors, simulating the proximal colon, which is the primary site of fermentation in the gastrointestinal tract [29]. The first PolyFermS model validated by Zihler Berner *et al.* [16] included an inoculum reactor operated under conditions of the first section of the proximal colon which was used to continuously feed control and test reactors with fermented effluent, operated with conditions selected to mimic the distal section of the proximal colon. Therefore in this model a combination of two stages, IR and CR or TR, was used to model the proximal colon. In contrast, test reactors in the porcine PolyFermS tested in this study were operated in conditions of the proximal colon and constantly inoculated with only 10% fecal microbiota produced in IR, while 90% of the feed was fresh medium to simulate the dynamic process of chyme inflow. This design is likely more suitable for studying the fate of dietary treatments on the fermentative capacity of the gut microbial community, since the complete response of the proximal colon microbiota, constantly supplied with fresh substrates from the small intestine, is of interest. In our study, the combination of the reactor set-up and the nutritive medium, adapted to simulate swine chyme, allowed the establishment of six self-contained parallel fermentations for the porcine proximal colon with a high

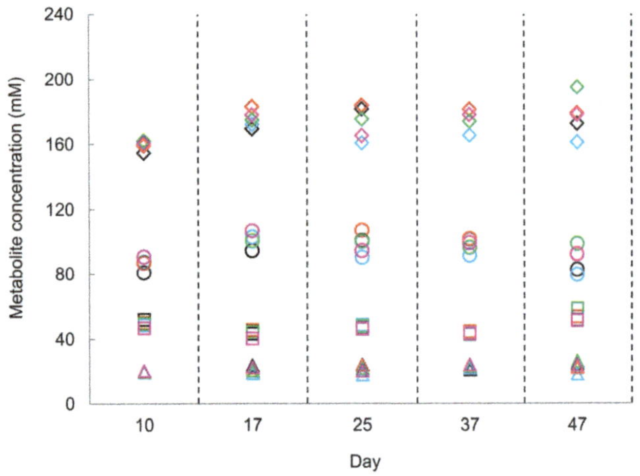

Figure 4. Main metabolites in CR and TR1-4 on the last day of each stabilization period. (◇) total SCFA; (○) acetate; (□) propionate; (Δ) butyrate; (black) CR; (blue) TR1; (red) TR2; (green) TR3; (pink) TR4.

metabolic and compositional stability, diversity and reproducibility throughout 54 days of fermentation.

The stable metabolite concentrations obtained during the continuous fermentation, indicate balanced microbial growth and the maintenance of gut microbiota functional capacity. In addition, metabolite ratios for IR and CR (61:27:12 and 57:29:14, acetate:propionate:butyrate) after initial stabilization were very similar and also in agreement with values reported for the pig proximal colon (60:25:15 [30]) and for fecal SCFA ratios (63:25:12 [31]) *in vivo*. Small but significant time effects were recorded for all detected metabolites, except total SCFA concentrations that remained constant during the entire stabilized fermentation. The ability to detect even small time changes is directly related to the sensitivity of the analysis, permitted by continuous operation of a stabilized fermentation system and achieved with a high number of time points analyzed [32].

The slight but significant increase of propionate concentrations during continuous fermentation may directly be related to the observed increase of the family *Succinivibrionaceae*, which was already detected in the fecal inoculum but at a low relative abundance of 0.1% by 454 pyrosequencing. *Succinivibrionaceae* belong to the γ-subclass of the phylum Proteobacteria and play an important role in starch digestion in sheep and cattle [33]. While Kim *et al.* [34] assigned the genus *Succinivibrio* to the group of the less abundant genera in the swine fecal microbiome, other studies in contrast have grouped the genus as a member of the core microbiota of the porcine proximal colon [35] or porcine cecum [36]. Different carbohydrate sources can be metabolized by *Succinivibrionaceae* resulting in the main fermentation products acetate and succinate [33]. Further decarboxylation of succinate can lead to propionate [37], likely due to cross-feeding reactions in the complex intestinal environment.

Microbial composition and diversity determined in the model effluents by qPCR and 454 pyrosequencing showed no major changes in the bacterial groups between days 11 and 54 in both, IR and CR, after the initial stabilization period of 10 days. The establishment of a microbial pseudo steady-state is an important factor to gain reliable data on the modulating potential of a specific treatment in order to avoid false positive conclusions related to the microbiota adaptation to *in vitro* conditions [38]. Compared to the fecal inoculum all bacterial groups targeted by qPCR were reduced in reactor effluents from IR and CR, except for *Enterobacteriaceae* that displayed higher copy numbers. Changes in microbiota composition and diversity may reflect the transfer from *in vivo* (feces) to *in vitro* (proximal colon) conditions, the adaptation to a new environment, which depends on the conditions in the host during collection of the fecal sample, and the lack of host effects *in vitro* [16,39]. Additionally, the shift in microbial composition may favor more robust species due to a competitive advantage in adaptation and may open niches, possibly occupied by *Enterobacteriaceae*, which can explain their increase. Using 454 pyrosequencing, *Clostridiaceae* exhibited the most remarkable decrease in relative abundance between the fecal inoculum and reactor effluents, whereas *Prevotellaceae* increased markedly in reactor effluents. The high prevalence of *Prevotellaceae* was directly linked to an increased Bacteroidetes and decreased Firmicutes ratio in reactor effluents compared to the fecal inoculum. The high occurrence of Bacteroidetes in intestinal *in vitro* models, has already been reported previously in the Twin-SHIME [39] and TNO intestinal model [40] and may be the result of the higher micromolar levels of oxygen and the less adhesive capacity of Bacteroidetes.

Studies on the swine fecal and cecal microbiome reported Firmicutes, Bacteroidetes and Proteobacteria as the most prevalent phyla in swine [1,34,36,41–43], which is in accordance with our study. Firmicutes and Bacteroidetes represented between 80–90% of assigned reads in all effluent samples, except for CR on day 52/53, when Proteobacteria increased to 29% and Firmicutes/Bacteroidetes decreased to 70%, respectively. On family level, *Prevotellaceae* was the most abundant family detected in all reactor samples, which is in accordance with other studies [1,35,36,41,44] and supported by our qPCR data. On genus level the predominant genera in the porcine PolyFermS (mean relative abundance in IR and CR >1%) were *Prevotella, Escherichia/Shigella, Roseburia, Ruminococcus, Oscillibacter, Succinivibrio, Blautia* and *Bacteroides*, genera that have previously been described as members of the cecal [36,43,45] and fecal [1,44] gut microbiota. We thus conclude that the microbial composition is representative for the porcine proximal colon, as demonstrated by the high accordance of bacterial genera between our *in vitro* study and previous *in vivo* studies on the cecal porcine microbiota. Furthermore, a dominant fraction of glycan degraders was identified, consisting of the predominant genera *Prevotella, Ruminococcus* and *Roseburia*. The high prevalence of fibrolytic bacteria and their associated carbohydrate utilization systems in swine has already been reported previously [1,46] and is possibly related to the high amount of complex polysaccharides found in the pig diet. Starch degraders play a key role in gut microbial ecosystems as they represent the top of the trophic chain by providing simple sugars from the breakdown of complex carbohydrates, thus directly rendering them accessible to other members of the microbial community [47].

Assessing the reproducibility of *in vitro* gut fermentation models is a permanent challenge [4] and is difficult to achieve with classical models. In this study, we used a constant 10% inoculation rate from IR to the subsequent reactors, which allowed reproducing similar and parallel evolving self-contained ecosystems in IR and the second-stage reactors. Due to the consecutive treatment periods with in-between washing of the test reactors, a fast re-establishment of reproducible environment in the test reactors was required. This is of particular importance for studying and more importantly comparing treatment-related responses of the microbial community *in vitro*, which presumes comparable experimental conditions in the different reactors. While the bacterial groups in TR1-4 targeted by qPCR reached similar numbers to that of CR after only one day of re-stabilization, complete metabolic activity was recovered after a longer time of approximately three days. Such delay of functionality response has also been reported with the Twine-SHIME model [39]. The short re-stabilization period of PolyFermS compared to 5–8 days for Twin-SHIME points on the benefits of immobilized fecal microbiota to provide bacterial stability and diversity [4].

To conclude, in the present study we validated a novel PolyFermS continuous intestinal fermentation model of the swine proximal colon, inoculated with immobilized fecal microbiota. This model operated with a nutritive medium designed to mimic pig chyme allowed to stably reproduce the microbiota and metabolic activity of swine proximal colon for at least 54 days. The particular model set-up allows comparing different treatments and a control, run with the same inoculated microbiota, simultaneously. Furthermore, our data demonstrate a considerable interplay between functionality and taxonomic composition and highlight the stringent potential of the model for compositional as well as functionality related studies. This *in vitro* gut model can further be expanded to simulate multiple stages of the large intestine (proximal, transverse, distal) and a number of consecutive treatment periods. It should be particularly suitable to accurately investigate the effects of dietary factors such as pro-and prebiotics, as well as environmental parameters or drugs on the porcine gut microbiota in highly controlled settings.

Supporting Information

Figure S1 Microbial composition in the fecal inoculum (FI), IR and CR measured by 454 pyrosequencing on (A) phylum level and (B) genus level. Values <1% are summarized in the group "others".

Figure S2 Microbial composition on family level in CR and TR1-4 on day 25 (last day of 3rd stabilization period) measured by 454 pyrosequencing. Values <1% are summarized in the group "others".

Figure S3 Mean main metabolite concentrations and total 16S rRNA gene copy numbers from TR1-4 during the last three days of each stabilization period. Data are depicted as mean values ± SD from TR1-4. (♦) total 16S rRNA gene copies mL^{-1} effluent; (○) acetate; (□) propionate; (Δ) butyrate.

Table S1 Composition of the nutritive medium simulating the swine ileal chyme.

Table S2 16S rRNA gene copy numbers (log_{10} mL^{-1} of effluent) of specific bacterial groups measured by qPCR in effluent samples from CR and TR1-4.

Acknowledgments

We kindly thank the Genetic Diversity Center (GDC), ETH Zurich, for providing equipment for quantitative real-time PCR analysis and Eun-Hee Doo, Lukas Meile and Simon Galenda for technical assistance.

Author Contributions

Conceived and designed the experiments: SAT AZB FG CC CL. Performed the experiments: SAT AZB ER. Analyzed the data: SAT AZB ER CC CL. Wrote the paper: SAT AZB CC CL.

References

1. Lamendella R, Domingo JW, Ghosh S, Martinson J, Oerther DB (2011) Comparative fecal metagenomics unveils unique functional capacity of the swine gut. BMC Microbiol 11: 103.
2. Macfarlane GT, Macfarlane S (2007) Models for intestinal fermentation: association between food components, delivery systems, bioavailability and functional interactions in the gut. Curr Opin Biotechnol 18: 156–162.
3. Russell WR, Hoyles L, Flint HJ, Dumas ME (2013) Colonic bacterial metabolites and human health. Curr Opin Microbiol 16: 246–254.
4. Payne AN, Zihler A, Chassard C, Lacroix C (2012) Advances and perspectives in *in vitro* human gut fermentation modeling. Trends Biotechnol 30: 17–25.
5. Jha R, Bindelle J, Van Kessel A, Leterme P (2011) *In vitro* fibre fermentation of feed ingredients with varying fermentable carbohydrate and protein levels and protein synthesis by colonic bacteria isolated from pigs. Anim Feed Sci Technol 165: 191–200.
6. Jonathan MC, van den Borne JJGC, van Wiechen P, da Silva CS, Schols HA, et al. (2012) *In vitro* fermentation of 12 dietary fibres by faecal inoculum from pigs and humans. Food Chem 133: 889–897.
7. Lin B, Gong J, Wang Q, Cui S, Yu H, et al. (2011) *In-vitro* assessment of the effects of dietary fibers on microbial fermentation and communities from large intestinal digesta of pigs. Food Hydrocolloid 25: 180–188.

8. Zhu WY, Williams BA, Konstantinov SR, Tamminga S, De Vos WM, et al. (2003) Analysis of 16S rDNA reveals bacterial shift during *in vitro* fermentation of fermentable carbohydrate using piglet faeces as inoculum. Anaerobe 9: 175–180.
9. Williams BA, Bosch MW, Awati A, Konstantinov SR, Smidt H, et al. (2005) In vitro assessment of gastrointestinal tract (GIT) fermentation in pigs: Fermentable substrates and microbial activity. Anim Res 54: 191–201.
10. Ricca DM, Ziemer CJ, Kerr BJ (2010) Changes in bacterial communities from swine feces during continuous culture with starch. Anaerobe 16: 516–521.
11. Messens W, Goris J, Dierick N, Herman L, Heyndrickx M (2010) Inhibition of *Salmonella* typhimurium by medium-chain fatty acids in an *in vitro* simulation of the porcine cecum. Vet Microbiol 141: 73–80.
12. Pinloche E, Williams M, D'Inca R, Auclair E, Newbold CJ (2012) Use of a colon simulation technique to assess the effect of live yeast on fermentation parameters and microbiota of the colon of pig. J Anim Sci 90: 353–355.
13. Cinquin C, Le Blay G, Fliss I, Lacroix C (2004) Immobilization of infant fecal microbiota and utilization in an *in vitro* colonic fermentation model. Microb Ecol 48: 128–138.
14. Cinquin C, Le Blay G, Fliss I, Lacroix C (2006) New three-stage *in vitro* model for infant colonic fermentation with immobilized fecal microbiota. FEMS Microbiol Ecol 57: 324–336.
15. Le Blay G, Chassard C, Baltzer S, Lacroix C (2010) Set up of a new *in vitro* model to study dietary fructans fermentation in formula-fed babies. Br J Nutr 103: 403–411.
16. Zihler Berner A, Fuentes S, Dostal A, Payne AN, Vazquez Gutierrez P, et al. (2013) Novel Polyfermentor Intestinal Model (PolyFermS) for controlled ecological studies: validation and effect of pH. PLoS One 8: e77772.
17. Macfarlane GT, Macfarlane S, Gibson GR (1998) Validation of a three-stage compound continuous culture system for investigating the effect of retention time on the ecology and metabolism of bacteria in the human colon. Microb Ecol 35: 180–187.
18. Nyannor EK, Williams P, Bedford MR, Adeola O (2007) Corn expressing an *Escherichia coli*-derived phytase gene: a proof-of-concept nutritional study in pigs. J Anim Sci 85: 1946–1952.
19. Lee TT, Huang YF, Chiang CC, Chung TK, Chiou PW, et al. (2011) Starch characteristics and their influences on *in vitro* and pig prececal starch digestion. J Agric Food Chem 59: 7353–7359.
20. Wang JP, Hong SM, Yan L, Cho JH, Lee HS, et al. (2011) The evaluation of soybean meals from 3 major soybean-producing countries on productive performance and feeding value of pig diets. J Anim Sci 89: 2768–2773.
21. Kararli TT (1995) Comparison of the gastrointestinal anatomy, physiology, and biochemistry of humans and commonly used laboratory animals. Biopharm Drug Dispos 16: 351–380.
22. Michel C, Kravtchenko TP, David A, Gueneau S, Kozlowski F, et al. (1998) *In vitro* prebiotic effects of *Acacia* gums onto the human intestinal microbiota depends on both botanical origin and environmental pH. Anaerobe 4: 257–266.
23. Sakata T, Kojima T, Fujieda M, Takahashi M, Michibata T (2003) Influences of probiotic bacteria on organic acid production by pig caecal bacteria *in vitro*. Proc Nutr Soc 62: 73–80.
24. Bindelle J, Buldgen A, Boudry C, Leterme P (2007) Effect of inoculum and pepsin-pancreatin hydrolysis on fibre fermentation mecasured by the gas production technique in pigs. Anim Feed Sci Technol 132: 111–122.
25. von Heimendahl E, Breves G, Abel HJ (2010) Fiber-related digestive processes in three different breeds of pigs. J Anim Sci 88: 972–981.
26. Dostal A, Fehlbaum S, Chassard C, Zimmermann MB, Lacroix C (2013) Low iron availability in continuous *in vitro* colonic fermentations induces strong dysbiosis of the child gut microbial consortium and a decrease in main metabolites. FEMS Microbiol Ecol 83: 161–175.
27. Jost T, Lacroix C, Braegger CP, Chassard C (2012) New insights in gut microbiota establishment in healthy breast fed neonates. PLoS One 7: e44595.
28. Pluske JR (2013) Feed- and feed additives-related aspects of gut health and development in weanling pigs. J Anim Sci Biotechnol 4: 1.
29. Guarner F, Malagelada JR (2003) Gut flora in health and disease. Lancet 361: 512–519.
30. Grieshop CM, Reese DE, Fahey Jr GC (2000) Nonstarch polysaccharides and oligosaccharides in swine nutrition. In: Lewis AJ and Southern LL, editors. Swine Nutrition. Boca Raton, FL: CRC Press. pp. 107–130.
31. Bird AR, Vuaran M, Crittenden R, Hayakawa T, Playne MJ, et al. (2009) Comparative effects of a high-amylose starch and a fructooligosaccharide on fecal bifidobacteria numbers and short-chain fatty acids in pigs fed *Bifidobacterium animalis*. Dig Dis Sci 54: 947–954.
32. Mozzetti V, Grattepanche F, Moine D, Berger B, Rezzonico E, et al. (2012) Transcriptome analysis and physiology of *Bifidobacterium longum* NCC2705 cells under continuous culture conditions. Benef Microbes 3: 261–272.
33. Hippe H, Hagelstein A, Kramer I, Swiderski J, Stackebrandt E (1999) Phylogenetic analysis of Formivibrio citricus, Propionivibrio dicarboxylicus, Anaerobiospirillum thomasii, Succinimonas amylolytica and Succinivibrio dextrinosolvens and proposal of Succinivibrionaceae fam. nov. Int J Syst Bacteriol 49 Pt 2: 779–782.
34. Kim HB, Borewicz K, White BA, Singer RS, Sreevatsan S, et al. (2012) Microbial shifts in the swine distal gut in response to the treatment with antimicrobial growth promoter, tylosin. Proc Natl Acad Sci U S A 109: 15485–15490.
35. Li RW, Wu S, Li W, Navarro K, Couch RD, et al. (2012) Alterations in the porcine colon microbiota induced by the gastrointestinal nematode *Trichuris suis*. Infect Immun 80: 2150–2157.
36. Buzoianu SG, Walsh MC, Rea MC, O'Sullivan O, Cotter PD, et al. (2012) High-throughput sequence-based analysis of the intestinal microbiota of weanling pigs fed genetically modified MON810 maize expressing *Bacillus thuringiensis* Cry1Ab (Bt maize) for 31 days. Appl Environ Microbiol 78: 4217–4224.
37. Hosseini E, Grootaert C, Verstraete W, Van de Wiele T (2011) Propionate as a health-promoting microbial metabolite in the human gut. Nutr Rev 69: 245–258.
38. Possemiers S, Verthe K, Uyttendaele S, Verstraete W (2004) PCR-DGGE-based quantification of stability of the microbial community in a simulator of the human intestinal microbial ecosystem. FEMS Microbiol Ecol 49: 495–507.
39. Van den Abbeele P, Grootaert C, Marzorati M, Possemiers S, Verstraete W, et al. (2010) Microbial community development in a dynamic gut model is reproducible, colon region specific, and selective for *Bacteroidetes* and *Clostridium* cluster IX. Appl Environ Microbiol 76: 5237–5246.
40. Rajilic-Stojanovic M, Maathuis A, Heilig HG, Venema K, de Vos WM, et al. (2010) Evaluating the microbial diversity of an *in vitro* model of the human large intestine by phylogenetic microarray analysis. Microbiology 156: 3270–3281.
41. Looft T, Johnson TA, Allen HK, Bayles DO, Alt DP, et al. (2012) In-feed antibiotic effects on the swine intestinal microbiome. Proc Natl Acad Sci U S A 109: 1691–1696.
42. Riboulet-Bisson E, Sturme MH, Jeffery IB, O'Donnell MM, Neville BA, et al. (2012) Effect of *Lactobacillus salivarius* bacteriocin Abp118 on the mouse and pig intestinal microbiota. PLoS One 7: e31113.
43. Poroyko V, White JR, Wang M, Donovan S, Alverdy J, et al. (2010) Gut microbial gene expression in mother-fed and formula-fed piglets. PLoS One 5: e12459.
44. Kim HB, Borewicz K, White BA, Singer RS, Sreevatsan S, et al. (2011) Longitudinal investigation of the age-related bacterial diversity in the feces of commercial pigs. Vet Microbiol 153: 124–133.
45. Haenen D, Zhang J, da Silva CS, Bosch G, van der Meer IM, et al. (2013) A diet high in resistant starch modulates microbiota composition, SCFA concentrations, and gene expression in pig intestine. J Nutr 143: 274–283.
46. Leser TD, Amenuvor JZ, Jensen TK, Lindecrona RH, Boye M, et al. (2002) Culture-independent analysis of gut bacteria: the pig gastrointestinal tract microbiota revisited. Appl Environ Microbiol 68: 673–690.
47. Chassard C, Lacroix C (2013) Carbohydrates and the human gut microbiota. Curr Opin Clin Nutr Metab Care 16: 453–460.
48. Guo X, Xia X, Tang R, Zhou J, Zhao H, et al. (2008) Development of a real-time PCR method for *Firmicutes* and *Bacteroidetes* in faeces and its application to quantify intestinal population of obese and lean pigs. Lett Appl Microbiol 47: 367–373.
49. Ramirez-Farias C, Slezak K, Fuller Z, Duncan A, Holtrop G, et al. (2009) Effect of inulin on the human gut microbiota: stimulation of *Bifidobacterium adolescentis* and *Faecalibacterium prausnitzii*. Br J Nutr 101: 541–550.
50. Furet JP, Firmesse O, Gourmelon M, Bridonneau C, Tap J, et al. (2009) Comparative assessment of human and farm animal faecal microbiota using real-time quantitative PCR. FEMS Microbiol Ecol 68: 351–362.
51. Bartosch S, Fite A, Macfarlane GT, McMurdo ME (2004) Characterization of bacterial communities in feces from healthy elderly volunteers and hospitalized elderly patients by using real-time PCR and effects of antibiotic treatment on the fecal microbiota. Appl Environ Microbiol 70: 3575–3581.
52. Cleusix V, Lacroix C, Dasen G, Leo M, Le Blay G (2010) Comparative study of a new quantitative real-time PCR targeting the xylulose-5-phosphate/fructose-6-phosphate phosphoketolase bifidobacterial gene (*xfp*) in faecal samples with two fluorescence *in situ* hybridization methods. J Appl Microbiol 108: 181–193.
53. Stevenson DM, Weimer PJ (2007) Dominance of *Prevotella* and low abundance of classical ruminal bacterial species in the bovine rumen revealed by relative quantification real-time PCR. Appl Microbiol Biotechnol 75: 165–174.

Microbial Community Structure of Relict Niter-Beds Previously Used for Saltpeter Production

Takashi Narihiro[1]*, Hideyuki Tamaki[1], Aya Akiba[1], Kazuto Takasaki[2], Koichiro Nakano[2], Yoichi Kamagata[1,3], Satoshi Hanada[1], Taizo Maji[4]

1 Bioproduction Research Institute, National Institute of Advanced Industrial Science and Technology (AIST), Tsukuba, Ibaraki, Japan, **2** FASMAC Co., Ltd., Atsugi, Kanagawa, Japan, **3** Bioproduction Research Institute, National Institute of Advanced Industrial Science and Technology (AIST), Sapporo, Hokkaido, Japan, **4** Shubun University, Ichinomiya, Aichi, Japan

Abstract

From the 16th to the 18th centuries in Japan, saltpeter was produced using a biological niter-bed process and was formed under the floor of gassho-style houses in the historic villages of Shirakawa-go and Gokayama, which are classified as United Nations Educational, Scientific and Cultural Organization (UNESCO) World Heritage Sites. The relict niter-beds are now conserved in the underfloor space of gassho-style houses, where they are isolated from destabilizing environmental factors and retain the ability to produce nitrate. However, little is known about the nitrifying microbes in such relict niter-bed ecosystems. In this study, the microbial community structures within nine relict niter-bed soils were investigated using 454 pyrotag analysis targeting the 16S rRNA gene and the bacterial and archaeal ammonia monooxygenase gene (*amoA*). The 16S rRNA gene pyrotag analysis showed that members of the phyla *Proteobacteria*, *Actinobacteria*, *Bacteroidetes*, *Chloroflexi*, *Firmicutes*, *Gemmatimonadetes*, and *Planctomycetes* were major microbial constituents, and principal coordinate analysis showed that the NO_3^-, Cl^-, K^+, and Na^+ contents were potential determinants of the structures of entire microbial communities in relict niter-bed soils. The bacterial and archaeal *amoA* libraries indicated that members of the *Nitrosospira*-type ammonia-oxidizing bacteria (AOB) and "*Ca.* Nitrososphaera"-type ammonia-oxidizing archaea (AOA), respectively, predominated in relict niter-bed soils. In addition, soil pH and organic carbon content were important factors for the ecological niche of AOB and AOA in relict niter-bed soil ecosystems.

Editor: A. Mark Ibekwe, U. S. Salinity Lab, United States of America

Funding: This research was supported by a Grant-in-Aid for Young Scientists (B) (Grant Number 23760512) from JSPS KAKENHI. http://www.jsps.go.jp/english/e-grants/. The funders had no role in study design, data collection and analysis, decision to publish, or preparation of the manuscript.

Competing Interests: TN, HT, AA, YK, SH and TM have neither financial nor occupational affiliations with FASMAC Co., Ltd. at the time of the study. KT and KN were employees of FASMAC Co., Ltd. at the time of the study. There are no patents, products in development or marketed products to declare.

* Email: t.narihiro@aist.go.jp

Introduction

Until the early 20th century, saltpeter (i.e., niter or potassium nitrate [KNO_3]), which is known as the main source of gunpowder, was produced via an artificial bioprocess, the so-called "niter-bed" [1,2]. The basic concept of saltpeter production via niter-bed involves the nitrification of ammonia-containing wastes (e.g., plant residue, horse manure, and human urine) mixed with a certain amount of soil [1,2]. This nitrification reaction was thought to be performed by ammonia-oxidizing and nitrite-oxidizing microbes [1,2]. The first description of a niter-bed was found in the military technological manual "Bellifortis," which was written by Konrad Kyeser and published in 1405. This manual contains brief instructions describing how to construct a niter-bed [3,4], and this process was employed for saltpeter production in European countries [1,5,6], India [7,8], and Japan [9,10] until the discovery of a large natural deposit in Chile [11]. In Finland, the export of saltpeter produced by niter-bed reached a maximum level (433 tons) in 1816 during the last stages of the Napoleonic wars [6]. In India, a large amount of saltpeter was produced annually (20,000 tons/year) using the niter-bed methodology in the early 20th century [7]. Thus, the niter-bed process is a historically important industry in several nations.

From the 16th to the 18th centuries in Japan, saltpeter was produced for military purposes using the niter-bed formed under the floor of gassho-style houses [10], which is a specific architectural style used in the historic villages of Shirakawa-go and Gokayama, which are listed as UNESCO World Heritage Sites [12]. Because these villages were located in the mountainous area of the former Kaga domain in Japan (currently the Toyama and Ishikawa prefectures), there was limited land for the production of agricultural crops. Hence, people in this area faced economic difficulties. To overcome this issue, the people practiced sericulture to produce silk as an alternative income source in the loft spaces of gassho-style houses [10]. Moreover, according to the historical records, the people produced saltpeter in the underfloor space of gassho-style houses using ammonia-containing wastes such as plant residues (e.g., the stems of Japanese millet, buckwheat, and tobacco), silkworm feces from sericulture, and human urine as primary substrates [10]. These substrates were mixed with soil under the floor of the gassho-style house to create a niter-bed, and nitrate (NO_3) was produced via a microbial nitrification reaction. The resulting nitrate contained in the niter-bed soil was further extracted with water and purified by adding wood ash to remove unnecessary minerals (e.g., Ca, Mg, K, and Na). Purified saltpeter was used as the raw material for gunpowder production, and this gunpowder also became an important source of income for people who lived in this area [9,10]. Thus, saltpeter production via the niter-bed process was a cornerstone of the lifestyle in this area.

Although saltpeter production in the villages of Shirakawa-go and Gokayama ended more than 100 years ago, the relict niter-beds are now conserved in the underfloor space of gassho-style houses, where they are isolated from rainfall and other destabilizing factors [10]. Surprisingly, bulk soils collected from relict niter-beds in Shirakawa-go village retained nitrification activity [13]. This finding implies that nitrifying microbes have persisted in the relict niter-bed soils; however, there is no information available regarding the microbial community structure responsible for nitrification in the relict niter-bed ecosystem. In this study, we investigated the microbial community structure of nine soil samples obtained from the underfloor areas of three gassho-style houses by using 454 pyrotag analysis targeting the 16S rRNA gene and the ammonium monooxygenase subunit A gene (*amoA*) of ammonia-oxidizing bacteria (AOB) and ammonia-oxidizing archaea (AOA). In addition, a previously reported *amoA* pyrotag dataset [14] was used to elucidate the ecological niche of AOB and AOA in soil environments.

Materials and Methods

Soil sample

A total of nine soil samples were collected from three relict niter-beds under the floors of gassho-style houses located in Shirakawa Village, referred to as OVA, OVB, and OVC. We collected five, two, and two soil samples from the OVA, OVB, and OVC houses; these samples are referred to as OVA1–5, OVB2 and OVB3, and OVC1 and OVC2, respectively. Each 500–1,000 g of soil sample was collected from relict niter-beds (for OVA and OVC, a depth of ca. 10 cm from the surface; for OVB, a depth of ca. 20 cm from the surface) by permission from the education board of Shirakawa-Village, Gifu, Japan. Detailed physicochemical parameters, including pH, organic carbon content, and ion contents, are summarized in Table S1 [13,15–17]. Pebbles were removed from the samples using a 30-mesh filter, and the samples were subjected to DNA extraction.

DNA extraction, PCR, and pyrosequencing

DNA was extracted from the soil samples using the FastDNA SPIN Kit for Soil (MP Biomedicals, Carlsbad, CA, USA) according to the manufacturer's instructions. Approximately 50 ng of template DNA was used for PCR amplification in a reaction volume of 50 μL. The 16S rRNA, bacterial *amoA*, and archaeal *amoA* genes were amplified with the forward/reverse primer sets Univ519F/Univ926R [18], amoA-1F/amoA-2R [19], and Arch-amoAF/Arch-amoAR [20], respectively. The forward and reverse primers were barcoded with fusion A (FA; 5′-CGTATCGCCTCCCTCGCGCCATCAG-3′) and fusion B (FB; 5′-CTATGCGCCTTGCCAGCCCGCTCAG-3′) adaptor sequences, respectively. The primers also included Multiplex Identifier (MID) sequences. Detailed information regarding the primer sequences is provided in Table S2. These genes were amplified with PrimeSTAR HS DNA Polymerase (TaKaRa, Otsu, Japan) using the following thermocycling conditions: the appro-

Table 1. Pyrosequencing results of 16S rRNA, bacterial *amoA*, and archaeal *amoA* genes and observed diversity indexes.

Sample ID	16S rRNA						Bacterial *amoA*						Archaeal *amoA*					
	Total reads	OTU	Chao 1	Shannon	Simpson	Coverage (%)	Total reads	OTU	Chao 1	Shannon	Simpson	Coverage (%)	Total reads	OTU	Chao 1	Shannon	Simpson	Coverage (%)
OVA1	7,770	514	757.8	4.57	37.42	97.52	N.D.[a]						4,240	6	7.00	1.16	1.06	99.95
OVA2	17,480	641	1,042.8	4.63	47.06	98.59	N.D.[a]						3,455	9	10.00	1.70	1.34	99.91
OVA3	16,615	990	1,485.7	4.92	33.83	97.80	7,826	6	6.00	0.07	1.02	99.99	3,872	6	7.00	2.18	1.83	99.95
OVA4	19,266	1,496	2,248.9	5.60	84.90	96.99	10,197	5	5.00	0.15	1.07	99.99	2,705	9	10.50	2.37	1.87	99.89
OVA5	18,029	1,360	2,124.7	5.50	77.00	97.06	14,558	6	6.00	0.22	1.10	99.99	3,670	8	8.33	2.02	1.72	99.95
OVB2	19,304	1,020	1,411.7	5.06	44.17	98.23	4,606	6	6.00	0.32	1.19	99.98	2,957	4	4.00	2.02	1.96	100.00
OVB3	22,764	1,028	1,538.0	4.99	36.75	98.42	6,896	9	10.50	0.32	1.17	99.96	3,695	8	8.33	2.13	1.94	99.95
OVC1	23,172	1,144	1,816.1	4.56	17.29	98.09	8,016	3	3.00	0.40	1.23	100.00	3,550	10	10.00	2.13	1.55	99.97
OVC2	28,616	1,238	1,824.5	4.66	20.59	98.42	7,305	5	5.00	0.75	1.89	100.00	4,943	12	18.00	1.85	1.48	99.92
Total	173,016	3,146					59,404	13					33,087	16				

[a]N.D., not detected.

priate number of cycles (25 cycles for 16S rRNA, 40 cycles for bacterial *amoA*, and 35 cycles for archaeal *amoA* genes) of denaturation at 98°C for 10 sec, annealing at 55°C for 15 sec, and extension at 72°C for 1 min. PCR amplicons were purified using a Wizard SV Gel and PCR Clean-Up System (Promega, Fitchburg, WI, USA) and pooled for subsequent 454 pyrotag analysis. Pyrosequencing was performed using the GS-FLX Titanium platform (Roche/454 Life Sciences, Branford, CT, USA) at the FASMAC Co., Ltd. (Atugi, Japan). Sequencing was performed using 1/2-region gaskets. For the 16S rRNA gene, we used pyrotag reads obtained from both fusion sides (A and B) because the 454 platform used in this study generated full-length amplicons (approximately 370 bp). For bacterial and archaeal *amoA* genes, we used pyrotag reads obtained from only the fusion A side because the length of each sequenced read did not cover the entire region of the amplicon.

Operational taxonomic unit (OTU) analysis for 16S rRNA gene pyrotags

Raw 16S rRNA gene pyrotag sequences were qualitatively screened and trimmed with QIIME 1.7.0 [21], maintaining the sequence length at ≥300 nt. The sequences had an average quality score of ≥25. A denoising function was applied to the filtered sequence data [22]. The denoised sequence data were grouped into OTUs with the UCLUST algorithm [23]. The sequences exhibiting >97% identities were grouped into one OTU. The representative sequences of each OTU were aligned with PyNAST [24], and chimeric sequences were detected and removed using ChimeraSlayer [25]. The taxonomic assignment of each OTU was performed with the RDP classifier trained by a dataset from Greengenes at a minimum confidence level of 0.8 [26]. The sequence data were compared with the 16S rRNA database of RDP release 10 [27] using the BLAST+ program [28].

OTU analysis for *amoA* pyrotags

The procedures for quality screening, trimming, denoising, and OTU selection were same as those described above for 16S rRNA gene pyrotag analysis, except that sequence identity (>85%) was determined according to the threshold of Pester *et al.* [14]. The filtered sequences were manually checked after aligning with MEGA program [29]. Chimeric sequences were detected and removed using the UCHIME algorithm [30] on the FunGene platform [27]. Sequence data were compared with the GenBank database using the BLAST+ program [28].

Statistical and phylogenetic analysis

After taxonomy assignments, α-diversity indices (e.g., Chao1, Shannon, and Simpson) of the 16S rRNA and *amoA* genes pyrotag libraries were calculated by QIIME [21] and EstimateS (version 9.1.0) [31] software, respectively. The coverage values were calculated using equation [1− (n/N)], where n is the number of OTUs in a single read (singleton) and N is the total number of reads analyzed [32]. The weighted UniFrac distances were calculated and used for principal coordinate analysis (PCoA) [33]. Phylogenetic trees for the bacterial and archaeal *amoA* libraries for Fast UniFrac analysis were constructed with the

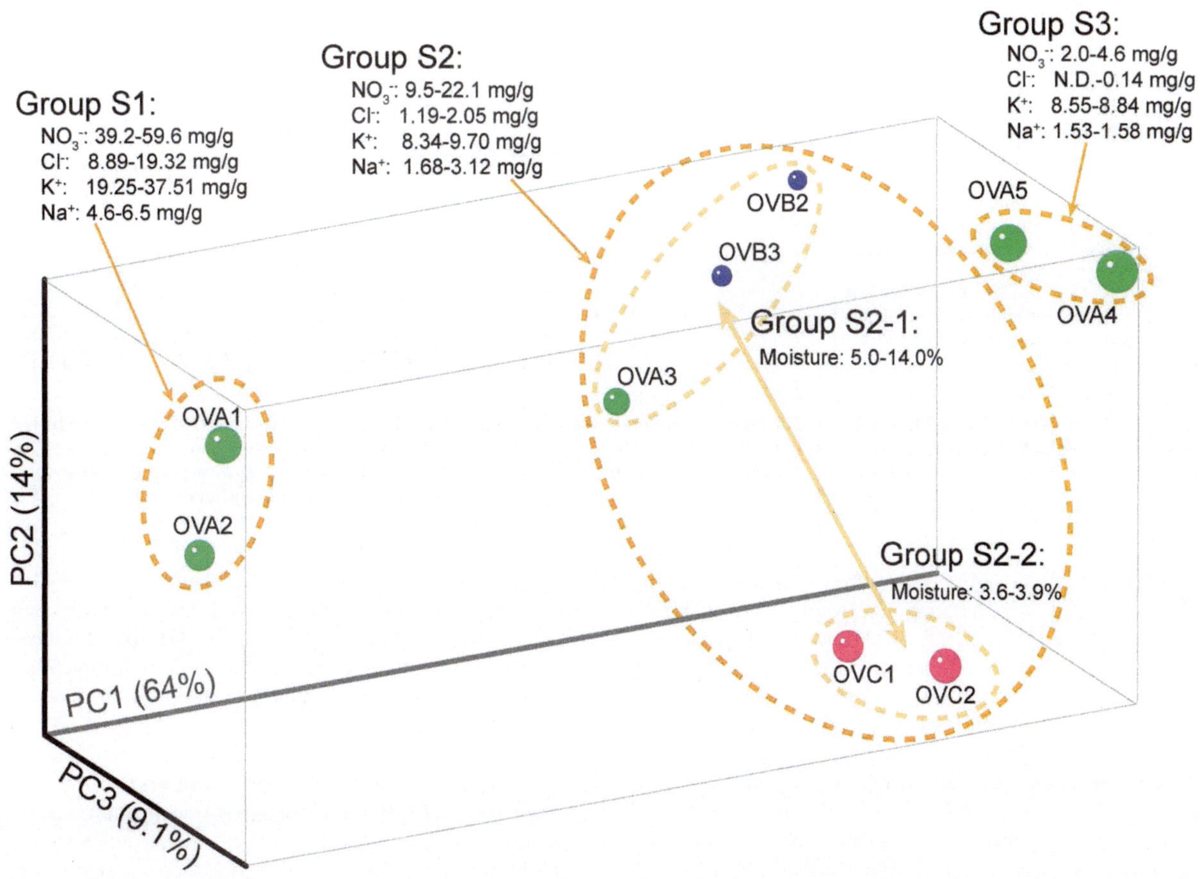

Figure 1. Principal coordinate analysis based on the abundances of 16S rRNA gene OTUs (weighted UniFrac) and separating soils according to their ion contents. For this analysis, observed 16S rRNA gene OTUs were normalized to 7,770 reads per soil.

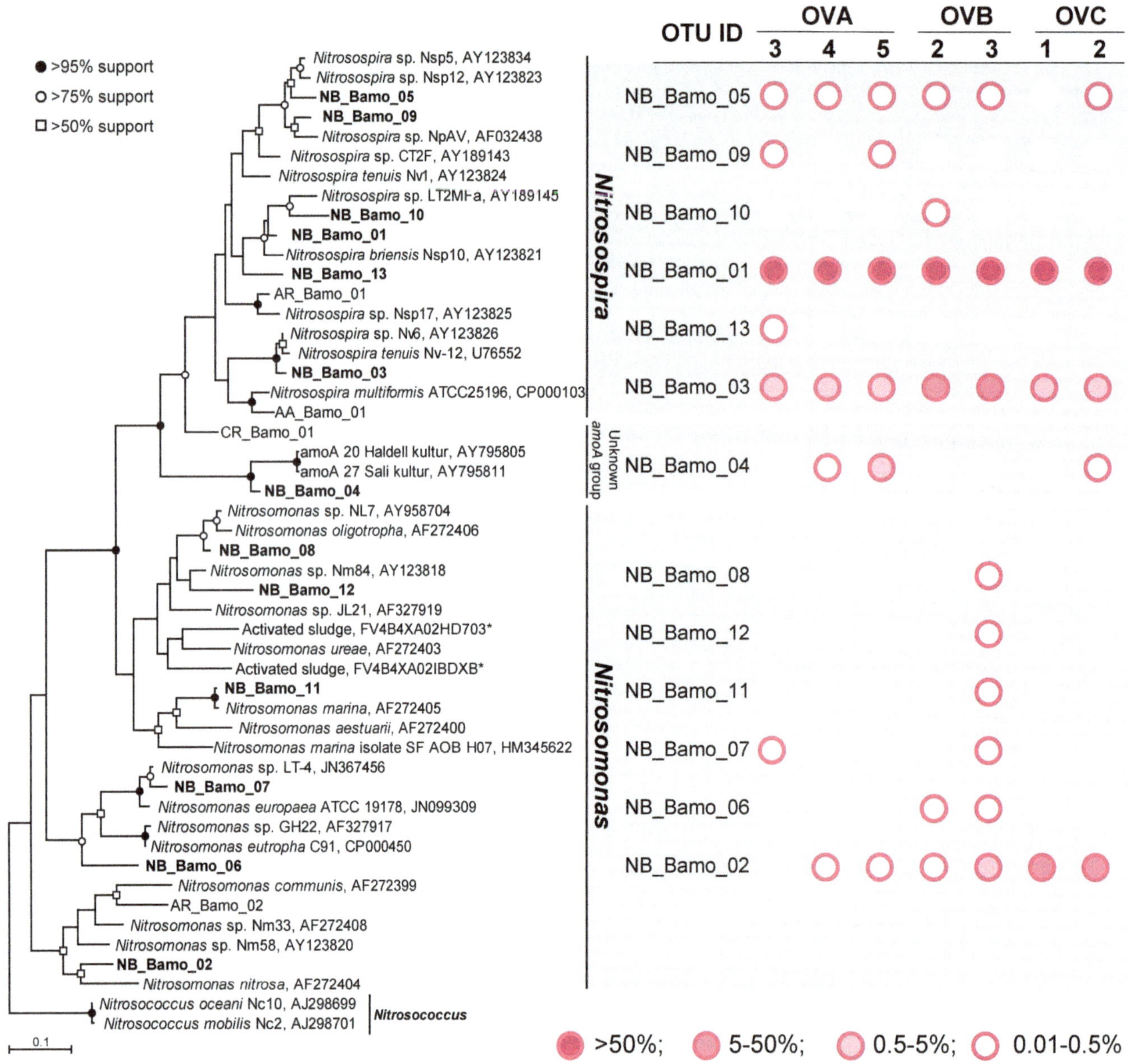

Figure 2. Distance matrix tree of bacterial *amoA* gene sequences retrieved from niter-bed soils based on the maximum likelihood method. Boldface indicates the sequences obtained in this study. The *Nitrosococcus amoA* gene sequences (AJ298699 and AJ298701) were used as outgroups. The bar indicates 10% base substitution. Branching points supported probabilities >95%, >75%, and >50% by bootstrap analyses (based on 1,000 replicates, estimated using the maximum likelihood method) are indicated by solid circle, open circles, and open square, respectively.

MEGA program [29] based on the neighbor-joining algorithm [34]. PCoA and jackknife clustering analyses were performed on weighted and normalized data with QIIME and the Fast UniFrac program [33], respectively. To identify OTUs that exhibited significant differences in abundance between different soil samples, Welch's *t*-test (confidence interval method: Welch's inverted, [$p < 0.05$] for two groups [definition of each group was described in the Results and Discussion]) having possibly unequal variances was performed with STAMP software [35,36].

The distance matrix tree of the bacterial and archaeal *amoA* genes, which is based on the maximum likelihood method, was constructed with the MEGA program [29]. The topology of the trees was estimated by 1,000 bootstrap replicates [37]. Previously reported *amoA* pyrotag sequences retrieved from soil environments in Austria, Costa Rica, Greenland, and Namibia were downloaded from the Sequence Read Archive (SRA) at NCBI (accession no. SRA047303) and used for the statistical and phylogenetic comparisons [14]. Subclusters within the "*Ca.* Nitrososphaera" and "*Ca.* Nitrosopumilus" were defined according to Pester *et al.* [14].

Nucleotide sequence accession numbers

The pyrosequence data obtained in this study have been deposited under DDBJ/EMBL/GenBank accession no. DRA001202.

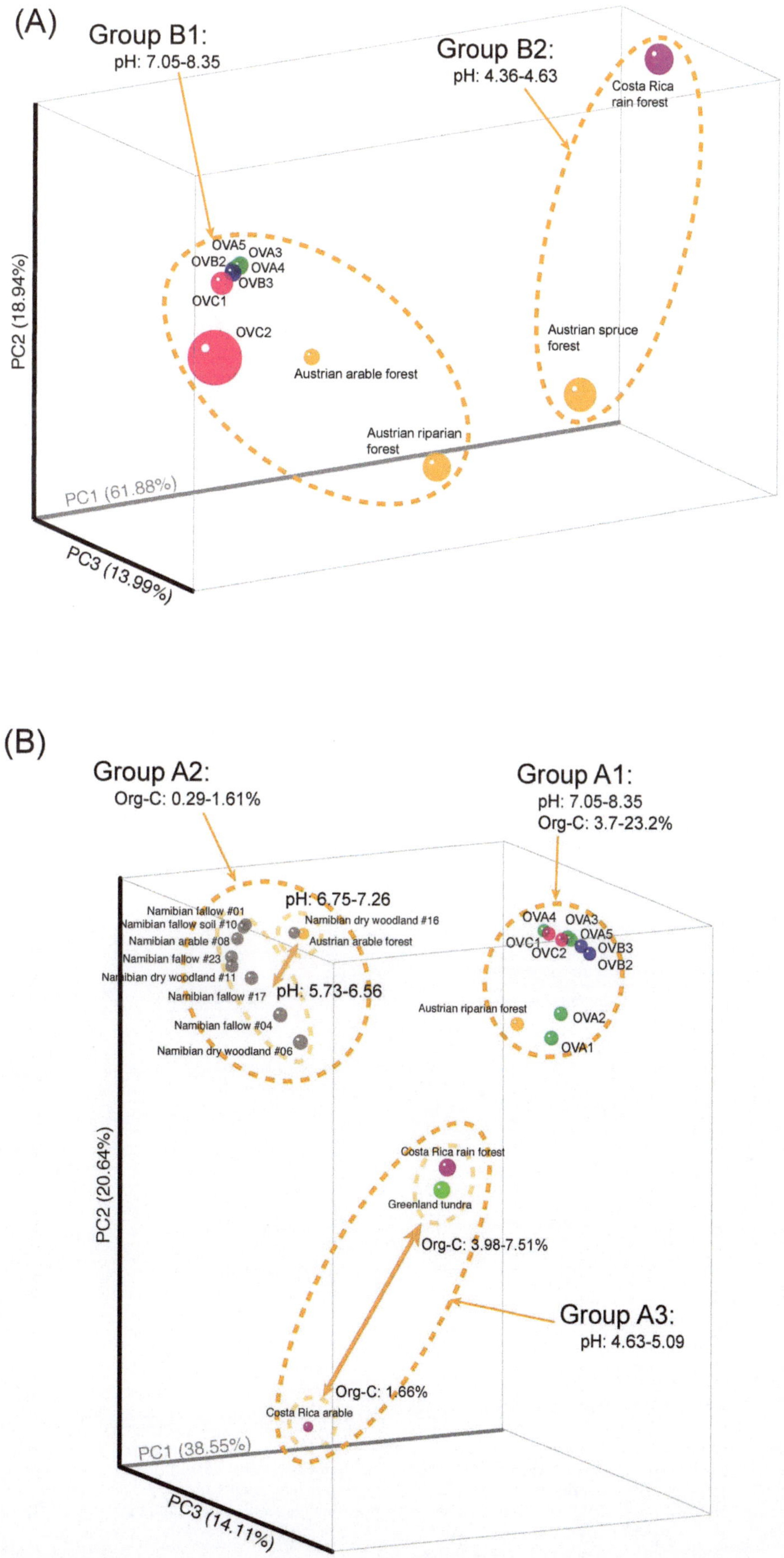
(A)
Group B1:
pH: 7.05-8.35
Group B2:
pH: 4.36-4.63
Costa Rica rain forest
OVA5
OVA3
OVB2
OVA4
OVB3
OVC1
OVC2
Austrian arable forest
Austrian spruce forest
Austrian riparian forest
PC1 (61.88%)
PC2 (18.94%)
PC3 (13.99%)
(B)
Group A2:
Org-C: 0.29-1.61%
Group A1:
pH: 7.05-8.35
Org-C: 3.7-23.2%
pH: 6.75-7.26
Namibian fallow #01
Namibian fallow soil #10
Namibian arable #08
Namibian fallow #23
Namibian dry woodland #11
Namibian fallow #17
Namibian fallow #04
Namibian dry woodland #06
Namibian dry woodland #16
Austrian arable forest
pH: 5.73-6.56
OVA4
OVA3
OVA5
OVC1
OVC2
OVB3
OVB2
Austrian riparian forest
OVA2
OVA1
Costa Rica rain forest
Greenland tundra
Org-C: 3.98-7.51%
Group A3:
pH: 4.63-5.09
Org-C: 1.66%
Costa Rica arable
PC1 (38.55%)
PC2 (20.64%)
PC3 (14.11%)

Figure 3. Principal coordinate analysis based on the abundances of (A) bacterial and (B) archaeal *amoA* gene OTUs (weighted UniFrac) and separating soils according to soil pH and organic carbon contents. For this analysis, *amoA* pyrotag libraries reported by Pester et al. [14] were used for the comparison, and bacterial and archaeal *amoA* OTUs observed here were normalized to 2,320 and 2,130 reads per soil, respectively.

Results and Discussion

16S rRNA gene pyrotag analysis

In this study, nine soil samples from three gassho-style houses, referred to as OVA (samples OVA1–5), OVB (samples OVB2 and OVB3), and OVC (samples OVC1 and OVC2), were used for microbial community analysis using 454 pyrosequencing. We retrieved a total of 173,016 16S rRNA gene pyrotags and found 3,146 OTUs with the criterion of >97% sequence identity (Table 1). Although the rarefaction curves showed that the accumulation of the 16S rRNA gene OTUs was insufficient to achieve the goal of saturation (Fig. S1A), the coverage values were calculated to be 96.99–98.59%, suggesting that the OTUs retrieved here were sufficient for estimating the microbial diversity in the relict niter-bed ecosystems. According to the Chao1 nonparametric estimators, the soil samples contained approximately 1.38–1.63-fold more OTUs than what was detected in our analyses.

The OTUs belonging to the phyla *Proteobacteria* (especially the classes *Alphaproteobacteria* and *Gammaproteobacteria*), *Actinobacteria*, *Bacteroidetes*, *Chloroflexi*, *Firmicutes*, *Gemmatimonadetes*, and *Planctomycetes* predominated and were detected in all samples (Table S3 and S4). Such microbial taxa are frequently found in soil environments [38]. In each sample, only a few number of the 16S rRNA gene pyrotags (<0.8% of the total populations) were associated with previously known ammonia-oxidizing microorganisms at the genus level: for AOB, three OTUs (nos. 1267, 2435, and 2701) of the *Nitrosococcus*, one OTU (no. 2080) of the *Nitrosomonas*, and one OTU (no. 673) of the *Nitrosospira*; for AOA, four OTUs (nos. 1207, 1897, 1946, and 3119) of the "*Ca.* Nitrososphaera" (Table S5). Given that soil samples from the relict niter-beds have retained microbial nitrification activity [13], these

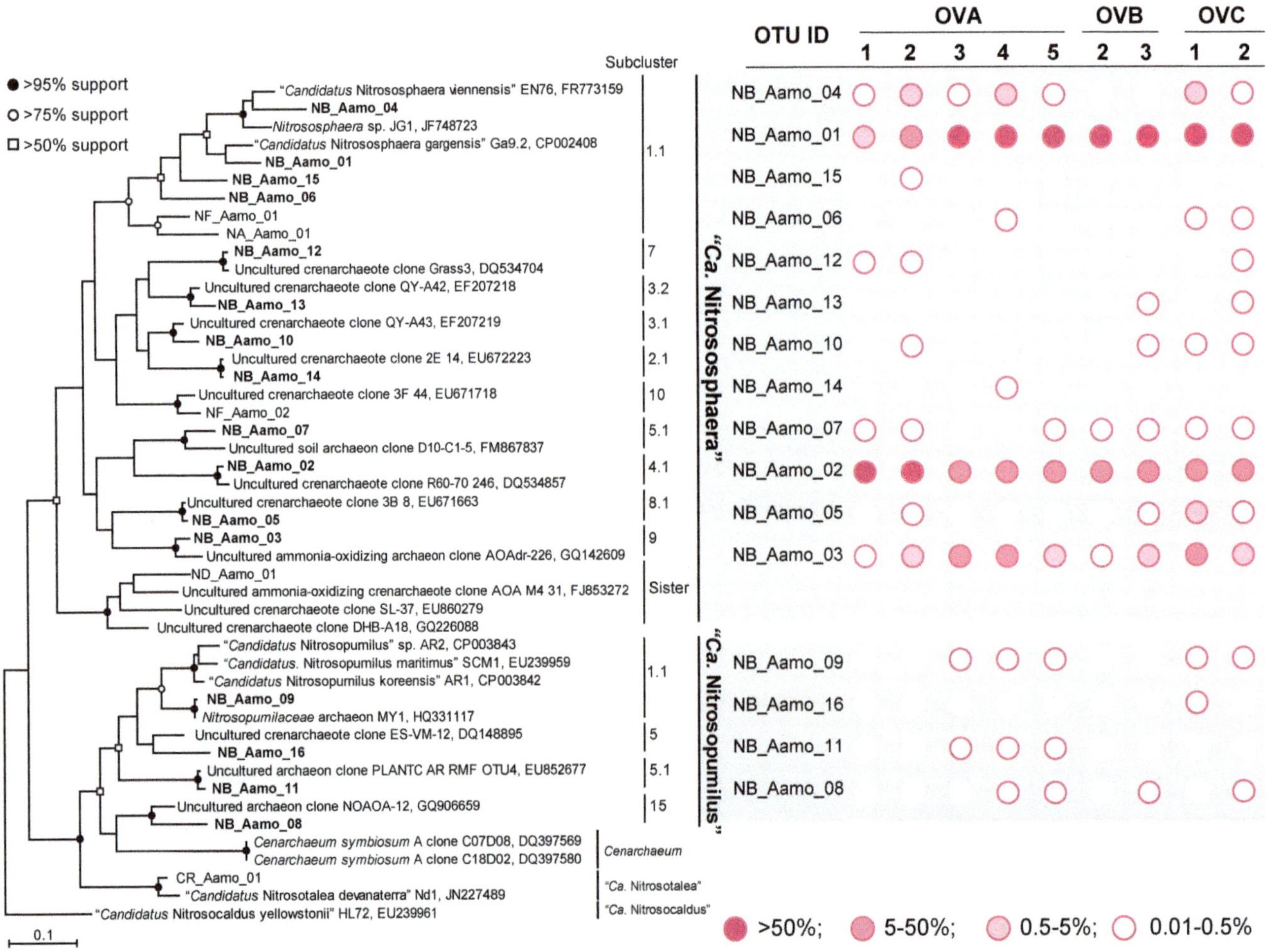

Figure 4. Distance matrix tree of archaeal *amoA* gene sequences retrieved from niter-bed soils based on the maximum likelihood method. Boldface indicates the sequences obtained in this study. The *amoA* gene sequence of the "*Ca.* Nitrosocaldus yellowstonii" strain HL72 (EU239961) was used as outgroup. The bar indicates 10% base substitution. Branching points supported probabilities >95%, >75%, and >50% by bootstrap analyses (based on 1,000 replicates, estimated using the maximum likelihood method) are indicated by solid circle, open circles, and open square, respectively.

nitrifying microbes may contribute to the nitrification at low population densities; alternatively, there may have been unidentified nitrifying microbes in the relict niter-bed soils. Although the relative abundances of nitrifying microbes were in good agreement with the microbial compositions of several types of soil ecosystems [38], we could not fully elucidate the biodiversity and ecological niche of ammonia-oxidizing microbes in the relict niter-bed ecosystems based on the results of 16S rRNA gene pyrotag analysis (e.g., there are no significant relationship between the ratio of the abundance ratio of AOB to AOA and clustering of the communities in PCoA and jackknife analysis described below) (Table S5). These observations emphasize the need to implement further analysis targeting *amoA* gene that enable specific detection of ammonia-oxidizing microbes.

The similarity of microbial compositions among the nine soil samples was compared on the basis of 16S rRNA gene pyrotag sequences. According to the PCoA and jackknife analyses, the microbial community structures of the relict niter-bed soil were classified into three groups (Fig. 1 and S2A). Group S1, consisting of OVA1 and OVA2, was distinct from the other two groups. Based on previously reported physicochemical characterization [15–17], relatively high contents of NO_3^-, Cl^-, K^+, and Na^+ were found in OVA1 and OVA2 compared with the samples clustered into groups S2 and S3 (Table S1). Because the soils in group S1 were not subjected to the extraction step of saltpeter production, the ion contents were higher than those in the other relict niter-bed soil samples [17]. In addition, the NO_3^- and Cl^- contents were relatively low in the soils clustered in group S3 compared with those in group S2. Based on Welch's *t*-test, we found a total of 21 OTUs that were significantly different in relative abundances among the groups and were associated with dominant taxa (e.g., *Alphaproteobacteria*, *Gammaproteobacteria*, *Actinobacteria*, *Bacteroidetes*, *Chloroflexi*, *Firmicutes*, and *Gemmatimonadetes*) (Fig. S3). Among them, *Gammaproteobacteria*-type OTU (NB_16S_507) and *Chloroflexi*-type OTU (NB_16S_1415) were involved in the differentiation of groups S1, S2, and S3. The samples in group S2 were further separated according to their moisture contents (groups S2-1 and S2-2), and the relative abundances of six OTUs (NB_16S_1093, 1144, 1409, 1415, 2046, and 2058) were significantly affected by the moisture contents (Fig. S3). Several types of ion species and moisture contents have been recognized as potential factors in the determination of soil microbial constituents [39–42]. Thus, the NO_3^-, Cl^-, K^+, Na^+, and moisture contents may affect the microbial community structures of relict niter-bed ecosystems.

Bacterial *amoA* gene pyrotag analysis

As described above, the proportions of 16S rRNA gene pyrotags associated with AOB and AOA were very low in the relict niter-bed soils. To further elucidate the microbial constituents responsible for ammonia oxidation, we performed pyrotag analysis targeting bacterial and archaeal *amoA* genes. As a result, we obtained a total of 59,404 bacterial *amoA* pyrotags and found 13 OTUs in relict niter-bed soils (Table 1). However, the bacterial *amoA* gene fragment could not be amplified from OVA1 and OVA2 using the primer set described in this study, suggesting that there were either low levels of AOB in the samples or there were some AOB with significantly different types of *amoA* genes. Instead, the archaeal *amoA* gene fragments were successfully amplified from OVA1 and OVA2 samples, suggesting that AOA may be a major player in ammonia oxidation in these soils (details described below). Although the rarefaction curves for the bacterial *amoA* libraries shorted to reach a plateau (except for OVC1 and OVC2), the coverage values were calculated to be 99.96–100%, indicating that the OTUs retrieved here were sufficient for estimating the biodiversity of AOB in seven relict niter-bed soils (Table 1 and Fig. S1B).

The most abundant AOB belonged to the genus *Nitrosospira* (6 OTUs, 68.13–99.96% of the total reads in each sample), and *Nitrosospira briensis*-type OTU (NB_Bamo_01) predominated in all samples (except for OVA1 and OVA2, in which there were no bacterial *amoA* gene amplification) (Table S6). This result is in good accordance with 16S rRNA gene pyrotag analysis that found a *Nitrosospira briensis*-type OTU (no. 673) across the samples. An OTU (NB_Bamo_01) related to the *Nitrosospira briensis* strain Nsp10 (95.5% *amoA* gene sequence identity) was detected as a major AOB population in all niter-bed soils (Fig. 2 and Table S6). *Nitrosospira briensis* are enriched in and have been isolated from soil samples taken from sandy and agriculturally uncultivated areas of Europe [43]. To the best of our knowledge, only two reports have described the clear appearance of *Nitrosospira briensis* populations: water columns of the Chesapeake Bay [44] and soil samples of the Cascade Mountains in Oregon [45] in the U.S. Interestingly, Bollmann *et al.* [46] reported that under ammonia-starved conditions, the cells of *Nitrosospira briensis* had the ability to immediately recover ammonia oxidation activity upon exposure to ammonia-rich conditions. Although the mechanism behind this phenomenon is unclear, *Nitrosospira briensis*-type AOB may reserve their activity in ammonia-limited environments, such as the relict niter-bed soils. Within the genus *Nitrosospira*, an OTU (NB_Bamo_03) that is closely related to the *Nitrosospira tenuis* strain Nv-12 (96.2% sequence identity) was also detected in all bacterial *amoA* gene pytotag libraries; however, its ecophysiological traits (other than ammonia oxidation) remain unclear.

In addition to the *Nitrosospira* populations, six *Nitrosomonas*-type OTUs were found in seven niter-bed soils (Fig. 2 and Table S6). In contrast to the *Nitrosospira*-type OTUs, 16S pyrotags related to *Nitrosomonas* were found in OVA3 and OVA4; however, the proportions were quite low (0.005–0.006% of total 16S rRNA gene pyrotags). This finding may also support the utility of *amoA*-based pyrotag analysis for relict niter-bed soils. An OTU (NB_Bamo_02) associated with *Nitrosomonas nitrosa* (87.2% sequence identity) was observed with significant abundance in OVC1 and OVC2 (5.6% and 31.8%, respectively) (Table S6). According to previous reports, *Nitrosomonas nitrosa*-type AOB have frequently been found in estuary soils [47], nitrification bioreactors [48,49], and freshwater ecosystems [50]. In addition to these taxa, an OTU (NB_Bamo_04) related to the unidentified *amoA* gene cluster consisting of clones retrieved from potash-polluted marsh soil [51] was found in three relict niter-bed soils (Fig. 2 and Table S6). This cluster has a significantly low identity with previously reported bacterial *amoA* gene sequences (e.g., NB_BamoA_04 had 77.4% identity with *Nitrosospira multiformis* strain C-71, accession no. X90822), suggesting that phylogenetically novel AOB may play a role in ammonia oxidation in some niter-bed ecosystems.

The similarity of AOB compositions in relict niter-bed soils was compared with previously reported *amoA* gene-based microbial community analysis [14]. PCoA and jackknife analyses showed that soil bacterial *amoA* libraries were separated into two groups according to their pH: seven niter-bed soils were clustered into group B1 with two Austrian arable and riparian forest soils (pH 7.05–8.35), and the remaining two soils, from an Austrian spruce forest and a Costa-Rican rain forest, were clustered into group B2 (pH 4.36–4.63) (Fig. 3A). pH is generally to be a key parameter in defining the abundance of AOB populations in soil ecosystems [52,53] and bioreactors [54,55]. The niter-bed soils in group B1 were further divided into two subgroups (Fig. S2B and S3A);

however, there was no clear explanation for the differentiation between OVC2 and the other soils based on the physicochemical parameters analyzed herein. In short, soil pH is one of the important factors affecting the AOB populations in relict niter-bed ecosystems.

Archaeal *amoA* gene pyrotag analysis

A total of 33,087 archaeal *amoA* gene sequences were retrieved from nine relict niter-bed soils, and we found 16 OTUs with the criterion of >85% sequence identity. The archaeal *amoA* diversity was almost fully covered according to the rarefaction curves, coverage values, and Chao1 diversity index (Table 1 and Fig. S1C).

Members of the genus "*Ca.* Nitrososphaera" were detected as major AOA populations (12 OTUs, 99.23–100% of the total reads in each sample), whereas those of "*Ca.* Nitrosopumilus" were absent or few (<0.28% of the total reads in each sample) in all samples tested (Fig. 4 and Table S6). Based on the 16S rRNA gene pyrotag analysis, only a few OTUs (accounting for 0.16–1.1% of the total prokaryotic population) associated with "*Ca.* Nitrososphaera" were identified as AOA populations in relict niter-bed soils (Table S5). Thus, the results from *amoA*-based pyrotag analysis expanded the biodiversity of AOA in relict niter-bed soils. The most abundant OTU in OVA1 and OVA2 soils was NB_Aamo_02, which was categorized into subcluster 4.1 of "*Ca.* Nitrososphaera" and was distantly related to previously characterized AOA (77.8% sequence identity with "*Ca.* Nitrososphaera viennensis" strain EN76, accession no. FR773159), whereas the most abundant OTU in the other seven samples (NB_Aamo_01) was closely related to "*Ca.* Nitrososphaera gargensis" strain Ga9.2 (90.6% sequence identity, accession no. CP002408) and categorized into subcluster 1.1. This marked difference in a major AOA population within OVA soils was clearly associated with the maximum nitrification rate of bulk soil (Table S1) [13]. This finding also implies that the nitrification activity and/or affinity for ammonia may differ between subclusters 1.1 and 1.4. The *amoA* fragments associated with "*Ca.* Nitrososphaera" subcluster 1.4 have been observed in diverse environments, including soil [20,56,57], salt marsh sediment [58], and marine sediment [20,59]. In addition to the major AOA, several subclusters of "*Ca.* Nitrososphaera" were detected as minor AOA populations. For example, an OTU (NB_Aamo_03) assigned to subcluster 9 was commonly observed in the relict niter-bed soils (Fig. 4). Although information regarding the ecological traits of "*Ca.* Nitrososphaera" subcluster 9 organisms is limited, these organisms were widely distributed in natural environments, such as marine sediment [60] and soil [20]. Further studies using metagenomic and single-cell genomic approaches are necessary to elucidate the ecological functions of unidentified AOA in relict niter-bed ecosystems.

PCoA and jackknife analyses clearly showed that archaeal *amoA* pyrotag libraries were separated into three clusters according to soil pH and organic carbon concentration (Fig. 3B and S2C). The niter-bed soils formed group A1 with an Austrian riparian forest soil. Group A1 soils had a weak alkaline pH (7.05–8.35) and relatively high organic carbon content (3.7–23.2%). In addition, relict niter-bed soils were further separated into two subgroups according to their NO_3^-, Cl^-, K^+, and Na^+ contents in addition to the 16S rRNA gene-based comparison (Fig. S4B). The samples clustered into group A2 were mainly composed of Namibian soils, which had relatively low organic carbon contents (0.29–1.61%) and a weakly acidic pH (5.7–6.8). The samples categorized into group A3 (soil samples from a Costa-Rican rain forest, a Costa-Rican arable forest, and Greenland tundra) were defined as having a relatively high organic carbon content (1.66–7.51%) and an acidic soil pH (4.63–5.09). Interestingly, there was a significant correlation between the Chao1 diversity index and organic carbon content, with higher richness at lower levels of organic carbon (Fig. S5A). Moreover, a relatively weak correlation between the Chao1 index and soil pH was also observed, with the highest diversity at a pH of approximately 6 (Fig. S5B). These results were in good agreement with the previously reported *amoA* gene-based pyrotag analysis [14], and they provided further evidence of the importance of soil pH and organic carbon for defining the ecological niche of the AOA community in soil ecosystems, including relict niter-beds.

Conclusions

16S rRNA and *amoA* gene-based pyrotag analyses revealed that microbes responsible for ammonia oxidation (both AOB and AOA) still persist in relict niter-bed soils, although saltpeter production ceased more than 100 years ago. Our study provided the first insights into the microbial diversity in relict niter-bed ecosystems. Our main conclusions are as follows. (i) Based on 16S rRNA gene pyrotag analysis, members of the phyla *Proteobacteria*, *Actinobacteria*, *Bacteroidetes*, *Chloroflexi*, *Firmicutes*, *Gemmatimonadetes*, and *Planctomycetes* were major microbial constituents in the niter-bed soil samples. The NO_3^-, Cl^-, K^+, Na^+, and moisture contents were potential factors for the microbial compositions of the relict niter-bed soils. (ii) Members of *Nitrosospira* and "*Ca.* Nitrososphaera" were the major AOB and AOA populations, respectively, in the niter-bed soils. Soil pH, organic carbon content, and NO_3^-, Cl^-, K^+, and Na^+ contents were important factors for the ecological niche of AOB and AOA in the relict niter-bed soils.

Supporting Information

Figure S1 Rarefaction curves of (A) 16S rRNA, (B) bacterial *amoA*, and (C) archaeal *amoA* gene sequences of nine niter-bed soils.

Figure S2 Jackknife clustering of (A) 16S rRNA, (B) bacterial *amoA*, and (C) archaeal *amoA* gene pyrotag libraries from the relict niter-bed soils (weighted UniFrac, normalized to 7,770, 2,320, and 2,130 reads per soil, respectively).

Figure S3 Extended error bar plot showing the OTUs that have significantly (p<0.05) different abundances in the group. Groups were defined on the basis of principle coordinate analysis (Fig. 1) and Jackknife clustering of 16S rRNA gene pyrotag analysis (Fig. S2A). The OTUs overrepresented in upper group have a positive difference between relative abundances, and those overrepresented in bottom group have a negative difference between relative abundances. Only OTUs with mean proportions more than 0.01% in the total populations were shown in the plot.

Figure S4 Principal coordinate analysis based on the abundances of (A) bacterial and (B) archaeal *amoA* gene OTUs (weighted UniFrac) only for niter-bed soils. For this analysis, bacterial and archaeal *amoA* OTUs observed here were normalized to 2,320 and 2,130 reads per soil, respectively.

Figure S5 Correlation analysis of organic carbon contents and soil pH to Chao1 diversity indexes. Previously reported archaeal *amoA* pyrotag data from various soil environments [14] were combined with those from niter-bed soils.

Table S1 Physicochemical parameters of niter-bed soil samples.

Table S2 Primer sequences for pyrotag analysis.

Table S3 The 16S rRNA gene pyrotag libraries of relict niter-bed soil.

Table S4 The OTU table of 16S rRNA gene pyrotag libraries.

Table S5 The relative abundances of ammonia-oxidizing bacteria and archaea in 16S rRNA gene pyrotag libraries.

Table S6 Phylogenetic identification of OTUs derived from bacterial and archaeal *amoA* gene sequences.

Acknowledgments

The relict niter-bed soils were kindly provided by the education board of Shirakawa-Village, Gifu, Japan. We are grateful to Ms. Kiriko Ikeba (National Institute of Advanced Industrial Science and Technology, Japan) for technical assistance.

Author Contributions

Conceived and designed the experiments: TN HT YK SH TM. Performed the experiments: TN AA. Analyzed the data: TN KT KN. Contributed reagents/materials/analysis tools: TN TM. Contributed to the writing of the manuscript: TN HT YK SH.

References

1. Williams AR (1975) The production of saltpetre in the middle ages. Ambix 22: 125–133.
2. Lockyer WJS (1908) Modern nitre beds. Nature 77: 513–515.
3. Kyeser K (1967) Umschrift und Ubersetzung nebst Erlauterungen (German translation). Düsseldorf, Germany: Verlag des Vereins Deutscher Ingenieure.
4. Kyeser K (1967) Facsimile-Ausgabe der Pergamenthandschrift, Cod. Ms. philos. 63 der Universititsbibliothek, Gottingen (German translation). Düsseldorf, Germany: Verlag des Vereins Deutscher Ingenieure.
5. Multhauf RP (1971) The French Crash Program for Saltpeter Production, 1776–94. Technol Cult 12: 163–181.
6. Kunnas J (2007) Potash, saltpeter and tar - Production, exports and use of wood in Finland in the 19th century. Scand J Hist 32: 281–311.
7. Anonymous (1912) The Indian saltpetre industry. Nature 88: 330–331.
8. Thorpe TE (1917) Indian saltpetre. Nature 99: 447–448.
9. Maji T (2009) Transaction on saltpeter of Shirakawa Village in the second half of the 18th century described by old letters found from Shinshoji-temple. Annu Rep Fac Edu, Gifu Univ, Humanit Soc Sci (in Japanese with English abstract) 57: 121–126.
10. Maji T, Maji A (2007) The Unknown History of Shirakawago: The Circulation of Nitrogen in Relation with Urine, Saltpeter and Rice, and The Daily Life in the Edo Era. Nagoya, Japan: Fubaisha Press.
11. Anonymous (1912) The condition of the Chilean saltpeter industry. J Ind Eng Chem-Us 4: 687–687.
12. Jimura T (2014) The impact of world heritage site designation on local communities - A case study of Ogimachi, Shirakawa-mura, Japan. Tourism Manage 32: 288–296.
13. Maji T (2009) Efficiency of nitrification and activity of nitrite production by soil in niter-bed. Sci Rep Fac Ed Gifu Univ Natur Sci (in Japanese with English abstract) 33: 103–108.
14. Pester M, Rattei T, Flechl S, Grongroft A, Richter A, et al. (2012) *amoA*-based consensus phylogeny of ammonia-oxidizing archaea and deep sequencing of *amoA* genes from soils of four different geographic regions. Environ Microbiol 14: 525–539.
15. Maji T (2006) A study of saltpeter production at Shirakawa Village in Edo era. Part 3: an estimation of contribution of urinay nitrogen to nitrate-N in niter-bed. Kagakushi (in Japanese with English abstract) 33: 1–14.
16. Maji T (2005) A study of saltpeter production at Shirakawa Village in Edo era. Part 2: productivity of saltpeter from artificial niter-bed soil: an estimation from extraction yield of nitrite. Kagakushi (in Japanese with English abstract) 32: 137–143.
17. Maji T (2005) A study of saltpeter production at Shirakawa Village in Edo era. Part 1: an analysis of constituents of soil gathered from niter-bed and the starting materials. Kagakushi (in Japanese with English abstract) 32: 75–84.
18. Baker GC, Smith JJ, Cowan DA (2003) Review and re-analysis of domain-specific 16S primers. J Microbiol Methods 55: 541–555.
19. Rotthauwe JH, Witzel KP, Liesack W (1997) The ammonia monooxygenase structural gene amoA as a functional marker: Molecular fine-scale analysis of natural ammonia-oxidizing populations. Appl Environ Microbiol 63: 4704–4712.
20. Francis CA, Roberts KJ, Beman JM, Santoro AE, Oakley BB (2005) Ubiquity and diversity of ammonia-oxidizing archaea in water columns and sediments of the ocean. Proc Natl Acad Sci USA 102: 14683–14688.
21. Caporaso JG, Kuczynski J, Stombaugh J, Bittinger K, Bushman FD, et al. (2010) QIIME allows analysis of high-throughput community sequencing data. Nat Methods 7: 335–336.
22. Reeder J, Knight R (2010) Rapidly denoising pyrosequencing amplicon reads by exploiting rank-abundance distributions. Nat Methods 7: 668–669.
23. Edgar RC (2010) Search and clustering orders of magnitude faster than BLAST. Bioinformatics 26: 2460–2461.
24. Caporaso JG, Bittinger K, Bushman FD, DeSantis TZ, Andersen GL, et al. (2010) PyNAST: a flexible tool for aligning sequences to a template alignment. Bioinformatics 26: 266–267.
25. Haas BJ, Gevers D, Earl AM, Feldgarden M, Ward DV, et al. (2011) Chimeric 16S rRNA sequence formation and detection in Sanger and 454-pyrosequenced PCR amplicons. Genome Res 21: 494–504.
26. Wang Y, Qian PY (2009) Conservative fragments in bacterial 16S rRNA genes and primer design for 16S ribosomal DNA amplicons in metagenomic studies. PLoS One 4: e7401.
27. Cole JR, Wang Q, Cardenas E, Fish J, Chai B, et al. (2009) The Ribosomal Database Project: improved alignments and new tools for rRNA analysis. Nucleic Acids Res 37: D141–D145.
28. Camacho C, Coulouris G, Avagyan V, Ma N, Papadopoulos J, et al. (2009) BLAST+: architecture and applications. BMC Bioinformatics 10: 421.
29. Tamura K, Peterson D, Peterson N, Stecher G, Nei M, et al. (2011) MEGA5: Molecular Evolutionary Genetics Analysis Using Maximum Likelihood, Evolutionary Distance, and Maximum Parsimony Methods. Mol Biol Evol 28: 2731–2739.
30. Edgar RC, Haas BJ, Clemente JC, Quince C, Knight R (2011) UCHIME improves sensitivity and speed of chimera detection. Bioinformatics 27: 2194–2200.
31. Colwell RK (2013) EstimateS: Statistical estimation of species richness and shared species from samples. Version 9 and earlier. User's Guide and application.
32. Good IJ (1953) The population frequencies of species and the estimation of population parameters. Biometrika 40: 237–264.
33. Hamady M, Lozupone C, Knight R (2010) Fast UniFrac: facilitating high-throughput phylogenetic analyses of microbial communities including analysis of pyrosequencing and PhyloChip data. ISME J 4: 17–27.
34. Saitou N, Nei M (1987) The neighbor-joining method - a new method for reconstructing phylogenetic trees. Mol Biol Evol 4: 406–425.
35. Welch BL (1947) The generalization of students problem when several different population variances are involved. Biometrika 34: 28–35.
36. Parks DH, Beiko RG (2010) Identifying biologically relevant differences between metagenomic communities. Bioinformatics 26: 715–721.
37. Felsenstein J (1985) Confidence-limits on phylogenies - an approach using the bootstrap. Evolution 39: 783–791.
38. Mao YJ, Yannarell AC, Mackie RI (2011) Changes in N-transforming archaea and bacteria in soil during the establishment of bioenergy crops. PLoS One 6: e24750.
39. Xue K, Wu LY, Deng Y, He ZL, Van Nostrand J, et al. (2013) Functional gene differences in soil microbial communities from conventional, low-input, and organic farmlands. Appl Environ Microbiol 79: 1284–1292.
40. Grossman JM, O'Neill BE, Tsai SM, Liang BQ, Neves E, et al. (2010) Amazonian anthrosols support similar microbial communities that differ distinctly from those extant in adjacent, unmodified soils of the same mineralogy. Microb Ecol 60: 192–205.
41. Corneo PE, Pellegrini A, Cappellin L, Roncador M, Chierici M, et al. (2013) Microbial community structure in vineyard soils across altitudinal gradients and in different seasons. FEMS Microbiol Ecol 84: 588–602.

42. Van Horn DJ, Van Horn ML, Barrett JE, Gooseff MN, Altrichter AE, et al. (2013) Factors controlling soil microbial biomass and bacterial diversity and community composition in a cold desert ecosystem: role of geographic scale. PLoS One 8: e66103.
43. Watson SW (1971) Reisolation of *Nitrosospira briensis* Winogradsky, S and Winogradsky, H 1933. Arch Mikrobiol 75: 179–188.
44. Ward BB, Eveillard D, Kirshtein JD, Nelson JD, Voytek MA, et al. (2007) Ammonia-oxidizing bacterial community composition in estuarine and oceanic environments assessed using a functional gene microarray. Environ Microbiol 9: 2522–2538.
45. Mintie AT, Heichen RS, Cromack K, Myrold DD, Bottomley PJ (2003) Ammonia-oxidizing bacteria along meadow-to-forest transects in the oregon cascade mountains. Appl Environ Microbiol 69: 3129–3136.
46. Bollmann A, Schmidt I, Saunders AM, Nicolaisen MH (2005) Influence of starvation on potential ammonia-oxidizing activity and *amoA* mRNA levels of *Nitrosospira briensis*. Appl Environ Microbiol 71: 1276–1282.
47. Li XR, Xiao YP, Ren WW, Liu ZF, Shi JH, et al. (2012) Abundance and composition of ammonia-oxidizing bacteria and archaea in different types of soil in the Yangtze River estuary. J Zhejiang Univ-Sc B 13: 769–782.
48. Fukushima T, Whang LM, Chiang TY, Lin YH, Chevalier LR, et al. (2013) Nitrifying bacterial community structures and their nitrification performance under sufficient and limited inorganic carbon conditions. Appl Microbiol Biotechnol 97: 6513–6523.
49. Shore JL, M'Coy WS, Gunsch CK, Deshusses MA (2012) Application of a moving bed biofilm reactor for tertiary ammonia treatment in high temperature industrial wastewater. Bioresour Technol 112: 51–60.
50. Avrahami S, Jia ZJ, Neufeld JD, Murrell JC, Conrad R, et al. (2011) Active autotrophic ammonia-oxidizing bacteria in biofilm enrichments from simulated creek ecosystems at two ammonium concentrations respond to temperature manipulation. Appl Environ Microbiol 77: 7329–7338.
51. Eilmus S, Rosch C, Bothe H (2007) Prokaryotic life in a potash-polluted marsh with emphasis on N-metabolizing microorganisms. Environ Pollut 146: 478–491.
52. Gubry-Rangin C, Nicol GW, Prosser JI (2010) Archaea rather than bacteria control nitrification in two agricultural acidic soils. FEMS Microbiol Ecol 74: 566–574.
53. Shen JP, Zhang LM, Zhu YG, Zhang JB, He JZ (2008) Abundance and composition of ammonia-oxidizing bacteria and ammonia-oxidizing archaea communities of an alkaline sandy loam. Environ Microbiol 10: 1601–1611.
54. Tarre S, Shlafman E, Beliavski M, Green M (2007) Changes in ammonia oxidiser population during transition to low pH in a biofilm reactor starting with *Nitrosomonas europaea*. Water Sci Technol 55: 363–368.
55. Yasuda T, Waki M, Kuroda K, Hanajima D, Fukumoto Y, et al. (2013) Responses of community structure of amoA-encoding archaea and ammonia-oxidizing bacteria in ammonia biofilter with rockwool mixtures to the gradual increases in ammonium and nitrate. J Appl Microbiol 114: 746–761.
56. Leininger S, Urich T, Schloter M, Schwark L, Qi J, et al. (2006) Archaea predominate among ammonia-oxidizing prokaryotes in soils. Nature 442: 806–809.
57. Jia ZJ, Conrad R (2009) *Bacteria* rather than *Archaea* dominate microbial ammonia oxidation in an agricultural soil. Environ Microbiol 11: 1658–1671.
58. Moin NS, Nelson KA, Bush A, Bernhard AE (2009) Distribution and diversity of archaeal and bacterial ammonia oxidizers in salt marsh sediments. Appl Environ Microbiol 75: 7461–7468.
59. Dang HY, Zhang XX, Sun J, Li TG, Zhang ZN, et al. (2008) Diversity and spatial distribution of sediment ammonia-oxidizing crenarchaeota in response to estuarine and environmental gradients in the Changjiang Estuary and East China Sea. Microbiology-SGM 154: 2084–2095.
60. Park HD, Wells GF, Bae H, Criddle CS, Francis CA (2006) Occurrence of ammonia-oxidizing archaea in wastewater treatment plant bioreactors. Appl Environ Microbiol 72: 5643–5647.

Representation of Dormant and Active Microbial Dynamics for Ecosystem Modeling

Gangsheng Wang[1,2]*, Melanie A. Mayes[1,2], Lianhong Gu[1,2], Christopher W. Schadt[1,3]

1 Climate Change Science Institute, Oak Ridge National Laboratory, Oak Ridge, Tennessee, United States of America, **2** Environmental Sciences Division, Oak Ridge National Laboratory, Oak Ridge, Tennessee, United States of America, **3** Biosciences Division, Oak Ridge National Laboratory, Oak Ridge, Tennessee, United States of America

Abstract

Dormancy is an essential strategy for microorganisms to cope with environmental stress. However, global ecosystem models typically ignore microbial dormancy, resulting in notable model uncertainties. To facilitate the consideration of dormancy in these large-scale models, we propose a new microbial physiology component that works for a wide range of substrate availabilities. This new model is based on microbial physiological states and the major parameters are the maximum specific growth and maintenance rates of active microbes and the ratio of dormant to active maintenance rates. A major improvement of our model over extant models is that it can explain the low active microbial fractions commonly observed in undisturbed soils. Our new model shows that the exponentially-increasing respiration from substrate-induced respiration experiments can only be used to determine the maximum specific growth rate and initial active microbial biomass, while the respiration data representing both exponentially-increasing and non-exponentially-increasing phases can robustly determine a range of key parameters including the initial total live biomass, initial active fraction, the maximum specific growth and maintenance rates, and the half-saturation constant. Our new model can be incorporated into existing ecosystem models to account for dormancy in microbially-driven processes and to provide improved estimates of microbial activities.

Editor: Jonathan H. Badger, J. Craig Venter Institute, United States of America

Funding: This research was funded by the Laboratory Directed Research and Development (LDRD) Program of the Oak Ridge National Laboratory (ORNL) and by the U.S. Department of Energy Biological and Environmental Research (BER) program. ORNL is managed by UT-Battelle, LLC, for the U.S. Department of Energy under contract DE-AC05-00OR22725. The funders had no role in study design, data collection and analysis, decision to publish, or preparation of the manuscript.

Competing Interests: The authors have declared that no competing interests exist.

* E-mail: wangg@ornl.gov

Introduction

Ecologically-important processes such as soil organic carbon and nutrient cycling largely depend on the active fraction of microbial communities [1]. At any given time in a given environment, microorganisms can be in active, dormant, or dead states [2]. Dormancy is considered an evolutionary strategy designed to maintain the genetic code until conditions improve to allow replication [3]. When environmental conditions are unfavorable for growth, e.g., resource limitation, microbes may enter a reversible state of low to zero metabolic activity to alleviate the loss of biomass and metabolic functions [4,5]. The maintenance coefficient (i.e., maintenance cost of C per unit microbial biomass C per unit time) can be two to three orders of magnitude lower in dormant microbes than in metabolically active microbes [6,7]. Many soils have slow organic matter turnover rates with seasonal changes in substrate supply, temperature, and moisture. The complexity of soils in space and time may result in uneven distributions of multiple potentially limiting resources, leading to significant rates of dormancy even when some resources are abundant. When spatial and temporal complexity is combined with differential resource partitioning among species in a community, high rates of dormancy could be a prominent feature in soil systems. Thus it is essential to understand dormancy in order to more accurately predict how active microorganisms contribute to ecosystem processes such as decomposition and nutrient turnover [1].

A complicating factor in studying microbial dormancy is that no single approach can be easily employe to simultaneously measure individual microbial states (active, dormant or dead), and a combination of different techniques is required. Differential staining is often used to segregate physiological states with direct microscopic counting of bacteria and fungi. 'Life-indicating' stains that require the presence of 'standard' physiological abilities, such as the esterase activity needed for fluorescein diacetate cleavage, may distinguish active from dormant+dead cells [8]. When combined with general-purpose stains, these strains can distinguish dormant cells by difference [9]. Combining membrane-permeant with membrane-impermeant nucleophilic stains (e.g., SYTO-9 and propidium iodide respectively) may distinguish live from dead, but not active from dormant [10,11]. Active microbes may or may not be 'viable' with common culture-based techniques, which complicates classification and measurement of dormancy phenomena [5]. Methods such as direct plating, serial dilution and most probable number (MPN) techniques will not distinguish between active and dormant organisms [12]. Substrate Induced Respiration (SIR) or Substrate Induced Growth Response (SIGR) method [13,14] can distinguish active and dormant communities if growth respiration curves are modeled (using initial exponentially-increasing respiration); however, the technique often needs to be combined with microscopy or chloroform fumigation/extraction in order to obtain total live microbial biomass [15,16].

Despite limitations in distinguishing active, dormant and dead microbial biomass, abundant evidence indicates that the majority of environmental microorganisms in a given community may be dormant under natural conditions [1,17]. Alvarez *et al.* [18] reported that only 3.8–9.7% of the total biomass is active in a Typic Argiudoll soil from the Argentinean Pampa. Khomutova *et al.* [19] showed that the fraction of active microbial biomass ranged from 0.02% to 19.1% in the subkurgan paleosoils of different age and 9.2–24.2% in modern background soils. Microbial biomass measured through SIR or SIGR is thought to reflect only the active portion because the maintenance respiration of dormancy biomass is negligible in the initial exponentially-increasing phase [13,16,20]. Through a mathematical analysis of respiration curves, Van de Werf & Verstraete [21] examined 16 soils and found that 4–49% of the total biomass was in an active state; and the active component in undisturbed natural ecosystems (18.8±8.8%, mean±standard deviation) was about 70% of that in arable agricultural soils (25.7±14.8%). Stenström *et al.* [22] showed that the fraction of active biomass typically varied from 5% to 20% in soils without recent addition of substrates. Lennon & Jones [5] found much lower active fractions in soils (18±15%) than in marine (65±19%) and fresh (54±11%) water environments. From the above studies it seems conservative to extrapolate that the active fraction is very likely below 50% of live microbes under most natural soil conditions.

Microbially-mediated processes have been incorporated into ecosystem models [23–28] although continued development is still required to bring microbial processes into global climate models [29–31]. However, these recent models do not consider physiological state changes and assume that measures of microbial biomass constitute the active biomass. The exclusion of dormancy from the microbially-driven ecosystem processes could result in incorrect estimates of total live microbial biomass, which further leads to deficiencies in model parameterization and predictions of soil organic carbon and nutrients.

Generally, there are two strategies to represent physiological states in microbial-ecology models: one is to explicitly separate the total live biomass into two pools, i.e., active and dormant [4,32]; the other is to directly regard the active fraction (i.e., ratio of active biomass to total live biomass) as a state variable [33,34]. Both of these two approaches predict the total live biomass, active and dormant biomass, and the flux between the active and dormant components. Apparently the introduction of the 'active fraction' as a state variable in the latter approach simplifies the model structure since the adaptive variation of microbial composition might be represented by one single variable (active fraction) [34,35]. However, another state variable indicating the microbial biomass pool size (e.g., total live biomass, active biomass or dormant biomass) is still essential for ecosystem modeling since the carbon and nutrient fluxes are pool-size dependent. For example, if we define active fraction and total microbial biomass as state variables, the active and dormant biomass could be determined by them, and the net flux between active and dormant fractions and other related fluxes could also be computed according to the active and dormant biomass constrained by mass balance. The above-mentioned modeling efforts have shown that adequate representation of dormancy and the transitions between the dormant and active states is crucial for modeling important microbially-mediated ecosystem processes.

Here, we review state-of-the-art microbial dormancy modeling approaches and discuss the rationales of these models with a focus on transformation processes between active and dormant states. We propose an improved synthetic microbial physiology model based on accepted assumptions and examine the model behavior with theoretical and experimental analyses. In this paper, the 'total microbial biomass' refers to the 'total live microbial biomass' unless otherwise stated. Our objective is to clarify the applicability of existing microbial dormancy models and provide a new theoretical basis for representing microbial activity and dormancy in ecosystem models.

Dormancy In Microbial Models

Transformation between active and dormant states

Although Buerger *et al.* [36] argued that dormant microbial cells could reactivate stochastically and might be independent of environmental cues, environmental factors such as substrate availability are often thought to control the transformation between active and dormant states [5]. Most models (see Appendix S1) distinguish the active biomass pool from the dormant pool and define them as two state variables (B_a and B_d) (Fig. 1). Only active microbes (B_a) can uptake substrate and produce new cells. The connection between the active and dormant states is a reversible process including two directional sub-processes, i.e., dormancy (from active to dormant) and reactivation (or resuscitation, from dormant to active). Losses from active biomass include growth respiration and maintenance (maintenance respiration, mortality, enzyme synthesis, etc.) [23]. Dormant microbes still require energy for maintenance and survival although at a lower metabolic rate [5].

The net transformation rate ($B^N_{a\to d}$) from active to dormant state is the difference between the flux from active to dormant ($B_{a\to d}$) and the flux from dormant to active state ($B_{d\to a}$), i.e., $B^N_{a\to d} = B_{a\to d} - B_{d\to a}$. The models of Hunt [37] (Equation S1-1 in Appendix S1) and Gignoux *et al.* [38] (Equation S1-2 in Appendix S1) directly formulate the net flux ($B^N_{a\to d}$) without explicit components for $B_{a\to d}$ and $B_{d\to a}$. The direction of the net flux depends on the maintenance requirement relative to the substrate availability. If the available substrate is less than the maintenance requirement, there is a positive net flux from active to dormant pool, and vice versa. In addition, Hunt [37] assumed a 'buffer zone' for the change of states: when the maintenance requirement surpasses the substrate supply but the deficit is within a small fraction (1% d^{-1}) of B_a, there is no flux between the two states.

Some models define rates for both dormancy and reactivation. In the model of Ayati [39] (Equation S1-3 in Appendix S1), the

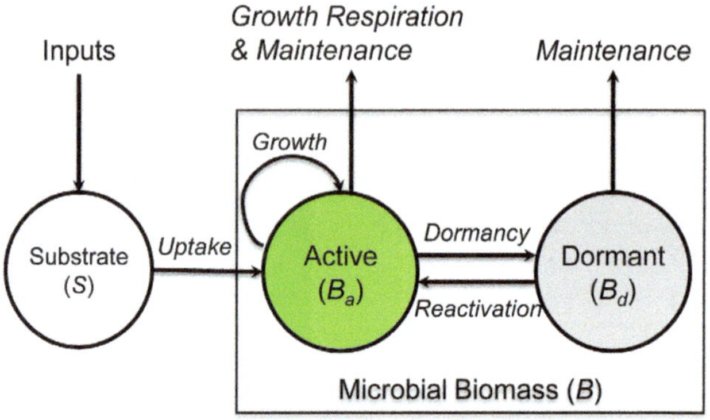

Figure 1. Active and dormant microbial biomass pools in microbial physiology models (modified from Fig. 2 in Lennon & Jones, 2011).

dormant rate ($\gamma_{a\to d}$) increases with declining substrate concentration, and the reactivation ($\gamma_{d\to a}$) only occurs when substrate concentration is higher than the half-saturation constant (K_s). Konopka [32] modified the potential rates for dormancy and reactivation by the relative growth rate (μ/μ_{max}, ratio of true specific growth rate to maximum specific growth rate), i.e, the two rates are multiplied by ($1-\mu/\mu_{max}$) and μ/μ_{max}, respectively (Equation S1-4 in Appendix S1). Similarly, Jones & Lennon [40] postulated two complementary rates (1-*R* and *R*) for dormancy and resuscitation (Equation S1-5 in Appendix S1).

Two other models also explicitly formulate the two conversion rates between states but do so using concepts of probability. Bär *et al.* [41] used two complementary factors (1-$\mathcal{J}$ and $\mathcal{J}$) to represent the probability for the transition between active and dormant state in addition to an identical potential rate constant for the two processes (Equation S1-6 in Appendix S1). The conceptual model of Locey [42] applies a deterministic dormant rate and a stochastic resuscitation rate (Equation S1-7 in Appendix S1). The potential resuscitation rate is modified by ($1-p$), where p is the probability that a disturbance in the active pool will result in the immigration of one individual from the metacommunity. The probability ($\mathcal{J}$) in Bär *et al.* [41] is explicitly calculated from the environmental cues (e.g., soil moisture), while the cause of the probability (p) in Locey [42] is not elucidated.

Switch function model

In addition to the dormancy and reactivation processes, a key concept in the model developed by Stolpovsky *et al.* [4] is 'switch function' (Equation S1-8 in Appendix S1). The switch function (θ) determines the fraction of active cells taking up dissolved organic carbon (DOC). This function refers to the growth fraction in active biomass (B_a) that consumes substrate and thus is not the same as the active fraction (r) in total biomass (B). Furthermore, the dormancy and reactivation fluxes are set to be proportional to ($1-\theta$) and θ, respectively. θ is formulated by the Fermi-Dirac statistics [4]. Another feature of this model is the consideration of 'depth' of dormancy in reactivation, where the reactivation rate is negatively dependent on the duration of dormancy. The switch function model includes at least 15 model parameters and it is difficult to compute the Gibbs energy change of the oxidation of DOC [4].

The switch function (θ) sets it apart from the conventional Michaelis-Menten (M-M) or Monod kinetics because of its new perspective of thermodynamics. According to the M-M kinetics [43], the substrate saturation level represents the fraction of enzyme-substrate complex (ES) in active enzyme (E_0), where the substrate saturation level is formulated by $S/(K_s + S)$ with S and K_s being the substrate concentration and the half-saturation constant [43]. When the M-M kinetics is applied to describe microbial uptake of substrate, the substrate (or combined with TEA) saturation level (i.e., $\mu(S, TEA)$ in Equation S1-8a of Appendix S1) is a measure of the actively growing fraction in the active microbial community. The switch function is also determined by the saturation levels of substrate and terminal electron acceptor (TEA), i.e., $\mu(S, TEA)$ (Equation S1-8e in Appendix S1). Mathematically the inclusion of both the switch function (θ) and the M-M kinetics (i.e., $\mu(S, TEA)$) might result in double counting of the impact of substrate and TEA. We would recommend using the switch function (θ) to modify the microbial uptake rate if the Gibbs energy change of the oxidation of substrate (ΔG) is tractable and the thermodynamic threshold (G_0) and the steepness of the step function (st) are identifiable.

Physiological state index models

As an alternative to models with two microbial biomass pools (i.e., active and dormant), a further state variable indicating the dormant or active fraction in total biomass has been proposed. Wirtz [44] developed a simple index ($r_d = 0.5$–1.0) representing the dormant microbial biomass as a fraction of the steady-state total biomass (B_{stat}) under the condition of $B_d << B_a$. In case of a net loss of total biomass ($dB/dt<0$), the dormant biomass $B_d = B_{stat} \cdot r_d$; otherwise ($dB/dt>0$), $B_d = B_{stat} \cdot (1-r_d)$. This model has a sudden change of dormant biomass at the transition point (i.e., $dB/dt = 0$) since $r_d>0.5$.

Different from the dormant index of Wirtz [44], the concept of an active index (i.e., index of physiological state) of soil microbial community has been employed in soil carbon and nutrient cycling models [33,34]. The index of physiological state (r), referring to the activity state, is often defined as the ratio of metabolically active microbial biomass to the total soil microbial biomass [22,33,34].

In the Synthetic Chemostat Model (SCM), the rate of change of the state variable r is described as follows [33,45]:

$$\frac{dr}{dt} = \frac{1}{B}\frac{dB}{dt} \cdot (\phi - r) = \mu \cdot (\phi - r) \tag{1}$$

with

$$\phi = \phi(S) = S^n/(K_r + S^n), \text{or} \phi = S/(K_r + S) \tag{2}$$

where $r = B_a/B$, representing the fraction (hereinafter referred to as 'active fraction') of active biomass in total biomass; μ is the specific growth rate of total biomass; ϕ denotes the saturation level of substrate (S); the simple power ($n = 1$) has been widely used [35] and, in this case ($n = 1$), K_r is called the half-saturation constant.

Blagodatsky & Richter [34] used the expression $\mu(S) = \mu_{max} \cdot \phi(S)$ in their model development. This expression was not derived in the original definition of the specific growth rate (see Equation 3) by Panikov [45] and because its validity cannot be inferred, the concepts will not be addressed here.

According to Panikov's derivation [45], the specific growth rate (μ) follows the general definition [46,47]:

$$\mu = \frac{1}{B}\frac{dB}{dt} \tag{3}$$

Based on Equations 1 and 3, we can derive (see Equation S2-1 in Appendix S2):

$$dB_a/dt = \phi \cdot (dB/dt) \tag{4}$$

$$dB_d/dt = (1-\phi) \cdot (dB/dt) \tag{5}$$

We find that the model described by Equation 1 is not applicable under low substrate availability, as described below. Generally, the rates of change in biomass pools (B, B_a, and B_d) can be expressed as

$$dB/dt = g^{\pm}(S, B_a) - f^{+}(S, B_d) \tag{6}$$

$$dB_a/dt = g^{\pm}(S,B_a) - B^N_{a\to d} \quad (7)$$

$$dB_d/dt = -f^{+}(S,B_d) + B^N_{a\to d} \quad (8)$$

where $B^N_{a\to d}$ denotes the net dormancy flux; $g^{\pm}(S, B_a)$ is a function that represents the difference between the substrate uptake and the maintenance requirements of B_a, i.e., the net growth of B_a; and $f^{+}(S, B_a)$ is a function denoting the maintenance and survival energy costs of B_d. The superscript '±' in $g^{\pm}$ indicates the function value of g could be positive at high S or non-positive when the substrate uptake cannot satisfy the maintenance requirements of B_a at low S. The superscript '+' in f^{+} implies $f \geq 0$. Note that the function $f^{+}(S, B_a)$ is not necessarily dependent on S [4].

From Equations 4, 6 and 7, we can obtain

$$B^N_{a\to d} = (1-\phi)\cdot g^{\pm}(S,B_a) - \phi\cdot f^{+}(S,B_d) \quad (9)$$

The two terms in the right side of Equation 9 may be regarded as the conversion of B_a to B_d (i.e., $B_{a\to d}$) and the transformation of B_d to B_a (i.e., $B_{d\to a}$), respectively. At high S resulting in $g \geq 0$, Equation 9 may be one of the possible expressions for $B_{a\to d}$ and $B_{d\to a}$. However, at low S leading to $g<0$ and $B_{a\to d}<0$, i.e., no active cells become dormant under insufficient substrate, which is inconsistent with the strategy of dormancy for microorganisms when faced with unfavorable environmental conditions [5].

Based on the above analysis, we conclude that the physiological state index model (Equation 1) needs to be improved. In other words, the empirical assumption that the steady state active fraction (r^{ss}) approaches the substrate saturation level (ϕ^{ss}) may not be necessary because this assumption could lead to impractical flux (Equation 9) between dormant and active states under low substrate availability.

A Synthetic Microbial Physiology Model

Based on the aforementioned review and analysis, we have developed a synthetic microbial physiology model component relating to substrate availability. As indicated by Fig. 1, the growth and maintenance functions of active microbes (B_a) are characterized by the maximum specific growth rate (μ_G) and maintenance rate (m_R); whereas the dormant microbes (B_d) cost energy to maintain their basic cellular functions at a much lower specific maintenance rate (denoted by $\beta\cdot m_R$, where $\beta<1$) [48].

General assumptions

First we define the substrate saturation level (ϕ) as

$$\phi = S/(K_S + S) \quad (10)$$

where the parameter K_s is the half saturation constant for substrate uptake as indicated by the M-M kinetics [43].

Based on the above review of existing dormancy models, the following assumptions are accepted in our new model: (1) the dormancy rate is proportional to the active biomass and the reactivation rate is proportional to the dormant biomass, i.e., $B_{a\to d} \propto B_a$ and $B_{d\to a} \propto B_d$; (2) under very high substrate concentration ($S>>K_s$), $\phi\to 1$, $B_{a\to d}\to 0$ and $B_{d\to a}\geq 0$; (3) under very low substrate ($S<<K_s$), $\phi\to 0$, $B_{a\to d}\geq 0$ and $B_{d\to a}\to 0$; (4) based on the assumptions (1–3), we derive that $B_{a\to d} \propto (1-\phi)\cdot B_a$ and $B_{d\to a} \propto \phi\cdot B_a$; (5) further we assume that the maximum specific maintenance rate for active microbes (m_R with units of h^{-1}) controls both transformation processes since the maintenance energy cost is the key factor regulating the dormancy strategy [5,37,38]. As a result we postulate that

$$B_{a\to d} = (1-\phi)\cdot m_R\cdot B_a \quad (11a)$$

$$B_{d\to a} = \phi\cdot m_R\cdot B_d \quad (11b)$$

Model Description

Equations 11a and 11b only describe the transformation between the active and dormant states. They need to be linked to a microbial growth and maintenance model for depicting microbial dynamics. Our recent work to develop the Microbial-ENzyme-mediated Decomposition (MEND) model [23] suggested that it might be adapted to serve this purpose due to its focus on microbial processes for which we have developed a firm theoretical basis [47,49]. Combining Equations 11a and 11b with the MEND model [23,47], we express the microbial physiology component (see Fig. 1) as a group of differential equations

$$dS/dt = I_s - \frac{1}{Y_G}\cdot\frac{\phi}{\alpha}m_R\cdot B_a \quad (12a)$$

$$dB/dt = d(B_a + B_d)/dt = (\phi/\alpha - 1)\cdot m_R\cdot B_a - (\beta\cdot m_R)\cdot B_d \quad (12b)$$

$$dB_a/dt = (\phi/\alpha - 1)\cdot m_R\cdot B_a - (1-\phi)\cdot m_R\cdot B_a + \phi\cdot m_R\cdot B_d \quad (12c)$$

$$dB_d/dt = -(\beta\cdot m_R)\cdot B_d + (1-\phi)\cdot m_R\cdot B_a - \phi\cdot m_R\cdot B_d \quad (12d)$$

where t is the time scale; ϕ is defined by Equation 10; I_s is the input to substrate pool; Y_G is the true growth yield; m_R denotes the specific maintenance rate at active state (h^{-1}); $\alpha = m_R/(\mu_G+m_R)$ is the ratio of m_R to the sum of maximum specific growth rate (μ_G) and m_R, $\alpha\in(0,0.5)$ since usually $m_R\leq\mu_G$; and β (0–1) is the ratio of dormant maintenance rate to active maintenance rate, i.e., (β m_R) denotes the maximum specific maintenance rate at dormant state.

In summary, there are five parameters (α, β, m_R or μ_G, Y_G, K_s) in the proposed model (hereinafter referred to as the MEND model). From Equations 12b and 12c, we can derive the change rate of active fraction (r) (see Equation S2-2 in Appendix S2)

$$dr/dt = m_R\cdot[(\phi - r) + (\phi/\alpha + \beta - 1)\cdot r\cdot(1-r)] \quad (12e)$$

This equation for r is more complicated than Equation 1 but still practical, given currently available data. Additionally, it implies that r needs not approach ϕ at steady state in our model, whereas $r\equiv\phi$ at steady state is required by the model of Panikov [45].

Steady state analysis

Assuming the input (I_s) is time-invariant, we can obtain the steady state solution to the above new MEND model (see Equations S2-3(a–e) in Appendix S2). Fig. 2 shows the steady

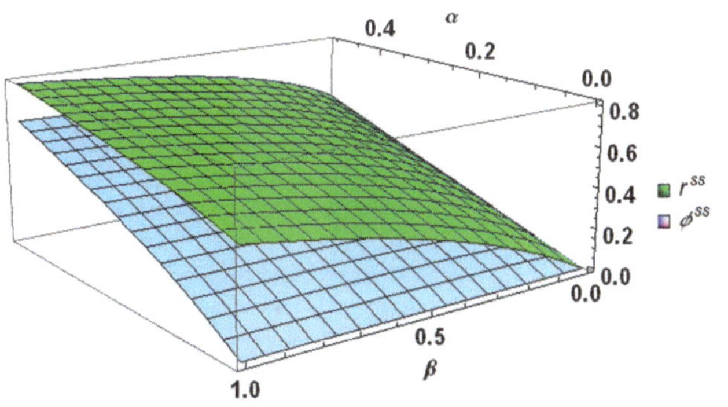

Figure 2. Steady state active fraction (r^{ss}) and substrate saturation level (ϕ^{ss}) as a function of α and β, $\alpha = m_R/(\mu_G + m_R)$, μ_G and m_R (h^{-1}) are maximum specific growth rate and specific maintenance rate for active microbial biomass, respectivly; β denotes the ratio of dormant specific maintenance rate to m_R.

state active fraction (r^{ss}) and substrate saturation level (ϕ^{ss}) as a function of the two physiological indices, i.e., α (0–0.5) and β (0–1). Both r^{ss} and ϕ^{ss} positively depend on α and β and $r^{ss} \geq \phi^{ss}$ for any combinations of α and β. If we consider two extreme values of $\beta \to 0$ or $\beta \to 1$, the r^{ss} and ϕ^{ss} (see Equations S2-4 and S2-5 in Appendix S2) can be simplified to

$$r^{ss}_{\beta \to 0} = \phi^{ss}_{\beta \to 0} = \alpha \tag{13a}$$

$$\begin{cases} r^{ss}_{\beta \to 1} = \frac{1+\sqrt{1+8\alpha}}{4} \\ \phi^{ss}_{\beta \to 1} = \frac{4\alpha - 1 + \sqrt{1+8\alpha}}{3+\sqrt{1+8\alpha}} \end{cases} \tag{13b}$$

Equation 13 and Fig. 2 indicate that: (1) the steady state active fraction (r^{ss}) is equal to ϕ^{ss} and they are identical to $\alpha = m_R/(\mu_G + m_R)$ only under the condition of $\beta \to 0$; (2) the upper bound of r^{ss} is approximately 0.8 at $\alpha \to 0.5$ and $\beta \to 1$; and (3) with $\alpha \leq 0.5$, the maximum r^{ss} is ca. 0.5 if the magnitude of β is around 0.001–0.01 [6]. This threshold value (0.5) of r^{ss} is a reasonable estimate that can explain how the measured active fraction of microbes in undisturbed soils is usually considerably less than the total biomass [5,21,22].

Model simplification under sufficient substrate condition

As mentioned in the Introduction, SIR or SIGR method can distinguish active from dormant composition and the data from these experiments have been widely used to estimate the active microbial biomass and the maximum specific growth rate [13,14]. The simplification of the microbial model under excess substrate has also been employed to estimate maximum specific growth rate (μ_G), active microbial biomass (B_a), and/or total microbial biomass (B) using the SIR or SIGR data [1,13,35]. Here we show the simplification of our model (Equation 12) for conditions appropriate to SIGR or SIR experiments, e.g., the short-term period of exponentially-increasing respiration of active biomass following substrate addition. We will test our reduced and full model with the SIGR data of Colores *et al.* [13] in the next section.

Under sufficient substrate (i.e., $S >> K_s$ in Equation 10 thus $\phi \to 1$), Equations 12(a–e) can be simplified and integrated for initial conditions, i.e., $S = S_0$, $B = B_0$ and $r = r_0$ at $t = 0$ (see Equations S2-6 and S2-7 in Appendix S2):

$$S(t) = S_0 - \frac{B(t) - B_0}{Y_G(1-\alpha)} \tag{14a}$$

$$B(t) = B_0 r_0 \cdot e^{(1/\alpha - 1)m_R t} + B_0(1 - r_0) \cdot [\alpha \cdot e^{(1/\alpha - 1)m_R t} + (1-\alpha) \cdot e^{-m_R t}] \tag{14b}$$

$$r(t) = \frac{[r_0 + \alpha(1-r_0)] \cdot e^{(m_R/\alpha) \cdot t} - \alpha(1-r_0)}{[r_0 + \alpha(1-r_0)] \cdot e^{(m_R/\alpha) \cdot t} + (1-\alpha)(1-r_0)} \tag{14c}$$

The CO_2 production rate, $v(t)$, during the exponential growth stage is derived as an explicit function of t (see Equation S2-7d in Appendix S2):

$$v(t) = \frac{dCO_2}{dt} = \frac{B_0(1-Y_G)}{Y_G} \left\{ [(m_R/\alpha) \cdot r_0 + m_R \cdot (1-r_0)] \cdot e^{(1/\alpha - 1)m_R t} - [m_R \cdot (1-r_0)] \cdot e^{-m_R t} \right\} \tag{14d}$$

The respiration rate, $v(t)$, is associated with two exponential items, i.e., $e^{(1/\alpha - 1)m_R t} = e^{\mu_G t}$ and $e^{-m_R t}$. Considering an extreme case that $m_R << \mu_G$ (i.e., $\alpha \to 0$), Equations 14(b–d) can be further simplified to Equations S2-8(b–d) (see Appendix S2). Equations S2-8b and S2-8c (denoting $B(t)$ and $r(t)$, respectively) are similar to Equations 11 and 10 in Panikov & Sizova [35], respectively. However, Equation S2-8d (for $v(t)$) is different from Equation 13 of Panikov & Sizova [35], where a constant 'A' was added to the exponential. Equation S2-8d is identical to Equation 7 derived for SIGR experiments in Colores *et al.* [13].

Panikov & Sizova [35] used their Equation 13 to fit respiration rates during the exponentially-increasing (i.e., no substrate limitation) phase (see Fig. 2 in Panikov & Sizova [35] for data and curve fittings). However, these data are based on glucose-induced respiration that includes both basal respiration of native SOC and respiration due to the addition of glucose [13]. The basal respiration rate may be regarded as a constant in certain cases (see Colores *et al.* [13] and data in Fig. 1 of Blagodatsky *et al.* [50]). The constant 'A' representing the basal respiration rate was included in Equation 13 of Panikov & Sizova [35] in order to fit the combined respiration from the addition of glucose and basal respiration. However, this constant 'A' cannot be derived from such governing equations as Equations S2-6(a–c) (see Appendix S2) that assume respiration is the sole result of substrate addition. In other words, the equations do not include basal respiration. Certainly, our predicted respiration could include basal respiration as long as (i) a basal respiration rate is added to Eq. 14d *ad hoc* or (ii) Equations S2-6(a–c), or more commonly Equations 12(a–e), are linked to a soil organic matter (SOM) decomposition model, which can produce decomposed native soil C in addition to the respiration of substrate addition. Because Equation 13 of Panikov & Sizova [35] is not linked to a native C decomposition model, fitting the model to combined native C and substrate respiration data is not appropriate.

Model test I: substrate-induced respiration

In this section, we used the respiration data from ^{14}C-labeled glucose SIGR experiments by Colores et al. [13] to calibrate our

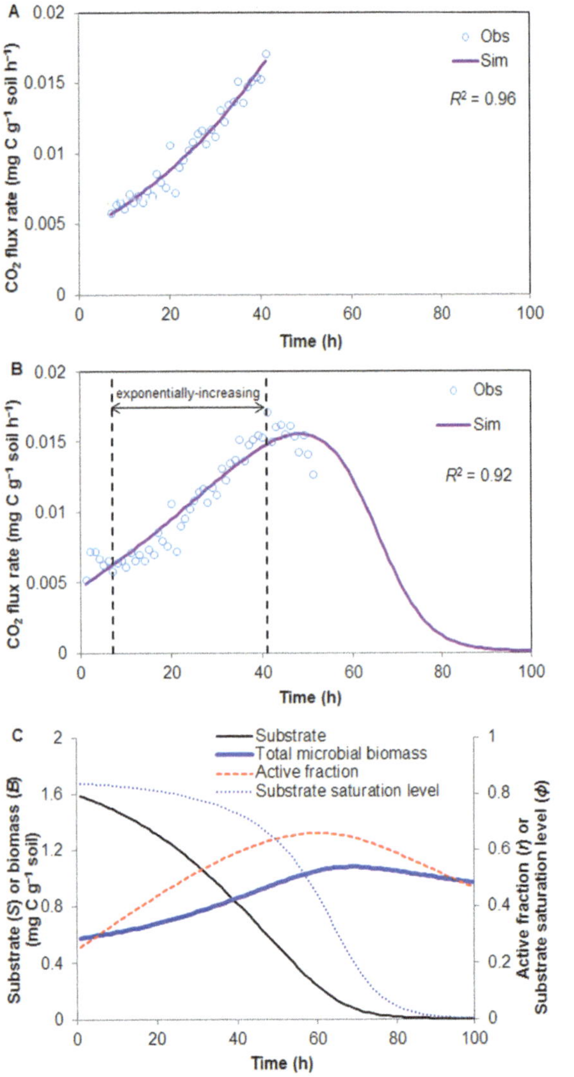

Figure 3. MEND model simulations against the respiration rates due to added ^{14}C-labeled glucose in Colores et al. [13]. (a) Fitting of the respiration rates in the exponentially-increasing phase using Equation 14, 'Obs' and 'Sim' denote observed and simulated data, respectively. (b) Fitting of the respiration rates in both exponentially-increasing and non-exponentially-increasing phases using Equation 12. (c) Simulated substrate (*S*), total live microbial biomass (*B*), active fraction (*r*) and substrate saturation level (ϑ) based on Equation 12.

MEND model. The respiration data only represented the CO_2 production from the added substrate and did not include basal respiration from the native C.

First we employed Equation 14d to fit the respiration rates during the exponentially-increasing stage and the result is shown in Fig. 3a (see original data in Fig. 3 of Colores et al. [13]). The true growth yield (Y_G) was set to 0.5 according to Colores *et al.* [13]. There are four undetermined parameters (B_0, r_0, μ_G, α) in Equation 14d (with $m_R = \mu_G \cdot \alpha/(1-\alpha)$). We found that only the maximum specific growth rate (μ_G) could be determined with high confidence (coefficient of variation (CV) = 5%) from the exponentially-increasing respiration rates. The CVs of the other three optimized parameters (B_0, r_0, α) were as high as 55–77% (Table 1). However, the initial active microbial biomass ($B_{a0} = B_0 \times r_0$) had a lower uncertainty (CV = 20%) compared to B_0 and r_0. The above results indicate that the exponentially-increasing respiration rates can only be used to obtain μ_G and B_{a0}.

We then conducted numerical simulations in terms of all data including both exponentially-increasing and non-exponentially-increasing respiration rates (Fig. 3b). The non-exponentially-increasing respiration rates include the lag period before the exponentially-increasing phase and the respiration at longer times after the rates cease to increase exponentially [13]. The latter phase is likely because of the substrate saturation levels (ϕ) become limiting to respiration. We used Equations 12a, 12b, 12e and the corresponding expression for CO_2 flux rate, to allow the substrate saturation level (ϕ) to change with time. Additionally, we used the ranges of μ_G determined above. We used the SCEUA (Shuffled Complex Evolution at University of Arizona) algorithm [51,52] to determine model parameters. The SCEUA is a widely used stochastic optimization algorithm for calibrating hydrological and environmental models [51].

When exponentially-increasing and non-exponentially-increasing data are included together, the CVs of all parameters (B_0, r_0, μ_G, α, K_s, β) are within 25% except β with a high CV of 76% (Table 1). The optimized μ_G values (0.030 ± 0.001 h^{-1}) are almost the same as obtained by Colores *et al.* [13]. Model estimates of $\alpha = 0.228\pm0.031$ indicate that the maximum specific maintenance rate of active microbes (m_R) is about 30% of μ_G and thus cannot be ignored. The initial active biomass (B_{a0}) is 0.145 ± 0.004 mg C g^{-1} soil (Table 1), which is lower than the values (0.194 ± 0.004 mg C g^{-1} soil) using the SIGR method [13]. This is likely due to the inclusion of maintenance respiration (characterized by m_R, see Equation 14d) in our model even for the exponentially-increasing stage; thus a lower B_{a0} could produce similar CO_2 flux to the case with higher B_{a0} that does not include the contributions from maintenance respiration. Our results also show that the initial active fraction (r_0) is 28.5 ± 6.4% and β is 0.025 ± 0.019. The magnitude of β is comparable to the estimation by Anderson & Domsch [6,7]. In addition, the half-saturation constant (K_s) was estimated as 0.275 ± 0.038 mg C g^{-1} soil, which is very close to the values derived from 16 soils by Van de Werf & Verstraete [21]. This K_s value indicates the substrate saturation level (ϕ) is higher than 0.7 before the transition from exponentially-increasing to non-exponentially-increasing phase (Fig. 3c). The changes of substrate (*S*), total microbial biomass (*B*) and active fraction (*r*) with time are also shown in Fig. 3c. In conclusion, the five parameters (B_0, r_0, μ_G, α, K_s) can be effectively determined using both exponentially-increasing and non-exponentially-increasing respiration rates, whereas β may also be determined but with a relatively high uncertainty (CV = 76%) than the other parameters.

Through this experimental analysis, we identified the need for isotopic data to discriminate between basal and substrate-induced respiration. We also discovered that the exponentially-increasing period due to substrate addition can be used to identify only a select set of model parameters (i.e., μ_G and B_{a0}) as also demonstrated by the method of Colores *et al.* [13]. These parameters, however, can be further applied to longer-term respiration experiments to enable fitting to obtain the remainder of model parameters by using our MEND model. Thus, we have found a new and unique solution to identify different parameters as a function of time, and to effectively use isotopic labeling to yield a specific set of model parameters.

Model test II: intermittent substrate supply

In order to further validate this additional physiological component in the MEND model, we also tested it against a laboratory experimental dataset with intermittent substrate supply [4]. In addition to the substrate, another limiting factor (i.e.,

Table 1. MEND model parameters values used for simulation of respiration rates due to added ^{14}C-labeled glucose in Colores et al. [13].

Parameter	Exponentially-increasing respiration*			All data†			Description
	Mean	SD‡	CV§	Mean	SD	CV	
B_0	0.504	0.279	55%	0.525	0.080	15%	Initial microbial biomass, (mg C g^{-1} soil)
r_0	0.394	0.263	67%	0.285	0.064	23%	Initial active fraction
μ_G	0.027	0.001	5%	0.030	0.001	3%	Maximum specific growth rate (h^{-1})
α	0.185	0.142	77%	0.228	0.031	13%	$m_R/(\mu_G+m_R)$, m_R is maximum specific maintenance rate for active microbes (h^{-1})
K_s	—	—	—	0.275	0.038	14%	Half-saturation constant for substrate (mg C g^{-1} soil)
β	—	—	—	0.025	0.019	76%	Ratio of dormant maintenance rate to m_R
Y_G	0.5	—	—	0.5	—	—	True growth yield, constant
B_{a0}	0.135	0.027	20%	0.145	0.004	3%	Initial active biomass (mg C g^{-1} soil), calculated by $B_0 \times r_0$

*Only the respiration rates during exponentially-increasing phase are used.
†All data including both exponentially-increasing and non-exponentially-increasing respiration.
‡SD: standard deviation.
§CV: Coefficient of variation.

oxygen, O_2) was included in this study. For this reason, we also introduced one more parameter (K_o: half saturation constant for O_2) to represent the limitation of O_2 on the microbial processes sketched in Fig. 1. Similar to substrates, the saturation level of O_2 is computed as $O_2/(O_2+K_o)$, where O_2 denotes the concentration of oxygen. The simulated oxygen concentrations by Stolpovsky et al. [4] were used as an input to our model. We used the SCEUA algorithm to determine the six model parameters in addition to the initial value for active fraction (r_0).

A summary of the seven parameters (one of them is r_0) and their fitted values is presented in Table 2. The initial active fraction (r_0) has a median of 0.925 with the 95% confidence interval (CI) of [0.628–1.000]. It means that a high r_0 is required for this experiment, but not necessary to be 1.0 set by Stolpovsky et al. [4]. The model and data are not sensitive to β since its 95% CI covers a wide range from 0.001 to 1. The reason is that the experiment only lasts for a very short time (33 h) so the influence of low metabolic rate at dormant state is insignificant.

Fig. 4 shows that the simulated total biomass (B) and substrate (S) concentrations agree very well with the observations (the coefficients of determination are 0.98 and 0.78 for biomass in Fig. 4a and substrate in Fig. 4b, respectively). Our simulation results indicate that, under limited O_2 between 12h and 24 h of the experiment, the active biomass decreases and the dormant biomass increases. As a result, the active fraction (r) declines from ca. 0.9 to 0.7 (Fig. 4a). For the same period Stolpovsky et al. [4] predicted a decrease of r from 1.0 to ca. 0, which means that all active biomass becomes dormant. Although there were not adequate measurements to confirm either prediction, our predicted changes in the active fraction (r) appear to be more reasonable during such a short experimental time period. This demonstration also shows that our model is capable of producing reasonable change in total, active, and dormant microbial biomass in response to substrate supply as well as an important forcing function (O_2).

Conclusions

We show that the physiological state index model (Equation 1) of Panikov [33] can be improved by eliminating the assumption that the steady state active fraction (r^{ss}) approaches the substrate saturation level (ϕ^{ss}). In particular, the model of Panikov [33] indicates that no active cells become dormant under insufficient substrate, which disregards the general nature of the strategy of dormancy in microorganisms when faced with unfavorable environmental conditions [5]. Our analysis also implies that the estimate of respiration rates under sufficient substrate by the

Table 2. MEND model parameter values used for simulation of the experiment described in Fig. 3 of Stolpovsky et al. (2011).

Parameter	Fitted Value*	Initial Range	Description
m_R	0.032, [0.011–0.048]	0.001–0.1	Specific maintenance rate for active biomass (h^{-1})
α	0.099, [0.045–0.181]	0.001–0.50	$m_R/(\mu_G+m_R)$, μ_G is specific growth rate (h^{-1})
K_s	3.110, [1.387–5.652]	0.1–9.0	Half-saturation constant for substrate (mg L^{-1})
Y_G	0.573, [0.463–0.600]	0.2–0.6	Growth yield factor (-)
K_o	0.0008, [0.0007–0.001]	0.005–0.1	Half-saturation constant for dissolved oxygen (mM)
β	0.351, [0.001–1.000]	0.001–1	Ratio of dormant maintenance rate to m_R
r_0	0.925, [0.628–1.000]	0–1	Initial fraction of active biomass (-)

*Medians and 95% confidence intervals of the fitted values from 100 optimization runs, i.e., 100 different random seeds are used for the stochastic optimization algorithm.

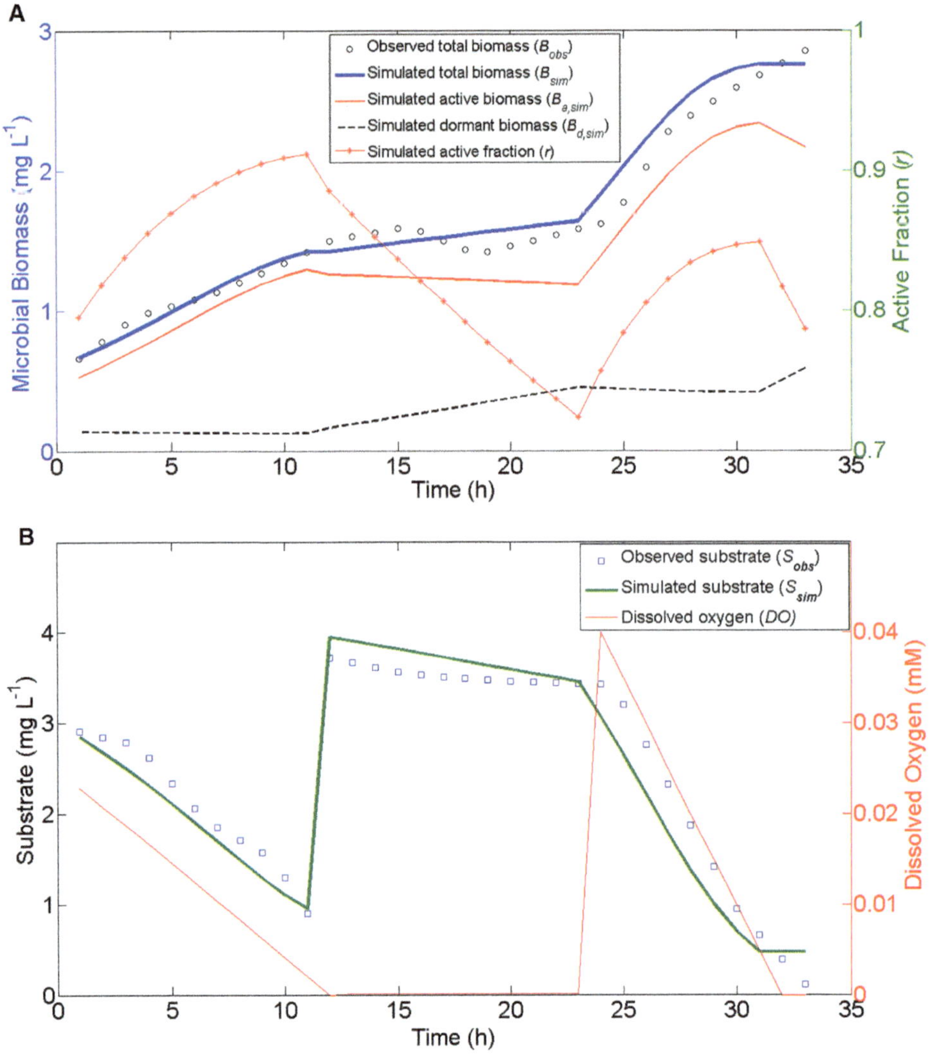

Figure 4. MEND model simulations against the experimental dataset used by Stolpovsky et al. (2011). (a) total live biomass, active and dormant biomass, and active fraction; (b) observed and simulated substrate concentration and prescribed O_2 concentration. There are three manipulations on the substrate and oxygen: (1) at time 0, the substrate (3 mg/L) and O_2 (0.025 mM) are added to the system; (2) after 12 h, the same amount of substrate is injected; (3) at 24 h, additional O_2 (0.04 mM) is injected to the system. The observed concentrations of substrate and total biomass are hourly data interpolated from the original observations in Stolpovsky et al. (2011). We scaled the substrate concentrations (with units of mM in original data) to match the magnitude of biomass concentration in units of mg/L.

physiological state index model is deficient. Pertaining to the switch function model, we argue that the switch function (θ) is also determined by the substrate (or combined with other impact factors) saturation level thus we would recommend using the switch function to modify the microbial uptake rate if the Gibbs energy change of the oxidation of substrate (ΔG) is tractable and the thermodynamic threshold (G_0) and the steepness of the step function (st) are identifiable. Based on the generally accepted assumptions summarized from existing dormancy models, we postulate a synthetic microbial physiology component to account for dormancy. Both the steady state active fraction (r^{ss}) and substrate saturation level (ϕ^{ss}) can be expressed as functions of two physiological indices: α and β. The index $\alpha = m_R/(\mu_G+m_R)$ is composed of μ_G and m_R denoting the maximum specific growth and maintenance rates, respectively, for active microbes. The index β represents the ratio of dormant to active maintenance rates. The value of r^{ss} is no less than ϕ^{ss}, and is equal only under the condition of $\beta \to 0$, where they are both identical to α. The upper bound of r^{ss} is ca. 0.8 at $\alpha \to 0.5$ and $\beta \to 1$. The maximum r^{ss} is ca. 0.5 if β (≤ 0.01) following the estimation of Anderson & Domsch [6]. It is evident that r^{ss} could be attenuated further by other limiting factors. The application of the MEND microbial physiology model to an experimental dataset with intermittent substrate supply shows satisfactory model performance (the determination coefficients are 0.98 and 0.78 for microbial biomass and substrate, respectively). The case study on the SIGR dataset indicate that the exponentially-increasing respiration rates can only be used to determine μ_G and B_{a0} (initial active biomass), while the major parameters (B_0, r_0, μ_G, α, K_s) can be effectively determined using both exponentially-increasing and non-exponentially-increasing respiration rates.

In conclusion, the microbial physiology model presented here can be incorporated into existing ecosystem models to account for dormancy in microbially-mediated processes. We have illustrated the impacts of substrate and oxygen availabilities on the physiological states through this study. Other environmental factors, such as soil temperature and soil water potential, could also be introduced into this framework to affect the transformation processes between the two microbial compositions. The changes in the physiological states of microbes could further alter the microbially-driven carbon and nutrient dynamics in ecosystems. Traditional measures of microbial biomass include the entire microbial population, even though dormancy is an important evolutionary strategy for preservation of microbial genetics and function until conditions for growth and replication improve. Parameterizing microbial ecosystem models assuming the entire population is active could therefore lead to significant errors. The approach described here provides a tractable and testable method to include dormancy as a response to external forcing.

Acknowledgments

The authors thank Dr. Sindhu Jagadamma for her helpful comments. Thanks also go to the two anonymous reviewers for their constructive comments.

Author Contributions

Conceived and designed the experiments: GW MAM. Analyzed the data: GW MAM CWS. Wrote the paper: GW MAM LG CWS. Mathematical analysis: GW LG.

References

1. Blagodatsky SA, Heinemeyer O, Richter J (2000) Estimating the active and total soil microbial biomass by kinetic respiration analysis. Biology and Fertility of Soils 32: 73–81.
2. Mason C, Hamer G, Bryers J (1986) The death and lysis of microorganisms in environmental processes. FEMS microbiology letters 39: 373–401.
3. Price PB, Sowers T (2004) Temperature dependence of metabolic rates for microbial growth, maintenance, and survival. Proceedings of the National Academy of Sciences of the United States of America 101: 4631–4636.
4. Stolpovsky K, Martinez-Lavanchy P, Heipieper HJ, Van Cappellen P, Thullner M (2011) Incorporating dormancy in dynamic microbial community models. Ecological Modelling 222: 3092–3102.
5. Lennon JT, Jones SE (2011) Microbial seed banks: the ecological and evolutionary implications of dormancy. Nature Reviews Microbiology 9: 119–130.
6. Anderson TH, Domsch KH (1985) Determination of ecophysiological maintenance carbon requirements of soil microorganisms in a dormant state. Biology and Fertility of Soils 1: 81–89.
7. Anderson TH, Domsch KH (1985) Maintenance carbon requirements of actively-metabolizing microbial populations under *in situ* conditions. Soil Biology and Biochemistry 17: 197–203.
8. Adam G, Duncan H (2001) Development of a sensitive and rapid method for the measurement of total microbial activity using fluorescein diacetate (FDA) in a range of soils. Soil Biology and Biochemistry 33: 943–951.
9. Jones KH, Senft JA (1985) An improved method to determine cell viability by simultaneous staining with fluorescein diacetate-propidium iodide. Journal of Histochemistry & Cytochemistry 33: 77–79.
10. Boulos L, Prévost M, Barbeau B, Coallier J, Desjardins R (1999) LIVE/DEAD® BacLight^{TM}: application of a new rapid staining method for direct enumeration of viable and total bacteria in drinking water. Journal of Microbiological Methods 37: 77–86.
11. Stocks S (2004) Mechanism and use of the commercially available viability stain, BacLight. Cytometry Part A 61: 189–195.
12. Schulz S, Peréz-de-Mora A, Engel M, Munch JC, Schloter M (2010) A comparative study of most probable number (MPN)-PCR vs. real-time-PCR for the measurement of abundance and assessment of diversity of alkB homologous genes in soil. Journal of Microbiological Methods 80: 295–298.
13. Colores GM, Schmidt SK, Fisk MC (1996) Estimating the biomass of microbial functional groups using rates of growth-related soil respiration. Soil Biology and Biochemistry 28: 1569–1577.
14. Anderson JPE, Domsch KH (1978) A physiological method for quantitative measurement of microbial biomass in soils. Soil Biology & Biochemistry 10: 215–221.
15. Jenkinson DS, Powlson DS (1976) The effects of biocidal treatments on metabolism in soil-V: A method for measuring soil biomass. Soil Biology and Biochemistry 8: 209–213.
16. Lodge DJ (1993) Nutrient cycling by fungi in wet tropical forests. In: Isaac S, editor. Aspects of Tropical Mycology: Symposium of the British Mycological Society Cambridge, UK: Cambridge University Press. pp. 37–57.
17. Yarwood S, Brewer E, Yarwood R, Lajtha K, Myrold D (2013) Soil microbe active community composition and capability of responding to litter addition after 12 years of no inputs. Applied and Environmental Microbiology 79: 1385–1392.
18. Alvarez C, Alvarez R, Grigera M, Lavado R (1998) Associations between organic matter fractions and the active soil microbial biomass. Soil Biology and Biochemistry 30: 767–773.
19. Khomutova T, Demkina T, Demkin V (2004) Estimation of the total and active microbial biomasses in buried subkurgan paleosoils of different age. Microbiology 73: 196–201.
20. Orwin KH, Wardle DA, Greenfield LG (2006) Context-dependent changes in the resistance and resilience of soil microbes to an experimental disturbance for three primary plant chronosequences. Oikos 112: 196–208.
21. Van de Werf H, Verstraete W (1987) Estimation of active soil microbial biomass by mathematical analysis of respiration curves: calibration of the test procedure. Soil Biology and Biochemistry 19: 261–265.
22. Stenström J, Svensson K, Johansson M (2001) Reversible transition between active and dormant microbial states in soil. FEMS Microbiology Ecology 36: 93–104.
23. Wang G, Post WM, Mayes MA (2013) Development of microbial-enzyme-mediated decomposition model parameters through steady-state and dynamic analyses. Ecological Applications 23: 255–272.
24. Moorhead DL, Lashermes G, Sinsabaugh RL (2012) A theoretical model of C- and N-acquiring exoenzyme activities, which balances microbial demands during decomposition. Soil Biology and Biochemistry 53: 133–141.
25. Sinsabaugh RL, Manzoni S, Moorhead DL, Richter A (2013) Carbon use efficiency of microbial communities: stoichiometry, methodology and modelling. Ecology Letters 16: 930–939.
26. Schimel JP, Weintraub MN (2003) The implications of exoenzyme activity on microbial carbon and nitrogen limitation in soil: a theoretical model. Soil Biology & Biochemistry 35: 549–563.
27. Lawrence CR, Neff JC, Schimel JP (2009) Does adding microbial mechanisms of decomposition improve soil organic matter models? A comparison of four models using data from a pulsed rewetting experiment. Soil Biology & Biochemistry 41: 1923–1934.
28. Wang G, Post WM, Mayes MA, Frerichs JT, Jagadamma S (2012) Parameter estimation for models of ligninolytic and cellulolytic enzyme kinetics. Soil Biology and Biochemistry 48: 28–38.
29. Todd-Brown KE, Hopkins FM, Kivlin SN, Talbot JM, Allison SD (2012) A framework for representing microbial decomposition in coupled climate models. Biogeochemistry 109: 19–33.
30. Treseder KK, Balser TC, Bradford MA, Brodie EL, Dubinsky EA, et al. (2012) Integrating microbial ecology into ecosystem models: challenges and priorities. Biogeochemistry 109: 7–18.
31. Wieder WR, Bonan GB, Allison SD (2013) Global soil carbon projections are improved by modelling microbial processes. Nature Clim Change 3: 909–912.
32. Konopka A (1999) Theoretical analysis of the starvation response under substrate pulses. Microbial Ecology 38: 321–329.
33. Panikov NS (1996) Mechanistic mathematical models of microbial growth in bioreactors and in natural soils: explanation of complex phenomena. Mathematics and Computers in Simulation 42: 179–186.
34. Blagodatsky S, Richter O (1998) Microbial growth in soil and nitrogen turnover: a theoretical model considering the activity state of microorganisms. Soil Biology and Biochemistry 30: 1743–1755.
35. Panikov NS, Sizova MV (1996) A kinetic method for estimating the biomass of microbial functional groups in soil. Journal of Microbiological Methods 24: 219–230.
36. Buerger S, Spoering A, Gavrish E, Leslin C, Ling L, et al. (2012) Microbial scout hypothesis, stochastic exit from dormancy, and the nature of slow growers. Applied and Environmental Microbiology 78: 3221–3228.
37. Hunt HW (1977) A simulation model for decomposition in grasslands. Ecology 58: 469–484.
38. Gignoux J, House J, Hall D, Masse D, Nacro HB, et al. (2001) Design and test of a generic cohort model of soil organic matter decomposition: the SOMKO model. Global Ecology and Biogeography 10: 639–660.
39. Ayati BP (2012) Microbial dormancy in batch cultures as a function of substrate-dependent mortality. Journal of Theoretical Biology 293: 34–40.

40. Jones SE, Lennon JT (2010) Dormancy contributes to the maintenance of microbial diversity. Proceedings of the National Academy of Sciences 107: 5881–5886.
41. Bär M, Hardenberg J, Meron E, Provenzale A (2002) Modelling the survival of bacteria in drylands: the advantage of being dormant. Proceedings of the Royal Society of London Series B: Biological Sciences 269: 937–942.
42. Locey KJ (2010) Synthesizing traditional biogeography with microbial ecology: the importance of dormancy. Journal of Biogeography 37: 1835–1841.
43. Wang G, Post WM (2013) A note on the reverse Michaelis–Menten kinetics. Soil Biology and Biochemistry 57: 946–949.
44. Wirtz KW (2003) Control of biogeochemical cycling by mobility and metabolic strategies of microbes in the sediments: an integrated model study. FEMS Microbiology Ecology 46: 295–306.
45. Panikov NS (1995) Microbial growth kinetics. London, UK: Chapman & Hall. 378 p.
46. Pirt SJ (1965) Maintenance energy of bateria in growing cultures. Proceedings of the Royal Society of London Series B-Biological Sciences 163: 224–231.
47. Wang G, Post WM (2012) A theoretical reassessment of microbial maintenance and implications for microbial ecology modeling. FEMS Microbiology Ecology 81: 610–617.
48. Hoehler TM, Jorgensen BB (2013) Microbial life under extreme energy limitation. Nat Rev Micro 11: 83–94.
49. Beeftink HH, Van der Heijden RTJM, Heijnen JJ (1990) Maintenance requirements: energy supply from simultaneous endogenous respiration and substrate consumption. FEMS microbiology letters 73: 203–209.
50. Blagodatsky S, Yevdokimov I, Larionova A, Richter J (1998) Microbial growth in soil and nitrogen turnover: model calibration with laboratory data. Soil Biology and Biochemistry 30: 1757–1764.
51. Duan QY, Sorooshian S, Gupta V (1992) Effective and efficient global optimization for conceptual rainfall-runoff models. Water Resources Research 28: 1015–1031.
52. Wang G, Xia J, Chen J (2009) Quantification of effects of climate variations and human activities on runoff by a monthly water balance model: A case study of the Chaobai River basin in northern China. Water Resources Research 45: W00A11.

23

Gut Microbiome in Down Syndrome

Elena Biagi[1], Marco Candela[1]*, Manuela Centanni[1], Clarissa Consolandi[2], Simone Rampelli[1], Silvia Turroni[1], Marco Severgnini[2], Clelia Peano[2], Alessandro Ghezzo[3], Maria Scurti[3,4], Stefano Salvioli[3,4], Claudio Franceschi[3,4], Patrizia Brigidi[1]

1 Department of Pharmacy and Biotechnology, University of Bologna, Bologna, Italy, **2** Institute of Biomedical Technologies - Italian National Research Council, Milan, Italy, **3** Department of Experimental, Diagnostic and Specialty Medicine, University of Bologna, Bologna, Italy, **4** Interdepartmental Centre "L. Galvani" (CIG), University of Bologna, Bologna, Italy

Abstract

Background: Premature aging seriously compromises the health status of Down Syndrome (DS) persons. Since human aging has been associated with a deterioration of the gut microbiota (GM)-host mutualism, here we investigated the composition of GM in DS.

Methods: The observational study presented involved 17 adult DS persons. We characterized the GM structure by 454 pyrosequencing of the V4 region of the 16S rRNA gene. DS microbiome was compared with that of age-matched healthy non-trisomic adults enrolled in the same geographic area.

Results and Conclusions: The dominant GM fraction of DS persons showed an overall mutualistic immune-modulatory layout, comparable to that of healthy controls. This makes GM a possible factor counteracting the genetic determined acceleration of immune senescence in DS persons. However, we also found detectable signatures specific for DS among subdominant GM components, such as the increase of *Parasporobacterium* and *Sutterella*. In particular, the abundance of this last microorganism significantly correlated with the Aberrant Behavior Checklist (ABC) total score, allowing us to hypothesize a possible role for this microbial genus in behavioral features in DS.

Editor: Francesco Cappello, University of Palermo, Italy

Funding: The research leading to these results has received funding from the European Union's Seventh Framework Programme (FP7/2007–2011) under grant agreement n. 259679 (IDEAL), and the Italian Ministry of Health, Progetto Ricerca Finalizzata 2008, convenzione 35: "An integrated approach to identify functional, biochemical and genetic markers for diagnostic and prognostic purposes in the elderly, in the centenarians and in people with dementia, Alzheimer's disease, mild cognitive impairment" to CF. The funders had no role in study design, data collection and analysis, decision to publish, or preparation of the manuscript.

Competing Interests: The authors have declared that no competing interests exist.

* Email: marco.candela@unibo.it

Introduction

Resulting from the trisomy of the 21st chromosome, Down Syndrome (DS) represents the most common genetic cause of intellectual disability. The incidence of DS is of 14.47 per 10,000 live births, with higher incidence for increasing maternal age [1]. Even if lifespan of persons with DS is significantly increased in the past decades - from 12 years of age in the 1940s to a mean life expectancy of 60 years today – DS persons exhibit sign of premature aging [2]. Thus, the overall increase in life expectancy in DS comes with a parallel increase of risk for age-related diseases, such as Alzheimer's disease, increased rate of infections, hypertension and obesity [3,4]. This results in a considerable need of medical, rehabilitative and social services, as well as a compromised health status and premature death.

The acceleration of biological aging in DS principally involves the increase in the oxidative stress [5], as well as a premature immunosenescence in the immune system [6]. In particular, adult DS persons show typical signs of immunosenescence which are normally found in much older non-trisomic individuals, including alterations of B and T cell subpopulations [7,8], and they have higher local inflammation with respect to age-matched non-trisomic subjects [9,10]; however it is not clear whether DS is characterized by the age-related increase of pro-inflammatory status - known as inflamm-aging - found in the elderly [11–13].

Human aging and immunosenescence are accompanied by a deterioration of the mutualistic relationship with the gut microbiota (GM) [14–16]. Encompassing the reduction of key immunomodulating groups as members of the butyrate producers *Clostridium* clusters IV and XIVa [17], and the concomitant increase of pro-inflammatory pathobionts [18], age-dependent GM dysbioses result in an overall pro-inflammatory configuration of the ecosystem capable to boost a self-sustained pro-inflammatory loop favoring immunosenescence [14]. In this context, we investigated - to our knowledge for the first time - the GM structure in DS persons, under the hypothesis that the GM structure of these subjects could present alterations typical of an aged-type microbiota which favors immunosenescence.

Autism Spectrum Disorder (ASD) features are also quite common in DS persons, among which behavioral problems are prominent [19]. In particular, higher total score of the Aberrant

Behavior Checklist (ABC), a rating scale specifically designed for individuals with severe intellectual disability [20], occurs in DS persons with ASD [21]. Since in recent years an association between ASD and specific GM genera has been outlined [22,23], the characterization of the GM ecosystem in DS persons allows us to further explore the association between GM and ASD.

Here we characterized the GM structure in 17 adult DS persons by 454 pyrosequencing of the V4 region of the 16S rRNA gene. In order to highlight deviations from a healthy-like GM structure, the gut microbial ecosystem of DS was compared with the previously published GM profiles of 16 age-matching healthy adults [24]. Furthermore, in order to look for any alterations of the DS microbiota ecosystem towards an aged-type configuration, a GM meta-analysis also including elderly and centenarians from Rampelli *et al.* [25] was performed. According to our data, the dominant GM fraction of DS persons showed an overall immune-modulatory layout, comparable to that of healthy adults. However, signatures specific for DS gut microbiome were detectable among the subdominant GM components and associated with the ABC total score. This suggests that such components are correlated to behavioral features in DS as already demonstrated in subjects with diagnosed ASD.

Materials and Methods

Subject enrolment and sample collection

Persons with DS were enrolled prospectively from 2008, in the framework of an open-label study on DS. The study was approved by the local Ethics Committee (S. Orsola Hospital, University of Bologna; ethical clearance #126/2007/U/Tess, released on December 18, 2007). A written informed consent to participate to the study was obtained from DS persons and from their parents or relatives (brothers/sisters). DS persons were recruited with the help of CEPS, OPIMM and ANFFAS, three local non-profit associations dealing with DS persons operating in the eastern part of Emilia-Romagna Region (Bologna and Ferrara provinces). Participation to the study was totally on a voluntary basis, with no reward for the participants or their families. A standard questionnaire to collect information regarding the health status, drugs use, clinical anamnesis, and lifestyle was administered. In order to take account of the dietary habits of DS persons, 24-hour dietary recalls were provided for each of the enrolled subjects for 3 days. Subjects affected by malignant neoplasia and/or in therapy with immunosuppressive drugs like cyclosporin, methotrexate, glucorticoids, anticoagulant drugs, and who recently (at one month) used antibiotics and/or probiotics were excluded from the study. Stool samples from 17 adult DS persons randomly selected from this group were used for this study. Feces were collected and stored at −80°C and analyses were performed within 3 months.

Neuropsychiatric assessment and adaptive behavior assessment

The cognitive level of 13 DS persons was evaluated with Wechsler Scale and resulted to be within the range of Cognitive Impairment (IQ scores below 70), as expected. For the remaining 4 DS persons, the cognitive level was assessed by using the mental age values extrapolated from VABS (see below). DS persons were classified within 4 levels of Cognitive Impairment (mild, moderate, severe and profound) according to DSM IV criteria.

For neuropsychiatric assessment, the ABC (Aberrant Behavior Checklist) [20] test was used. The italian version of ABC was used during verbal interview to care givers (parents and/or siblings, or educational personnel in one case). ABC is a 58-item caregiver report checklist that assesses maladaptive behaviors in individuals with developmental disabilities using a simple four-point rating scale (0–3) with higher scores reflecting more problems. ABC items are grouped into five subscales: (1) Irritability, Agitation, Crying, (2) Lethargy, social withdrawal, (3) Stereotypic behavior, (4) Hyperactivity, non-compliance, and (5) Inappropriate speech.

For adaptive behavior assessment, the Vineland Adaptive Behavior Scales (VABS) [26,27] were used. The test assesses personal autonomy and social responsibility of individuals from birth to adulthood through questions designed to assess the skills actually possessed by the subject at the time of evaluation. The scales are used with both non-disabled and disabled persons. The questionnaire in its complete form consists of 540 items that include the following areas: communication, daily living skills, socialization, motor skills. The questionnaire was directly administered to parents, or whoever takes care of the DS person.

Microbial DNA extraction from feces

Total bacterial DNA from fecal material was extracted using QIAamp DNA Stool Mini Kit (QIAGEN) with a modified protocol [28]. Briefly, 250 milligrams of feces were suspended in 1 mL of lysis buffer (500 mM NaCl, 50 mM Tris-HCl pH 8, 50 mM EDTA, 4% SDS). Four 3-mm glass beads and 0.5 g of 0.1-mm zirconia beads (BioSpec Products) were added, and the samples were treated in FastPrep (MP Biomedicals) at 5.5 movements per second for 1 min for 3 times, with 5-min intervals in ice. Samples were heated at 95°C for 15 min, then centrifuged for 5 min at full speed to pellet stool particles. Supernatants were collected and 260 μL of 10 M ammonium acetate was added, incubated in ice for 5 min and centrifuged at full speed for 10 min. One volume of isopropanol was added to each sample and incubated in ice for 30 min. The precipitated nucleic acids were collected by centrifugation for 15 min at full speed and washed with 70% ethanol. Pellets were resuspended in 100 μL of TE buffer and treated with 2 μL of DNase-free RNase (10 mg/mL) at 37°C for 15 min. Proteins were removed by proteinase K treatment, and DNA was subsequently purified following the manufacturer's instructions. Final DNA concentration was measured by using NanoDrop ND-1000 (NanoDrop Technologies).

16S rRNA gene amplification

For the amplification of the V4 region of the 16S rRNA gene the primer set 520F (5′-AYTGGGYDTAAAGNG-3′) and 802R (5′-TACNVGGGTATCTAATCC-3′) (with Y = C/T, D = A/G/T, N = any base, V = A/C/G) [29] was utilized. Primers included at their 5′ end the adaptor sequence used in the 454-sequencing library preparation protocol, linked to a unique MID tag barcode of 10 bases that allowed the identification of the different samples. PCR mixtures contained 0.5 μM of each forward and reverse primer, 100 ng of template DNA, 2.5 U of GoTaq Flexi Polymerase (Promega), 200 μM of dNTPs and 2 mM of $MgCl_2$ in a final volume of 50 μL. Samples were initially denatured at 95°C for 5 min, followed by 35 cycles of 94°C for 50 s, 40°C for 30 s, and 72°C for 60 s, with a final extension step at 72°C for 5 min [24]. PCR amplifications were carried out in a Biometra Thermal Cycler T Gradient (Biometra).

Pyrosequencing

16S rRNA gene V4 hypervariable region amplicons were purified using MinElute PCR Purification Kit (QIAGEN) and quantified using the Quant-iT PicoGreen dsDNA Kit (Invitrogen). Amplicons were pooled in equal amounts (creating one 9-plex and one 8-plex pools) and again purified by 454-Roche Double Ampure size selection protocol with Agencourt AMPure XP DNA purification beads (Beckman Coulter Genomics GmbH) to remove

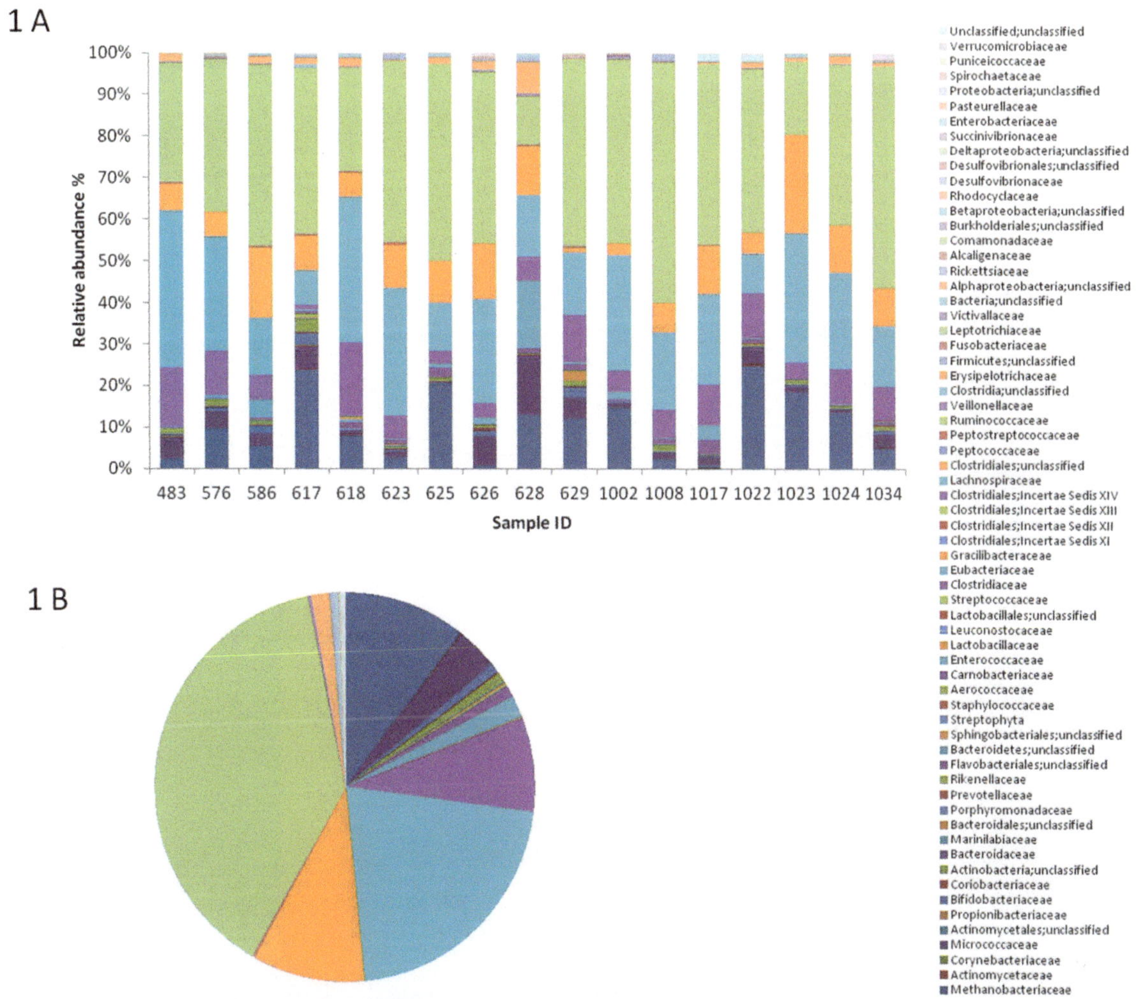

Figure 1. Individual (A) and average (B) gut microbiota profiles of the enrolled DS persons. Relative abundance of family-level assigned OTUs is reported. Colors were assigned for all families detected.

primer dimers, according to the manufacturer's instructions (454 LifeSciences, Roche). Before emulsion PCR, the purified amplicon pools were quantified using a quantitative Real Time PCR (qPCR) by KAPA Library Quant Kits (KAPA Biosystems). Afterwards pools were fixed to microbeads to be clonally amplified by emulsion PCR following the GS-FLX protocol Titanium emPCR LIB-A (454 LifeSciences, Roche). Beads were enriched in order to keep only those carrying identical PCR products on their surface, and loaded onto a picotiter plate for pyrosequencing reactions, following the GS-FLX Titanium sequencing protocol. Each pool was sequenced in one eighth of a plate. Amplicon sequences were deposited in MG-RAST under the project ID 10557 (http://metagenomics.anl.gov/linkin.cgi?project=10557).

Bioinformatic analysis and statistical data analysis

Sequencing reads were analyzed using the QIIME pipeline as described previously [24]. Briefly, V4 sequences were filtered according to the following criteria: (i) read length not shorter than 150 bp and not longer than 350 bp; (ii) no ambiguous bases (Ns); (iii) a minimum average quality score over a 50-bp rolling window of 25; (iv) exact match to primer sequences and maximum 1 error in barcode tags. RDP-classifier (version 2.2) with 50% as confidence value threshold was used for bacterial taxonomy assignment. Trimmed reads were clustered into OTUs at 97% identity level and further filtered for chimeric sequences using ChimeraSlayer (http://microbiomeutil.sourceforge.net/#A_CS). Alpha-diversity and rarefaction plots were computed using Shannon, PD whole tree, Chao1 and observed species metrics. Weighted and unweighted UniFrac distances were used to perform Principal Coordinates Analysis (PCoA). PCoA and heatmap were built using R packages Made4 [30], and Vegan (http://cran.r-project.org/package=vegan).

The R packages Stats and Vegan were used to perform statistical analysis. In particular, to compare the GM structure between DS persons and healthy adults, Wilcoxon signed rank test was used. Data separation in the PCoA was tested using a permutation test with pseudo F-ratios (function Adonis in the Vegan package). Significant differences in phylum- or genus-level abundance between DS persons and healthy controls were assessed by Mann-Whitney U tests, and corrected for multiple comparisons using the Benjamini-Hochberg method when appropriate. PCoA of the Euclidean distances of GM genera between DS persons, 16 healthy Italian adults from Schnorr *et al.* [24], and 5 elderly and 3 centenarians from Rampelli *et al.* [25] was carried out by using Vegan. The Kendall correlation test between ABC score and the relative abundance of genera in DS microbiome was

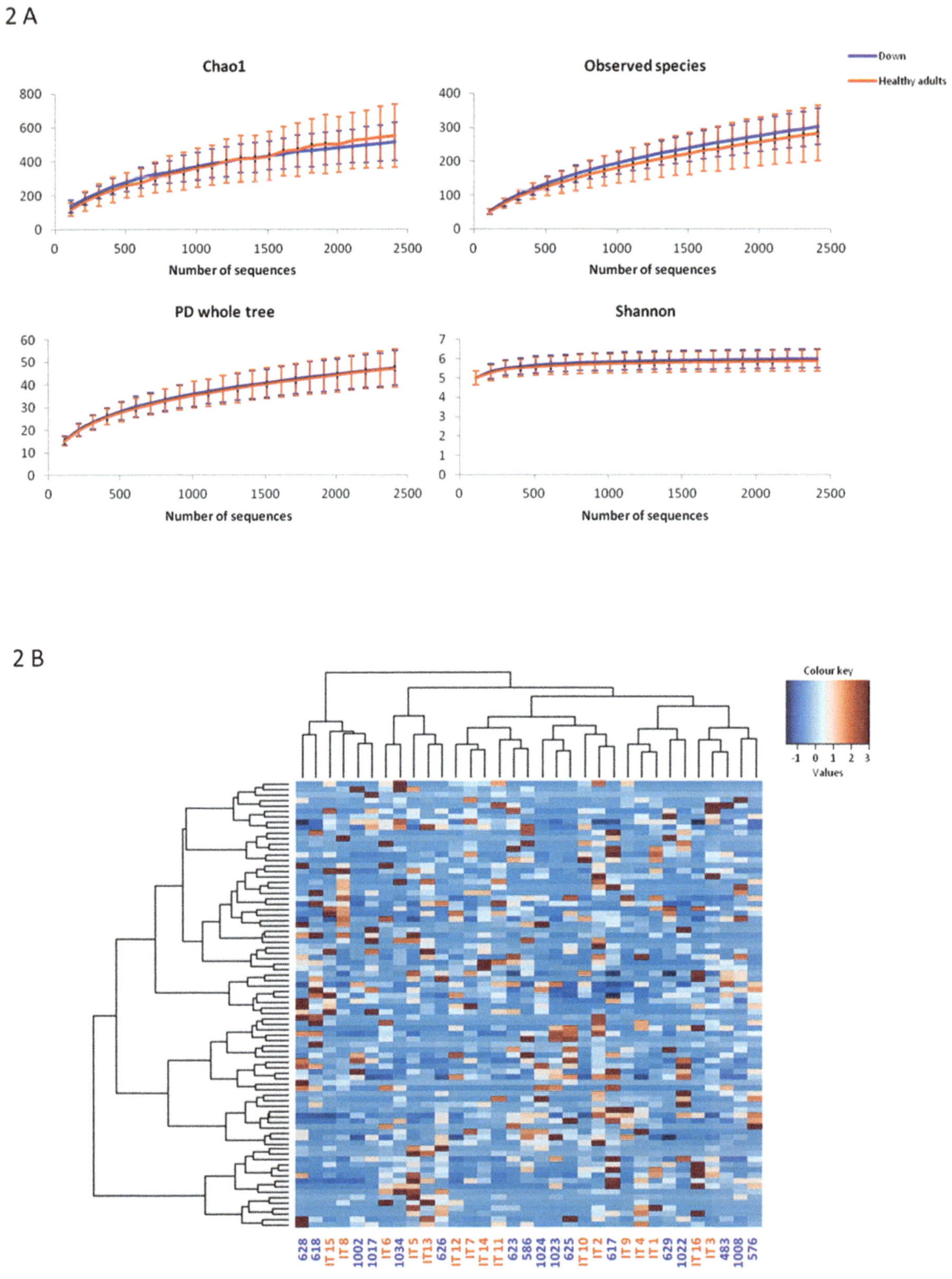

Figure 2. Similarity between the gut microbiota of DS persons and healthy non-trisomic adults for diversity and overall community composition. (A) Superimposition of the rarefaction curves of different α-diversity metrics (Faith's phylogenetic diversity (PD whole tree), observed OTUs, the Chao1 measure of microbial richness, and the Shannon index of biodiversity). (B) Hierarchical Ward-linkage clustering based on the Spearman correlation coefficients of genus proportion, showing no separation between DS persons and healthy adults. Blue, DS persons; red, healthy adults [24].

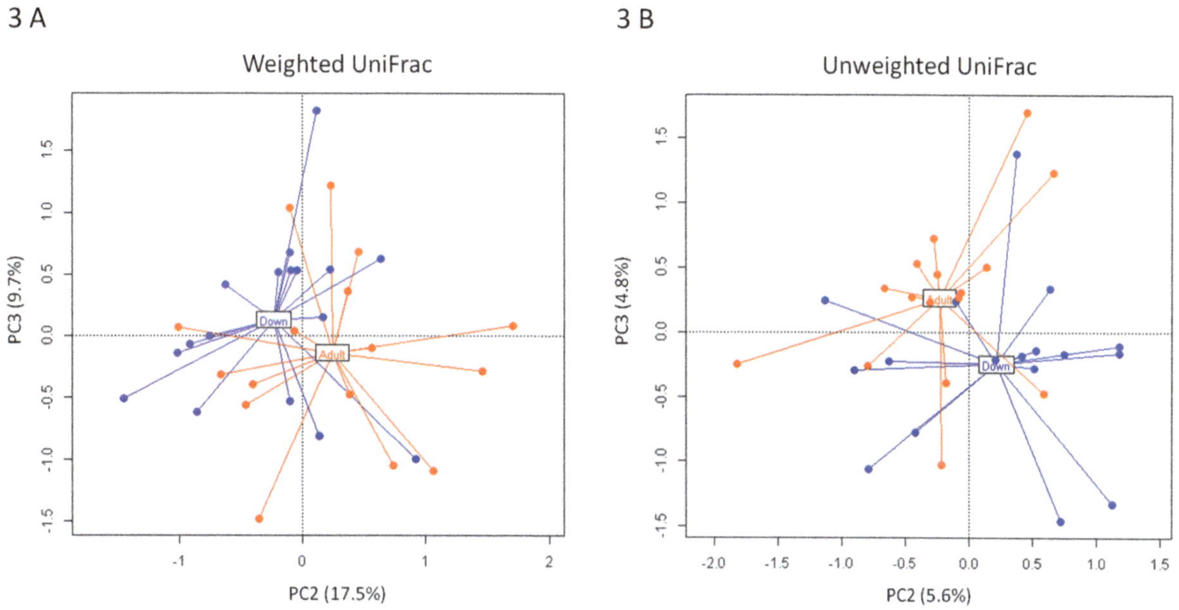

Figure 3. Weighted (A) and unweighted (B) UniFrac distance PCoA of the fecal microbiota of DS persons and healthy non-trisomic adults. Percentage of variance in the dataset shown by the second and third principal component (PC) is reported. Blue, DS persons; red, healthy adults [24].

achieved using function cor.test of the R package Stats. In all statistical tests, $P<0.05$ was considered as statistically significant.

Results

Intestinal microbiota in DS persons

Seventeen adult DS persons (19–35 years, 7 males and 10 females) randomly selected from a larger cohort participated to this study. The main characteristics of these subjects are provided in Table S1. As expected, they were in general overweight (mean BMI 27.05), while problems in alvus or eating were reported for 7 out of 17 subjects (41%). Stools were collected and GM structure was characterized by means of pyrosequencing of the V4 region of the bacterial 16S rRNA gene (Figure 1A). A total of 67,029 high-quality reads were obtained with a mean of 3,943 reads per subject. Reads were clustered in 7,457 operational taxonomic units at 97% of identity. OTU rarefaction curves based on number of unique OTUs were obtained with different phylogenetic diversity metrics, such as Shannon, Chao1, PD whole tree and observed species; curves reached the plateau, approximating the saturation level, after 2,400 reads (see Figure S1).

According to our data, the GM of DS persons was largely dominated by Firmicutes (relative abundance (rel. ab.) 83%±2), Actinobacteria (rel. ab. 9%±1.6) and Bacteroidetes (rel. ab. 8%±1.8). With rel. ab. values below 5%, Proteobacteria, Verrucomicrobia and Fusobacteria were largely subdominant phyla. The most represented families in DS gut microbial communities were *Ruminococcaceae* (rel. ab. 39%±3), *Lachnospiraceae* (rel. ab. 20%±1.4), *Clostridiales* (rel. ab. 9%±1), *Clostridiales Incertae Sedis XIV* (rel. ab. 8%±1.2), *Bifidobacteriaceae* (rel. ab. 8%±1.5) and *Bacteroidaceae* (rel. ab. 5%±1.4) (Figure 1B). Subdominant families showing rel. ab. values below 5% were *Eubacterium*, *Porphyromonadaceae*, *Rikenellaceae*, *Erysipelotrichaceae*, *Veillonellaceae*, *Alcaligenaceae*, and *Verrucomicrobiaceae*.

No separation of the DS microbiota structure according to BMI [31] was found ($P>0.05$) (see Figure S2).

Comparison between GM communities in DS persons and healthy non-trisomic adults

In order to highlight possible GM signatures of DS, the microbiota structure from the 17 DS persons enrolled in the present study was compared with that previously published from 16 healthy young adults aged between 20 and 40 years [24]. To exclude any country- and age-related effect on microbiota structure, as well as biases due to a study effect [32], age-matched Italian adults were selected for comparison. Both DS persons (see Table S2) and healthy adults [24] adhered to the standard Mediterranean diet, abundant in plant foods, fruit, pasta, bread and olive oil, and low-to-moderate in dairy and red meat.

According to our data, DS persons and healthy adults showed comparable levels of GM diversity and an overall similarity in the composition of the GM community (Figure 2). The visualization of community structures using PCoA of weighted UniFrac distances showed only a weak segregation between the two groups ($P = 0.28$), confirming the overall similarity of the two GM ecosystems (Figure 3A). However, the PCoA of unweighted UniFrac distances tended to separate the two subject groups (Figure 3B) ($P = 0.09$). This suggests that GM differences between DS persons and healthy controls principally involved subdominant GM components. Indeed, within the dominant GM families no significant differences in rel. ab. between the two groups of subjects were detected. Differently, DS persons were characterized by a higher abundance in several subdominant GM genera, such as *Parasporobacterium* ($P = 0.036$) and *Sutterella* ($P = 0.012$), as well as by a reduction in the rel. ab. of the family *Veillonellaceae* ($P = 0.005$).

In order to explore possible alterations of the GM of DS persons matching an aged-type ecosystem, we performed a meta-analysis of the GM structure also including 5 healthy Italian elderly and 3 centenarians from Rampelli *et al.* [25]. The PCoA of the Euclidean distances of the genus-level GM composition in DS persons, healthy adults, elderly (mean age 66.4 years) and centenarians (mean age 100.7 years) is provided as Figure S3.

Data shows that DS persons and heathy adults segregate together, suggesting that the gut microbial ecosystem of DS persons does not show the typical deviations of an aged-type gut microbial community.

Correlation between ABC score and signature genera of DS microbiome

Since microbiota dysbioses have been recently associated with certain neuro-behavioral disorders [33], here we sought Kendall correlations between ABC or VABS total score in DS persons and the three GM groups characterizing DS microbiome: *Parasporobacterium*, *Sutterella* and *Veillonellaceae*. The values of ABC and VABS scores for the enrolled DS persons are reported in Table S3. While no significant correlations were found between signature DS microorganisms and VABS score, which is indicative of the adaptive behavior of DS persons, *Sutterella* was positively correlated with the ABC total score (tau = 0.40, $P<0.05$). On the other hand, no correlation was found between ABC and *Parasporobacterium* as well as *Veillonellaceae*.

Discussion

Hypothesizing that the layout of the GM in DS persons could be of some relevance for their health, here we investigated the GM structure of 17 DS persons. In order to highlight signatures of an aged-type microbiota in DS, the GM profile of DS persons was compared with that of age-matched healthy non-trisomic Italian adults, and healthy Italian elderly and centenarians. According to our finding, DS persons and healthy adults possessed a similar complement of dominant groups in the GM, showing an overall healthy-like microbiota profile without any signs of premature microbiota aging [34].

The GM of DS persons was found to be well forged to provide the host with short-chain fatty acids (SCFA), butyrate, propionate and acetate, from fermentation of indigestible polysaccharides in the gut. Indeed, DS microbiota was largely dominated by the butyrate producers *Ruminococcaceae* (*Clostridium* cluster IV) and *Lachnospiraceae* (*Clostridium* cluster XIVa) [35], propionate-producing Bacteroidetes [36], and the acetate producer *Bifidobacterium* [37,38]. SCFA are pivotal metabolites in the microbiota-host immunological cross-talk, and their specific role as immune system modulators has been recently proved. In particular, butyrate is crucial to promote the extrathymic differentiation of regulatory T cells (Treg), whereas acetate is functional to drive their accumulation in the colon. Propionate exerts a dual function, capable to boost peripheral Treg generation and subsequent homing in the colon [39]. Thus, it is tempting to speculate that - by providing the host with SCFA from dietary polysaccharides - the gut microbiome of DS persons could be strategic to counteract the pro-inflammatory immune defects driving to immunosenescence [40]. This could explain why signs of premature immunosenescence found at the level of lymphocyte subpopulations (decreased B cells and T cells, in particular CD8+ T cells, and increased NK cells) are not apparently paralleled by the other major event of immunosenescence, *i.e.* the increased level of pro-inflammatory molecules. In any case, due to the limited number of DS persons enrolled in the present study, these assumptions need to be taken with extreme caution. Moreover, it is to note that the relatively young age of the subjects studied here (max 35 years) does not allow us to exclude that later on this event takes place anyway. Thus, further studies enrolling more DS persons and including older subjects are needed to prove these hypotheses, however, should it turn to be confirmed, it will make GM a strategic factor to counteract the genetically determined acceleration of the process of immunosenescence of DS persons. Within this context, it will become of extreme relevance to control diet in DS persons, supporting the maintenance of their healthy profile of the dominant GM. For instance, the consumption of a fiber-containing diet, which favors immunomodulating GM actors as *Clostridium* clusters IV and XIVa [41], should be encouraged.

While the general GM structure of DS persons was comparable to the one of healthy subjects, we observed significant differences among the subdominant groups. In particular, according to our findings, DS persons were enriched in *Parasporobacterium* and *Sutterella*, and reduced in *Veillonellaceae*. The increase of *Sutterella* and the reduction of *Veillonellaceae* have been described in autistic children with gastrointestinal symptoms [22,23,42]. In our study, the abundance of *Sutterella* in the GM was positively correlated with the ABC total score in DS persons, suggesting a possible link between this microorganism and ASD in DS [21,43]. As a Gram-negative opportunistic pathogen, *Sutterella* has been proposed to aggravate ASD through LPS-induced inflammation in the brain [44]. Conversely, the connection between *Veillonellaceae* and autistic symptoms was apparently less obvious. However, according to Kang *et al.* [42], the reduction of this group in the gut was more closely linked to the presence of autistic symptoms, rather than to the severity of gastrointestinal symptoms or the specific dietary regimen.

The correlation between the relative abundance of *Sutterella* in the GM and ABC total score in DS persons further strengthens the hypothesis in favor of a possible role for this gut microorganism in maladaptive behavior, opening the way to further investigation on the role of GM dysbioses in adults with ASD.

Supporting Information

Figure S1 Alpha-diversity rarefaction curves for the 16S rRNA V4 region pyrosequencing reads. The OTU table was rarefied up to 2,500 reads per sample and analyzed using various diversity metrics for each enrolled DS person. Metrics used were the Chao1 index of microbial richness, observed species, Faith's phylogenetic diversity index (PD whole tree), and the Shannon index of biodiversity. The individual rarefaction curves are color-coded according to the list of DS persons on the right.

Figure S2 PCoA of the unweighted Unifrac distances of the fecal microbiota of DS persons grouped according to body mass index (BMI). The BMI cutoff point of ≥ 25 kg/m^2 for overweight was used [31]. Green, BMI <25 kg/m^2; yellow, BMI ≥ 25 kg/m^2.

Figure S3 PCoA of the fecal microbiota in DS persons, healthy adults, elderly and centenarians. PCoA was based on the Euclidean distances of the relative abundance of gut microbiota genera in DS persons from the present study (green), healthy adults from Schnorr *et al.* [24] (blue), healthy elderly (red) and centenarians from Rampelli *et al.* [25] (orange).

Table S1 Demographic and clinical features of the enrolled Down Syndrome persons.

Table S2 Percent contribution of food categories to the diet of Down Syndrome persons.

Table S3 Level of cognitive impairment, Aberrant Behavior Checklist (ABC) and Vineland Adaptive Behavior Scale (VABS) scores in the enrolled Down Syndrome persons.

Acknowledgments

The authors are grateful to "ANFFAS", "CEPS" and "OPIMM" no-profit associations and all the DS persons and their families who participated to the study. The authors also acknowledge Dr. Daniela Follo for her precious contribution in contacting DS persons' families.

Author Contributions

Conceived and designed the experiments: CF PB. Performed the experiments: EB M. Centanni CC CP SS AG M. Scurti. Analyzed the data: SR M. Candela M. Severgnini. Contributed reagents/materials/analysis tools: CF PB. Contributed to the writing of the manuscript: M. Candela SS EB ST.

References

1. Parker SE, Mai CT, Canfield MA, Rickard R, Wang Y, et al; National Birth Defects Prevention Network (2010) Updated National Birth Prevalence estimates for selected birth defects in the United States, 2004–2006. Birth Defects Res A Clin Mol Teratol 88(12): 1008–1116.
2. Zigman WB (2013) Atypical aging in Down syndrome. Dev Disabil Res Rev 18(1): 51–67.
3. Nakamura E, Tanaka S (1998) Biological ages of adult men and women with Down's syndrome and its changes with aging. Mech Ageing Dev 105(1–2): 89–103.
4. Bittles AH, Glasson EJ (2004) Clinical, social, and ethical implications of changing life expectancy in Down syndrome. Dev Med Child Neurol 46(4): 282–286.
5. Patterson D, Cabelof DC (2012) Down syndrome as a model of DNA polymerase beta haploinsufficiency and accelerated aging. Mech Ageing Dev 133(4): 133–137.
6. Trotta MB, Serro Azul JB, Wajngarten M, Fonseca SG, Goldberg AC, et al. (2011) Inflammatory and Immunological parameters in adults with Down syndrome. Immun Ageing 8(1): 4.
7. Cossarizza A, Monti D, Montagnani G, Ortolani C, Masi M, et al. (1990) Precocious aging of the immune system in Down syndrome: alteration of B lymphocytes, T-lymphocyte subsets, and cells with natural killer markers. Am J Med Genet Suppl 7: 213–218.
8. Cossarizza A, Ortolani C, Forti E, Montagnani G, Paganelli R, et al. (1991) Age-related expansion of functionally inefficient cells with markers of natural killer activity in Down's syndrome. Blood 77(6): 1263–1270.
9. Griffin WS, Stanley LC, Ling C, White L, MacLeod V, et al. (1989) Brain interleukin 1 and S-100 immunoreactivity are elevated in Down syndrome and Alzheimer disease. Proc Natl Acad Sci U S A 86(19): 7611–7615.
10. Broers CJ, Gemke RJ, Weijerman ME, van der Sluijs KF, van Furth AM (2012) Increased pro-inflammatory cytokine production in Down Syndrome children upon stimulation with live influenza A virus. J Clin Immunol 32(2): 323–329.
11. Franceschi C, Bonafè M, Valensin S, Olivieri F, De Luca M, et al. (2000) Inflamm-aging. An evolutionary perspective on immunosenescence. Ann N Y Acad Sci 908: 244–254.
12. Franceschi C, Capri M, Monti D, Giunta S, Olivieri F, et al. (2007) Inflammaging and anti-inflammaging: a systemic perspective on aging and longevity emerged from studies in humans. Mech Ageing Dev 128(1): 92–105.
13. Franceschi C, Campisi J (2014) Chronic inflammation (inflammaging) and its potential contribution to age-associated diseases. J Gerontol A Biol Sci Med Sci (Suppl 1): S4–S9.
14. Biagi E, Nylund L, Candela M, Ostan R, Bucci L, et al. (2010) Through ageing, and beyond: gut microbiota and inflammatory status in seniors and centenarians. PLoS One 5(5): e10667.
15. Biagi E, Candela M, Franceschi C, Brigidi P (2011) The aging gut microbiota: new perspectives. Ageing Res Rev 10(4): 428–429.
16. Biagi E, Candela M, Turroni S, Garagnani P, Franceschi C, et al. (2013) Ageing and gut microbes: perspectives for health maintenance and longevity. Pharmacol Res 69(1): 11–20.
17. Atarashi K, Tanoue T, Oshima K, Suda W, Nagano Y, et al. (2013) Treg induction by a rationally selected mixture of Clostridia strains from the human microbiota. Nature 500(7461): 232–236.
18. Kamada N, Chen GY, Inohara N, Núñez G (2013) Control of pathogens and pathobionts by the gut microbiota. Nat Immunol 14(7): 685–690.
19. Roizen NJ, Patterson D (2003) Down's syndrome. Lancet 361(9365): 1281–1289.
20. Aman MG, Singh NN, Stewart AW, Field CJ (1985) The aberrant behavior checklist: a behavior rating scale for the assessment of treatment effects. Am J Ment Defic 89: 485–491.
21. Ji NY, Capone GT, Kaufmann WE (2011) Autism spectrum disorder in Down syndrome: cluster analysis of Aberrant Behaviour Checklist data supports diagnosis. J Intellect Disabil Res 55(11): 1064–1077.
22. Williams BL, Hornig M, Parekh T, Lipkin WI (2012) Application of novel PCR-based methods for detection, quantitation, and phylogenetic characterization of Sutterella species in intestinal biopsy samples from children with autism and gastrointestinal disturbances. MBio 3(1). pii: e00261-11.
23. Wang L, Christophersen CT, Sorich MJ, Gerber JP, Angley MT, et al. (2013) Increased abundance of Sutterella spp. and Ruminococcus torques in feces of children with autism spectrum disorder. Mol Autism 4(1): 42.
24. Schnorr SL, Candela M, Rampelli S, Centanni M, Consolandi C, et al. (2014) Gut microbiome of the Hadza hunter-gatherers. Nat Commun 5: 3654.
25. Rampelli S, Candela M, Turroni S, Biagi E, Collino S, et al. (2013) Functional metagenomic profiling of intestinal microbiome in extreme ageing. Aging (Albany NY) 5: 902–912.
26. Sparrow SS, Balla DA, Cicchetti DV (1984) Vineland adaptive behavior scales. Circles Pines, MN: American Guidance Service.
27. Balboni G, Pedrabissi L (2003) Adattamento italiano delle Vineland Adaptive Behavior Scales. Firenze: O.S. Organizzazioni Speciali.
28. Salonen A, Nikkilä J, Jalanka-Tuovinen J, Immonen O, Rajilić-Stojanović M, et al. (2010) Comparative analysis of fecal DNA extraction methods with phylogenetic microarray: effective recovery of bacterial and archaeal DNA using mechanical cell lysis. J Microbiol Methods 81(2): 127–134.
29. Claesson MJ, O'Sullivan O, Wang Q, Nikkila J, Marchesi JR, et al. (2009) Comparative analysis of pyrosequencing and a phylogenetic microarray for exploring microbial community structures in the human distal intestine. PLoS One 4(8): e6669.
30. Culhane AC, Thioulouse J, Perrière G, Higgins DG (2005) MADE4: an R package for multivariate analysis of gene expression data. Bioinformatics 21: 2789–2790.
31. Physical Status: the use and interpretation of anthropometry. Report of a WHO Expert Committee. (1995) World Health Organ Tech Rep Ser 854: 1–452.
32. Lozupone CA, Stombaugh J, Gonzalez A, Ackermann G, Wendel D, et al. (2013) Meta-analyses of studies of the human microbiota. Genome Res 23(10): 1704–1714.
33. Benach JL, Li E, McGovern MM (2012) A microbial association with autism. MBio 14: 3(1).
34. Biagi E, Candela M, Fairweather-Tait S, Franceschi C, Brigidi P (2012) Aging of the human metaorganism: the microbial counterpart. Age (Dordr) 34(1): 247–267.
35. Flint HJ, Bayer EA, Rincon MT, Lamed R, White BA (2008) Polysaccharide utilization by gut bacteria: potential for new insights from genomic analysis. Nat Rev Microbiol 6(2): 121–131.
36. Fischbach MA, Sonnenburg JL (2011) Eating for two: how metabolism establishes interspecies interactions in the gut. Cell Host Microbe 10(4): 336–347.
37. Fukuda S, Toh H, Hase K, Oshima K, Nakanishi Y, et al. (2011) Bifidobacteria can protect from enteropathogenic infection through production of acetate. Nature 469(7331): 543–547.
38. Fanning S, Hall LJ, Cronin M, Zomer A, MacSharry J, et al. (2012) Bifidobacterial surface-exopolysaccharide facilitates commensal-host interaction through immune modulation and pathogen protection. Proc Natl Acad Sci U S A 109(6): 2108–2113.
39. Arpaia N, Campbell C, Fan X, Dikiy S, van der Veeken J, et al. (2013) Metabolites produced by commensal bacteria promote peripheral regulatory T-cell generation. Nature 504(7480): 451–455.
40. Li C, Jin L, Bai Y, Chen Q, Fu L, et al. (2012) Genome-wide expression analysis in Down syndrome: insight into immunodeficiency. PLoS One 7(11): e49130.
41. David LA, Maurice CF, Carmody RN, Gootenberg DB, Button JE, et al. (2014) Diet rapidly and reproducibly alters the human gut microbiome. Nature 505(7484): 559–563.
42. Kang DW, Park JG, Ilhan ZE, Wallstrom G, Labaer J, et al. (2013) Reduced incidence of Prevotella and other fermenters in intestinal microflora of autistic children. PLoS One 8(7): e68322.
43. Carter JC, Capone GT, Gray RM, Cox CS, Kaufmann WE (2007) Autistic-spectrum disorders in Down syndrome: further delineation and distinction from other behavioral abnormalities. Am J Med Genet B Neuropsychiatr Genet 144B (1): 87–94.
44. Adams JB, Romdalvik J, Levine KE, Hu LW (2008) Mercury in first-cut baby hair of children with autism versus typically-developing children. Toxicol Environ Chem 90: 739–753.

PERMISSIONS

All chapters in this book were first published in PLOS ONE, by The Public Library of Science; hereby published with permission under the Creative Commons Attribution License or equivalent. Every chapter published in this book has been scrutinized by our experts. Their significance has been extensively debated. The topics covered herein carry significant findings which will fuel the growth of the discipline. They may even be implemented as practical applications or may be referred to as a beginning point for another development.

The contributors of this book come from diverse backgrounds, making this book a truly international effort. This book will bring forth new frontiers with its revolutionizing research information and detailed analysis of the nascent developments around the world.

We would like to thank all the contributing authors for lending their expertise to make the book truly unique. They have played a crucial role in the development of this book. Without their invaluable contributions this book wouldn't have been possible. They have made vital efforts to compile up to date information on the varied aspects of this subject to make this book a valuable addition to the collection of many professionals and students.

This book was conceptualized with the vision of imparting up-to-date information and advanced data in this field. To ensure the same, a matchless editorial board was set up. Every individual on the board went through rigorous rounds of assessment to prove their worth. After which they invested a large part of their time researching and compiling the most relevant data for our readers.

The editorial board has been involved in producing this book since its inception. They have spent rigorous hours researching and exploring the diverse topics which have resulted in the successful publishing of this book. They have passed on their knowledge of decades through this book. To expedite this challenging task, the publisher supported the team at every step. A small team of assistant editors was also appointed to further simplify the editing procedure and attain best results for the readers.

Apart from the editorial board, the designing team has also invested a significant amount of their time in understanding the subject and creating the most relevant covers. They scrutinized every image to scout for the most suitable representation of the subject and create an appropriate cover for the book.

The publishing team has been an ardent support to the editorial, designing and production team. Their endless efforts to recruit the best for this project, has resulted in the accomplishment of this book. They are a veteran in the field of academics and their pool of knowledge is as vast as their experience in printing. Their expertise and guidance has proved useful at every step. Their uncompromising quality standards have made this book an exceptional effort. Their encouragement from time to time has been an inspiration for everyone.

The publisher and the editorial board hope that this book will prove to be a valuable piece of knowledge for researchers, students, practitioners and scholars across the globe.

LIST OF CONTRIBUTORS

Clara Belzera
Department of Pathology, Brigham & Women's Hospital, Boston, Massachusetts, United States of America
Department of Pathology, Harvard Medical School, Boston, Massachusetts, United States of America

Georg K. Gerber, Andrew Onderdonk and Lynn Bry
Department of Pathology, Brigham & Women's Hospital, Boston, Massachusetts, United States of America
Department of Pathology, Harvard Medical School, Boston, Massachusetts, United States of America
Center for Clinical and Translational Metagenomics, Brigham & Women's Hospital, Harvard Digestive Diseases Center, Harvard Medical School, Boston, Massachusetts, United States of America

Mary Delaney, Andrea DuBois, Qing Liu, Vera Belavusava and Vladimir Yeliseyev
Department of Pathology, Brigham & Women's Hospital, Boston, Massachusetts, United States of America
Center for Clinical and Translational Metagenomics, Brigham & Women's Hospital, Harvard Digestive Diseases Center, Harvard Medical School, Boston, Massachusetts, United States of America

Guus Roeselers and Colleen Cavanaugh
Department of Organismic and Evolutionary Biology, Harvard University, Cambridge, Massachusetts,United States of America

Andres Housemanc
Department of Biostatistics, Harvard School of Public Health, Boston, Massachusetts, United States of America

Ximei Zhang
State Key Laboratory of Forest and Soil Ecology, Institute of Applied Ecology, Chinese Academy of Sciences, Shenyang, China

Xingguo Han
State Key Laboratory of Forest and Soil Ecology, Institute of Applied Ecology, Chinese Academy of Sciences, Shenyang, China
State Key Laboratory of Vegetation and Environmental Change, Institute of Botany, Chinese Academy of Sciences, Beijing, China

Quansheng Chen
State Key Laboratory of Vegetation and Environmental Change, Institute of Botany, Chinese Academy of Sciences, Beijing, China

Qian Wang, Yun-tao Jiang, Rui Ma, Zi-sheng Tang, Zheng Liu, Jingping Liang and Zheng-wei Huang
Department of Endodontics, Ninth People's Hospital, Shanghai Jiao Tong University School of Medicine, Shanghai Key Laboratory of Stomatology, Shanghai, China

Zi-yang Shao
Division of Radiation Oncology, Department of Oral and Maxillofacial Surgery, Ninth People's Hospital, Shanghai Jiao Tong University School of Medicine, Shanghai, China

Yue-jian IIu
Department of Endodontics, Ninth People's Hospital, Shanghai Jiao Tong University School of Medicine, Shanghai Key Laboratory of Stomatology, Shanghai, China
Department of Endodontics, The Affiliated Stomatology Hospital of Tongji University, Tongji University School of Stomatology, Shanghai, China

Inés Martínez, Catherine E. Muller and Jens Walter
Department of Food Science and Technology, University of Nebraska, Lincoln, Nebraska, United States of America

Varun Mazumdar
Bioinformatics Program, Boston University, Boston, Massachusetts, United States of America

Salomon Amar
Bioinformatics Program, Boston University, Boston, Massachusetts, United States of America
Center for Anti-Inflammatory Therapeutics; Boston University Goldman
School of Dental Medicine, Boston, Massachusetts, United States of America

Daniel Segré
Bioinformatics Program, Boston University, Boston, Massachusetts, United States of America
Department of Biology and Department of Biomedical Engineering, Boston University, Boston, Massachusetts, United States of America

Irene Dorado-García, Guillermo Herrera, Marco J. Cabrerizo and Presentación Carrillo
Instituto del Agua, Universidad de Granada, Granada, Spain

Juan Manuel Medina-Sánchez
Departamento de Ecología, Facultad de Ciencias, Universidad de Granada, Granada, Spain

Béatrice A. Frank-Fahle
Alfred Wegener Institute, Helmholtz Centre for Polar and Marine Research, Research Unit Potsdam, Potsdam, Germany

Hugues Lantuit
Alfred Wegener Institute, Helmholtz Centre for Polar and Marine Research, Research Unit Potsdam, Potsdam, Germany
University of Potsdam, Potsdam, Germany

Étienne Yergeau and Charles W. Greer
National Research Council, Montréal, Quebec Canada

Dirk Wagner
GFZ German Center for Geosciences, Section 4.5 Geomicrobiology, Potsdam, Germany

Guilherme Martins and Isabelle Masneuf-Pomarède
USC Oenologie-INRA, Université Bordeaux Segalen, ISVV, Villenave d'Ornon, France
Bordeaux Science Agro, Gradignan cedex, Bordeaux, France

Béatrice Lauga
Bordeaux Science Agro, Gradignan cedex, Bordeaux, France

Cécile Miot-Sertier, Anne Mercier, Aline Lonvaud, Marie- Louise Soulas and Guy Soulas
Equipe Environnement et Microbiologie UMR IPREM 5254 IBEAS, Universitéde Pau et des Pays de l'Adour, Pau, France

Louise C. Andresen and Anders Michelsen
Department of Biology, University of Copenhagen, Copenhagen, Denmark

Jennifer A. J. Dungait
Sustainable Soils and Grassland Systems Department, Rothamsted Research-North Wyke, Okehampton, United Kingdom

Roland Bol
Sustainable Soils and Grassland Systems Department, Rothamsted
Research-North Wyke, Okehampton, United Kingdom
Institute of Bio- and Geosciences, Forschungszentrum Jülich, Jülich, Germany

Merete B. Selsted and Per Ambus
Chemical and Biochemical Engineering, Technical University of Denmark, Kongens Lyngby, Denmark

John M. Blair and Michael A. Herman
Kansas State University, Ecological Genomics Institute, Division of Biology, Manhattan, Kansas, United States of America

Joseph D. Coolon
Kansas State University, Ecological Genomics Institute, Division of Biology, Manhattan, Kansas, United States of America
University of Michigan, Department of Ecology and Evolutionary Biology, Ann Arbor, Michigan, United States of America

Kenneth L. Jones
Kansas State University, Ecological Genomics Institute, Division of Biology, Manhattan, Kansas, United States of America
University of Colorado School of Medicine, Department of Biochemistry and Molecular Genetics, Aurora, Colorado, United States of America

Timothy C. Todd
Kansas State University, Ecological Genomics Institute, Division of Biology, Manhattan, Kansas, United States of America
Kansas State University, Department of Plant Pathology, Manhattan, Kansas, United States of America

Kelly Saeed, Huai Li, Richard Isom and Abdul Shakur Abdullah
CosmosID, College Park, Maryland, United States of America

Nur A. Hasan and Seon Young Choi
CosmosID, College Park, Maryland, United States of America
Maryland Pathogen Research Institute, University of Maryland, College Park, Maryland, United States of America

Thomas A. Cebula
CosmosID, College Park, Maryland, United States of America
Department of Biology, Johns Hopkins University, Baltimore, Maryland, United States of America

Brian A. Young, Angela T. Minard-Smith, Esley M. Heizer, Nancy J. McMillan, Daniel M. Bornman, Seth A. Faith and Michael L. Dickens
Battelle, Columbus, Ohio, United States of America

Rita R. Colwell
CosmosID, College Park, Maryland, United States of America

Maryland Pathogen Research Institute, University of Maryland, College Park, Maryland, United States of America
University of Maryland Institute for Advanced Computer Studies, University of Maryland, College Park, Maryland, United States of America
Bloomberg School of Public Health, Johns Hopkins University, Baltimore, Maryland, United States of America

Etienne Slabbert and Karin Jacobs
Department of Microbiology, Stellenbosch University, Stellenbosch, Western Cape, South Africa

Shayne Martin Jacobs
Department of Microbiology, Stellenbosch University, Stellenbosch, Western Cape, South Africa
Department of Conservation Ecology and Entomology, Stellenbosch University, Stellenbosch, Western Cape, South Africa

Luca Cocolin, Valentina Alessandria, Cristian Botta, Roberta Gorra and Kalliopi Rantsiou
Dipartimento di Scienze Agrarie, Forestali e Alimentari, Universita` degli Studi di Torino, Grugliasco, Italy

Francesca De Filippis and Danilo Ercolini
Dipartimento di Agraria, Universitádegli Studi di Napoli Federico II, Portici, Italy

Sean Rosenbaum and Ignatia Van den Veyver
Department of Obstetrics & Gynecology, Division of Maternal-Fetal Medicine, Baylor College of Medicine and Texas Children's Hospital, Houston, Texas, United States of America

Kjersti Aagaard, Kevin Riehle and Jun Ma
Department of Obstetrics & Gynecology, Division of Maternal-Fetal Medicine, Baylor College of Medicine and Texas Children's Hospital, Houston, Texas, United States of America
Bioinformatics Research Laboratory, Baylor College of Medicine and Texas Children's Hospital, Houston, Texas, United States of America

Toni-Ann Mistretta, Sabeen Raza and James Versalovic
Department of Pathology & Immunology, Baylor College of Medicine and Texas Children's Hospital, Houston, Texas, United States of America

Cristian Coarfa and Aleksandar Milosavljevic
Bioinformatics Research Laboratory, Baylor College of Medicine and Texas Children's Hospital, Houston, Texas, United States of America

Nicola Segata and Curtis Huttenhower
Harvard School of Public Health, Harvard University, Cambridge, Massachusetts, United States of America

Dirk Gevers
Broad Institute, Massachusetts Institute of Technology, Cambridge, Massachusetts, United States of America

Joseph Petrosino
Department of Molecular Virology and Microbiology, Baylor College of Medicine and Texas Children's Hospital, Houston, Texas, United States of America

Ximei Zhang and Xingguo Han
State Key Laboratory of Forest and Soil Ecology, Institute of Applied Ecology, Chinese Academy of Sciences, Shenyang, China
State Key Laboratory of Vegetation and Environmental Change, Institute of Botany, Chinese Academy of Sciences, Beijing, Chin

Guangming Zhang and Quansheng Chen
State Key Laboratory of Vegetation and Environmental Change, Institute of Botany, Chinese Academy of Sciences, Beijing, Chin

Heather Maughan and Julio Diaz Caballero
Department of Cell and Systems Biology, University of Toronto, Toronto, Ontario, Canada

David S. Guttman
Department of Cell and Systems Biology, University of Toronto, Toronto, Ontario, Canada
Centre for the Analysis of Genome Evolution and Function, University of Toronto, Toronto, Ontario, Canada

Pauline W. Wang, Pauline Fung, Yunchen Gong, Sylva L. Donaldson and Lijie Yuan
Centre for the Analysis of Genome Evolution and Function, University of Toronto, Toronto, Ontario, Canada

Shaf Keshavjee
Latner Thoracic Surgery Laboratories, University Health Network, University of Toronto, Toronto, Ontario, Canada

Yu Zhang and David M. Hwang
Latner Thoracic Surgery Laboratories, University Health Network, University of Toronto, Toronto, Ontario, Canada
Department of Pathology, University Health Network, University of Toronto, Toronto, Ontario, Canada

Yvonne C. W. Yau and Valerie J. Waters
Research Institute, The Hospital for Sick Children, Toronto, Ontario, Canada

D.Elizabeth Tullis
Adult Cystic Fibrosis Clinic, St. Michael's Hospital, Toronto, ON, Canada

KathiJo Jankowski, Daniel E. Schindler and M. Claire Horner-Devine
University of Washington, School of Aquatic and Fisheries Sciences, Seattle, Washington, United States of America

Yann Surget-Groba
Key Laboratory of Tropical Forest Ecology, Xishuangbanna Tropical Botanical Garden (XTBG), Chinese Academy of Sciences, Kunming, Yunnan, P. R. China

Shi Lingling
Key Laboratory of Tropical Forest Ecology, Xishuangbanna Tropical Botanical Garden (XTBG), Chinese Academy of Sciences, Kunming, Yunnan, P. R. China
University of the Chinese Academy of Sciences, Beijing, P. R. China

Charles H. Cannon
Key Laboratory of Tropical Forest Ecology, Xishuangbanna Tropical Botanical Garden (XTBG), Chinese Academy of Sciences, Kunming, Yunnan, P. R. China
Texas Tech University, Lubbock, Texas, United States of America

Peter O. Alele
Key Laboratory of Tropical Forest Ecology, Xishuangbanna Tropical Botanical Garden (XTBG), Chinese Academy of Sciences, Kunming, Yunnan, P. R. China
University of the Chinese Academy of Sciences, Beijing, P. R. China
Great Nile Conservation Centre (GNCC), Lira, Uganda
Institute of Tropical Forest Conservation (ITFC), Mbarara University of Science and Technology (MUST), Kabale, Uganda

Douglas Sheil
Department of Ecology and Natural Resource Management, Norwegian University of Life Sciences, Ås, Norway
Center for International Forestry Research (CIFOR), Bogor, Indonesia
Department of Ecology and Natural Resource Management, School of Environment, Science and Engineering, Southern Cross University, Lismore, New South Wales, Australia
Institute of Tropical Forest Conservation (ITFC), Mbarara University of Science and Technology (MUST), Kabale, Uganda

Wei Wang and Weile Chen
Department of Ecology, College of Urban and Environmental Sciences and the Key Laboratory for Earth Surface Processes of the Ministry of Education, Peking University, Beijing, China

Xinyue Zhang
Department of Ecology, College of Urban and Environmental Sciences and the Key Laboratory for Earth Surface Processes of the Ministry of Education, Peking University, Beijing, China
Key Laboratory for Cyclic Economy, School of Urban Planning and Design, Shenzhen Graduate School, Peking University, Shenzhen, China

Hui Zeng
Key Laboratory for Cyclic Economy, School of Urban Planning and Design, Shenzhen Graduate School, Peking University, Shenzhen, China

Naili Zhang
State Key Laboratory of Vegetation and Environmental Change, Institute of Botany, Chinese Academy of Sciences, Beijing, China

Sabine A. Tanner, Annina Zihler Berner, Eugenia Rigozzi, Franck Grattepanche, Christophe Chassard and Christophe Lacroix
Laboratory of Food Biotechnology, Institute of Food, Nutrition and Health, ETH Zurich, Zurich, Switzerland

Takashi Narihiro, Hideyuki Tamaki, Aya Akiba and Satoshi Hanada
Bioproduction Research Institute, National Institute of Advanced Industrial Science and Technology (AIST), Tsukuba, Ibaraki, Japan

Yoichi Kamagata
Bioproduction Research Institute, National Institute of Advanced Industrial Science and Technology (AIST), Tsukuba, Ibaraki, Japan
Bioproduction Research Institute, National Institute of Advanced Industrial Science and Technology (AIST), Sapporo, Hokkaido, Japan

Kazuto Takasaki and Koichiro Nakano
FASMAC Co., Ltd., Atsugi, Kanagawa, Japan

Taizo Maji
Shubun University, Ichinomiya, Aichi, Japan

Gangsheng Wang, Melanie A. Mayes and Lianhong Gu
Climate Change Science Institute, Oak Ridge National Laboratory, Oak Ridge, Tennessee, United States of America
Environmental Sciences Division, Oak Ridge National Laboratory, Oak Ridge, Tennessee, United States of America

Christopher W. Schadt
Climate Change Science Institute, Oak Ridge National Laboratory, Oak Ridge, Tennessee, United States of America
Biosciences Division, Oak Ridge National Laboratory, Oak Ridge, Tennessee, United States of America

Elena Biagi, Marco Candela, Manuela Centanni, Simone Rampelli, Silvia Turroni and Patrizia Brigidi
Department of Pharmacy and Biotechnology, University of Bologna, Bologna, Italy

Clarissa Consolandi, Marco Severgnini and Clelia Peano
Institute of Biomedical Technologies - Italian National Research Council, Milan, Italy

Alessandro Ghezzo
Department of Experimental, Diagnostic and Specialty Medicine, University of Bologna, Bologna, Italy

Maria Scurti, Stefano Salvioli and Claudio Franceschi
Department of Experimental, Diagnostic and Specialty Medicine, University of Bologna, Bologna, Italy
Interdepartmental Centre "L. Galvani" (CIG), University of Bologna Bologna, Ital

Index